How Genes Influence Behavior

How Genes Influence Behavior

SECOND EDITION

Jonathan Flint

Semel Institute for Neuroscience and Human Behavior, University of California, Los Angeles, California, USA

Ralph J. Greenspan

Kavli Institute for Brain and Mind, University of California, San Diego, California, USA

Kenneth S. Kendler

Virginia Institute for Psychiatric and Behavioral Genetics, Virginia Commonwealth University, Richmond, Virginia, USA

OXFORD
UNIVERSITY PRESS

Great Clarendon Street, Oxford, OX2 6DP,
United Kingdom

Oxford University Press is a department of the University of Oxford.
It furthers the University's objective of excellence in research, scholarship,
and education by publishing worldwide. Oxford is a registered trade mark of
Oxford University Press in the UK and in certain other countries

Published in the United States of America by Oxford University Press
198 Madison Avenue, New York, NY 10016, United States of America

British Library Cataloguing in Publication Data
Data available

Library of Congress Control Number: 2019952173

ISBN 978–0–19–871687–7

Printed and bound by CPI Group (UK) Ltd, Croydon, CR0 4YY

About the Online Resources

The online resources that accompany this book provide a number of useful teaching and learning materials and are free of charge.

The site can be accessed at:

www.oup.com/he/flint2e

Please note that instructor resources are available only to registered adopters of the textbook. To register, simply visit the site and follow the appropriate links.

Student resources are openly available to all, without registration.

FOR STUDENTS

Multiple choice questions for you to check your threshold knowledge

Data sets for you to manipulate, so that you can apply what you have learned

For registered adopters of the book:

Figures and tables from the book, in ready-to-download format.

Acknowledgements

The fields of psychiatric, behavioral, and molecular genetics have changed dramatically since the first edition of our book was published in 2009. The second edition is longer, partly because there is more to cover, and partly because we have expanded our remit to tackle behavior genetics more broadly. In so doing, we stray from our original focus on subjects we work on ourselves, and deal with more subject matter at second hand. As a result, our book is less personal. However, where possible, we have tried to maintain the informal tone that characterized the first edition.

We owe our first thanks to Martha Bailes, our editor at Oxford University Press. She did a marvelous job keeping us on track through the many months of work on this volume. She was consistently cheerful, firm when she needed to be, and shared with the authors our vision of what this text was striving to be: an up-to-date and interesting-to-read text for the many students curious about the rapid developments in our understanding of how genes impact on behavior.

Each of us (J.F., R.J.G., and K.S.K.) has been blessed through our now rather long academic careers with many stimulating and supportive colleagues and students. This is not the place to try to list them all. However, we want to acknowledge here the depth of our gratitude. At its best, academic life is about as good as it gets, reflecting the motto of the college attended by one of us (K.S.K.): the pursuit of truth in the company of friends.

A number of our colleagues read substantial sections of this book and provided a range of helpful comments and averted a number of embarrassing errors. We list them here as an expression of our gratitude: Daniel Geschwind, Alcino Silva, Zoe Donaldson, Peter Visscher, Maria Karayiorgou, Marcus Munafo, Laurent Keller, Eleazar Eskin, David Krantz, Adam Whitley, Daniel Benjamin, and David Anderson.

As with our first edition, we dedicate this book to our wives, Alison, Dani, and Susan, for a decade more of their love, support, and tolerance.

Contents

1

Introduction

1.1 Genetics: the universal research language

Genetics has dominated all of biomedical research for some time now. From the 1980s onwards, there has been a steady increase in the number of diseases subjected to molecular genetic analysis. But things really took off with the completion of the human genome project and the sequencing of the genomes of model organisms. These prodigious achievements are expected to be succeeded by equally prodigious advances in medicine. Enormous numbers of people, half a million in the UK alone, were enrolled in studies of genetic effects on health (BYCROFT et al. 2018). Plans are afoot to sequence either the expressed portion of the human genome (called the exome) or the entire genome of hundreds of thousands of individuals. Some have claimed, as part of an initiative called 'precision medicine', that doctors will soon be prescribing medicines based on the specific genetic composition of their patients. It's hard to argue with science on this scale: billions of dollars, touching the lives of millions of people.

Psychiatry has not been spared the genetic assault on disease. Media reports provide a stream of apparent successes: not just genes that make you depressed but also happiness genes, gay genes, God genes, and,

from a newspaper headline: 'Britain's biggest study into the **genetic basis** of promiscuity … One in five women cheat—and it's **genetic** … **Infidelity** linked to **genetic** make-up.'

However, there has been strong opposition to geneticists' claims to understand and explain human behavior. *The Bell Curve*, a book about the genetics of intelligence (HERRNSTEIN AND MURRAY 1994), aroused such passion that protestors picketed book stores. *Not In Our Genes* is one of a number of books arguing that heritability (a measure of the extent to which behavior and other traits are genetically influenced) is not only a flawed concept but one that is ideologically suspect, developed by conservative academics and seized on by right-wing politicians (LEWONTIN et al. 1984). Genes get political because it can be claimed (falsely) that if genes are shown to have important effects on behavior, for example on criminality, there would be no point in rehabilitation programs. As criminals are genetically programmed to be criminals, incarceration is society's only recourse. The same would be true for drug addicts, although perhaps once we know which genes make a person devote their entire life to the pursuit of drug-induced euphoria, we would then be able to manufacture a **biological** therapy (obviously no point in changing the environment, if the disorder is due to faulty genes).

So writing about genes and behavior promises to be controversial. Although controversy is not something that any of us willingly seeks out, it would be disingenuous to ignore the contentious nature of psychiatric genetics and sanctimonious to assert that all we want to do is educate our readers. In fact, it is the elusive but irresistible nature of the quest, and the stages of understanding that it has led us through, that provide the motivation and framework for this book.

1.2 What will you get in the subsequent pages?

This book is a collaboration among three scientists who have spent a good proportion of their adult lives trying to understand how genes influence behavior. We joined up in this effort because we enjoy talking to each other about our mutual interests, which we approach from complementary areas of expertise. Two of us, Kenneth Kendler and Jonathan Flint, are trained in medicine and psychiatry but have taken rather different career paths: Kenneth Kendler has focused on psychiatric genetic studies in humans while Jonathan Flint has spent most of his time examining the impact of genes on behavior in rodents. Ralph Greenspan, by contrast, works on the genetics of behavior in the common fruit fly (*Drosophila melanogaster*, an organism that seems to devote its life to hovering around fruit bowls and wine glasses).

Our goal in this book is to illustrate the key issues and results in behavior genetics, and to describe how they were obtained. Along the way, we hope to convey a feel for the chase and how the search for these elusive genetic influences has changed the way scientists think about the relationship between genes and behavior. The problems are not just ones of empirical knowledge, of the results we obtain, and how we place them within the canon of accepted facts. It's a truism that there is no such thing as theory-free data, but theories themselves have to emerge from a particular viewpoint, a set of assumptions or, in short, a philosophy about how the world works. When we consider how genes contribute to the production of behavior, we have to determine to what extent we can reduce the biological basis of behavior to a more fundamental level. Reductive explanations are attractive and very powerful. They explain complex phenomena very successfully. For example, physics is full of simple

laws that explain much of our world, like $F = ma$ (force equals mass times acceleration, which we owe to Isaac Newton) and $E = mc^2$ (energy equals mass times the square of the speed of light, which we owe to Albert Einstein).

Do similar explanations work for behavior? We know that genes encode information about an organism: the entire set of proteins that it can make and many aspects of its developmental trajectory, from egg to adult and all the intervening steps. Metabolic pathways that fuel the process, and the machinery for the immense physical transformations involved, are determined by the linear structure of DNA. In some instances, you can even observe developmental order reflected in the physical order of genes along a chromosome (Lewis 1996). Many aspects of the problem of how an organism develops are hence reduced to the problem of how genes are turned on and off in a coordinated fashion. Is the same true for behavior? Can the problem of how a nervous system, with all its connections between neurons, branching dendrites, cellular architecture, and anatomical organization into a brain, gives rise to behavior? Can all of this be reduced to a problem of which genes specify which neural components? Even more challenging, can we reduce the nature of our mental experiences—part of the 'internal' aspects of our own behavior—our emotions, perceptions, wishes, and fears down to the simple level of genes and their associated biology? Our views on these questions will make more sense to you after you have journeyed with us through the whole book. There is a tremendous amount that we do not know and a lot we are still learning. But we think we now know enough to see the broad shape of how genes influence behavior. We hope to explain this still incompletely formed vision to you. We start with a brief historical overview, which sets the scene for a thread that runs throughout the book: the contrast between Mendelian and biometric genetics.

1.3 An historical sketch: genetics as a four-act play

1.3.1 Act I

One of the two major roots of modern psychiatric and behavioral genetics can be traced back to Francis Galton (1822–1911) and his book *Hereditary Genius*, published in 1869 (GALTON 1869). This book explored the tendency for genius of all sorts (e.g. in jurisprudence, music, literature, and science) to cluster within families, noting that the number of eminent relatives was greater with a closer degree of kinship. Galton, a cousin of Charles Darwin, first popularized the concepts of nature and nurture, a phrase taken from the play *The Tempest* by William Shakespeare. Galton, with his student Karl Pearson, developed the correlation coefficient as a way of summarizing the global resemblance of close relatives for classical quantitative traits like height. Geneticists trained in this way of thinking, later known as biometricians, understood genetics to be best studied by examining correlations of traits, especially in different kinds of relatives.

One useful application of the biometric method is the classical twin study where (with some added bells and whistles) researchers compare the correlations of traits in genetically identical (or monozygotic) twins and fraternal (or dizygotic) twins, who are as genetically similar as regular brothers or sisters. Act 1 in our drama, where the biometricians ruled the roost in human genetics, lasted from 1869 to 1900.

I.3.2 Act 2

In 1900, several researchers rediscovered the key 1866 publication by an Augustinian friar living and working in what is now the Czech Republic: Gregor Mendel (1822–1884). Mendel's studies showed that inheritance arose from the action of distinct units, which we now call genes. For the next two decades, an intense intellectual battle about the nature of genetic transmission raged between the followers of Mendel (called Mendelians) and the biometricians. Were genetic effects transmitted continuously in what was described as a 'blended inheritance' that should be studied by correlations, or in discrete units that produced clearly distinct phenotypes (as was true for Mendel's tall and short pea plants, and his smooth and wrinkled peas)?

Investigators set out to understand human behavioral and psychiatric using simple Mendelian models. In Germany, Ernst Rüdin conducted a large and systematic study of siblings of individuals with schizophrenia to see if the pattern of transmission fitted with a simple Mendelian pattern. It did not, but that finding did not stop Rüdin from proposing a two-gene recessive model that roughly fitted his results. In the USA, Charles Davenport studied pedigrees trying to find the Mendelian basis for a range of behavioral traits including personality, 'the wandering impulse', and 'feeble-mindedness'. He believed that, for these traits, he had discovered simple inheritance patterns consistent with Mendel's theories.

While these rather shaky results were emerging from studies of human psychiatric and behavioral traits, Mendelian models were having dramatic success in rare human conditions, including Albinism, certain metabolic abnormalities, and rare neurological disorders like Huntington's disease. Even more impressive was the discovery of Mendelian genetic variants in the fruit fly (*Drosophila melanogaster*) in the laboratory of Thomas Hunt Morgan at Columbia University in New York (MORGAN 1911).

Gradually, during this act, the Mendelians gained ground on the biometricians. The results were beginning to turn their way. But this left the field with a major problem. Quantitative traits like human height were known to be similar in close relatives. Everyone knew that when a tall and short parent married, their children could **not** be divided into tall and short groups in ratios predicted by Mendel's theories. What could be going on to explain these results?

I.3.3 Act 3

Act 3 begins with the entry of a statistical genius: Ronald Fisher. In 1918, he published a paper that, although statistically demanding to read to this day, dramatically changed the intellectual landscape of genetics. The message of his paper is provided in its title: 'The correlation between relatives on the supposition of Mendelian inheritance' (FISHER 1918). Fisher showed how the biometric theory of genetic transmission could be understood from Mendelian principles.

Fisher's key insight, which permitted him to unify Mendelian and biometric worldviews, was what he termed the 'infinitesimal model'. Mendel had focused solely on traits affected by a single gene. Models had subsequently been developed in which two, or even three, independent genes influence traits. Fisher took an intellectual leap and asked what would happen if a trait was influenced by a very large number of Mendelian genes, each of very small effect. The answer was that one would observe the correlations seen by the biometricians. In one stroke, these two contrasting and sometimes conflicting schools of thought were brought together within a single conceptual rubric.

One more piece of intellectual workmanship is needed to complete this picture. Psychiatric disorders are not quantitative traits. Rather, they are things you either have or do not have. How could biometrical approaches that were developed for quantitative traits be applied to such conditions? Here the key insight was from Karl Pearson, who articulated what we would now call a liability-threshold model.

We will illustrate the liability-threshold model with the example of height. Instead of studying height as a quantitative trait, you could instead study a categorical trait called 'tallness'. We can define the trait of 'tallness' as anyone who is more than 6 feet tall. Below the 6-foot threshold, you do not have the trait, and above it you do. In this way a continuous, quantitative trait (height) is converted into a categorical, dichotomous trait (tallness). Pearson developed an algebraic approach that made it possible to extrapolate the underlying dimension of height from the dichotomous trait of tallness. So by examining the 'correlation' in tallness in pairs of relatives, you could infer with considerably accuracy their real correlation in height.

In our tallness example, you know what the underlying trait is (height) and you could, if you wanted to, measure it. But what about schizophrenia? The approach Pearson developed for tallness would apply perfectly well to schizophrenia. You could study pairs of relatives and calculate their 'correlation', but now the underlying trait would be unmeasured and would reflect what we would call the 'liability to schizophrenia'. All of the twin studies that we will review in this book are based on this model (the liability-threshold model) where the calculated heritability refers *not* to the measured dichotomous trait (be it depression, anxiety, or schizophrenia) but instead to its *underlying liability*.

With Fisher's 1918 paper, the Mendelian and biometric views on genetic transmission were theoretically unified, but for many decades for psychiatric genetics that did not matter much. Psychiatric genetics focused largely on methods consistent with the biometrical methods—especially family, twin, and, somewhat later, adoption studies—because that is what they could do. The occasional paper would be published proposing a single-gene model for schizophrenia, but there was no practical way to test such models.

One prescient paper published in 1967 deserves attention. It was written by a pair of behavior geneticists/twin researchers, Irv Gottesman and Jerry Shields,

and was entitled 'A polygenic theory of schizophrenia' (GOTTESMAN AND SHIELDS 1967). They proposed that the genetic risk for schizophrenia can best be described by a combination of Fisher's infinitesimal model of many genes of very small effect (often now referred to as a **polygenic model**) and Pearson's liability-threshold model. So instead of accepting the standard biometrical model because it was convenient, they argued that it was true.

1.3.4 Act 4

While there were sputterings of molecular work on psychiatric and behavioral traits from the 1950s through to the early 1980s using the then-available methods (e.g. looking a blood groups, color blindness, and so on), the defining moment for the emergence of Act 4, the one we are still currently in, occurred in 1983 with the first use of molecular markers to map a human genetic disease by linkage analysis. The disorder was Huntington's chorea (GUSELLA et al. 1983).

Act 4 began with two competing models for human genetics. One was genetic epidemiology, derived from the biometricians' viewpoint, which applies liability-threshold models to various forms of family, twin, and adoption designs. Molecular makers are not measured and the Fisherian infinitesimal or polygene model is assumed. The other research approach, in a direct lineage from the early Mendelians, assumed that the genetic influences on major psychiatric and behavioral traits resulted from 'Mendelian' loci and used the methods of linkage analysis to detect these loci. In these years, the Mendelian researchers were often critical of the genetic epidemiologists, noting that their polygenic models were uninformative and more 'statistical' than 'biological'.

You will see in Chapter 6 that the rise of genome-wide association studies resulted in a dramatic realignment of the two classical models of genetic effect. Ironically, instead of the tremendous advances in molecular genetic methodologically bringing the triumph of the Mendelian genetic models, just the opposite has occurred. Genome-wide association studies have shown that the Fisherian polygenic model of many genes of quite small effect, assumed first by the early biometricians and later by the more modern genetic epidemiologists, was the correct one and not the major gene models advocated by the Mendelians.

1.4 Genetic terminology

The last section of this chapter is devoted to introducing terminology that will recur throughout the rest of the book. Figure 1.1 gives a summary overview. Genetics is full of terminology, which is not always used consistently in the literature.

In this book, by locus we mean a position on a chromosome. If we consider more than one locus, we write loci. You might think, why use a new word for place, or position, when we have perfectly good words already. This is because, in genetics, locus means more than just position. It refers to a contiguous region of DNA that encodes a specific function, and it is often used interchangeably with 'gene', the primary functional unit of DNA. However, for all practical purposes, remember that locus means a place in the genome.

By allele, we mean one of the alternative DNA sequence variants that occur at a locus. As each human (and most of the organisms in this book) has pairs of chromosomes (two copies of chromosome 1, two copies of chromosome 2, and so on) there will be two alleles at each locus, except in males on the sex chromosomes, because males have one X and one Y chromosome.

A genotype is the combination of alleles present at a locus. Note that genotype can also refer to genotypes of all loci in the genome, so that, together with the environment, it shapes an individual's phenotype, a term that is just another way of referring to a trait, feature, or characteristic that we measure for a genetic study. In a way, all of genetics can be reduced to questions about how genotypes influence phenotypes.

At a locus where sequence variants are found (thus, by definition, a polymorphic locus), genotypes can be heterozygous (not the same allele) or homozygous (the same allele). The number of alleles in a population determines the possible number of genotypes that can occur, so that for two alleles there will be three genotypes, for three alleles there will be six genotypes, for four alleles there are nine genotypes, and so on (for n alleles at a locus, the number of possible genotypes is the sum of integers from 1 to n). For most purposes in this book, we work with loci that have two alleles. For example, using A to represent one allele and a for the second, we expect to find in a population the homozygous genotypes aa and AA and the heterozygous genotype Aa.

A haplotype refers to the combination of alleles at different loci but on the same chromosome. Where locus 1 has the genotype AT and locus 2 the genotype CG, the two haplotypes could be either A-C and T-G, or A-G and T-C.

The definition of another term, gene, is not so easy and is explained in more detail in Chapter 3 (section 3.6). For now, consider a gene as a region of DNA that encodes RNA, which in turn encodes a protein.

A gamete is a male or female germ cell that exists as a single copy of DNA (a single set of chromosomes, rather than many pairs) and can join another gamete of the opposite sex to form a zygote.

Haploid regions of the genome are single copy DNA, while diploid regions consist of two copies. By definition, gametes are haploid cells.

We define additional terms later in the book. Chapter 3 gives an overview of molecular biology, and introduces the important, and easily misunderstood, difference between sequencing and genotyping (see Chapter 3, section 3.8). Chapter 4 introduces even more molecular terminology, most of it due to the application of next-generation sequencing to biology. The Appendix provides an introduction to basic quantitative genetics, another area rich in jargon. In general, you can regard this book as an introduction to the synthesis of molecular and quantitative genetics. Current practitioners of behavioral genetics need to be fluent in both fields. Good luck!

Figure 1.1 Genetic terminology. The figure shows two (very short!) chromosomes, one in blue and one in black. Genotypes are shown at two loci. At the top locus, there is a single allele, T, and a homozygous genotype (TT). At the lower locus there are two alleles (G and C) and a heterozygous genotype (GC). The two loci form two haplotypes: T-G and T-C.

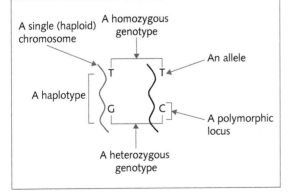

2

The genetic epidemiology of schizophrenia

[On studies with families and what they tell us about the inheritance of schizophrenia]

2.1 What is schizophrenia?

Q. Are you feeling ill?

A. You see as soon as the skull is smashed and one still has flowers with difficulty, so it will not leak out constantly. I have a sort of silver bullet which held me by my leg, that one cannot jump in, where one wants, and that ends beautifully like the stars. Former service, then she puts it on her head and will soon be respectable, I say, O, God, but one must have eyes.

KRAEPELIN (1915).

Schizophrenia is the most debilitating of adult psychiatric disorders. Onset is typically in early adulthood. Only a small minority of affected individuals truly recover. For most, their ability to live a full life—to have a meaningful job, to sustain lasting love relationships—is impaired. Schizophrenia strikes around one out of 150 individuals. In the 1950s prior to the deinstitutionalization movement that emptied many large public psychiatric hospitals, schizophrenia accounted for more than half of all the hospital beds in the USA.

Schizophrenia is the psychiatric disorder that comes closest to epitomizing what we understand as madness. Delusions, hallucinations, and thought disorder (as illustrated in the quote above) are all characteristic symptoms. In addition to the biomedical view of schizophrenia—as a poor-prognosis neuropsychiatric disorder—people with schizophrenia have also been described as the shamans or, in our culture, those in touch with the magical and divine nature of our world. In reality, individuals with schizophrenia are more likely to be impaired, low-status individuals living on the periphery of both traditional and modern western societies.

More intriguing is the possible relationship between schizophrenia and creativity. Anecdotally, the number of highly creative unconventional individuals who have relatives with schizophrenia has been noted. For example, in the twentieth century, Albert Einstein, James Joyce, and James Watson, all known for their genius, have had children with schizophrenia. Joyce's ill daughter was named Lucia. The Republic of Ireland, to which our story turns, designated July 26 as 'Lucia Day' to promote schizophrenia awareness.

Figure 2.1 shows pictures of his cat painted by the artist Louis Wain as he slid into a schizophrenic

Figure 2.1 The nineteenth- and twentieth-century artist Louis Wain, who was fascinated by cats, painted these pictures over a period of time in which he developed schizophrenia.

Images courtesy Henry Boxer Gallery.

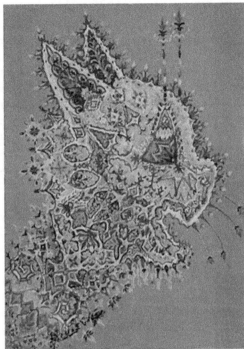

psychosis. These pictures convey a hint of what the world might begin to look like for those who suffer from schizophrenia. But lest we get carried away by a romantic view of madness, we should note that a very large study of the relationship between intelligence quotient (IQ) and schizophrenia in Sweden showed that the smarter you were, the lower your risk for schizophrenia (KENDLER et al. 2015d). No increase in risk was seen in those with the highest IQs.

2.2 Does schizophrenia run in families?

The idea that schizophrenia ran in families is far from new. The first modern describer of this disorder, Emil Kraepelin, writing toward the end of the nineteenth century, noted the apparent increased risk of schizophrenia in close relatives of his affected patients. This impression was validated by more systematic and increasingly sophisticated studies over the course of the twentieth century. We report one such study here.

Awareness of the disease in Ireland spawned one of the most informative studies to date of the genetics of schizophrenia. The Irish public health agencies had been keeping, in certain counties, detailed case registries: records of all individuals showing up at hospitals, clinics, or doctors' offices for any kind of psychiatric treatment. These registries documented large families that were ideal for a genetic study: the more family members recorded, the better the chance of demonstrating a genetic link. Beyond this, Ireland has the advantage of a relatively homogeneous population. Immigration to Ireland was a rare event prior to the last few years. With all of these attractions, and because of the long and complicated relationship between England and Ireland, in 1987 the US National Institutes of Health, rather than an agency in the UK, funded a study of the genetics of schizophrenia in the west of Ireland.

2.2.1 The Roscommon Family Study

Working with a number of Irish researchers, one of us (K.S.K.) came to know well (at times, perhaps, too well) the bed and breakfast houses and psychiatric hospitals of the western part of the Republic. During a 5-and-a-half year period, we first tried to track down all individuals in the county of Roscommon born after 1 January 1930 who had been treated for schizophrenia. There were a total of 285 such individuals. At the same time, we identified another 100 individuals who had been hospitalized with depression and other severe mood disorders. These individuals are known as probands, meaning an individual through whom a family with a genetically influenced disorder is ascertained. Finally, we identified an additional 150 people, selected at random from the voter registration rolls, for comparison as our control group.

The first steps in the study were to follow up and try to interview the probands. We succeeded in 88% of the cases where they were alive and traceable. When we found them, we conducted a psychiatric assessment that included detailed questions about the kinds of emotional and psychiatric problems they might have experienced in their lives. We also searched for hospital records, and these were systematically reviewed and abstracted.

The next step was to identify, locate, and attempt to interview all the first-degree relatives of the controls, those with mood disorder, and those with schizophrenia—that is their parents, full siblings, and offspring. By the time we were done, we had used up a lot of index cards (laptops did not exist when we started this work) identifying a total of 2,043 relatives who were still alive and hadn't migrated to somewhere inaccessible.

Next, we tried to interview these relatives; most were still in Ireland but many were abroad, in Liverpool, Birmingham, London, and other parts of England. And, as before, we interviewed everyone we could find, 1,753 in all, speaking to them mostly at home but occasionally in parked cars, pubs, libraries, or even standing in fields. Members of our interviewing team were 'blinded' meaning that they did not know (and asked the respondent not to tell them) whether they were interviewing the relative of a control, or of a proband with a mood disorder or schizophrenia. Similarly, when senior psychiatrists (including K.S.K.) reviewed the interviews and abstracts of hospital records if they had been hospitalized, they were

similarly blind. This approach helps protect research from subjective bias. Like most people, scientists can make the mistake of seeing the world as they want to, rather than as it really is. Blinding the interview process makes it much less likely that we impose our biases on our results.

At the end of all this interviewing (it took our team over 5 years), we had enough data to answer definitively the question of whether first-degree relatives of an individual diagnosed with schizophrenia are more likely to be schizophrenic than someone picked at random from the same population. Our key result is shown in Figure 2.2 as a life table (KENDLER et al. 1993b). The figure shows the cumulative risk for schizophrenia on the y-axis and age on the x-axis. The difference in risk for schizophrenia in the relatives of our schizophrenic versus matched control probands was striking.

We found that if you are in the immediate family of someone with schizophrenia, the chances are 13 times higher that you too will develop schizophrenia, compared with those who are related to a randomly chosen individual from the electoral rolls (KENDLER et al. 1993a). The odds are not actually all that high in absolute terms, 65 out of 1,000 compared with five out of 1,000, but given the nature of this disease for those

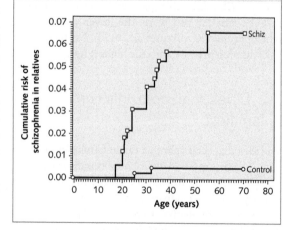

Figure 2.2 Graph showing cumulative risk for schizophrenia in relatives against the age of probands. Schiz, individuals with schizophrenia.

Reproduced with permission from K. S. Kendler, M. McGuire, A. M. Gruenberg, A. O'Hare, M. Spellman, and D. Walsh. The Roscommon Family Study. I. Methods, diagnosis of probands, and risk of schizophrenia in relatives. *Arch Gen Psychiatry* 50 (7):527–40, 1993. Copyright © American Medical Association.

affected by it and their families, any number of cases is a tragedy. Either way, a family history is a better predictor of future disease than any other factor. Other studies came to a similar conclusion: at least ten other family studies of schizophrenia conducted since 1980, using similar methods, all showed that close relatives of individuals with schizophrenia have approximately ten times the risk of suffering from the illness themselves.

Roscommon is a predominantly rural region consisting of small villages connected by winding narrow lanes. One summer evening during one of many home visits, an elderly grandmother took me aside after the interview had ended. *'Now doctor,'* she asked, *'what is this you've been doing, driving all about the place, asking all these questions of my relatives?'* I explained to her that we were studying mental illness and trying to understand if it ran in families, and to do so we needed to study all kinds of families. *'Do you mean to tell me,'* she said with a slight sense of indignation, *'that you are spending all this time just to see if being daft runs in families? Why everybody knows that! Take the O'Donnells for example. They are mad as can be and it goes back generations!'* She proceeded to give me a detailed genealogy of the O'Donnell family and the characteristics of each member, describing how, when she was a little girl, the grandfather had taken off in the family tractor down to the main town shouting and carrying on, talking about God and the Devil and who knows what else, and ended up in the local mental hospital (St Pat's as it was known). She told me about his children and grandchildren many of whom had had psychiatric difficulties. Then she told me about the McGuires with a similar level of detail. That was enough for her. She went back into the house shaking her head and laughing gently to herself. *'What will these crazy Americans think of next? You'd think they have all the time and money in God's creation.'* I felt a bit foolish. After all, was it really necessary to go to all this trouble to demonstrate what so many people take for granted, that madness does indeed run in families?

2.3 Traditional assumptions about heredity, mental illness, and behavior

After the discovery of the principles of inheritance in the late nineteenth and early twentieth centuries, it was immediately assumed that there were hereditary factors (genes) for each human disease and trait.

One of the main proponents of this view was Charles Davenport, who sent out staffers to knock on doors all over central Long Island (in the state of New York) to interview families in pursuit of evidence for a gene that determined 'nomadism'. In a 1915 monograph with the title 'Nomadism, or the wandering impulse, with special reference to heredity' (DAVENPORT 1915), he reported 100 'family histories' showing the *'distribution of the nomadic tendency'*. In contrast to the Roscommon study described in the previous section, Davenport's methods were not what we would consider rigorous. His diagnostic criteria were vague and subjective, and he had personally interviewed only a very small proportion of his sample. To give you a flavor of his clinical methods, here are a few of the 'remarks' we see about members of his first 25 pedigrees: *'Mother has 2 sisters of good repute. Mother's father was a Western desperado who married a good woman'*, *'A bear-hunter of note'*, *'A stage-driver'*, and *'At one time an itinerant tinkerer.'* On the basis of these descriptions, Davenport concluded that nomadism was a sex-linked recessive trait.

We might see this and similar work as simply a product of its time. After all, at that time the science of statistics, psychology, and sociology was still immature, and survey methods were primitive. Few alternative explanations were available, and all traits were assumed to be hereditary. However, this work has a more ominous side: results from this and other studies, especially of idiocy or feeble-mindedness, were used by the US Congress in subsequent decades to restrict immigration into the USA (KEVLES 1985).

2.4 Diagnosing schizophrenia

Is schizophrenia any more precisely defined than the 'wandering impulse'? Has our work really progressed from the days of Davenport? Up until the early 1970s, definitions of mental illness, including schizophrenia, consisted largely of clinical theories articulated by German psychiatrists formulated in the late nineteenth and early twentieth centuries. During the 1960s, psychiatry in general, and psychiatric diagnoses specifically, were under attack by the 'anti-psychiatry' movement, associated with such names as Szasz, Laing, and Foucault (LAING AND ESTERSON 1964; FOUCAULT 1965; SZASZ 1974). Psychiatric diagnoses at that time were not that reliable (but neither were diagnoses in most of the rest of medicine). Diagnostic manuals contained vague descriptions of psychiatric disorders, open to varying interpretations. The situation wasn't helped by several high-profile court cases featuring 'expert' psychiatric testimonies that completely disagreed with each other.

But the problems were not intractable, and the field of psychiatry responded. A first critical step was the realization that much of the difficulty resulted from differences in training and in concepts about the nature of the illness. This insight arose from efforts to understand why admission rates for schizophrenia were higher in the USA than in the UK. In the 1960s, the US–UK Diagnostic Project compared 250 patients admitted to the Brooklyn State Hospital in New York and a comparable number admitted to Netherne Hospital in the south of London. In order to assure uniform criteria, they developed a structured interviewing procedure. When this standardized form of assessment was used, they found that there was no significant difference in the way key diagnostic symptoms and signs were rated on either side of the Atlantic (COOPER et al. 1972). (Physicians make a distinction between a symptom which is something a patient tells you about and a sign, which is something that a doctor observes about a patient.) What did differ was the way those symptoms and signs were assembled into a diagnosis.

The solution to this problem, which had two major components, arose over a decade on both sides of the Atlantic. The first was the development of clear operationalized diagnostic criteria for psychiatric illness, first proposed in a famous article published by John Feighner and colleagues in 1972 (FEIGHNER et al. 1972). Instead of vague diagnostic descriptions, psychiatric diagnostic manuals came to contain clearly specified individual criteria. Second, structured psychiatric interviews were developed first in England by John Wing (WING et al. 1967) and later by a number of psychiatrists and epidemiologists in North America. Instead of the typical free-flowing interview, these were carefully scripted, so that no matter who was giving the interview, the same questions would be asked.

One of the first major studies that utilized the first generation of structured psychiatric interviews was called the International Pilot Study of Schizophrenia. Researchers in nine centers worldwide, including hospitals in Africa, India, South America, and Russia, interviewed patients clinically diagnosed with schizophrenia (SARTORIUS et al. 1972; SARTORIUS et al. 1977). The goal was to address this simple but important question: were the main features of schizophrenia similar in patients from varying world cultures? The answer was clear: yes. The most consistent and salient features of schizophrenia were seen at all the centers: delusions of control, thought insertion, thought withdrawal, thought broadcasting, and third-person auditory hallucinations.

Psychiatric interviews have now been standardized to the point that it is possible to run them on a computer. You can get a feel for the way the interview works from these questions, taken from the start of one of the most recently developed instruments—the CIDI (which stands for Composite International Diagnostic Interview):

1. Now I want to ask you about some ideas you might have had about other people. Have you ever believed people were spying on you?
   NO (SKIP TO 2)
   YES

      A. How did you know that was happening?
      IS EXAMPLE IMPLAUSIBLE?

2. Was there ever a time when you believed people were following you?
   NO (SKIP TO B)
   YES

      A. How did you know people were following you?
      IS EXAMPLE IMPLAUSIBLE?

      B. Have you thought that people you saw talking to each other were talking about you or laughing at you?
   NO (SKIP TO 3)
   YES

      C. What made you think it was you they were discussing or laughing at?
      IS EXAMPLE IMPLAUSIBLE?

3. Have you ever believed you could actually hear what another person was thinking, even though he or she was not speaking?
   NO (SKIP TO 4)
   YES

      A. How was it possible for you to hear what a person thought if that person didn't say anything?

4. Have you ever been convinced that you were under the control of some power or force, so that your actions and thoughts were not your own?

Schizophrenia, then, can be reliably diagnosed by the presence, and absence, of a number of well-characterized psychological and behavioral phenomena. These include auditory hallucinations (typically hearing voices), delusions (very distorted and unrealistic beliefs), and thought disorder, as illustrated by the opening quote of this chapter.

The reliability of diagnoses was carefully assessed in the Roscommon Family Study. Two well-trained psychiatrists (one of them K.S.K.) reviewed 69 cases weighted toward those with symptoms. They each used 12 different diagnostic categories and agreed 83% of the time. While few studies have quite this high level of reliability, many have shown that using modern methods, schizophrenia and other forms of severe psychiatric illness can be assessed at least as reliably as many other medical conditions. For example, 933 physicians, when asked to match a set of taped heart sounds to 15 possible diagnoses, agreed in about 80% of cases (BUTTERWORTH AND REPPERT 1960), pretty much the same concordance as achieved for the diagnosis of schizophrenia. We should, however, recognize that being able to make a consistent diagnosis may not reflect anything about the underlying disorder. What if different underlying causes can produce the same set of symptoms? This takes us back to the Irish woman's question: why ask if madness runs in families?

2.5 Adoption, heredity, and environment

In human families, relatives can be similar because they share environments or because they share genes, or both. Relatives share all kinds of environments. They live in the same neighborhoods, often attend religious services together, eat a similar diet, are exposed to the same level of harmony or conflict in the home, and might all live close to an industrial plant that pumps out pollution into the air or water.

All of these reasons and more could explain the resemblance among members of the same family. If we could find cases where genetically related individuals are exposed to different environments, we might be able to sort out the relative importance of genetic and environmental factors in the origins of a disease such as schizophrenia. Studying what happens to adopted children provides one way to separate genes

from environment, although, as we shall see, the approach has its limitations.

Adoption studies are one of the two main 'experiments of nature' that can be used in human populations to separate out the effects of nature and nurture (twin studies are the other). The design of adoption studies requires subjects to be adopted at an early age by individuals to whom they are genetically unrelated. The power of the design arises from the fact that adopted individuals share genetic but not environmental factors with their biological relatives, and share environmental but not genetic factors with their adoptive relatives.

Two different adoption designs have been used most commonly, both of which have been used to study the genetics of schizophrenia. The first, and simplest, begins with ill mothers who have given children up for adoption and a matched group of well mothers who similarly have had their offspring adopted. The risk of illness is then compared in the two groups of children. If genetic factors influence the transmission of the disease within families, then the adopted-away offspring of the ill mothers should have a higher rate of illness than the adopted-away offspring of the control mothers.

The second design begins with ill adoptees and requires an evaluation of their biological and their adoptive relatives. If genetic factors are responsible for the familial nature of the disorder, then the biological relatives of the ill adoptees will have higher rates of illness than the biological relatives of the control adoptees. If the disorder runs in families for environmental reasons, then the adoptive relatives of the ill adoptees will have higher rates of illness than the adoptive relatives of the control adoptees (see Chapter 11, section 11.2 for a further description of adoption studies). Examples will make clear how the two adoption designs work to distinguish genetic from environmental influences on schizophrenia.

2.5.1 An adoption study in Oregon

While still completing his psychiatric training at Oregon State Hospital in the 1960s, Leonard Heston noted that pregnant mothers with schizophrenia were typically urged to give their children up for adoption, usually never to see them again. Reviewing records of births from 1915 to 1945, Heston identified 58 adoptees from schizophrenic mothers who, within 3 days of birth, were placed in either adoptive or foundling homes and had little or no further contact with their mother or with her biological relatives. He then found a matched set of control adoptees from the same agencies whose biological mothers and fathers had no record of psychiatric hospitalization (HESTON 1966).

Heston made exhaustive efforts to track down all of the adoptees, just as we had had to do in order to find people for our study in Ireland. Fourteen adoptees had died in infancy or childhood, and there was no information at all about a further five. This left 47 high-risk and 50 control adoptees whom he interviewed personally. Reasonably good records from a variety of sources were available for the other adoptees.

All of the information about each adoptee was organized into a dossier, omitting all information about the mental health of the biological mother, and was evaluated by two psychiatrists as well as by Heston. A fourth psychiatrist was brought in to resolve diagnostic disagreements. As seen in Figure 2.3, five of the 97 subjects were diagnosed as having schizophrenia. All five were adopted-away offspring of schizophrenic mothers; none came from the control adoptees. In fact, the rate of schizophrenia in the adopted-away offspring of schizophrenic mothers (10.6%) was indistinguishable from the rates of illness seen in prior studies of children who were born of and reared by schizophrenic parents. Although the number of subjects was small, these results had a dramatic

effect on the field at their time of publication because of the compelling evidence that schizophrenia ran in families for genetic reasons.

2.5.2 The Danish Studies and schizophrenia spectrum disorders

The second kind of adoption design used in schizophrenia begins with adoptees who developed schizophrenia (rather than with mothers who have schizophrenia). Seymour Kety and Fini Schulsinger, in a very influential series of studies, utilized this design. They began by contacting the Department of Justice in Denmark, which records every legal adoption. They identified almost 6,000 adoptees born in the Greater Copenhagen area between 1924 and 1947. To identify those who had gone on to develop schizophrenia, they searched through the National Psychiatric Registry, which documents every psychiatric hospitalization in Denmark. This is not a bad place to start, as we know from a number of studies that in developed countries almost every individual affected with schizophrenia is admitted to a psychiatric hospital at some point in their life. In this 'Copenhagen sample', the investigators identified 34 adoptees who were hospitalized for schizophrenia, 34 control adoptees, and all of the biological and adoptive relatives. Over a decade later, they did something rarely accomplished in human studies: they replicated the entire project, searching

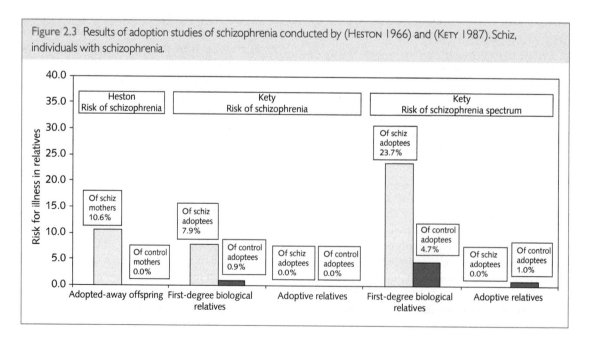

Figure 2.3 Results of adoption studies of schizophrenia conducted by (HESTON 1966) and (KETY 1987). Schiz, individuals with schizophrenia.

the adoption records for cases that occurred elsewhere in Denmark (the 'Provincial sample'). From a total of 8,944 adoptees, they identified 41 adoptees later hospitalized for broadly defined schizophrenia and a matched set of control adoptees (KETY 1987).

When the findings of both studies (Copenhagen and Provincial) were collated, analyzed, and re-analyzed using modern diagnostic criteria, one of us (K.S.K.) along with a close colleague (Alan Gruenberg) found that the rate of schizophrenia in the first-degree biological relatives of the schizophrenic adoptees was 7.9% versus 0.9% in the control adoptees (Figure 2.3) (KENDLER et al. 1994a). Sharing genes, but not environment, with a schizophrenic individual produced a nearly tenfold increased risk for schizophrenia. Conversely, those reared in a family where an adoptee developed schizophrenia were not themselves at increased risk: no cases of strictly defined schizophrenia were diagnosed in nonbiological relatives in either the schizophrenic or the control groups. So sharing a rearing environment but not genes with an individual who has schizophrenia produced no increase in risk of schizophrenic illness.

The Danish adoption studies have turned out to be crucial in several ways. In addition to explaining why schizophrenia ran in families (showing that it was due to genes), they also helped to sort out some of the inconsistencies nagging the diagnosis of schizophrenia.

The international study of schizophrenia, alluded to previously, had found that no matter where the patients were or who saw them, there was agreement on the diagnosis of schizophrenia in two-thirds of cases (SARTORIUS et al. 1972; SARTORIUS et al. 1977). That meant that one-third of the patients had symptoms that left room for doubt. Some of these cases were marginal and unlikely to be diagnosed as schizophrenia outside of two particular centers. The remaining 20%, on the other hand, manifested some aspects of schizophrenia but not a complete enough set of symptoms to meet a strict definition. This raised the possibility that there might be a number of distinct conditions, with different causes, hidden under the label 'schizophrenia'. Psychiatrists talk about schizophrenia spectrum disorders, a set of conditions that possess some but not all of the features of narrowly defined schizophrenia. Sometimes the symptoms are not so severe, or don't last as long and don't interfere with the patient's life style as obviously as the frank delusions and hallucinations that are characteristic of the full-blown psychosis.

One of the most interesting disorders within the schizophrenia spectrum is known as 'schizotypal personality disorder'. For the last 100 years, clinicians have noted that some relatives of individuals with schizophrenia seemed to possess, in mild form, some of the key symptoms and signs of schizophrenia. Such individuals tended to be socially odd, suspicious, are often both physically and socially awkward, commonly have odd ideas (e.g. thinking that people are looking at them or making fun of them in public places), and are prone to unusual patterns of communication.

The Danish adoption studies discovered that almost a quarter of the first-degree biological relatives of schizophrenic adoptees (those with the strictly defined disease) could be diagnosed by this broader definition of schizophrenia. By contrast, the rate was less than 5% in the first-degree relatives of the control subjects, 1.0% in the adoptive relatives of the controls, and 0% in the adoptive relatives of the schizophrenic subjects (Figure 2.3). In other words, whatever genetic predisposition increased the susceptibility to schizophrenia, strictly defined, also increased the likelihood of developing other syndromes within the schizophrenia spectrum, particularly schizotypal personality disorder. This led to an important conclusion: the spectrum of symptoms in schizophrenia and the variability of their appearance are all influenced by the same genetic liability. We discuss further the relationship between personality types and schizophrenia in Chapter 12, section 12.3.1.

There is almost a recursive property to this finding: we not only learn about the nature of the gene action, but the gene action also tells us something about the boundaries of the disorder itself. The relationship thus revealed between genes and phenotype has implications for how we go on to interpret the molecular findings described later. The recursive nature of genetic studies of behavior, in which we learn about the phenotype, just as much as about the genotype, is a recurring theme of this book. As an example, genetic studies of autism identified evidence for autism spectrum disorders (see Chapter 8, section 8.2).

2.5.3 Consistency and pitfalls in adoption studies

Altogether, there have been five major adoption studies of schizophrenia performed in the USA, Denmark, and Finland (HESTON 1966; KETY et al. 1968; ROSENTHAL et al. 1971; TIENARI 1991; KETY et al. 1994); all found strong and consistent evidence for substantial genetic effects on the risk for schizophrenia. But even the best adoption studies can have important limitations.

In the ideal adoption study, there should be no correlation between the characteristics of the biological parents who are giving the child up and the adoptive family into which the child is going. This is critical in order to separate the effects of genes and family environment. In reality, however, adoption agencies are sometimes asked to match characteristics of the adoptive family with the biological parents. In the USA, adoption agencies are often affiliated with religious organizations, where Catholic, Protestant, and Jewish babies are given to homes of the same religious background. Adoptive parents sometimes ask for adoptive children with a certain eye color, or from parents with musical or athletic ability, or of a certain height or particular complexion.

Another concern in adoption studies is the possible impact of the intrauterine environment and early (pre-adoption) childhood experiences. Some resemblance between the biological mother and her child might be due to what happens during the 9 months of pregnancy when the child and mother inhabit the same body. Furthermore, many children are not adopted until **after** they have spent the earliest years of life with either of their biological parents. If being raised as an infant and toddler by an individual with a severe psychiatric illness contributes to future risk of illness (although the evidence to date does not suggest that this is the case), it could confound adoption studies.

Finally, biological or adoptive parents of adoptees are not generally representative of the population at large. Adoption agencies see it as their task to try to find ideal homes for their adoptees. Because of this, adoptive parents have higher socioeconomic status and lower rates of drug, alcohol, and psychiatric problems than are found in the general population. The biological relatives of adoptees are not a random sampling either, but the nature of the bias varies with social circumstances and historical period. Immediately after World War II, for example, a large proportion of adoptions in Europe took place as a result of poverty. Most adoptees came from intact poor rural families. In general, their rates of psychopathology were not elevated. However, in the late 1990s the large majority of adoptees were the children of teenage mothers in urban areas, born out of wedlock. Rates of psychiatric illness, as well as drug and alcohol abuse, were much higher than normal in these parents. In addition, the identity of the father might be impossible to determine in some cases. In other words, either biological or adoptive parents may be quite unrepresentative of the population, thus biasing the findings.

2.6 Twins

None of the completed adoption studies of schizophrenia was free of some methodological limitations (indeed, this is true for all human genetic studies, which can never approach laboratory sciences in the degree of the control of the experimental variables). To increase our confidence in the validity of these findings, it would be best to try to replicate these adoption results in a completely different kind of human study that could also separate out the effects of genetic and environmental influences. Fortunately, there is one such other method—twin studies. The potential value of twins for genetic research was first recognized by one of the founders of genetics and statistics, Francis Galton, whom we introduced in Chapter 1. The title of his 1876 paper, 'The history of twins, as a criterion of the relative powers of nature and nurture', says it all (GALTON 1876).

2.6.1 Two types of twins

Human twins come in two different kinds. One kind of twin, called monozygotic (MZ) or identical, is a result of a single egg being fertilized by a single sperm. In the first 2 weeks of development, the fertilized egg splits apart into two genetically identical organisms that then grow up into two adults who are always of the same sex. We still do not really understand what causes splitting, but it happens at a similar rate (approximately three per 1,000 births) regardless of ethnic background or maternal age. The second kind of twin is called dizygotic (DZ) or fraternal, and results from the female producing two eggs instead of one during ovulation. We understand more about the biology of dizygotic twinning. It is, in essence, a misfiring of the female reproductive cycle due to 'overdrive' of the ovaries by hormones secreted from the pituitary gland. Dizygotic twinning is most common in African and least common in Asian populations. The rates are also higher in older women. As the two eggs are nearly always fertilized by separate sperm from the same father, dizygotic twins are, from a genetic perspective, exactly like regular brothers and sisters except that they are fertilized at the same time, share a common uterine environment, are born at the same time, and grow up together at the same age in the family.

For all intents and purposes, therefore, monozygotic twins are genetically identical: they share 100% of their genetic variants at birth. Fraternal twins, like regular siblings, on average share 50% of their genetic variants identical by descent. For traits where the resemblance is due solely to inherited factors, identical twins ought to resemble one another twice as much as fraternal twins (although, because of experimental error, this will never be exactly true). On the other hand, if we make the reasonable assumption that the environments of identical and fraternal twins are approximately similar, you would predict that if the causes of resemblance for a trait are due to environmental factors, the level of resemblance between identical and fraternal twins ought to be approximately the same.

Twin studies are based on the idea that resemblance between twins results from two distinct sets of factors: genes and shared environment. We start by working out how similar each twin is to their twin sibling. In this case, we are measuring similarity in terms of a diagnosis (schizophrenia), but it could equally well be height, personality, or intelligence. We record similarity as a correlation, which can range from 0 (meaning that the two traits are completely unrelated to each other) to 1.0 (meaning the two traits are perfectly related to each other). Recall our discussion of the liability threshold model in Chapter 1, section 1.3.3 that is applied to twin studies of dichotomous traits such as schizophrenia. The standard twin study assumes that shared environment contributes equally to the correlation in MZ and DZ pairs. This is not true, however, for genetic factors. As MZ twins are genetically identical while DZ twins share half their genes, the genetic component of the correlation in the MZ twins should be twice the value of that seen in DZ twins.

By comparing the degree of correlation for a trait in MZ and DZ pairs, it is possible to infer the role of genetic versus shared environmental factors. Take religious affiliation: for example, whether people self-identify as Catholic, Jewish, Baptist, Muslim, and so on. Twins strongly resemble one another in their religion, but the degree of resemblance is virtually the same in MZ and DZ twins. These results suggest (consistent with common sense) that genes have nothing to do with religious identity. Resemblance in twins for religious affiliation results solely from shared environmental experiences (EAVES et al. 1990).

Contrast this to the pattern seen for height, where the correlation in MZ twin pairs is typically very high (around 0.90) and the correlation for height in DZ twin pairs is moderate (around 0.45). When correlations in MZ twins are approximately twice that seen in DZ twins, we can conclude that twin resemblance is driven primarily by genetic factors. Of course, there are traits that are influenced both by genes and shared environment. In this case, the correlations in MZ twins are greater than those seen in DZ twins, but the difference between them is less than the 2:1 ratio expected if only genetic factors are operating.

These three rules are worth repeating:

1. If the correlation in identical (MZ) twins is at least twice as large as in fraternal (DZ) twins, twin resemblance is likely to be due entirely to genetic factors.

2. If the correlations are approximately the same magnitude in both types of twins, twin resemblance is likely to be due to shared environmental effects.

3. If the magnitude of the correlation in DZ twins is between 50 and 100% of that seen in MZ pairs, then it is likely that both genetic and shared environmental effects are operative.

The methods used to analyze twin data in behavior genetic research are described in more detail in the Appendix, section A5, together with an explanation of the mathematical modeling used to assess both genetic and environmental contributions to behavior.

2.6.2 Heritability and twin studies

Just how useful are these rules? Any conclusions about the relative influence of genetics versus environment in studies of this sort rely on the statistical concept of heritability. Because heritability is an often misunderstood and sometimes maligned statistic, we will take time to explain what it is and what it is not.

Most importantly, heritability is based on the concept of variability. Assume that we are studying height in a population of 5,000 individuals and imagine that we could line up all these individuals from the tallest to the shortest. We would see a great deal of variation, with the largest number of people of middling height and a diminution in numbers at the extremes of the very tall and the very short.

We also know that, just as individuals differ in height, they also have differences in their genomes. The extent of DNA variation among humans is still not fully known, but it's definitely there. Heritability is nothing more than **the proportion of variation in height (or whatever phenotype we study) that is due to the genetic differences among individuals in the population**. This is important enough to express in a different way as the following ratio:

$$\text{Heritability} = \frac{\text{genetic variance}}{\text{total variance}}$$

In this formula, total variance in a trait is broken down into genetic and environmental variance. If height had a heritability of 100%, it would mean that all of its variability could be explained by genetic differences between individuals. In fact, the heritability of height is about 80%. A heritability of 0% would indicate that genes contribute nothing at all to the observed differences between individuals. This is close to what is seen when we study religious affiliation.

Heritability is inherently a population-based statistic. In fact, a disease in a single person does not, by definition, have a meaningful heritability. Diseases themselves also do not have heritabilities. Only given diseases in specific populations at specific times have a defined heritability. That is, heritabilities could change between populations or within a population over time if genetic variability changed or if environmental variability changed.

Imagine an isolated population, somewhere on the edge of Europe, where together with the environmental risks for schizophrenia, the population contains 100 different genetic variants, each of which independently influences the risk for the disease. Psychiatric geneticists carry out a twin study and determine that the heritability of schizophrenia is 70%. Now it so happens that a virus, passed from birds to humans, such as avian flu, arrives and randomly infects part of the population. This fictional infection does not kill, but it substantially increases the risk of developing schizophrenia.

Psychiatric geneticists, keen to add another publication to their curriculum vitae, rush back into the field and reassess the heritability. They find that the figure has now dropped to 50%, because the total variability in risk to schizophrenia has increased. Some people are infected, but not all. The increased variability in risk is due entirely to an environmental variable.

A little later, a group of immigrant laborers arrive, settle, and begin marrying the inhabitants. The immigrants carry ten new genetic variants hitherto unknown in the indigenous population, all of which increase susceptibility to schizophrenia. After a few generations, when these genetic variants have spread through the population and the virus is no longer present, psychiatric geneticists scent another publication. For the third time, they assess heritability. This time the estimate has increased, to 80%. New variability in risk of illness has been introduced into the population, but in this case the new risk is genetic in origin.

2.6.3 Controversies over heritability

How useful is heritability as determined by twin studies? The answer is controversial. Let us begin by summarizing the views of the psychologist Leon Kamin. Here he is in full flood on the subject of the inheritance of IQ (these and subsequent quotes are all from KAMIN 1974): *'There exist no data which should lead a prudent man to accept the hypothesis that I.Q. test scores are in any degree heritable … The fact that twin correlations are high, and that MZs resemble one another more than DZs, are wholly consistent with the expectations of an environmental view.'* Kamin argues strongly against the assumption that MZ and DZ twins grow up in similar environments: *'The various sampling problems, however, pale into relative insignificance compared to a fundamental objection … the essential assumption that relevant environmental differences impinging upon MZ and DZ pairs are equal is, on its face, absurd.'* There then follow a few pages debunking the idea that the environments could ever be thought equal (*'the quaint notion of a constant "within-family environmental variance" acting equally on all categories of twin pairs seems clearly untenable'*) and the final conclusion that *'our study of kinship correlations has revealed no evidence sufficient to reject the hypothesis of zero I.Q. heritability.'*

Kamin's book contributes to a long history of controversy surrounding twin studies of behavior and genetics. Much of the passion comes from the mixing of politics with science that always seems to accompany these studies. From the very beginning, Galton saw genetics as a means of improving the human race through selective breeding. This philosophy gave rise to the eugenics movement, which found one of its most ardent supporters in the American geneticist whom we discussed above: Charles Davenport.

His studies of hereditary 'feeble-mindedness' were inspired at least in part by his wish to eliminate what he saw as a genetic plague (DAVENPORT 1915). Through his efforts and those of others, the movement eventually succeeded in reinforcing the prejudices of US public policymakers in several infamous cases. Laws mandating the compulsory sterilization of mental patients were passed in many states, and in 1924 the U.S. Congress reversed its open door immigration policy with an ethnically restrictive quota system. The ultimate embodiment of eugenics came under the National Socialist (Nazi) program in Germany, starting with the compulsory sterilization of mental patients (modeled after US statutes) and ending with the Final Solution (KEVLES 1985).

2.6.4 Challenging the assumptions

How well do the results from twin studies withstand Kamin's criticisms? To answer, we can draw upon the large and remarkably rich literature on the quantitative genetic basis of individual differences, inferred from twin studies.

The first methodological concern, and one that Kamin raises, is whether the assumption that the environmental exposures of MZ and DZ twins are equally similar is valid. Standard twin analyses assume that the greater resemblance in MZ twins compared with DZ twins results entirely from sharing greater genetic similarity (the 'equal environment assumption' or EEA for short).

One approach to evaluate the EAA is to examine the physical similarity of the twins. Identical twins, on average, resemble each other much more than fraternal twins. Perhaps the world treats people more similarly if they look more similar. If so, then the physical similarity of identical twins would result in their greater behavioral similarity. So is it the case that greater similarity for psychiatric disorders is predicted by greater physical similarity? By taking photographs of twins, rating them on the level of physical similarity, and comparing the results with other measures, it is possible to test this hypothesis. With substantial, but not complete, uniformity, the answer has been no.

Another more direct way to evaluate the EEA is to ask twins about the similarity of their environmental experiences. Most of these studies have focused on childhood environments. Without doubt, monozygotic twins are more frequently dressed alike as children than are dizygotic twins. Other studies ask whether twins shared the same room at home, had similar playmates, or shared the same classroom when children. Putting all of this together, it is possible to create an index of 'childhood environmental similarity'. Such an index is nearly always greater in monozygotic than in dizygotic pairs, sometimes quite substantially so. The question can then be asked whether 'childhood environmental similarity' can explain the greater similarity in the trait being studied (e.g. schizophrenia or depression). In the vast majority of studies where this has been asked, the answer is no.

Sometimes, twins and their parents are misinformed about their true twin status; that is, some twins who are really MZ think they are DZ and some who are really DZ think they are MZ. This provides an ideal natural experiment to test whether the self-perception of twins about their zygosity, or the expectations of them by their family and social environment, or both, might actually influence the degree of resemblance. This test of the EEA—whether twins who are really MZ but think they are DZ are any less similar than twins who are truly MZ and also think they are MZ—has been applied in several different studies to personality measures, IQ, and risk for psychiatric disorders. In nearly all cases, the results have been consistent: in misidentified twins, self-perception of twin status is not a good predictor of resemblance.

Yet another method used to test the EEA examines what might be termed the 'parental philosophy' toward raising twins. Most parents of twins approach the task with one of two extremes: either by tending to treat the twins as a unit or by treating them as distinct individuals. Not surprisingly, the former approach is more common among parents of MZ twins than among parents of DZ twins. Perhaps MZ twins turn out more similarly than DZ twins, in part, because they are more frequently treated as a unit by their parents. One study examined this issue, and the outcome was that twin similarity for common psychiatric and drug-use disorders could not be explained by parental treatment.

The last approach comes from a famous study with the provocative title, 'Do parents create, or respond to, differences in twins?' The psychologist Hugh Lytton and his research team examined a detailed set of behavioral interactions between 2-and-a-half-year-old male twins and their parents in their natural home environment. They observed 48 different aspects of parent–child interactions in 17 MZ and 29 DZ twin pairs. Parents treated the MZ twins more similarly than the DZ twins in only seven of these behaviors, which included the mother's tendency to use material reward, such as candy, for reinforcing behavior, the degree to which the mother encouraged independent behavior by the child, and the frequency with which the father used reasoning to justify his commands and rules. Although only a fraction of parental behavior was involved, the results might, on the face of it, be thought to represent an obvious violation of the EEA: parents do treat identical twins more similarly than fraternal twins (LYTTON et al. 1977).

However, Lytton then asked whether parents treat MZ twins more similarly than DZ because MZ twins really do behave more similarly than DZ, or because the parents expect them to behave more similarly. To resolve this question, he was able to observe parental behaviors that were not directly elicited by a child. He divided all parent-initiated behaviors into four broad classes: 'command prohibitions', 'suggestions' (a milder form of parental control), 'positive actions' (which included parents doing such things as showing pleasure, approving, expressing affection, and playing with the child), and 'neutral actions' (involving emotionally neutral parental behaviors such as talking, approaching, or showing things to their twin sons). He then compared the similarity of parent-initiated actions in these four categories between identical and fraternal twins, separately for mothers and fathers. In only one of the comparisons—mother's suggestions—was there the expected finding of greater similarity in parent-initiated behaviors in identical than in fraternal pairs. In the other seven categories of parental behavior, there was not even a trend for parent-initiated actions to be more similar in identical versus fraternal twins. Lytton made the following observation about his data:

... the results ... lead to the conclusion that: a) parents do treat MZ twins more alike than DZ twins in some respects, but, b) they do not introduce systematic greater similarity of treatment for MZ twins in actions which they initiate themselves and which are not contingent on the child's immediate preceding behavior ... In other words, parents respond to, rather than create, differences between the twins ...

This study carries an important potential message: in measuring similarity of treatment or social environments in twin studies, it is important to consider that MZ twins, because of their more similar behavior, can elicit more similar treatment.

So, what can we say about the validity of the EEA for twin studies of psychological and psychiatric traits? Tests have with substantial but not perfect consistency supported its validity. The studies are not faultless and may not, for example, have identified the right kind of environment that truly impacts on twin resemblance (e.g. the impact of intrauterine environment). In sum, however, it seems unlikely that most twin studies are seriously biased by violations of the EEA.

There are, as you might imagine, other objections to the twin method. As twins constitute no more than roughly 1.5% of most human populations, what we learn about them may be irrelevant to the rest of us. A large literature addresses the question of the similarity between twins and singletons on a variety of medical, physical, psychological, and psychiatric traits. Twins are at increased risk for a few problems such as low birth weight and prenatal or perinatal complications. (This is because the reproductive system of the human female has largely evolved to have 'womb for one'.) However, the risk of twins for a very broad array of other medical conditions and for psychiatric disorders and symptoms is entirely typical of the general singleton population. Aside from traits directly related to the risks of twin pregnancy, it seems entirely justified to apply the results from twin studies to the general population.

Recall that we noted previously that heritability estimates are population specific. How good a job have we done of determining whether the heritability of schizophrenia is broadly consistent across different human populations? To date, twin studies of schizophrenia have been conducted in the USA, England, Japan, Denmark, Norway, Sweden, Finland, and Germany: nearly all studies are from Europe or North America. No large-scale twin (or adoption) studies of schizophrenia have been performed in populations of African descent and few (but an increasing number) in Asian populations. When we compare the various findings for the same trait across populations that have been tested, the results have generally been encouraging: consistent findings emerge in the main pattern of results for twin studies of schizophrenia. But major ethnic groups of humans have yet to be studied carefully.

2.6.5 Twin studies of schizophrenia

Given Kamin's criticisms, is there any reason to believe the results of twin studies of schizophrenia? In the last 80 years, more than a dozen studies have been conducted that assessed the presence of schizophrenia in both members of MZ and DZ twin pairs. These studies measure the concordance rate, that is, the presence of the same trait in both members of a pair of twins. Strictly speaking, concordance is the probability that a pair of individuals will both have a certain characteristic, given that one of the pair has the characteristic. The work cuts across many countries, involves many different investigators, and has used a variety of different methods. Two representative twin studies of schizophrenia are described below.

The first investigation, known as the Maudsley Twin Study, started shortly after World War II when any individual who registered for treatment at the Maudsley hospital in South London was asked whether or not they were a twin. If they said yes and gave consent, they were entered into the research project. The most recent update of this study covers the years from 1948 to 1993. Of the 224 individuals with symptoms of severe mental illness, 106 met the criteria for strictly defined schizophrenia. The correlation in MZ twins for the liability to schizophrenia was quite high (0.81), and slightly more than twice as high as in DZ twins (0.31) (CARDNO et al. 1999). By Rule 1 described earlier, genetic factors are very likely to be influencing the risk for schizophrenia, and shared environmental factors are probably making very little contribution.

The main strengths of this study were the systematic method used to locate the twins and the relatively high quality of the clinical information available for each individual. The diagnostic evaluations were done with care. However, the twins were obtained from a single facility and one that is a specialist center. How representative of the general population are such results likely to be?

The second study was conducted in Finland by a combined American–Finnish team. In some ways, its strengths and weaknesses are the mirror image of the Maudsley study: the quality of the diagnosis was variable, but the sample collection credentials were hard to fault. Because of the high quality of medical and social records in Finland, the investigators were able to examine virtually every twin born in Finland from 1940 to 1957—a total of 19,124 individuals. The researchers then matched these records to three registries that contained information about psychiatric illness: the national hospital discharge register, which contains diagnoses assigned during any hospital stay in all of Finland, the Pension Register, and the Free Medicine Register. The latter two contain information provided by physicians to justify eligibility for either a pension, access to state-subsidized medications, or both. We know from other studies that nearly all individuals with schizophrenia in a country like Finland would at some point come into medical care and seek a pension or receive state-subsidized medication or both, so it is unlikely that many cases of schizophrenic illness in twins would have been missed (CANNON et al. 1998).

The investigators looked in these three registries from the years 1969–1991 and identified 670 twins who had a diagnosis of schizophrenia in one or more of the databases. Diagnoses were recorded by a variety of different clinicians under a range of circumstances, which means that their quality almost certainly varied widely. Determining which twins were MZ and which were DZ was also difficult. A mailed questionnaire was sent out to members of the twin registry in 1975, but individuals with schizophrenia are not always cooperative with such research requests. In nearly a quarter of the twin pairs, no information was available about which type of twin they were. Nevertheless, the investigators obtained 508 usable pairs, a much larger sample than in the Maudsley Twin Study, and compared them with all other twins in the registry. This is a better method than that used in the Maudsley Twin Study, which had to assume an estimate of the population risk for schizophrenia. The correlation for the liability to schizophrenia in Finnish MZ and DZ pairs was estimated at 0.84 in MZ twins and 0.34 in DZ twins—remarkably close to the figures for the Maudsley series (0.81 and 0.31). Given the wide differences in method, the similarity in the results of these two studies—one from a hospital in South London using personal interviews and the other from the entire country of Finland relying on medical, pension, and pharmacy records—is striking.

While the Maudsley and Finnish studies both had limitations, each had been systematic in their collection of data from twins with schizophrenia. This has not always been true in other twin studies of schizophrenia.

One of us (K.S.K.), with colleagues Patrick Sullivan and Michael Neale, combined the results of all of the major studies that met reasonable methodologic criteria in a formal meta-analysis. Several key findings emerged. First, all but one of the 12 studies provided strong evidence that genetic risk factors were of critical importance in schizophrenia. Across these studies, the heritability of liability to schizophrenia was estimated to be 81%. Second, the results of the methodologically strongest studies did not differ from those of the more problematic ones. Third, familial environmental factors played a modest, but nevertheless detectable, role in causing schizophrenia, accounting for 11% of the variance in liability (SULLIVAN et al. 2003).

2.7 Should we rely on twin and adoption studies?

There are a few important conclusions we can now draw from the foregoing discussion. First, research using adoptees and twins is far from flawless. Like most other kinds of science, these studies can be done well, or poorly, and the confidence that we should have in the results relates directly to the quality of the methodology employed. Second, the main worry about the validity of the twin method is the EEA. Based on the available empirical studies, this assumption is probably not seriously violated in most twin studies. Other concerns with the twin method—for example, that twins are highly atypical—have little evidence in their favor. Like other epidemiological methods, however, twin and adoption studies are susceptible to unrepresentative sampling. Third, the evidence to date suggests no large differences in results across various countries, although far more studies are needed representing additional human populations.

What can we conclude from research efforts to disentangle the effects of genes and family environment (nature versus nurture) for schizophrenia? Most importantly, the various twin and adoption studies generally agree that genetic risk factors play a major role in the development of schizophrenia. However, we have also seen that there is no such thing as a flawless genetic study when investigating humans. Both twin and adoption studies have many possible methodological problems. Fortunately, the problems in the two kinds of studies are rather different. Therefore, if twin and adoption methods give broadly similar answers, the probability that the results are spurious is low. As one colleague put it, 'Nature would have to be particularly perverse to provide us with a set of different biases that would each produce quite similar findings in twin and adoption studies.' We do not claim that these methodological issues are entirely resolved, but the consistency of findings across the two different methods strongly suggests that genes are important in schizophrenia.

Another general point is worth making here. One way to divide sciences is into those that observe the world and those that go out and manipulate it, typically in laboratory situations. The first kind of science includes astronomy, geology, and human genetics. Laboratory genetics, by contrast, belongs in the second group. We cannot do laboratory genetic studies in humans. Human genetics researchers always have to be on their toes, looking for the unsuspected bias, because we cannot control everything in our studies. We can just look at the experiments that nature provides us with.

So if genes are important, how are they important? The foregoing results treat genetic effects, in some ways, like a huge black box. Trying to peer into that black box is the next critical step.

Summary

1. The first question that a psychiatric geneticist should always ask about a disorder is whether it runs in families.

2. The key features of a high-quality family study for a psychiatric disorder include representative ill and matched control probands, use of structured interviews, blind assessment of relatives, and blind final diagnosis.

3. As illustrated by the Roscommon Family Study, the first-degree relatives of patients with schizophrenia have an approximately tenfold increased risk of illness.

4. The next logical question that a psychiatric geneticist should ask is: why does the disorder run in families? To what degree is this familial aggregation a result of genes or shared family environment?

5. In humans, there are two quasi-experimental approaches to answering this question: adoption studies and twin studies.

6. There are two major adoption designs for psychiatric disorders that begin either with ill biological mothers who put their children up for adoption, or with ill adoptees.

7. Twin studies depend on a comparison of the level of resemblance for a trait or a disorder in monozygotic versus dizygotic twins.

8. The concept of heritability reflects the proportion of phenotypic variance for a particular trait or disorder in a population that results from genetic differences between individuals in that population.

9. The best current evidence from twin studies suggests that the heritability of schizophrenia is high—around 80%.

10. Twin studies have a number of potential methodological limitations, especially the equal environment assumption. The best evidence to date suggests that in most situations, these biases are modest and do not substantially distort results.

11. For estimates of the heritability of schizophrenia, added confidence is gained from the congruent findings about the importance of genetic factors from both adoption and twin studies.

3

Molecular biology of nucleic acids

[On the structure and function of nucleic acids, the nature of the gene, methods for sequencing and genotyping, and a description of genetic variation]

In this chapter, we provide an introduction to the molecular biology relevant to behavior genetics. We start with the molecules themselves—nucleic acids and proteins—and explain why understanding their structure and function has become so critical for biology. We explain the molecular technologies of sequencing and genotyping that are central to the discoveries that made DNA (and its acolytes) biology's cynosure, and we outline what is known about DNA sequence variation.

3.1 DNA

Everyone recognizes DNA, and while the names of the discoverers of DNA structure, James Watson and Francis Crick, aren't as famous as those of Einstein, they're as close to superstardom as biologists get. The Watson–Crick structure of DNA is among the most important discoveries of the twentieth century because it addressed a fundamental question in biology: what is life?

Three findings flowed from the structure of DNA: first, the structure elucidates how DNA contains the information about the form and function of cells; second, it explains how cells reproduce; and third, it places information processing at the center of biology. The first two findings meant that biology was subject to the laws of physics and chemistry, and that nothing new needed to be invoked to explain the origins of life. The third meant that instead of reducing biology to chemical reactions, biology is reduced to understanding how information is encoded and decoded. We are still living with the consequences of that monumental shift in thinking. We describe these findings, starting with the startling discovery that DNA, one of the two components of chromosomes, encoded hereditary information.

How is genetic information encoded on chromosomes and how could that information be inherited? When these questions were posed, it made sense to believe that genes would be made of the same stuff as chromosomes: nucleic acid (deoxyribonucleic acid, DNA) and proteins. The stunning, seemingly limitless

Figure 3.1 Chemical components of nucleic acids.

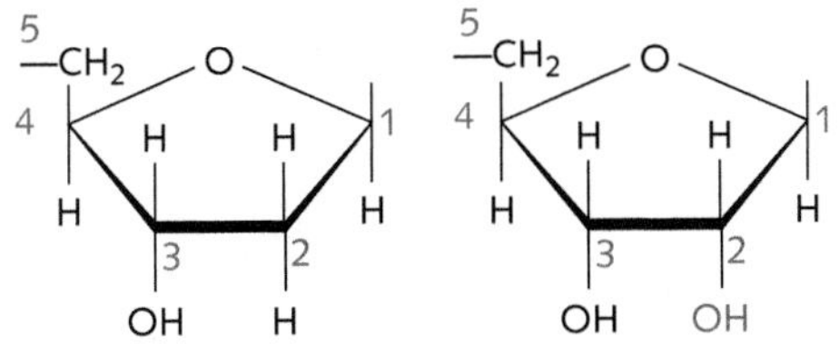

The pentose sugars: deoxyribose on the left is used in DNA; ribose on the right is used in RNA. The five carbon atoms are numbered. Bonds at carbon atoms 3 and 5 give rise to the chain of phosphates and sugars that forms the backbone of DNA, and are responsible for the naming convention. 5' and 3' (five prime and three prime)

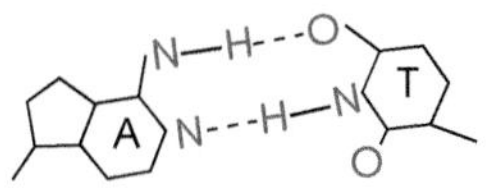

Two nitrogenous bases: adenine (A) bound to thymine (T)

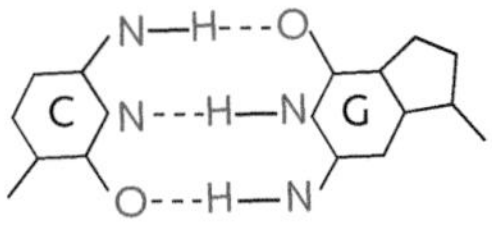

Two more nitrogenous bases: guanine (G) bound to cytosine (C)

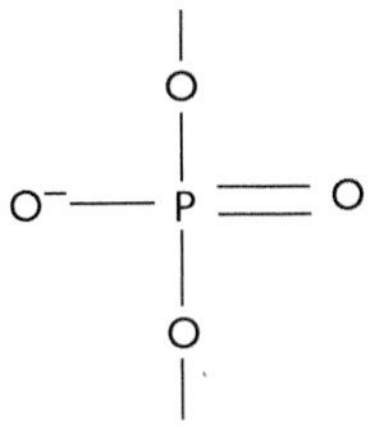

The phosphate group that joins the sugars

complexity of protein, together with its many known roles (e.g. structural proteins are responsible for the properties of skin and hair, proteins known as enzymes catalyze the biochemical reactions that power the cell, and protein hormones act as chemical messengers), made it seem evident that genes must be made of protein. Fred Sanger's finding that proteins consist of a linear arrangement of amino acids significantly added to the evidence in support of the view that genes were made of protein. Sanger was a little-known biochemist at Cambridge when he showed that proteins contain information, in the linear order of their amino acids, and the ability to contain information is one of the requirements of genes (SANGER AND TUPPY 1951a, b). But a gene also had to duplicate itself to pass on the information to the next generation, and that was a problem for the hypothesis that genes were made of protein (CRICK 1958). How could a protein duplicate itself? It has no template from which it could create a copy.

DNA appeared unlikely to be the material out of which genes were made. From what was then known, DNA could not contain sufficient information. Its composition argued against such a possibility. It is made of a sugar (deoxyribose), a phosphate, and a mixture of four nitrogenous bases: adenine and guanine

(known as purines), and thymine and cytosine (known as pyrimidines) (Figure 3.1). The amount of adenine (A) always equals the amount of thymine (T) and the amount of guanine (G) always equals the amount of cytosine. Imagine that the four bases are four letters (A, G, C and T). If each time you write A you have to write T and each time you write C you have also to write G then the code is reduced to just two letters, hardly sufficient to encode an entire organism.

The structure of DNA solves both the information problem and the inheritance problem (Figure 3.2). DNA consists of two strands, which are complementary to each other. Where one strand has an A, the other has T, and where one strand has C the other has G. This explains why there are equal amounts of A and T, and C and G, while at the same time allowing the DNA to contain information in a linear arrangement of the four bases. The existence of two strands solves the problem of inheritance because each strand acts as a template for its own duplication. Since adenine (A) can only bind to thymine (T), and guanine (G) to cytosine (C), when the two strands separate a complete copy of the original DNA molecule can be made by allowing nucleotide monomers (A, C, G, and T) to bind to the exposed bases. Joining them up through a

Figure 3.2 The structure of DNA.

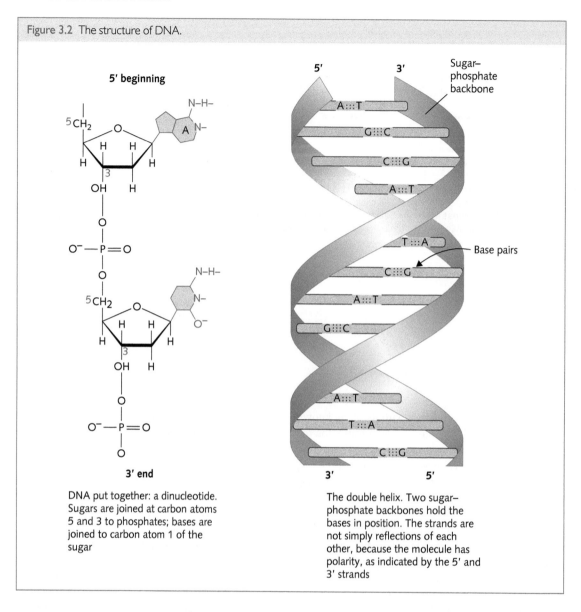

5′ beginning

DNA put together: a dinucleotide. Sugars are joined at carbon atoms 5 and 3 to phosphates; bases are joined to carbon atom 1 of the sugar

The double helix. Two sugar–phosphate backbones hold the bases in position. The strands are not simply reflections of each other, because the molecule has polarity, as indicated by the 5′ and 3′ strands

process called polymerization generates a new strand of DNA, a replica of the original.

One important point to note about the structure is that the two strands of DNA are not simply reflections of each other. The complement of the sequence 'ATTGCCAT' is not 'TAACGGTA' because the two strands have different polarities. They run in opposite directions, having a 5′ (five prime) beginning and a 3′ (three prime) end. The numbers 5 and 3 refer to the positions on the sugar molecules where the sugar's carbon atoms are attached to phosphate groups, as shown in Figure 3.2. Consequently the complement of

5′-ATTGCCAT-3′ is 5′-ATGGCAAT-3′. This important point is easily overlooked, leading to laboratory misfortune when a novice, forgetting the polarity of DNA, orders oligonucleotides for a polymerase chain reaction (I won't apologize for repeating this point below, in section 3.4, as it's really important).

The two publications in *Nature* that described DNA's structure and its *'genetical implications'* (WATSON AND CRICK 1953a, b) did not prove that heritable information was encoded in DNA. Sydney Brenner pointed out that *'even in 1958 (five years after the publication of the structure of DNA) the whole of DNA was still thought to be a flash in the pan—not right you know, not proven'* (BRENNER AND WOLPERT 2001). With hindsight, it is easy to forget the problems facing Watson and Crick's hypothesis, not the least of which was how the double helix could unwind, which clearly had to happen if replication were to occur in the way they proposed. It's like untwisting telephone cables: if you unwind it in one place, it winds up in another. How does all that uncoiling and coiling take place? There has to be a mechanism to deal with the build-up of tension. Brenner and Crick called this a 'don't worry' problem. *'A don't worry hypothesis is very important in biology. It is there to provide one with any plausible way that something can be done so you don't have to worry about it at that very moment ... one plausible mechanism that we postulated for unwinding the two DNA ... is that there will be enzymes to do it. Which of course there are. They are called DNA helicases'* (BRENNER AND WOLPERT 2001). As the structure of DNA turned out to be correct, the 'don't worry' hypothesis (at least in this instance) was justified.

3.2 RNA

Of the two types of nucleic acid that are found in cells, ribonucleic acid (RNA) is far more abundant than DNA. RNA, like DNA, is composed of a sugar (ribose), a phosphate, and a base. It differs from DNA in that it has uracil (U) instead of thymine; otherwise, it is a linear combination of four bases. Because of this structure, RNA is able to carry information, and, arguably, its most important role is as an intermediary, carrying information encoded in DNA to a place where it can be turned into protein (see Chapter 4, section 4.2.3, which describes the structure and function of other forms of RNA). The form of RNA that does this is called messenger RNA (mRNA), a central discovery in the canon of molecular biology.

Two structural differences between DNA and RNA explain much of RNA biology. First, DNA has a hydroxyl group on part of its sugar constituent, whereas RNA has a hydrogen atom (Figure 3.1). The result is that, in most biological environments, RNA is much more unstable than DNA. This makes sense because RNA mediates the expression of genetic information; its production and degradation have to be tightly controlled. Second, RNA normally exists as a single molecule, whereas DNA is typically a double helix. The single-stranded structure of RNA is well suited to its role in conveying information from DNA

to protein, through the processes of transcription (in which RNA is copied from a DNA template) and translation (in which the information in the RNA molecule is used to create a protein).

DNA is more than a stable repository of encoded protein sequence information; it also contains information that controls the transcription of mRNA. Disorder of the template function of DNA is the molecular basis of inherited disease. By contrast, gene expression (the transcription of DNA into mRNA and the translation of mRNA into protein) is not entirely genetically predetermined. It is highly regulated but in response to changes in the cellular environment, which reflect changes in the state of the organism, which is in turn impacted by its surrounding environment. Details of this process, together with other forms of RNA that mediate the relationship between genetic material and the environment, are described in Chapter 4.

A few years after the discovery of the structure of DNA, Crick put forward two hypotheses, the sequence hypothesis and the central dogma (CRICK 1958), which dealt with the problem of how information flowed from DNA to protein. The sequence hypothesis asserted that *'the specificity of a piece of nucleic acid is expressed solely by the sequence of its bases, and that this sequence is a (simple) code for the amino acid sequence of a particular protein'*. The central dogma was the idea that *'once "information" has passed into protein it cannot get out again. In more detail, the transfer of information from nucleic acid to nucleic acid, or from nucleic acid to protein may be possible, but transfer from protein to protein, or from protein to nucleic acid is impossible. Information means here the precise determination of sequence, either of bases in the nucleic acid or of amino acid residues in the protein.'* However, as Crick admitted, *'The direct evidence for both [hypotheses] is negligible.'*

The evidence to prove the role of RNA as the molecule that carried information came from studying how bacteria make protein. Most of the RNA in cells is concentrated in ribosomes, which localize at the site of protein synthesis, but in the 1960s, no one knew what ribosomes did. Ribosomes contain RNA, but ribosomal RNA wasn't the message; instead, there was a separate mRNA molecule that had escaped detection. Brenner described what they had to do in order to prove the role of mRNA (BRENNER AND WOLPERT 2001):

We went to California to do this experiment [this was in the 1960s]. These experiments were awful! They took a lot of expensive isotope and you didn't know the answer

Figure 3.3 The translation of messenger RNA (mRNA) into protein. A ribosome holds a single-stranded mRNA molecule whose nucleic acid sequence encodes the information for manufacturing protein. Transfer RNAs (tRNAs) carrying an amino acid (shown on the right) are loaded onto the ribosome, where they bind to the mRNA (shown in the center). Once bound, the amino acid is added to a growing protein, the tRNA is ejected (one is shown on the left, waiting to be recharged with an amino acid), and the mRNA is moved three nucleotides along and then waits for the next appropriate tRNA to arrive. Each tRNA has its own specific amino acid, and each recognizes, and binds to, a specific three base pairs of the mRNA. These triplets, or codons, constitute the genetic code. For a wonderful description of how ribosomes do their job, based on their structure, see RAMAKRISHNAN (2018).

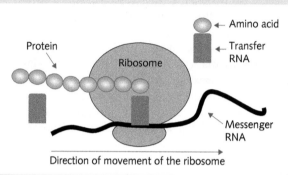

until you had run it in a centrifuge for a day. And if anything went wrong, if the centrifuge had broken down, it was really terrible ... and every experiment we did, didn't work ... Francois was so nervous he dropped the radioactive phosphate in the water-bath. We hid the water-bath behind the Coca-Cola machine so they wouldn't find it because it was now very radioactive ... In the middle of the run the centrifuge broke down. These gradients are tremendously mechanically stable, but they are not thermally stable. And so when the centrifuge broke down we had to move the rotor from one cold room to another cold room ... This day I will never forget.

Nevertheless, in the end, Brenner's experiment confirmed the existence of mRNA as the molecule that carries sequence information from DNA stored in the cell's nucleus to the ribosomal factories that make protein. Figure 3.3 depicts the process by which mRNA is translated into protein.

3.3 The genetic code

Everyone knows that DNA codes for something, although most of us have no idea what, and most of us confuse the genetic code with the sequence itself. The genetic code is generally shown as a table showing the relationship between DNA (and RNA) sequence and the 20 amino acids that constitute proteins (Figure 3.4) (CRICK 1968). The genetic code is **not** the genome, which refers to the complete DNA sequence of an

organism. The code is a triplet code, which is to say three bases define one of the 20 amino acids of which proteins are formed; for example, TCA codes for the amino acid serine (Ser for short) and AAA for lysine (Lys for short). The triplets are called codons and the sequence that matches each codon is an anticodon (the anticodon sequence must be complementary to the codon, so the anticodon for AAA is TTT).

Proof that a specific nucleic acid sequence codes for a specific protein came from a biochemical assay: add mRNA to the cellular machinery that makes protein and determine the output. This was the achievement of Marshal Nierenberg in developing a cell-free system. By adding a synthetic RNA of known sequence, he could unambiguously determine the relationship between the nucleic acid sequence and amino acids (NIRENBERG AND MATTHAEI 1961).

There are three important things to know about the genetic code. First, it is redundant. The four bases can be arranged to obtain 64 unique triplets (four choices for the first base, times four for the second, times four for the third), which means that some triplets encode more than one amino acid. So, while TCA codes for serine, so does TCC, TCT, and TCG. Second, the code contains messages for the start (ATG, which also codes for methionine (Met)) and the termination (TAG, TAA, and TGA) of translation. Mutations that alter mRNA sequence to form these codons result in the truncation (or elongation) of proteins. Third, and most strikingly, the code is universal. All

Figure 3.4 The genetic code. The code is in triplets (or codons), so to find the amino acid that corresponds to the sequence CAG, look along the row for the first letter 'C', the column for the second letter 'A' and within that box find the codon with the third letter 'G'. The translation of CAG is the amino acid glutamine (Gln). There are four special codons: AUG is the start codon, where translation should start, while UAG, UAA, and UGA are stop codons, where translation should stop. Note that it is mRNA that is translated, which uses 'U' instead of 'T', not DNA.

Second letter

	T	C	A	G	
T	TTT TTC } Phe TTA TTG } Leu	TCT TCC TCA TCG } Ser	TAT TAC } Tyr TAA Stop TAG Stop	TGT TGC } Cys TGA Stop TGG Trp	T C A G
C	CTT CTC CTA CTG } Leu	CCT CCC CCA CCG } Pro	CAT CAC } His CAA CAG } Gin	CGT CGC CGA CGG } Arg	T C A G
A	ATT ATC ATA } Ile ATG Met	ACT ACC ACA ACG } Thr	AAT AAC } Asn AAA AAG } Lys	AGT AGC } Ser AGA AGG } Arg	T C A G
G	GTT GTC GTA GTG } Val	GCT GCC GCA GCG } Ala	GAT GAC } Asp GAA GAG } Glu	GGT GGC GGA GGG } Gly	T C A G

First letter (left margin) / *Third letter* (right margin)

organisms—viruses, bacteria, yeast, insects, plants, and vertebrates—use the same code (there are some minor exceptions: vertebrate mitochondria use AGA and AGG as stop codons, but not UGA, which codes for tryptophan instead, and some unicellular protozoans, such as *Paramecium*, use stop codons for glutamate).

The discovery of a universal code is at the core of molecular genetics' reshaping of biology. Because the code is universal, DNA means the same thing in all organisms. Everything else can vary, but whether an organism uses sunlight to generate energy or metabolizes hydrogen or carbon compounds, whether it lives near a thermal vent at the bottom of the ocean or spends its life at the poles of the earth, whether it is the size of a blue whale or is no more than a single cell, the DNA codes for the same amino acids: TTT, for example, always encodes phenylalanine. Either this is because there is only one way to encode amino acids, because of chemical or physical constraints on protein synthesis, or the code was established once, at life's origins, and was passed on to all descendants

of that initial genesis. In fact, the universality of the code is **not** due to physical constraints; work from Jason Chin's laboratory has shown that it is possible to reprogram the code so that it generates new amino acids not found in nature (GREISS AND CHIN 2011). Instead, the code's conservation means that all life arose from the same source, that we are all descended from the same organisms that once, early in life's history, learned to use nucleic acids to encode proteins.

The discovery of a universal genetic code marked the point at which the analysis of information processing became central to understanding biology. The idea that biology involves solving codes is now so well accepted that there are many groups devoted to solving other codes in biology (the 'histone code', the 'neuronal code', and so on). But at the time of the discovery of DNA, the idea that biology had anything to do with information was unthinkable. Placing information at the center of biology represented a fundamental discontinuity in the way scientists were thinking, and its impact still reverberates. This was one of the intellectual paradigm shifts of the twentieth century, which has left a deep stamp ever since on the way biological problems are formulated and solved.

However, the full extent of the information that DNA contained and the implications of the informatics revolution in biology only became visible when DNA was sequenced. DNA sequencing lies at the center of the revolution in biology that followed from the discovery of the structure of DNA. But before we explain how sequencing works, we need to introduce the polymerase chain reaction.

3.4 The polymerase chain reaction

The polymerase chain reaction (PCR) is a biological chain reaction, a way of generating millions of copies of a piece of nucleic acid. Its development was key to later advances in molecular biology, and it currently underlies almost all the technologies described later. PCR makes it possible to amplify ridiculously small amounts of DNA. Starting from as little as one molecule, it's possible to make a solution that contains only the piece of DNA of interest (Figure 3.5).

PCR depends on the ability of two strands of nucleic acid to pair, using the Watson–Crick bonds that tie A to T and C to G. Heat (or alkali) will cause the two strands of DNA to separate, but as they cool (or as the pH returns to near neutral), the strands find their partners again, in a process called hybridization.

Hybridization means that if you manufacture single-stranded DNA and add it to other single-stranded sequences, it will find its match (assuming there is one). Hybridization is critical to the success of PCR, which uses a pair of short, single-stranded sequences (called oligonucleotides, or oligos for short) to find a match either side of a region to be amplified (Figure 3.5). An enzyme (DNA polymerase) combines DNA constituents to fill in the gap between the two oligonucleotides, thereby creating a copy of the target DNA. Polymerases can only add free nucleotides to the 3′ end of the newly forming strand (so they extend DNA (and RNA) in the direction 5′ to 3′). Having a single copy does not help much, but you can simply repeat the process, using the product of the first amplification as a template for another amplification and thus starting a chain reaction that doubles the amount of target in each reaction cycle (Figure 3.6). Thus, with PCR, it's possible to amplify and isolate one tiny fragment from the billions of bases in the genome, and to get sufficient quantities to measure it, to calculate its size (DNA fragments can differ in length as well as base composition), and, of course, to sequence it.

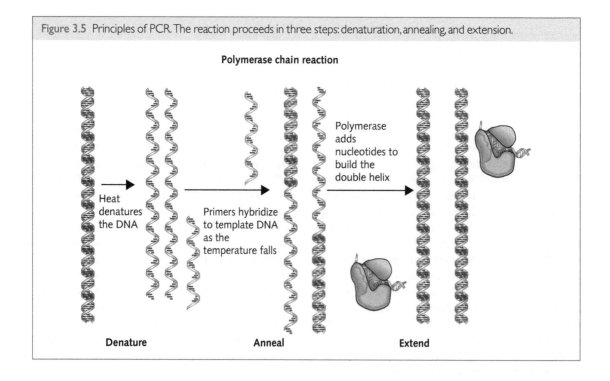

Figure 3.5 Principles of PCR. The reaction proceeds in three steps: denaturation, annealing, and extension.

Polymerase chain reaction

Heat denatures the DNA

Primers hybridize to template DNA as the temperature falls

Polymerase adds nucleotides to build the double helix

Denature   Anneal   Extend

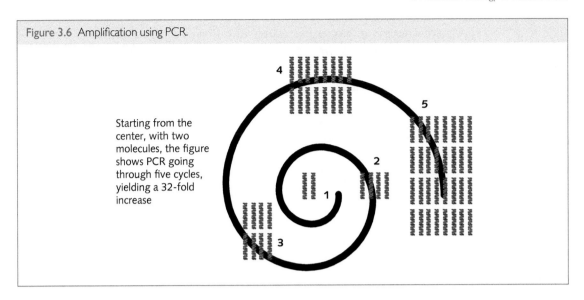

Figure 3.6 Amplification using PCR.

Starting from the center, with two molecules, the figure shows PCR going through five cycles, yielding a 32-fold increase

Because of the 5′ and 3′ ends, complementary DNA sequences, written in the standard format from 5′ to 3′, are deeply confusing. Suppose you are asked to design oligonucleotide primers for a PCR. The task sounds simple. Here is a piece of DNA sequence that we want to amplify (usually the sequence is much longer):

5′-CCCCCGGGATATGGGCCCAAGGCCTTAACCGG-3′

Given the rule that the complementary letter of A is T and of C is G, write down the complementary sequence of the last six base pairs of the sequence above (i.e. AACCGG). The answer you should get is CCGGTT. The difference between 5′-AACCGG-3′ and 5′-CCGGTT-3′ is the difference between DNA sequencing or PCR working or not working.

3.5 DNA sequencing

The size and complexity of the human genome was a formidable, apparently insuperable, barrier to obtaining its complete DNA sequence. The genome of the T4 bacteriophage, a virus that played such an important part in the early discoveries of molecular biology, contains 169,000 base pairs (bp), encoding 289 proteins. Its host bacterium, *Escherichia coli*, contains a circle of DNA that measures 4.6 million bp, encoding 4,288 protein-coding genes. In contrast, the human genome consists of 3.6 billion bp (3,600,000,000) encoding 20,300 genes (http://www.ensembl.org). We can translate this into physical sizes as follows: each base is about 3.4 angstroms (Å, where 1 Å is 0.0000000001 m), so the length of phage DNA is a fraction of a millimeter (0.06 mm), a bacterial genome is 1.5 mm, and the human genome is 1.2 m (DNA molecules have a diameter of 2 nm).

Decoding the vastness of the human genome required technologies that seemed unattainable when DNA's structure was solved. The chain-terminator method for DNA sequencing that Sanger developed (thereby earning him a second Nobel Prize) was incredibly laborious, and hardly suited to generating more than a few hundred base pairs of DNA (in the late 1970s and 1980s, you could get a PhD for sequencing a single gene). But with Leroy Hood's automation of the Sanger protocol (making liberal use of PCR) and the manufacture of the first DNA sequencing machines in the 1990s (SMITH et al. 1986; HUNKAPILLER et al. 1991) came the taming of first the kilobase (kb, 1000 bp) (SULSTON et al. 1992) and then the megabase (Mb, 1 million bp) (WILSON et al. 1994). Taming the gigabase (Gb, 1000 million bp) required the development of 'next-generation' sequencing, in which millions of sequencing reactions are carried out in parallel, using novel chemistry and technology, which has driven the cost of genome sequencing to less than a thousand dollars, compared with the US$300 million for the first draft of the human genome (Figure 3.7).

Genome sequencing of any organism is now (relatively) trivial, and thousands of genomes have been

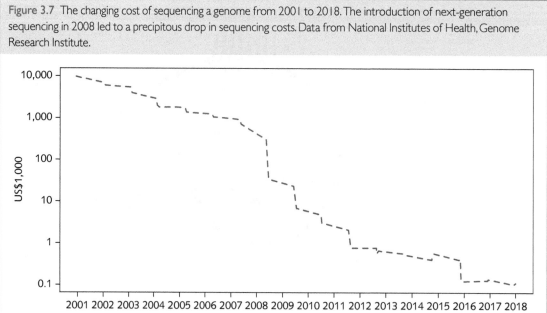

Figure 3.7 The changing cost of sequencing a genome from 2001 to 2018. The introduction of next-generation sequencing in 2008 led to a precipitous drop in sequencing costs. Data from National Institutes of Health, Genome Research Institute.

generated, many of which can be seen at http://www.genome.ucsc.edu or http://www.ensembl.org (be warned, these sites are daunting, even for the initiated). The introduction of automated sequencing, with the massive increase in data that this enabled, means that analyzing DNA-derived data is more time-consuming (for a scientist) than creating it, leading to the creation of a new profession, the bioinformatician.

Despite these impressive advances, the generation of genome sequences from the data spewing out of DNA sequencing machines still uses the same strategy that Sanger introduced, known as shotgun sequencing. This is how it is done. Sequencing machines generate a large number of random small segments of DNA sequence, called sequencing reads. For human studies, or any organism where we have a complete genome, the task is to match the sequence reads to the known genome (assembly of novel genomes is more complex). Assembling sequence reads to generate a genome is like taking many copies of one page of newspaper text, tearing the copies up into small sections, each with a few hundred letters, and randomly picking up a proportion of them.

Suppose each copy of the page is torn up into approximately 1,000 fragments, and you pick up 1,000 fragments. You wouldn't expect to be able to reconstruct the whole page from that selection. You will have picked up multiple copies of some words and no copies at all of others. But if you keep picking up more and more pieces, you'll increase your chances of getting every word. Picking up 1,000 fragments is the equivalent to picking up the same amount of text as a whole page, which in genome speak is called 1X (one 'X') coverage. Sequencing studies usually aim for about 30X coverage to obtain enough sequence to reconstruct a genome. However, if the amount of information in each fragment is too small, it's going to be very hard to find a single match—any fragment that contains only the word 'and' or 'the' won't be unique; longer fragments with five or six words won't have this problem and will be much more useful for assembling the sequence.

The length of a sequencing read is important in dealing with a problem that still afflicts genome sequencing: the amount of repetitive DNA that genomes contain. Longer reads are needed to span repeats and so acquire unique sequence that can be matched back to the known (or reference) genome. However, sequencing reads are currently too short to do this, and so DNA sequencing strategies that claim to generate 'whole genomes' actually only generate data on about 85% of a typical mammalian genome.

3.6 Genome sequence and the problem of repetitive DNA

One reason why the human genome (and other eukaryotic genomes) is so large is the unexpectedly complex structure of its sequence. You might think that the human genome is a model of efficiency, better designed than an Apple laptop, where each base pair plays a critical but finely judged role. As far as we can tell, nothing could be further from the truth. An enormous proportion of the genome consists of repetitive DNA: 60–70% (DE KONING et al. 2011). The simplest repeats are short nucleotide stretches, including anything from a single base (strings of As are common) and dinucleotides (e.g. CACACA) to complex longer sequences. What are these repeats doing? The function of some repeats is related to their position in the genome: repeats of the hexanucleotide TTACCC at the ends of chromosomes (the telomeres) act as a cap, protecting the chromosome from degradation; enormously long arrays of 120 bp repeats are found at the functional center of chromosomes (the centromere), extending in a head-to-tail arrangement for hundreds of kilobases. Their function is still unclear, although presumably they play some kind of structural role.

By far the largest contribution to repetitive DNA, up to half a vertebrate genome, is from degenerate copies of elements that have the remarkable property of being able to move from one location to another in the genome. Such elements are called **transposons** (discovered by Barbara McClintock in maize; MCCLINTOCK 1951), and work either by directly reinserting DNA copies of themselves or by transcription into RNA first before being reinserted (retrotransposed) as DNA back into the genome. The ability to jump (or to transpose) to almost any location in the genome makes functional transposable elements highly mutagenic.

Variation in the number of transposable elements in genomes is largely responsible for the remarkable differences in genome size between species (Figure 3.8). Lilies, with genomes 20 times as big as the human genome, do not have 20 times as many genes; they have 20 times as much repetitive DNA.

Why do half of our genomes consist of a (mostly degenerate) mutagen? One option is that most of our genome is 'junk', with no specific function. It could, as Lesley Orgel and Francis Crick argued (ORGEL AND CRICK 1980), be regarded as 'selfish'. Its existence depends on its ability to propagate parasitically, much like a virus. Genomes are thus a warzone: transposable elements invade and expand their occupation of the host genome, while host cells impose increasingly severe controls to limit the mutagenic potential of unwelcome immigrants. Alternatively, it is possible

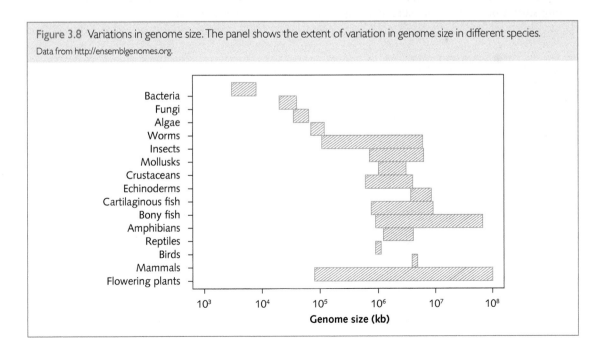

Figure 3.8 Variations in genome size. The panel shows the extent of variation in genome size in different species. Data from http://ensemblgenomes.org.

that repetitive sequences have unknown roles in regulating gene expression. Genome complexity may in part be explicable by population size, as large population sizes are able to purge their genomes of repetitive DNA (LYNCH AND CONERY 2003); metabolic effects may also be important (LANE AND MARTIN 2010).

3.7 What is a gene?

'DNA makes RNA makes protein' was James Watson's summary of Crick's central dogma. Genes are made of DNA and are transcribed into mRNA, which is translated into protein. But by the end of the twentieth century, you could take an entire course to learn what a gene was. The concept of the gene has a long history, going from Mendel's hereditary 'factor' and William Bateson's coining of the term 'gene' in 1905 up to the realization that genes are not as simple as they at first appeared to be.

The first complication is that there is not a one-to-one correspondence between DNA sequence and mRNA sequence. Bacterial genes are relatively well behaved in this respect: they consist of a linear arrangement of bases that can be transcribed into a corresponding RNA whose sequence encodes protein. The genes in eukaryotic cells (cells with nuclei to contain their DNA) don't have this comfortable arrangement (BERGET et al. 1977). DNA is first transcribed into a long stretch of RNA (often called heterogeneous RNA in the literature), some sections of which are then spliced out to leave a mature mRNA. Each gene is a patchwork of coding and noncoding segments in our genomes. Biochemical machinery was discovered whose function is to splice the primary RNA transcript, remove the noncoding **introns**, and join together the coding regions, the **exons**, to generate a mature transcript (Figure 3.9). The entirety of the coding regions of a genome is called the **exome**, so when you read about sequencing an exome, it means obtaining sequence information only for the coding portion of the genome. Almost, but not all, eukaryotic genes consist of introns and exons.

The lack of correspondence between DNA sequence and RNA became even more confusing when it was discovered that not all transcripts used identical exons. A single gene can produce multiple different transcripts (isoforms) by using different exons from the same primary transcript. This phenomenon is called **alternative splicing** (Figure 3.9).

Alternative splicing is one answer to the mystery of why there are relatively few genes in the human genome. One of the surprises to come out of the completion of genome projects, which included analyses of exomes, was the answer to a much-debated question about the number of genes in a species. Before we had genomic information, no one knew if humans had 10,000 genes or 100,000, or more. Table 3.1 summarizes the number of protein-coding genes known from sequencing studies.

Figure 3.9 Alternative splicing produces multiple transcripts from a single primary gene sequence. The figure shows two different processed mRNAs produced from the same pre-mRNA (an unprocessed transcript of DNA). Introns (the horizontal lines in the pre-mRNA) are spliced out, thus joining the exons (the boxes). The choice of which exons to include in the mature processed mRNA will depend on cell type, developmental stage, and environmental conditions. Alternative splicing means the cell can produce a variety of proteins from the same DNA sequence to meet specific and changing needs.

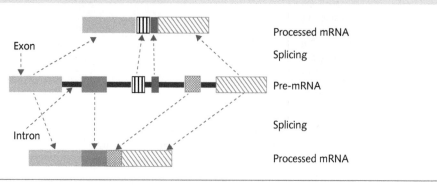

Table 3.1 Number of genes identified in different species		
Mammal	Human	20,310
	Gorilla	20,962
	Cat	19,493
	Panda	19,343
	Dog	19,856
	Mouse	22,615
	Pig	21,630
	Armadillo	22,711
Marsupial	Wallaby	15,290
Bird	Zebra finch	17,488
	Chicken	18,346
Fish	Fugu	18,523
	Zebra fish	25,832
Amphibian	Xenopus	18,442
Nematode worm	Caenorhabditis elegans	20,362
Insect	Drosophila melanogaster	13,918
Yeast	Saccharomyces cerevisiae	6,692

Data from http://www.ensembl.org/.

Look, my genome contains about 20,000 genes! Look further down the list and notice that the genome of a worm (*C. elegans*) contains more genes (20,362) than I do! Can this be right? I am more complex than a worm, so why don't I have more genes than a worm? The answer is that, while I might not have more genes, I do have more proteins, in part because alternative splicing generates multiple isoforms. So how many proteins do I have? Or to put it another way, how many mRNA transcripts are there? That's hard to answer, because different tissues have different transcripts; indeed, that's one way in which a tissue is defined. Brain tissues—mostly neurons—contain on average a greater diversity of transcripts than other tissues, and use more isoforms than other tissues.

The multiplicity of isoforms in brains addresses an outstanding issue in neurobiology: how are all the different cell types and connections in the brain specified? There clearly aren't enough genes to determine the billions of connections. Isoforms help (there are an extraordinary number of isoforms for a gene called

Dscam because this gene's function is to regulate connectivity; SCHMUCKER et al. 2000). But even counting all isoforms, there is still insufficient information in the exome to determine how the brain is wired. The total number of synapses in the human cortex is about 10^{15} (a typical neuron has approximately 5,000–200,000 synapses) (MUOTRI AND GAGE 2006). We still can't explain the brain's complexity from our understanding of the genome.

The original definition of the gene implies that it refers to DNA that is transcribed into RNA that is then translated into protein, so what should we call the large amount of transcribed RNA that is **not** turned into protein? Following the completion of the human genome project, it became possible to interrogate transcription from everywhere in the genome, which led to the unexpected finding that most of the mammalian genome is transcribed (MENDES SOARES AND VALCARCEL 2006). What is all that RNA doing? Some of the RNA species have roles in the regulation of gene expression (see Chapter 4, section 4.2.3), by interfering with transcription or by binding to transcripts (KATAYAMA et al. 2005), but there is another discovery that upsets the gene definition: RNA can act as a catalyst. Thomas Cech reported that some introns are self-splicing, and showed that RNA behaves as a catalyst for this reaction. As with proteins, the structure of the RNA molecule is critical for its function. So, in short, some DNA is transcribed into RNA that functions like a protein.

A further undermining of the original concept of the gene occurred when it was found that the sequence of some RNA species is not fully defined by DNA. Imagine Peter Seeburg's surprise when he discovered that his carefully characterized neurotransmitter receptor transcripts included a glutamine codon (CAG), while the genomic DNA predicted an arginine (CGG) (SOMMER et al. 1991). Seeburg had, inadvertently, discovered post-translational editing of adenosine residues. RNA editing turns out to be relatively widespread, occurring in thousands of transcripts, particularly in the brain (BLOW et al. 2004), leading some to argue for its role in the evolution of cognitive function (BARRY AND MATTICK 2012). One of the possibilities here is that expression of RNA editing enzymes depends on environmental cues, allowing a cell to reprogram its coding sequence. There is also some evidence that dysregulated RNA editing is, somehow, involved in autism (TRAN et al. 2019). Dan Geschwind's group has observed that, in the brains of patients with autism, there is a set of downregulated RNA editing sites, enriched in autism-susceptibility genes.

3.8 Genetic variation: genotyping versus sequencing

So far, we've explained the structure of genomes and genes, but we have not yet touched on the nature of genetic variation, the stuff that in part drives behavioral variation and which is central to the subject of this book. Before we describe the forms of sequence variation, there is one crucial piece of information you should know, which has to do with the way we obtain information on genetic variation, a process called genotyping. The important thing we want you to remember is the distinction between genotyping and sequencing.

Sequencing means obtaining the linear order of bases in some region of DNA (or an entire genome). Genotyping means obtaining information about sequence variation in the genome. Let's suppose there is an organism with just one very short chromosome, so short it consists of just nine nucleotides. Like you and other mammals, this organism has two copies of its single small chromosome, one from its mother and one from its father. The sequence of the chromosome it inherits from its father is:

CCACAGCTT

And the second, from its mother, is:

CCACGGCTT

There is one difference in the sequence of the two chromosomes, at the fifth position (A in the paternal chromosome and G in the maternal chromosome). The sequence of each chromosome is obtained in a sequencing reaction, which will return the linear order of the nucleotides, as written above. If we genotype the individual, we obtain information about the nucleotides on both chromosomes at a given position. For example, the genotype of the first position is CC. This is a homozygous genotype (the same nucleotides on both chromosomes). We can continue genotyping each position, continuing across the entire chromosome to obtain a set of genotypes:

CC, CC, AA, CC, AG, GG, CC, TT, TT

Eight of the nine positions are homozygous, but the fifth position is heterozygous: one chromosome is A and the other is G. We can thus construct genotypes from the sequence, but usually we are only interested in knowing the genotypes at positions known to be variant in a population (polymorphic loci), so it's more efficient (and a lot cheaper) to obtain the genotypes in another way than by sequencing. How does genotyping work? Well, there are lots of methods, but currently most genotyping for genome-wide association studies is carried out using microarrays.

Genotyping microarrays exploit hybridization between single-stranded oligonucleotides and genomic DNA. It works like this. Hybridization describes the ability of one strand of DNA to recognize, and bind to, its complementary partner, on the basis of Watson–Crick binding: A to T and C to G. I am going to repeat again that DNA sequences have polarity (they run from 5′ to 3′), so the sequence ACTGGCTGGC will bind to GCCAGCCAGT and not to TGACCGACCG:

```
5′-  A  C  T  G  G  C  T  G  G  C  -3′
     |  |  |  |  |  |  |  |  |  |
3′-  T  G  A  C  C  G  A  C  C  G  -5′
```

Thus, because DNA strands run from 5′ to 3′, if you are asked to write out the sequence to which 5′-ACTGGCTGGC-3′ hybridizes, the answer is 5′-GCCAGCCAGT-3′. Conventionally, sequences are always written in the 5′ to 3′ direction.

Hybridization between DNA strands is fussy enough to complain about a single mismatch so ACTGGCTGGC won't bind as well to GCCTGCCAGT. This perfectionism can be exploited to detect single-nucleotide polymorphisms (SNPs, pronounced 'snips'). Differences in hybridization signal can be turned into genotypes by using oligonucleotides that bind to either of the two alleles (in fact, multiple different oligonucleotides are used at each locus to improve signal detection). For mammalian genomes, we need oligonucleotides longer than the example of ten bases that I have used here. Short oligonucleotides hybridize to multiple positions in the genome, but an oligonucleotide of 20 bases in length is often sufficient to obtain a unique match. For genome-wide association studies, we have to interrogate hundreds of thousands of loci, which means making millions of oligonucleotides, an impossible objective before the advent of 'light-directed, spatially addressable parallel chemical synthesis' (FODOR et al. 1991). Using this technique, oligonucleotides are synthesized directly onto small glass plates, which are called genotyping arrays, or simply microarrays. They can also be used to interrogate the abundance of nucleic acids (both DNA and RNA) and were a key resource for detecting deletions and insertions in the genome until the advent of cheap next-generation sequencing (see Chapter 7 section 7.7 and Chapter 8, section 8.4 for the importance of deletions and insertions as a cause of psychiatric disease).

3.9 Genome variation: sequence variants, polymorphisms, mutations, and their functional consequences

Reading the genetics literature, you would not be alone if you are confused about the distinction between a sequence variant, a mutation, and a polymorphism. We'll clarify: a **sequence variant** is the generic term that covers all forms of variation, big and small, common or rare, regardless of the functional consequences. A **polymorphism** is a locus in the genome that is known to vary in a population (the definition derives from the work of the ecological geneticist E.B. Ford; FORD 1940). A **mutation** is any change in the sequence of the genome that is not normal, where normality is usually defined by frequency (mutations are rare events, whereas polymorphisms are relatively common).

The functional consequences of sequence variation can be minimal or catastrophic. As a general rule, the impact depends on the size of the variant: the larger the amount of DNA involved, the more likely it is that there will be a change in the phenotype. However, a single nucleotide altered in a critical coding sequence can have a far bigger impact than the duplication of many megabases of DNA, so size is not enough to predict the functional consequences; it's necessary to know which genes or regulatory elements are affected. The consequences of a mutation are best understood when they impact a coding sequence. We have already pointed out that the redundancy of the genetic code means that some coding mutations will not alter the amino acid sequence of a protein (synonymous mutations), whereas others will (nonsynonymous mutations). Table 3.2 lists and describes the different classes of coding mutations.

Deleterious mutations are almost always those that introduce stop sites early into the RNA sequence,

Table 3.2 Classification of coding variation

Type of coding variant	Definition	Example
Synonymous	A mutation that leads to no change in the protein sequence	TCA → TCC; both triplets code for serine, so there is no change to the protein
Nonsynonymous/ missense	A mutation that alters an amino acid in the translated protein	TCA → CCA; serine changes to proline, which may affect the protein structure and function
Stop gain/nonsense	A mutation that changes a codon for an amino acid into a stop codon	TCA → TAA; serine changes to a stop codon, bringing translation to a premature stop
Stop loss	A mutation that changes a stop codon into one that codes for an amino acid	TAA → TTA; a stop codon changes to leucine
Splice site loss	The loss of a component of the canonical splice site	e.g. … GT … AG … → … GG … AG …
Splice site gain	The gain of a component of the canonical splice site	e.g. … CT … AG … → … GT … AG …
Frameshift	A mutation that alters the reading frame of the RNA. Inserting or deleting one base pair will mean codons from that point on will be translated differently	Delete G: AA**G**,TTC → AAT,TC; or insert an A: AAG,TTC → AAG,A**TT**,C

Type	Definition
Table 3.3 Types of sequence variant	
Single-nucleotide polymorphism (SNP)	A single base is changed, such as A → C or T → G
Insertion/deletion (indel)	An insertion or deletion of one or more nucleotides (less than 50 bp affected). This can include small runs of repetitive DNA, such as CACACACA, often also called microsatellites.
Structural variant	A deletion, insertion, or inversion of a stretch of DNA, typically longer than 50 bp. Structural variants include copy-number variants.
Copy-number variant	Variation in the number of copies of a segment of DNA. Usually these are larger than 50 bp and have a complex sequence structure, rather than simple short repeats.

frameshifts, and splicing site changes. However, assessing pathogenicity is hard, as we will see later in the discussion of variants that cause autism and intellectual disability (see Chapters 8 and 9).

In some cases, a mutation that is pathogenic in one environment can be beneficial in another. There are a small number of well-documented examples in human populations of this phenomenon, among which are the mutations that cause inherited disorders of red blood cell function (sickle-cell disease and thalassemias). In northern Europe, the mutations are rare and are associated with disease, but where malaria is common, mutant allele frequencies can be very high, because the mutations confer resistance to the blood-borne parasites that cause malaria (KATO et al. 2018; TAHER et al. 2018). This is an example of the role of natural selection. While the importance of selection as a force in maintaining genetic variation is not in doubt, it is still not clear how to identify and quantify the role of selection in contributing to the genetic basis of complex traits such as schizophrenia (and other polygenic traits) (DE VLADAR AND BARTON 2011; DE VLADAR AND BARTON 2014). We return to this issue later in Chapter 19, section 19.1.

There are three questions about genome variation relevant to behavior genetics. First, what is the nature of genetic variation? Second, how much of the genome varies? And third, how is that variation distributed in the population?

3.9.1 The nature of genetic variation

We know the answers to these three questions from sequencing studies in human populations, such as the 1000 Genomes project (1000 GENOMES PROJECT CONSORTIUM 2010, 2015; ABECASIS et al. 2012) and the UK10K project (UK10K CONSORTIUM et al. 2015). The commonest form of sequence variation is the SNP. Each individual that UK10K sequenced had an average of 3.2 million SNPs (3,222,597 to be precise). Other forms of sequence variation are described in Table 3.3.

There are small (less than 50 bp) insertions or deletions, called indels. UK10K found that each individual has on average 705,684 indels. There are larger changes, more than 50 bp, of which each individual has on average 215. Finer categorization of the type of variation is possible. In particular, it's worth highlighting variation due to differences in the number of copies of a segment of DNA, called a **copy-number variant** (CNV), about which we'll have more to say when we describe some of the molecular changes known to alter the risk of psychosis and intellectual disability (see Chapter 7 section 7.7 and Chapter 9, section 9.3).

3.9.2 The amount of variation

Overall, you differ from any other randomly chosen person at about 4 million loci (that might sound like a lot, but in fact it's only about 0.13% of your genome). This figure is derived from a largely European population, and will vary depending on whether the comparison is with someone from another ethnic group (ABECASIS et al. 2012).

3.9.3 The frequency of variants

The frequency of variants in human populations is very skewed, with most variation being rare. In the UK10K project, each person had about 5,000 SNPs that were

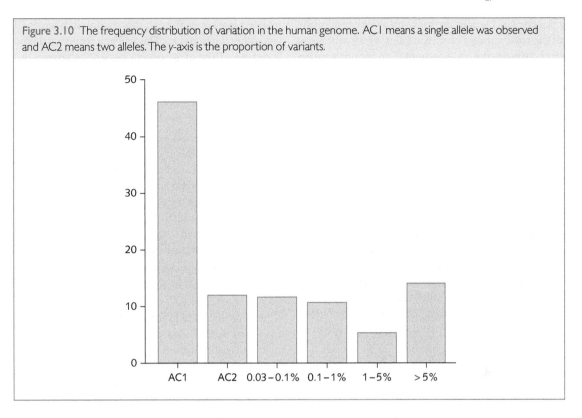

Figure 3.10 The frequency distribution of variation in the human genome. AC1 means a single allele was observed and AC2 means two alleles. The y-axis is the proportion of variants.

'private' (i.e. unique to an individual). Figure 3.10 illustrates the distribution of these. AC1 means that there is a single variant found (AC1 stands for allele count 1) and AC2 means that just two were found. The vertical scale is the proportion of variants: 46% (almost half!) were found as a single allele.

Almost all of the variation is noncoding. In the UK10K analysis, only 576 genes were found that contained variants predicted to result in the loss of function of a protein. It's useful to remember that human genomes typically contain about 100 genuine loss-of-function variants, and have 20 genes completely inactivated (MacArthur et al. 2012). That's to say you, reading this sentence now, lack about 20 genes.

A distinction is often drawn between 'common' variants and 'rare' variants, where a frequency of 1% (or sometimes 5%) is used as the cut-off between rare and common. In European populations—still the best studied—96.5% of variants with a frequency of 1% or greater in the genome are shared between ethnic groups, so that genotyping arrays can be designed that will interrogate almost all common variation in Europeans (the arrays capture much less variation in African populations).

For species other than our own, information on the amount, types, and frequency distribution of genetic variants is more limited, but there are two important general observations.

First, consistent with population genetic principles, species that have large population sizes, particularly when they have a large choice of breeding partners, have much more variation. Genetic variation in wild mice, excluding those commensal with humans, is much larger than in humans. Sequencing of just ten wild mice caught in India identified more than 50 million sequence variants (Halligan et al. 2010). Fruit flies (*Drosophila melanogaster*) have larger population sizes than mice but smaller genomes. When Trudy Mackay sequenced 168 inbred flies that were derived from a wild population, she reported 4,672,297 SNPs, 105,799 indels, and 36,810 other forms of variation (Mackay et al. 2012). Similar catalogs are not available for other species, but there is no reason to expect that the amount, and categorization, will be substantially different.

Second, the distribution of genetic variation in wild populations is similar to that seen in humans: most variation is rare. Common variants (those with a frequency of 5% or more) make up a small proportion of the total number of variants, but, as we will see in later chapters, the small amount of common variation makes a large contribution to variation in behavioral (and other) phenotypes. This will be a recurring theme in our discussion of the genetic basis of behavior.

Summary

1. Cells contain two types of nucleic acid: DNA and RNA. DNA is a double-stranded molecule that contains a linear arrangement of four bases, the 'letters' that encode hereditary information. RNA is a single-stranded molecule that is much more unstable than DNA. Its major role is to serve as an intermediary, carrying information encoded in DNA to a place where it can be turned into protein. The production of RNA from a DNA template is called transcription. Protein is made from an RNA template in a process called translation.

2. Three findings flow from the double-helix structure of DNA: first, it elucidates how DNA contains the information about the form and function of cells; second, it explains how cells reproduce; and third, it places information processing at the center of biology.

3. The genetic code shows the relationship between DNA (and RNA) sequence and the 20 amino acids that constitute proteins. Triplets of bases encode amino acids. The code is redundant (there are 64 triplets but only 20 amino acids), contains messages for the start and end of the amino acid, and is universal: all organisms—viruses, bacteria, yeast, insects, plants, and vertebrates—use the same code.

4. To obtain the sequence of a genome, DNA is broken up into small fragments that are sequenced on automated machines and then assembled on computers. Because 60–70% of the human genome is repetitive, most methods of genome sequencing usually generate only 80% of a human genome.

5. The human genome contains 20,310 protein-coding genes. Eukaryotic genes consist of exons that contain the coding sequence, and introns that are not coding and have to be spliced out of the RNA before translation occurs. Alternative splicing generates multiple transcripts from the same gene. Brain tissue has more alternatively spliced transcripts than other tissues.

6. Genotyping refers to methods that obtain information about sequence variation in the genome. Genotypes report the sequence at a specific locus in the genome from both chromosomes. Most genotyping methods exploit hybridization between single-stranded oligonucleotides and genomic DNA to genotype at the variant positions.

7. Genetic variants include single-nucleotide polymorphisms (SNPs), small insertion/deletions, and larger structural variants. Copy-number variants are a form of structural variant that have been implicated as risk factors in psychiatric disease.

8. The majority of genetic variants are rare. Each person has about 5,000 SNPs that are unique. Human genomes contain about 100 loss-of-function variants and have 20 genes completely inactivated.

9. Common variants are those with a frequency greater than 1%. In European populations, almost all common variants are shared.

4

Epigenetics, gene regulation, and 'omic technologies

[On the mechanisms underlying the regulation of gene expression and how these may explain behavioral variation and psychiatric disease]

4.1 What is 'epigenetics'?

In this chapter, we continue our presentation of molecular biology relevant to the genetic basis of behavior. We start with epigenetics, a subject that requires an understanding of what does, and what does not, regulate gene expression. After describing what is known about the molecular basis of gene expression, we introduce epigenetic mechanisms relevant to behavior, including genomic imprinting and inheritance that does not depend on DNA.

A developmental biologist (Conrad Waddington) introduced the term epigenetics in 1942 to mean the analysis of how genes are related to phenotypes through what was known about experimental embryology. For Waddington, epigenetic processes were those that led from DNA to a final adult organism (WADDINGTON 1942). More recently, epigenetics has come to mean the study of heritable changes in gene function, heritable in this case referring to cellular memory. The central question in epigenetic research is: how does a cell remember its characteristics? For example, how does a neuron know it is a neuron?

Epigenetics is the science of explaining why, when a liver cell divides, it produces two more liver cells, and not muscle cells or neurons.

Perhaps you think that understanding cellular memory is hardly relevant to the genetics of behavior. However, epigenetics applies not only to the consequences of mitotic division, where cells divide to reproduce themselves, usually as part of a tissue, but also to meiotic division, which produces germ cells (eggs and sperm). In the latter case, epigenetics could explain why some characteristics are inherited and others not. Epigenetic changes might be the reason why individuals who otherwise share identical DNA sequence have discordant phenotypes (FRAGA et al. 2005). On this basis, epigenetic explanations are put forward to explain why early life experience, and indeed maternal experience, can influence behavior through adulthood.

For example, differences occurring in the womb are associated with differences in postnatal intelligence and alterations of brain anatomy that persist at least into late adolescence (RAZNAHAN et al. 2012). There is also evidence that, during pregnancy, a mother's health or physical state affects her fetus, changing the offspring's physiology in an attempt to reduce long-term risk of disease (GLUCKMAN et al. 2008)—this is the 'fetal origins' hypothesis of adult disease (BARKER 1998; an idea that is not uncontested: see HUXLEY et al. 2002). We'll return to look at more examples later in this chapter, focusing, of course, on those where there are claims that epigenetics explains behavior (FRANCIS et al. 2003; MURGATROYD et al. 2009).

Later, the two uses of epigenetics are explained, namely changes in gene function persisting through mitosis (section 4.3.1) and heritable changes in meiosis (section 4.3.2). First, though, you have to learn some basic molecular biology, because it's impossible to understand an epigenetic effect, and assess whether it is has been adequately established, without familiarity with such terms as 'trans-acting factor', 'methylation', 'histones', and 'noncoding RNAs'. We'll try our best to clarify.

We suspect one of the reasons that epigenetics is popular is because so few know the basics of gene regulation (after all, biochemistry does not make easy reading). For example, 'for epidemiologists, invoking epigenetics is one way of establishing their up-to-the-minute "scientific" credibility as they make hand-waving statements about potential causal processes in disease, which can help dispel the impression of them being overpaid phlebotomists or (worse) moonlighting statisticians' (SMITH 2011). Most of us don't know how to argue against the claim that something is an epigenetic phenomenon, but by the time you've finished reading this chapter, we'll have provided you with sufficient information, and arguments, to assess the validity of 'epigenetic' explanations.

4.2 The regulation of gene expression

The science of gene regulation is one of those areas in biology lacking fundamental principles. It's all detail. Sydney Brenner tells of a physicist colleague, Leo Szilard, expressing his frustration about biology, 'When he was a physicist he could lie in the bath for three hours thinking about physics. But when he went into biology, no sooner did he get into the bath than he'd have to get out and look up another fact!' (BRENNER AND WOLPERT 2001). Mastering the literature on gene regulation means being able to navigate through more than the usual collection of acronyms.

In a simplified view, we can recognize five features known or suspected of being involved in the regulation of gene expression: DNA sequence features, trans-acting factors, regulatory RNAs, DNA methylation, and chromatin packaging (BIRNEY et al. 2007). All of these terms need unpacking. I warn you: the next bit of text is hard work, but I've tried to keep it short. Let's start with the DNA features.

4.2.1 Regulatory elements in DNA

There are two important DNA sequence features for regulation of gene expression, promoters and enhancers. Promoters are the regions in front of genes where a series of proteins bind to the DNA and to each other in order to turn on gene expression (Figure 4.1). The process by which DNA sequences are read out and converted into messenger RNA (mRNA) is called transcription, and the protein machinery that assembles at the promoter is referred to as a transcriptional complex. The most important component of the complex is an RNA polymerase, whose job is to make an RNA copy of the DNA, generating a primary RNA transcript (it is called primary because it has to undergo some additional processing before it becomes

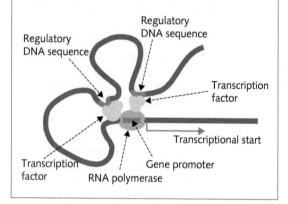

Figure 4.1 The elements of gene regulation. The blue line represents the DNA sequence. Regulatory elements are shown as gray segments. Enhancers are regulatory DNA sequences and are typically distant from the gene whose expression they regulate. The figures shows two transcription factors (in light blue) binding to regulatory elements and to an RNA polymerase, part of a transcriptional complex, which is located at the gene promoter, poised to start transcription from the transcriptional start site.

a mature mRNA, including the splicing out of introns (see Chapter 3, section 3.7 and Figure 3.9). Promoters can be recognized by the presence of (relatively) well-known sequence features. Enhancers are regulatory regions within DNA typically located at some distance from the genes they regulate (Figure 4.1). The classical definition of an enhancer is a segment of DNA whose effect on gene expression is independent of orientation and position. This means that if an enhancer is plucked out and placed somewhere else, facing a different direction, it will still alter the expression of the gene that it regulates. This is not true of promoters, which only function when they are in the correct orientation and placed at the front of the gene (technically the 5′ end of the gene; see Chapter 3, section 3.1 and Figure 3.1). It is currently almost impossible to recognize enhancers from sequence features alone, so the consequences of sequence variation within an enhancer can't yet be predicted from knowledge of the sequence (this is likely to change as artificial intelligence is applied, a hot topic in genetics and genomics research; LIBBRECHT AND NOBLE 2015). In addition to promoters and enhancers, there are also other, less well-characterized DNA elements that are involved in functions such as DNA packing.

Enhancer elements can be separated by up to hundreds of kilobases from the gene they regulate. So how do they act on their target genes? The current view is that they combine with the proximal elements (the promoters) to control gene transcription (Figure 4.1) (DEKKER et al. 2002; ZABIDI AND STARK 2016). A simple interpretation of what happens is that a DNA enhancer binds a protein called a *trans*-acting factor (described in more detail in section 4.2.2), which in turn binds to a transcriptional complex including an RNA polymerase that transcribes DNA. In this model, the DNA between the regulatory elements has to loop, so that the regulatory elements come into contact (JING et al. 2008; MITCHELL AND FRASER 2008). In Figure 6.11 in Chapter 6, we'll see that looping is critical to one idea of how many common genetic variants have their effect on phenotypes.

4.2.2 *Trans-acting factors*

Trans-acting factors are proteins that bind to DNA and alter transcription by forming complexes at promoters and enhancers (Figure 4.1). Steroid receptor proteins are an example: the stress hormone cortisol binds to its receptor (the glucocorticoid receptor) and

the resulting complex enters the cell's nucleus, where it binds to specific DNA sequences and initiates the transcription of a series of genes (BEATO 1989).

Although transcription factors have specific recognition sequences, these are usually relatively short motifs (sometimes only 4 bp), so there are always far more sequence matches in the genome than sites that actually bind the proteins. This means that sequence alone is not a good predictor of where the protein binds, nor of what it regulates. For example, the glucocorticoid response element binds to the 15-nucleotide sequence GGTACAnnnTGTTCT (where 'n' is any nucleotide); the glucocorticoid receptor binds to only a small fraction of these sequences in the genome. There have to be additional factors that determine transcription binding, but we don't know what they are.

4.2.3 **Regulatory RNAs**

Regulatory RNAs are just that: RNA species that regulate transcripts. There are three forms (possibly more): microRNAs (miRNAs), short interfering RNAs (siRNAs), and long noncoding RNAs (lncRNAs, pronounced 'link RNAs'). The role of lncRNAs is particularly hard to fathom. Some are well known and clearly have critically important functions, such as determining the inactivation of one of the two X chromosomes in mammals (BROWN et al. 1992) and interactions with proteins determining gene expression patterns (GUPTA et al. 2010). But there's still disagreement about how many have any effect at all (GUTTMAN AND RINN 2012; JANDURA AND KRAUSE 2017). It's possible that lncRNAs themselves don't directly regulate genes, but rather the act of lncRNA production does, by interfering with the expression of other genes (ENGREITZ et al. 2016a, b). Our understanding of the function of other regulatory RNAs is, fortunately, more complete.

The discovery that small pieces of RNA were part of the machinery of gene expression regulation was totally unexpected. Small really does mean small: 22 nucleotides is long enough to function as a regulatory RNA. Of the two forms of short RNAs, the distinction between miRNAs and siRNAs isn't entirely clear cut. They are similar in terms of their molecular characteristics and biogenesis. They both have targets; that is, their sequence is complementary to some genomic feature and binding to that feature is part of what they do. miRNAs and siRNAs both act by binding to mRNAs (the RNA species that carry protein-coding information) and thereby form an RNA interference system, an evolutionarily conserved, gene-silencing mechanism. However, they differ in the nature of their targets: miRNAs are relatively nonspecific, while siRNAs are relatively specific. In most cases, miRNAs match their targets imperfectly at multiple sites so that, as a class, miRNAs affect the expression of about one-third of all genes (Lewis et al. 2005). By contrast, siRNAs often bind a single site. The target mRNA is cut at the site of complementarity, thus disabling it. The specificity of siRNAs has led to their use in the laboratory as a versatile way of downregulating mRNAs. They now belong to the tools that molecular biologists use to interfere with gene expression.

4.2.4 DNA methylation

In eukaryotes, DNA methylation is generally associated with the silencing of gene expression (Mohn et al. 2008). DNA methylation refers to the addition of a methyl group (CH_3) to cytosine or adenine bases. In mammals, this happens almost exclusively on a cytosine followed by guanine (called CpG in the literature where 'CpG' means 5'-C-phosphate-G-3', to indicate the polarity of the molecule). But the pattern of methylation in mammalian DNA does not apply to all other species: for example, invertebrates don't use methylation in the same way at all.

In mammals, methylated CpGs attract their own specific binding proteins, of which one, methyl-CpG-binding protein 2 (MeCP2), has gained importance in the human behavior genetics literature because mutations in the gene cause Rett syndrome (Amir et al. 1999), a rare genetic condition in which a postnatal progressive neurodevelopmental disorder manifests in girls during early childhood (see Chapter 9, section 9.5.2). MeCP2 also has a role in chromatin remodeling, discussed in the following section.

About 70% of promoters in the human genome are associated with the phenomenon of CpG islands: regions of about 1,000 bases where there is an enrichment of CpG dinucleotides that are *not* methylated (Deaton and Bird 2011). These islands are known to bind a set of specific transcription factors, which in turn prevent DNA methylation (Thomson et al. 2010), even in the absence of ongoing transcription. Differences between low, medium, and high levels of CpG DNA methylation explain whether distinct transcription factors gain access to their binding sites. Proteins that bind to CpG islands recruit histone methyl transferases, whose role in regulating gene expression, through an effect on chromatin, is described next.

4.2.5 Chromatin and chromatin modifications

Chromatin plays the most complex and possibly the most important role in epigenetics (Felsenfeld 2014; Allis and Jenuwein 2016). Chromatin is the name that the German anatomist Walther Flemming gave to the mixture of DNA, RNA, and protein that is found in the cell nucleus. He called it chromatin because of its affinity to dyes used to stain the cell nucleus (Flemming 1882).

Chromatin in cell nuclei has two forms, which are distinguished by how intensely they stain. One form, euchromatin, is less intense, while heterochromatin stains intensely. The DNA sequence associated with these two forms has different features: heterochromatin is usually associated with long stretches of repetitive DNA, located at the center and the ends of chromosomes. Genes are rarely expressed in regions of heterochromatin. For example, the second (inactivated) X chromosome in human females is wrapped up in heterochromatin. Euchromatin is found in the rest of the genome, and is associated with regions of gene expression. These observations suggest a relationship between chromatin and gene expression, but initially it was not known whether chromatin was the passive consequence of processes that controlled gene expression, or whether it was actively involved in gene regulation. Indeed, for many years, the prevailing view was that chromatin was an inert packaging material.

The protein component of chromatin is composed of histones, which are used to wrap up the DNA so that it fits into the nucleus (each human cell has to fit a DNA string that is 1.2 m long into a cell nucleus with a radius of about 10 μm). The basic unit of the chromatin fiber is the nucleosome, which consists of two copies each of four histone proteins (H2A, H2B, H3, and H4; sorry, I don't know what happened to H1!), which package 147 bp of DNA (Kornberg 1974; Kornberg and Thomas 1974). Packaging is really only a serious problem for cells with nuclei (eukaryotes), so don't be surprised to learn that bacteria (prokaryotes, with no nuclei) don't have histones. If the job of histones were just to package DNA, then activating a gene would mean removing the histones and allowing transcription to proceed in much the same way that occurs in prokaryotes. That view has turned out to be far too simplistic.

There is now abundant evidence that histones are involved in gene regulation, by virtue of the fact that they can be biochemically modified (Figure 4.2).

Figure 4.2 DNA (shown as a black line) is wrapped around histones (the gray cylinders). Histones have 'tails' that are subject to molecular additions. The addition of acetyl groups (Ac) or methyl groups (Me3) determines whether transcription of the DNA is active or repressed. Histone methyl transferase (HMT) adds Me3 to histones and histone demethyl transferase (HDMT) removes them. Histone acetyl transferase (HAT) adds acetyl groups and histone deacetyl transferase (HDAC) removes them.

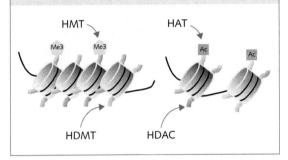

Michael Grunstein, a biochemist at the University of California, Los Angeles (UCLA), provided the evidence in the 1980s that nucleosomes repress transcription (HAN AND GRUNSTEIN 1988): he showed that without histones, every yeast gene that was normally repressed was activated. This was the experiment that convinced people that histones had to be involved in gene regulation. But how?

Histones bear multiple types of modification (ALLIS AND JENUWEIN 2016), but what the modifications did, if anything, was not known until the discovery in the 1990s of enzymes that add on and take off acetyl groups. Acetyl groups are really nothing much more than vinegar (acetic acid) (Figure 4.3; the wiggly line is where the acetyl group is joined to another molecule, such as a histone). The enzyme that adds acetyl groups to histones is called histone acetyltransferase (HAT) and is a transcriptional activator, revealing a link between histone acetylation and gene activity (BROWNELL et al. 1996). Furthermore, when the first histone deacetylase (HDAC)

was discovered, it turned out to be a transcriptional repressor (TAUNTON et al. 1996). Histone acetylation and deacetylation are thus coupled to 'on' and 'off' states of gene regulation.

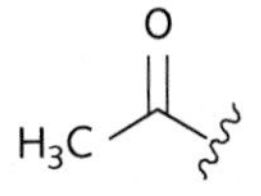

The second functional histone modification to be found was methylation (be careful not to confuse histone methylation with DNA methylation: both are involved in epigenetics, but they are completely different biochemical processes). An enzyme was discovered that adds methyl groups to the lysine residue in histones, a histone lysine methyltransferase (REA et al. 2000). Once methylated, the lysine residue provides a specific binding site for proteins that recognize the altered histone structure. The recruited proteins in turn alter chromatin structure from euchromatin to heterochromatin, which represses gene expression. Like the acetyl modification system, methylation marks can be removed as well as added (SHI et al. 2004), consistent with the role of histone methylation in dynamically regulating gene expression: there are histone demethylases that remove methyl groups from histones.

As additional methylation marks have been identified at different places on histone proteins, the literature has become replete with acronyms: trimethylation of H3K4 is enriched near active genes; monomethylation of H3K4 (H3K4me1) is enriched over active enhancers (RUTHENBURG et al. 2007); elevated levels of H3 lysine 27 methylation (H3K27me) correlate with gene repression mediated by a protein called *Polycomb* (TROJER AND REINBERG 2006). The acronyms refer to the location of the modification on the histone, so H3K4 means the fourth amino acid residue from the terminus of the histone H3. K is the single letter code for the amino acid lysine.

The complex nature of histone modifications has suggested the existence of a histone code, where some proteins lay down or 'write' histone modifications, while others 'read' modifications and perform actions that reflect the pattern of histone modification (STRAHL AND ALLIS 2000). Not everyone likes this idea (PTASHNE 2007), but it's one to be aware of, as it's claimed to play a part in explaining behavior. We turn to this next.

4.3 Epigenetic explanations of behavior

Now that we have told you about the five components of gene regulation (DNA regulatory sequences, *trans*-acting factors, DNA methylation, short RNAs,

Figure 4.3 Structure of an acetyl group.

$$H_3C \underset{\displaystyle }{\overset{\displaystyle O}{\|}} $$

and histone modifications), we can introduce the ways in which epigenetic mechanisms are said to explain behavior. As we explained earlier, there are two classes of mechanism: one that explains the transfer of information from a cell to its descendants (so a form of cellular memory) and one that explains transfer of information from an entire organism generation to its progeny (*trans*- or intergenerational memory).

4.3.1 Cellular memory

For some neuroscientists, psychiatric disease is best explained by epigenetic mechanisms. Consider the problem of explaining how a single, large stressor early in life is still causing misery, increasing susceptibility to anxiety and depression, in late adulthood. How does that work? Surely that represents a change in cellular memory that epigenetic processes can explain? And indeed that is precisely the argument that Eric Nestler, a neuroscientist at Mount Sinai Hospital in New York, has taken with two psychiatric disorders: addiction (NESTLER 2014b; PENA AND NESTLER 2018) and depression (KRISHNAN et al. 2007; SUN et al. 2013; NESTLER 2014a), both of which have environmental determinants (adversity in the case of depression and drugs in the case of addiction). His work exemplifies how each of the mechanisms we have described above in section 4.2 could deliver a mechanistic understanding of the biological basis of a psychiatric illness (TSANKOVA et al. 2007). I'll briefly explain his observations, but I should warn you that the next paragraph is technical; it contains a lot of the acronyms and neologisms that we introduced earlier in the chapter.

Almost all of the data come from experiments in mice in which the consequences of a behavioral intervention (such as repeated cocaine exposure: RENTHAL et al. 2007; MAZE et al. 2010; or a model of chronic stress: LEPACK et al. 2016) is associated with a change in one or more epigenetic factors. Nestler identified changes in expression of a transcription factor, called OTX2, as the key mediator of the transcriptional changes underlying stress-induced vulnerability to lifelong depression (PENA et al. 2017). He also proposed that addiction and depression are related to chromatin modifications. Chronic exposure to cocaine or stress decreases the function of an HDAC in a brain reward region (the nucleus accumbens), which results in increased histone acetylation and transcription of its target genes. Loss of the HDAC causes increased responses to chronic cocaine exposure or to stress (RENTHAL et al. 2007).

The amount of a histone variant (called H3.3) in the nucleus accumbens is thought to be involved in the regulation of stress-mediated gene expression and the precipitation of depressive-like behaviors in mice (LEPACK et al. 2016). There is also evidence that repeated cocaine administration prior to social defeat stress results in depressive-like behaviors in mice through decreased levels of histone H3 lysine 9 dimethylation in the nucleus accumbens (MAZE et al. 2010; COVINGTON et al. 2011). And Nestler has demonstrated that selectively modifying chromatin by molecular engineering of transcription factors can control drug- and stress-evoked transcriptional and behavioral responses (HELLER et al. 2014). Finally, DNA methylation is implicated through the finding that changes in the activity of an enzyme that methylates DNA (methyltransferase DNMT3a) alters both addictive and depressive behaviors in mice (LAPLANT et al. 2010).

Nestler's findings, and those of others (PENA AND NESTLER 2018), implicate one or more epigenetic mechanisms in the regulation of behavior. But strictly speaking, none of the examples we have given is epigenetic in Waddington's sense. Waddington's intent in introducing the term epigenetics was to advance our understanding of the pathway from the causal genetic variant to the altered phenotype. This requires the epigenetic change to be passed on, through cell division, to maintain a cellular state. None of Nestler's experiments demonstrates a cellular memory that extends through mitosis, for the simple reason that only a tiny fraction of neurons continue to divide in adulthood. So while these mechanisms are potentially important explanations of the molecular basis of behavior, and can explain how a cellular state is established and maintained (representing a cellular memory of environmental stress, for example), the experimental findings won't change our views of the importance of genetic factors, or even allow us to interpret how genetic variants are transmitted into behavioral change. In short, the sort of epigenetic experiments we have described above lie outside the purview of our book. They really don't tell us anything about the relationship between genes and behavior.

However, there are some findings that do invoke cellular memory passed on through cell division to explain behavioral differences. The prime example comes from twin studies: epigenetic cellular memory could explain why monozygotic (MZ) twins are discordant, differences that we would otherwise attribute to environmental effects. Suppose that, during development, the effect of the environment alters a cell's epigenetic marks (it acquires some extra DNA methylation or a change in histone modifications if the cell is stressed, for example by starvation or by a toxic substance). You can imagine that, during development, during cell division, the acquired epigenetic marks are passed on to the cell's descendants. In this way, while the DNA remains the same between two individuals, the epigenetic marks might not.

What evidence is there to support this view? In the first place, there are rare examples where discordance in MZ twin pairs has been chased down to an epigenetic difference. In 2002, a pair of female MZ twins was born of whom one twin had abnormalities of the lower spine, including a tumor. Following clues about which genes are involved in spinal abnormalities, it was found that one such gene was significantly more methylated in the affected than in the unaffected twin (OATES et al. 2006). The methylated site is probably an example of a 'meta-stable' allele; that is, the state of the epigenetic mark is unstable and hence shows variation. We'll describe a well-documented case of meta-stability in section 4.5. The important lesson of this single case is that, in the face of apparently identical sequence, methylation differences contributed to a phenotype. This means that epigenetic effects can have an impact, but it doesn't inform us as to how important a phenomenon that is.

Second, twin studies have assessed the heritability of methylation. The results are not easily summarized into a single statistic for methylation heritability, This is because different methylated sites of the genome have different heritabilities, which is what you would expect from the discovery that some sites are unstable. In general, DNA methylation differences in MZ twins are significantly lower in comparison with DZ twins (KAMINSKY et al. 2009), but, crucially, epigenetic profiles are not fully determined by DNA sequence; if that were the case, MZ twins would show no epigenetic differences. In other words, DNA methylation differences could explain differences between MZ twins (including behavioral differences).

Third, older twins show more differences in methylated DNA and histone acetylation sites than younger twin pairs (FRAGA et al. 2005). The epigenetic differences are more obvious in MZ twins who have had different lifestyles and have spent less of their lives together, consistent with the idea that environmental factors are responsible for changes in epigenetic marks. One way to think about these epigenetic effects is to consider them as a molecular record of what the environment has done to an organism, a sort of cellular record of the past. In this case, they are the consequence, rather than the cause, of behavioral differences.

In summary, so far, no examples have been shown where MZ twin differences in behavior or psychiatric disease can definitively be attributed to epigenetic variation, but it remains a possible explanation (DEMPSTER et al. 2011). This is an active area of research, with strong hints that important effects will be found, but as you can see from comparing our first and third examples, deciding whether epigenetic changes cause behavioral differences or not is going to be hard.

4.3.2 Transgenerational epigenetic inheritance

The discovery of the inheritance of epigenetic marks from one generation to the next, with an effect on behavior that can't be explained by DNA sequence variation, would overturn the central dogma presented in Chapter 3, which states that DNA is the sole repository of inherited information. It would imply the existence of a separate channel through which behavioral features could be passed on to the next generation. The idea has its proponents, and we will outline some of the evidence in the following section.

Do we believe that epigenetic marks can be passed on from generation to generation and that can they account for variation in inherited phenotypes? The short answer to that question is yes, but the circumstances in which this happens are limited and the impact is not as broad as some suggest. The best example of where an epigenetic modification persists through cell division, is passed through the germ line, and unarguably impacts on behavior is genomic imprinting at a locus on human chromosome 15q (BARLOW AND BARTOLOMEI 2014).

4.4 Prader–Willi and Angelman syndromes

Genomic imprinting lies at the root of a rare example in human genetics of inherited human behavior that is not dependent on DNA sequence. Two different conditions, Prader–Willi syndrome and Angelman syndrome, are caused, in the majority of cases, by the same 5–6 Mb deletion on chromosome 15, but have markedly different phenotypes. Just to be clear, the DNA deletion is exactly the same, but the phenotypes are radically different. The disorders are named after the doctors who first described the phenotypes: Prader–Willi syndrome occurs in approximately 1 in 10,000 births, and Angelman cases are even rarer. Children with Prader–Willi syndrome have recognizable physical characteristics, including small hands and feet, almond-shaped eyes, and a thin upper lip. The majority show obsessive–compulsive behavior, often focused on food consumption; a clinician told me of parents who had to put a lock on the freezer to prevent their son from eating frozen pizzas. Patients with Angelman syndrome are also recognizable by their physical appearance, which is different from Prader–Willi syndrome: they have small heads, a prominent mandible, and a wide mouth. Their behavior is also characteristic: they have severe developmental delay, minimal verbal skills, and abnormal hand-flapping movements. They laugh inexplicably, appear happy, and smile frequently.

In Prader–Willi syndrome, the deletion is paternally inherited, whereas in Angelman syndrome it is maternally inherited (NICHOLLS et al. 1989). Don't imagine that this means the parents are affected. Rather, the deletion arises for the first time in the germ cells, that is, in the egg or sperm. A deletion on chromosome 15 that occurs in the sperm gives rise to Prader–Willi syndrome, while one that happens in the egg produces Angelman syndrome.

About 70% of cases of Prader–Willi and Angelman syndromes are due to deletions. But about a quarter of the cases of Prader–Willi syndrome and 5% of Angelman syndrome are due to something even stranger. They have no deletion, not even a point mutation or any of the variants presented in Table 3.2 in Chapter 3. Instead, the disorder happens because they inherited two copies of chromosome 15 from one parent rather than one from the father and one from the mother (NICHOLLS et al. 1989). This is called uniparental disomy. Maternal uniparental disomy gives rise to Prader–Willi syndrome and paternal uniparental disomy gives rise to Angelman syndrome. You can see that the uniparental disomy and deletion both mean that the child has only DNA from one parent at this locus on chromosome 15. Why is it necessary to have DNA for this gene from both parents? To understand what is going on at the chromosome 15 locus, we need to explain what is known about the molecular biology of genomic imprinting.

4.4.1 Genomic imprinting

The implication of the 15q locus finding is that something in addition to DNA sequence is needed to ensure that we are all born healthy. That is also the conclusion of experiments that attempted to generate mouse embryos solely from mouse egg nuclei: healthy mice are only born from genomes that include both paternal and maternal contributions (MCGRATH AND SOLTER 1984; SURANI et al. 1984). Cells somehow distinguish between the parental origins of their genomes, and parental origin affects the function of DNA. This leads to two important questions: first, what is the nature of this imprint, and second, how does the imprint result in the functional changes?

We can state the following about the nature of the imprint. First, the action of the genomic imprint is limited to the chromosome on which it is found, otherwise cells could not recognize a specific chromosome. This is called *cis* action, in contrast to *trans* action in which a signal could diffuse from one chromosome to another. Because genomic imprinting was discovered in inbred mice whose genomes are clonal, with no sequence differences, the imprint has to be epigenetic—a mark added to the DNA sequence, rather than an alteration of the sequence itself. The functional consequence of the imprint is to change gene expression on one parental chromosome affecting one or more genes. Second, the parental imprints are placed on the DNA when the two parental chromosome sets are separate. This only happens during gamete formation (the production of eggs and sperm). Third, the imprint is sufficiently stable to remain on its chromosome of origin after fertilization, and it is capable of being passed on to the daughter cells, through mitosis, from the embryo to the adult. Yet it must also be possible to erase an imprint. During development, some of the embryonic cells, all of which carry parental imprints on their chromosomes, give rise to germ cells, the eggs and sperm of the newly forming organism. In order for the next generation to receive an imprint indicating their chromosome's parental origin, the old imprint must be stripped off and a new one added. Furthermore, after fertilization, another round of genome-wide reprogramming occurs, followed by *de novo* DNA methylation around the time when the fertilized cell implants in the womb (LI AND ZHANG 2014).

So far, only one epigenetic mark has been shown to fulfill the characteristics required of a genomic imprint. This is DNA methylation: the addition of a methyl group to cytosine, the only known heritable modification of DNA (LI AND ZHANG 2014). DNA methylation is acquired and maintained through the action of enzymes called methyltransferases, and mutations that disrupt the function of methyltransferases perturb gene expression at imprinted regions, consistent with the hypothesis that methylation is an imprinting mark that alters gene expression.

How does the imprint result in the functional changes? The answer is not fully understood and takes us into some complex biology, but in brief what we know is the following, most of which comes from mouse experiments. Mice contain about 150 imprinted genes organized into 17 clusters. Differential methylation controls imprinted expression at these clusters, and so these methylation regions are called imprint control elements. Currently, two major classes of *cis*-acting silencing mechanisms are believed to govern imprinting. In the insulator model, methylation prevents a specific protein from binding to DNA and as a result this disrupts interaction between DNA regulatory elements that are required to turn on gene expression. In the noncoding RNA model, imprinting results in the transcription of a lncRNA that somehow interferes with the transcription of the imprinted genes.

How important is genomic imprinting as a mechanism for understanding behavior? Probably not very important. Prader–Willi and Angelman syndromes are extraordinarily rare natural experiments that revealed the existence of imprinted loci on chromosome 15q, but they are almost unique examples. The mouse experiments identify just 17 imprinted clusters, mostly containing genes involved in metabolism. However, more subtle, less easily detectable effects of imprinting have been detected in a large-scale analysis of outbred mice, where it was shown that nonimprinted genes can generate parent-of-origin effects by interaction with imprinted loci (MOTT et al. 2014). So it is possible that the true extent of parent-of-origin effects may be larger than is currently imagined.

4.5 Epigenetic transgenerational inheritance

There are currently few well-characterized examples of transgenerational inheritance that do not depend on DNA. The best evidence comes from experimental

organisms, and they don't involve behavior. For example, the transmission of one form of coat color variation in mice, called 'agouti', depends on the phenotype of the mother, not the genotype (MORGAN et al. 1999). When a particular agouti allele called A^{vy} passes through the female germline, offspring are more likely to be agouti because an epigenetic modification is incompletely erased, not because of inheritance of the allele. The A^{vy} allele is a mutation due to a transposable element at the locus (transposable elements are segments of DNA that jump into and out of genomic DNA, a bit like a virus that parasitizes its host's genome; see Chapter 3, section 3.6). The allele is said to be 'meta-stable' and links epigenetic machinery to the need to repress transcription from invading DNA. A small number of other examples of 'meta-stable' alleles emerged in a systematic screen of the mouse genome (KAZACHENKA et al. 2018), so the A^{vy} allele is not unique, but the phenomenon is uncommon. There is also evidence for the role of RNA in mediating nongenomic transmission of information (RASSOULZADEGAN et al. 2006). However, the contribution of these unusual forms of inheritance to trait variation of any kind, let alone just behavior, is small.

Considering other examples of nongenomic transmission affecting behavior leads us into contentious territory. One story has garnered considerable attention since its publication in 1999 (FRANCIS et al. 1999). Starting from a finding about the consequences of rat parenting skills, Michael Meaney and colleagues argued that not all inheritance is DNA based. Meaney observed that a mother who cares for her pups by gathering them around her, licking and grooming them, is more likely to have offspring who grow up less fearful and more resistant to stress. The female offspring also learn their mother's caring style and thus transmit the behavioral pattern to subsequent generations. Inherited differences in maternal behavior and in fearfulness could be the outcome of variation in DNA, consistent with the patterns of inheritance we've seen elsewhere in this book, but in this case Meaney suggested that, *The variation in maternal behavior may thus constitute a mechanism for the nongenomic behavioral transmission of fearfulness from parent to offspring.* To decide between these alternatives, the researchers used a cross-fostering design: pups from the caring mother were moved to the noncaring mother and vice versa. The result was that pups took on the behavior of their foster mothers, not their biological parents, supporting the view for nongenomic transmission.

Meaney's group went on to ask, as rats breast-feed their pups for a week, how could such a short exposure result in a marked difference in behavior observed months (and years) later? Meaney suggested an answer: maternal care causes a stable change in gene expression in the brain, or at least of one gene (a glucocorticoid receptor), due to the cytosine methylation state of the gene's promoter (WEAVER et al. 2004, 2007). The authors have also argued that the same gene shows similar changes in samples from suicide victims with a history of childhood abuse (compared with controls) (MCGOWAN et al. 2009), and they claim to have discovered an example of environmental 'programming' manifest in the biology of the central nervous system.

One of the more intriguing claims for an epigenetic transmission of behavior is that a constituent of sperm mediates epigenetic inheritance (CHEN et al. 2016). A male's environment, either during development or during adulthood, has been reported to affect his children's phenotype (RANDO 2012). Altering a mouse's diet (e.g. by starvation) changes the metabolism of his offspring (CARONE et al. 2010; RADFORD et al. 2012). Stress a male mouse (e.g. by giving him mild foot shock) and his offspring will show behavioral changes, even when his sperm is used via artificial insemination to generate progeny. Claims for behaviors inherited via sperm include the response to a specific odor (the father is trained to associate the odor with foot shock and the offspring react specifically to that odor; DIAS AND RESSLER 2014), alterations in response to many different stressful experiences (GAPP et al. 2014), and changes in neuronal function and cognition in the offspring of mice given environmental enrichment (BENITO et al. 2018). What could be carrying this information? It is possible that it is an RNA species in sperm, such as the regulatory RNAs we discussed earlier.

4.6 Not every cellular mechanism is epigenetic

How should you, a reader of our book, respond to claims that an RNA species in sperm carries information about a father's life experience, such as a stressful life experience, than can alter the behavior of his children? Just because a claim is bizarre does not mean it is not true. But how do we know whether a result is true or not? This is a question that recurs throughout

our book. On the face of it, the relationship between an offspring's behavior and RNA in sperm is another association, like the genetic associations between DNA sequence variants and behavioral variation, or susceptibility to psychiatric disease, described in other chapters. However, there are two important differences that make that analogy inappropriate.

First, there is no known mechanism by which any sperm constituent, apart from DNA, can influence the behavior of the adult animal that it creates through its union with an egg. You could say that's not so different for the genetic studies, because we also don't know the pathway by which DNA variants contribute to behavior. That's true—we're not too knowledgeable about how sequence variants contribute to our personality, or our risk of schizophrenia, but one thing we do know for sure is that such a pathway must exist, otherwise behavior would not be heritable. By contrast, we don't know if such a pathway for an RNA species exists. All that we have is an association result. The existence of a pathway is a hypothesis, and it's an unlikely hypothesis given that the biochemical properties of RNA mean that the nucleic acid is soon degraded (see Chapter 3, section 3.2).

Of course, we could claim that the RNA species initiates an epigenetic modification, which is often what you will read about in papers that argue for the non-genomic transmission of inherited material, but what does that mean? As an informed reader of this book, you'll know by now that, strictly speaking, **epigenetic modifications are those that have the property of being maintained through cell division.** Unless that is demonstrated, then it's not correct to claim that the modification is epigenetic. Many biochemical processes are likely to involve histone modification, such as gene expression or DNA strand breakage repair, but proving that such modifications are maintained through cell division is difficult, largely, of course, because many cells in an organism don't divide (e.g. neurons don't). Please beware of the too easy resort to 'epigenetics' as a mechanism for an unexplained phenomenon (HEARD AND MARTIENSSEN 2014).

Second, for genetic associations we know the likely confounds, the effect sizes, the power required, and the appropriate tests to be used. Establishing all of those things took time and effort, and, as you will see in Chapter 5, when we didn't know what to expect, false-positive findings were commoner than weeds in a neglected garden. Lacking a mechanistic explanation, and in the absence of experimental evidence, we rely on statistical association, on tests whose choice, deployment, and interpretation we discuss in Chapter 21. Perhaps, just possibly, some of the results, the *P* values reported in high-profile journals, might be false positives.

How can we decide whether the findings are true? You will find in Chapter 21, section 21.9.2 a quantitative way to estimate the probability of a result being true. We show that the probability of a result being true is a function of the power to detect the effect, the prior probability that the effect is there at all, and the significance threshold applied. While that's an appealing way to proceed, the difficulty in applying this admirable method is the hand waving that inevitably accompanies any attempt to assess power (how can you accurately estimate power when you don't have a robust estimate of the effect size) and to assess the prior probability (those in favor of the hypothesis will say the prior probability is high, while those against will claim it is low).

Ideally, to resolve this debate, we'd like some experimental demonstration that the RNA species does the thing it is supposed to do, but before you rush out to the laboratory, it is worth quoting Max Delbruck's view of experiments: his sole test for an acceptable PhD thesis was that it be *'either interesting or original – not necessarily both'*. Reality was not on his list of criteria. After all, *'Max never really believed any experimental demonstration of anything'* (FISCHER 2007). At this point, we can only advise the application of common sense.

4.7 'Omics

At a behavioral genetics meeting in Ventura, just north of Los Angeles, Bambos Kyriacou, a fly geneticist who works on genes involved in periodicity, stood up and said, *'Today is my birthday and I'm not going to talk about any sort of 'omics, genomics, proteomics, metabolomics, or phenomics. No, today I am going to talk about Bambomics.'* 'Omics refers to the analysis of anything and everything to which you can add the suffix 'omics. For our purposes, however, 'omics refers to the information we can obtain from next-generation technologies (with the exception of metabolomics and proteomics).

Figure 4.4 summarizes information obtained about DNA function from the ENCODE project, a follow-up to the human genome project that is attempting to assign function to the noncoding section of the genome (ENCODE 2012). At the top of Figure 4.4 is a chromosome (far left) followed by nucleosomes (which package the DNA in chromatin), a region of DNA free of chromatin (called a DNase hypersensitive site, or DHS, because here an enzyme called DNase easily digests the DNA), then DNA wrapped in histones with two modifications, methylation (Me3) and acetylation (Ac), and then a region of DNA that is being actively transcribed into mRNA. We've discussed what is known about each of these features earlier in the chapter. Here, we briefly review the use of sequencing technology to acquire information about DNA function across the genome, hence the suffix

Figure 4.4 Sources of 'omics data from next-generation sequencing. The names of the technologies that acquire the 'omic information are shown below the molecular phenotypes that are being acquired (e.g. DNase-seq uses sequencing to determine the presence of DNase hypersensitivity sites). Other acronyms are defined in the text.

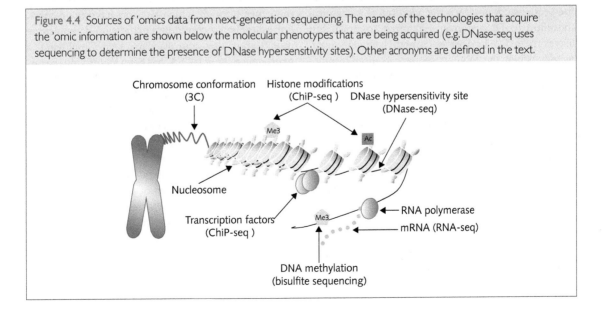

'omic. There are alternative methods, usually using microarrays, to obtain 'omics data, but their use is diminishing in importance as the costs of sequencing drop (see Chapter 3, Figure 3.7).

RNA-sequencing (RNA-seq) is a direct application of next-generation sequencing for 'omics. It refers to the sequencing of transcripts, including coding sequence variants and splicing structures, and has a large dynamic range (greater than 9,000-fold), so expression of the rarest and the most abundant species can be quantified (WANG et al. 2009). It generates the transcriptome, and the development of single-cell sequencing technologies allows RNA-seq to determine what each cell is expressing (DARMANIS et al. 2015; TASIC et al. 2016). It has become a standard tool in the functional characterization of genetic effects.

Obtaining information about transcription factor binding, and the modification states of chromatin, is slightly more complicated. By treating cells with compounds that create chemical bonds between DNA and proteins, it is possible to separate specific proteins, and the DNA bound to them, by employing antibodies that recognize a transcription factor, or a histone with a particular modification. The antibodies isolate from all the other cellular contents not just their targets but also the DNA that the targets are bound to. Thus, the DNA bound to a specific transcription factor, or the DNA associated with a specific histone mark, can be sequenced. The method is called ChiP-seq for chromatin immunoprecipitation and sequencing.

The DNA methylation state is obtained using bisulfite sequencing, in which the addition of a chemical (bisulfite) to DNA converts cytosine to uracil but leaves methylated cytosines (5-methylcytosine) unaffected. When the DNA is sequenced after bisulfite treatment, any bases in the sequence reads that are scored as 'C' must have been methylated. Thus, bisulfite sequencing generates single-nucleotide-resolution information about the methylation status of DNA.

Finally, next-generation sequencing obtains information about the relationship between regulatory regions and genes. Enhancers are likely to interact with the promoters of the genes they regulate (Figure 4.1), and this interaction occurs through proteins that bind to both places: proteins bind to the enhancers and then interact physically with the proteins at the promoter. The loop structure can be exploited to identify genes that the enhancers regulate.

The loop structure can be identified using sequencing in the following way.

Imagine pinching a loop of wool between two fingers and then slicing the loop so that you all have left are two ends at the base of the loop. It's possible to do this chemically, to isolate the two ends, and then sequence them. Sequencing identifies the fragments that were at the base of the loop: one section comes from one part of the genome and the other section comes from another, more distant location. If we know that one of the sequences is an enhancer (from a separate experiment where we sequenced DNA associated with histone modifications that indicate enhancers), then we can look to see what the enhancer is interacting with. This method is called chromosome conformation capture, or 3C for short. 3C has been used to show that distal enhancer elements, separated by up to hundreds of kilobases from the gene they regulate, combine with proximal elements to control gene transcription (DEKKER et al. 2002; SCHOENFELDER et al. 2010).

'Omics provides vast amounts of data, of which the largest compendia are part of the ENCODE project (ENCODE 2012). These data have been critical in showing that the great majority of the single-nucleotide polymorphisms associated with common psychiatric disease risk do not lie randomly in the genome but lie preferentially in regulatory regions (MAURANO et al. 2012). As this chapter has shown, regulatory regions affect the expression and splicing of genes, so that identifying the consequences of sequence variation within a regulatory element can be used to identify the genes through which risk loci have their effect. For example, by combining expression data, histone modification data and 3C data, Dan Geschwind, a neurogeneticist at UCLA, has been identifying the targets of schizophrenia risk loci (WON et al. 2016). While the importance of epigenetics for understanding the inherited components of psychiatric disease might not be as large as some claim, if by 'epigenetics' you mean 'gene regulation' then understanding how risk loci exert their effect on behavior is going to involve solving an epigenetic problem. We return to the subject of the molecular basis of psychiatric disease in Chapter 7 and following chapters, but before we do so, we explain in the next two chapters how genetic studies were developed to identify the molecular variants that contribute to psychiatric disease and to behavioral variation.

Summary

1. Epigenetics is the study of heritable changes in gene function; it is the science of explaining why, when a liver cell divides, it produces two more liver cells and not muscle cells or neurons, for example. Epigenetic modifications are those that have the property of being maintained through cell division.

2. There are five features involved in the regulation of gene expression: DNA sequence features, regulatory RNAs, *trans*-acting factors, DNA methylation, and chromatin packaging.

3. DNA regulatory elements include gene promoters and enhancers. *Trans*-acting factors are proteins that bind to DNA and alter transcription by forming complexes at promoters and enhancers. Short pieces of RNA are part of the machinery of gene expression: microRNAs are relatively nonspecific, while short interfering RNAs are relatively specific. In vertebrates, DNA methylation is generally associated with the silencing of gene expression. The protein component of chromatin is composed of histones, which are used to wrap up the DNA so that it fits into the nucleus. Biochemical modification of histones, such as methylation and acetylation, is involved in gene regulation.

4. Epigenetic differences have been found between individuals who have the same genome sequence. Their role in altering the risk of psychiatric disease is still not clear. In most cases, they may reflect the changes subsequent to disease rather than the changes that lead to disease.

5. Prader–Willi and Angelman syndromes are examples of epigenetic changes that result in disease. Identical DNA deletions result in different disorders depending on whether the deletion is on the maternal or paternal chromosome.

6. Examples of epigenetic transgenerational inheritance are rare.

7. 'Omics refers to the generation of large data sets, often from next-generation sequencing, that generate catalogs of near-complete cellular data. Examples include catalogs of RNA transcripts (transcriptomics), regulatory regions of the genome (the regulome), and metabolites (metabolomics).

5

Linkage and association

[On the use of genetic relationships within families to map genes, and the use of unrelated individuals to identify associations between genetic markers and behavioral variation]

5.1 Linkage analysis

DNA sequence variants are common. When the same piece of human DNA is sequenced in different people, the base pairs that make up the DNA code will be found to differ at a rate of about 0.1% (or one base in 1000) (1000 GENOMES PROJECT CONSORTIUM 2010, 2015; ABECASIS et al. 2012; UK10K CONSORTIUM et al. 2015). As far as we know, the vast majority of these sequence variants have no effect. They are not mutations, in the sense that they don't tamper with a gene's structure, and many are so common that half the population has the variant.

The realization that these apparently useless variants could be put to work, to discover the molecular basis of disease, occurred in the late 1970s. David Botstein described the origin of the critical insight. It took place in the middle of a typical academic argument—this one reviewing a genetics training grant in Utah. A student was presenting a plan of a genetic mapping study of hemochromatosis, a disease that involves abnormal deposits of iron all over the body but especially in the liver. This effort was focused on a special place in the human genome called the human leukocyte antigen (HLA) region, one of the most variable regions in the genome. At the HLA region, almost all human beings (except monozygotic (MZ) twins) have different copies of a series of immune genes. Botstein and his close colleague, Ron Davis, were defending the efforts of this student because the audience, mostly immunologists, did not understand the first thing about the principles of gene mapping. Quoting Botstein:

So finally I say something like, 'Look there is nothing special about HLA. What's good about HLA is that it has many alleles, and because it has many alleles, you can tell if you have linkage, and if you have many multiallelic markers all over the genome, you can map anything!' And as soon as the words were out of my mouth, I look at Davis, Davis looks at me, and we both understand that of course there are such markers, and we could make a map of the human genome tomorrow.

Botstein's paper outlining the insight that he had that day showed how the map could be used to find

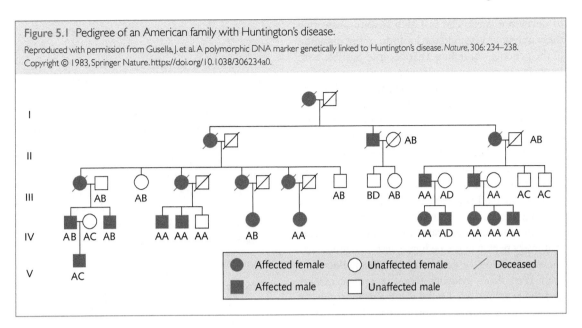

Figure 5.1 Pedigree of an American family with Huntington's disease.

Reproduced with permission from Gusella, J. et al. A polymorphic DNA marker genetically linked to Huntington's disease. *Nature*, 306: 234–238. Copyright © 1983, Springer Nature. https://doi.org/10.1038/306234a0.

the chromosomal location of mutations that give rise to a genetic disease (BOTSTEIN et al. 1980). The method required no prior knowledge of the gene involved, indeed no information about the biology at all, just some assumptions as to how the mutation was passed down through the generations, one of the well-understood patterns of Mendelian inheritance: recessive, dominant, and X-linked (see Chapter 8, Figure 8.3, for a simple review of inheritance patterns).

At the time, the idea was controversial. It was technically challenging, expensive, and there was no guarantee it would work. But the first success, proving the value of molecular biology for understanding genetic disease, came remarkably quickly. Within 3 years, the first genetic markers—the sequence variants that were to form the basis of the genetic map—had been used to identify the chromosomal location of a human disease gene.

Huntington's disease is an adult-onset neurological condition. First described by the American physician George Huntington in 1872, Huntington's disease became one of the earliest examples of a disease inherited as a Mendelian 'factor', after being so described by the English biologist William Bateson.

The first symptoms, usually mental disturbances and mild uncontrolled movements, typically emerge at an age when the sufferer has already started their own family, in their 40s. The condition worsens over a number of years, leading finally to a vegetative state in which the patient needs 24-hour care. The inheritance

pattern is a classic example of a Mendelian dominant mutation: one copy of the mutation is sufficient to cause a disease as severe as that due to two mutations (one on each chromosome).

Figure 5.1 shows the pedigree for five generations of an American family who have Huntington's disease (GUSELLA et al. 1983). Squares represent males, circles are females, and affected members are shown in blue (a line through them means they were dead when the pedigree was drawn up). About half of the offspring at each generation are affected individuals, and an affected individual always has an affected parent. If you see this pattern in a pedigree, then it's likely that there is a dominant mutation segregating. The other possibilities are X-linked inheritance, where the mutation affects only males (there are affected females in the pedigree, so that is excluded here), recessive inheritance (two copies of the mutation are required, so both parents have to carry at least one mutation), and more complex inheritance where many mutations are segregating (this can mimic the patterns of the other Mendelian segregation patterns). Mendelian patterns can also be obscured because sometimes an individual with a mutation shows little or even no manifestation of the trait. You will see this referred to as 'penetrance': a mutation can be fully penetrant (everyone with the mutation shows the trait) or variably penetrant. Huntington's disease is due to a single, dominant, fully penetrant mutation. What is that mutation, how does it cause the disease, and can we use that information to develop a cure? The first step toward answering these questions is to find the location of the mutation in the genome, something that became possible with Botstein's map of genetic variants.

The map provides genetic markers for each of the human genome's 22 chromosomes and the X and Y sex chromosomes. You pass on one of your two copies of each chromosome to your children, and which one they inherit is entirely random. It's like tossing a coin: a 50% chance of heads or tails, a 50% chance of passing on this or that copy. In the pedigree above, there are 22 affected offspring and seven unaffected (I'm not counting the marriage partners, as I am assuming they are unrelated and don't carry the mutation). If I find that every person in the pedigree with the disease has the same allele at a genetic marker from Botstein's map, then that's evidence that the genetic marker and the mutation are on the same chromosome. Of course, that could occur by chance. It's relatively straightforward to work out that probability: the lower the probability, the higher the chance that the marker and the disease are on the same chromosome, or to use genetic terminology, the higher the chance that the marker and the disease are linked. The complication is that chromosomes recombine, a process that occurs during the production of egg and sperm cells, when two copies of the same chromosome, one from our mother and the other from our father, swap DNA. Figure 5.2 explains how this happens. Recombination means that a disease locus can be on the same chromosome as a marker but not linked to it.

On an average-sized chromosome, there will be about two or three sites where chromosomal material is exchanged. Consequently, within a family, the allele of a genetic marker situated at one end of a chromosome won't be much use in predicting the allele of a marker at the other end of the chromosome (or therefore of a disease-causing mutation), as recombination will scramble the relationship. In this case, information at the two markers is said to segregate independently. If the two markers are close, let's say less than a quarter of the length of one of the chromosome arms, then the alleles at the two markers will tend to be inherited together; that is, they are linked. The closer they are, the less likely that recombination occurs between the two markers. You can see this in the figure, where the alleles at markers 1 and 2 are inherited together, but recombination breaks the relationship with the alleles at marker 3. Markers 1 and 2 are linked, but markers 1 and 3 are not linked.

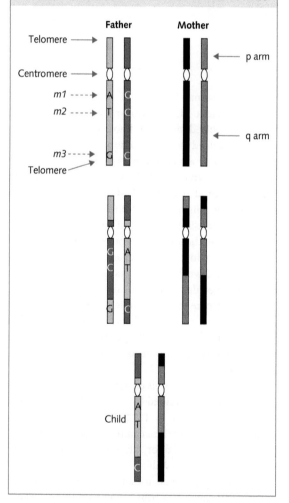

Figure 5.2 The process of recombination and linkage. Chromosomes have two ends (telomeres), a center (centromere), and two arms, the shorter of which is the p arm and the longer the q arm. In the sex cells where meiosis occurs, chromosomes find their partners and recombine to generate a mosaic of the paternal and maternal chromosomes, as shown in the second stage of the figure. In the final stage, at the bottom of the figure, one of these chromosomes is chosen at random to pass on to the next generation. Three genetic markers are shown (m1, m2, and m3) with their genotypes, on the paternal chromosomes. Because of recombination, the allele at m1 segregates with the allele at m2, but not with the allele at m3. Markers m1 and m2 are said to be linked.

An important concept related to linkage, and one that will turn up throughout this book, is linkage disequilibrium. In this chapter, we talk about linkage within pedigrees, and we mean that, within a family, genotypes at one marker predict, with some degree of accuracy, genotypes at a second marker. In other chapters, we deal with genotypes obtained from a population, rather than from a family, and again we are interested in the degree to which genotypes at one marker predict genotypes at a second. We use the term linkage disequilibrium to refer to the degree of correlation between genotypes obtained from two or more markers within a population. There is a further discussion of linkage disequilibrium in Chapter 6, section 6.4. For now, think of linkage as a feature of families, and linkage disequilibrium as a feature of populations.

In disease linkage mapping, we look for linkage between markers, like those in the figure, and disease status, assuming that there is a disease-causing mutation segregating according to one of the three Mendelian forms of inheritance: dominant, recessive, or sex linked. Disease status is treated just like any other marker, with one allele (the mutation) conferring risk, while the other allele is normal (the wild type). With sufficient numbers of markers to capture recombination events anywhere in the genome, it's possible to detect linkage with disease anywhere in the genome. How many markers? Recombination occurs two or three times in each generation on each chromosome, so a few hundred markers is enough to track linkage across the entire human genome. That's a tiny number by today's standards, but at the time the first attempts were made, it was a heroic undertaking.

In the early 1980s when Jim Gusella and David Housman of the Massachusetts Institute of Technology tried genetic linkage mapping for Huntington's disease, they imagined it would take years, possibly decades, before they would find linkage. Gusella began testing markers in the summer of 1982. With remarkable good fortune, the third marker he tried gave weak evidence of linkage, and when he tested the marker on a larger cohort of patients, the evidence in favor became overwhelming (GUSELLA et al. 1983). No one had imagined that success would come so quickly.

Finding linkage is not the same as finding the causative mutation. Linkage gets you into a rough area of the human genome. The information from a successful linkage experiment is like helping you find an individual's whereabouts in the USA by telling you in which state to look. There's a considerable amount of work ahead, as there are often thousands of genes in a linkage region. Progressing from linkage to mutation is a process called positional cloning, a Herculean

task in an age without a complete map of the genome, let alone a complete genome sequence. When the Huntington gene was finally identified, Peter Goodfellow, a Cambridge geneticist who had identified the gene that determines sex, admitted: *'In the satirical masterpiece Catch-22 by Joseph Heller, Major Major Major's father makes a living by not growing alfalfa. In a moment of uncalled-for unkindness stimulated by alcohol, I once accused a close colleague of making a living by not cloning the HD gene'* (GOODFELLOW 1993). A key to success was the establishment of a collaborative research group, held together primarily by Nancy Wexler, then a psychologist at Columbia University, whose father established a research foundation after his wife was diagnosed with Huntington's disease. *'Although everyone had agreed to collaborate, until then they had not realized just what they were in for,'* said Wexler (ROBERTS 1990). Tempers flared at a meeting in Boston in 1986, with charges of withheld data, unreturned phone calls, and probes lost in the mail. The ostensible problem was that they were all planning to do essentially the same experiment. But the real issues, says Wexler, were simply 'How much can I trust you? Were others being honest or were they going off to be the Lone Ranger?' *'At that time everyone thought it would be easy to find the gene,'* she said, *'and part of the tension was the fear that someone in the group would find and clone it before the others even heard about it. But it wasn't the case and the difficulty in tracking down the gene has helped pull the group together'* (ROBERTS 1990). Hard as it was to identify the Huntington mutation, the identification of disease-susceptibility genes in complex traits like schizophrenia was to prove considerably harder.

Work on Huntington's disease was the first demonstration that molecular mapping delivered results. In the decade following the first successful linkage studies of human genetic disease, there were attempts to apply the molecular strategy to disorders where the pattern of inheritance was not as clear cut as in Huntington's disease. Most diseases and most traits don't follow a Mendelian pattern of inheritance: there might be a family resemblance, but the inheritance pattern can't be explained by a single genetic variant passing through the family. In these cases, one way to apply the linkage paradigm is to find those (rare) families where the disease does follow a classical Mendelian pattern (e.g. dominant or recessive patterns). For instance, breast cancer doesn't typically show a Mendelian pattern of segregation, but clinicians knew that in some families the disease struck young and was often accompanied with other cancers, especially of the ovaries. Concentrating on these rare families, a single Mendelian genetic effect could be seen, mapped (Hall et al. 1990), and finally cloned (Miki et al. 1994). The same was true for rare forms of Alzheimer's dementia. While the vast majority of cases of Alzheimer's disease have only a modest tendency to run in families and follow no clear Mendelian pattern, in a very few families Alzheimer's disease is an autosomal-dominant condition. Interestingly, in these families, the age of onset of the disease is much younger than is typically seen—often in the 40s or 50s. Investigations in these families resulted in the discovery of three different genes that cause this rare form of Alzheimer's disease (Tanzi and Bertram 2001). These studies resulted in fundamental advances in our understanding of the causes of breast cancer and of dementia, with consequent improved diagnosis and the possible development of novel therapies. The genetic mapping and cloning experiments were extraordinary achievements, involving high-profile races between scientific groups to be the first to map and then identify the disease genes. (For a flavor of the drama of the scientific competition, read Mary-Claire King's description of the race to find the breast-cancer gene; King 2014.) Unfortunately, the approach didn't help with identifying the genes for psychiatric disease, as we shall see.

5.1.1 Linkage analysis of language: the KE family

Even in the 1980s, when gene hunting first hit the press, some psychiatrists were asking if molecular approaches would work in their discipline. After all, many patients with Huntington's disease have significant psychiatric symptoms; admittedly, the neurologists claimed it was their illness, but that was just an issue of demarcation. Alzheimer's disease was just as much a neurological condition as Huntington's disease, yet patients with Alzheimer's were often looked after by psychiatrists (Alzheimer was a psychiatrist). Another of the high-profile races was the hunt for the gene for fragile X syndrome, a cause of intellectual disability—another condition that belonged to psychiatrists (Fu et al. 1991; Oberle et al. 1991). Could other psychiatric conditions be tackled this way?

One of the most striking successes for linkage mapping was to come with a condition even closer to the psychiatric heartland: language disorder. Families become famous for various reasons, but the nature of the KE family's fame is unique. They suffer from a dominant mutation that affects their speech. Affected family members are almost incomprehensible but not just because of difficulties in pronunciation (Hurst et al. 1990). There is a defect in the production of correctly formed language. For example, the affected individuals would fail tests such as filling in the missing word: 'This creature is smaller than this one, but this creature must be the [smallest]', or filling in the missing nonword: 'This creature is ponner than this one, but this one must be the [ponnest].'

Simon Fisher, then a postdoctoral scientist working with the geneticist Tony Monaco in Oxford, mapped the gene to a region of chromosome 7 in 1998 (Fisher et al. 1998) and by 2001 confirmed that a mutation in a single gene (called *FOXP2*) gave rise to the speech disorder (Lai et al. 2001). Brain imaging of affected KE members has shown that there are abnormalities in language-related cortical regions, providing further evidence that their problems extend beyond uncoordinated motor activity (Vargha-Khadem et al. 2005).

The discovery of *FOXP2* advanced our understanding of the relationship between genes and behavior, but we're not dealing here with one of the usual conditions that psychiatrists treat. Does success with genetic analysis of a speech disorder imply that finding genes contributing to schizophrenia, bipolar disorder, autism, depression, or anxiety is possible? Could these genes be identified using the same molecular approaches? Finding more 'KE'-like families would be good, that is, finding families in which a single genetic mutation could be held responsible for schizophrenia or indeed any of the common psychiatric illnesses. Mary-Claire King's success with breast cancer gave hope that, among diseases with complex genetic inheritance (such as schizophrenia), there could be found rare cases of Mendelian inheritance, which would be tractable to the new molecular mapping approaches. If psychiatrists could identify families with Mendelian mutations, then there was a good chance that they would find mutations that give rise to psychiatric disease.

5.1.2 Linkage analysis of psychiatric disorders

In 1988, a London-based research group headed by Hugh Gurling published evidence in *Nature* for a susceptibility gene for schizophrenia on chromosome 5 (Sherrington et al. 1988). The results were reported for seven large British and Icelandic families containing multiple members affected with schizophrenia. Not surprisingly, the finding created a huge stir both in the scientific community and in the more general public. It appeared as if the genetic puzzle of schizophrenia had been solved. Moreover, in February 1987, a US team had reported that a locus on chromosome 11 was linked to bipolar disorder (EGELAND et al. 1987).

A few weeks prior to the publication of the chromosome 5 paper, one of us (K.S.K.) was being visited by a colleague of Hugh Gurling and recalled:

I had put him up in my home and we were driving into work. Twenty-one years later, I can still clearly recall exactly where on the highway he casually told me about Gurling's discovery. He commented that the data 'looked very strong'. Although typically a very careful driver, I nearly drove the car off the road in my astonishment! Nothing that I knew, or thought I knew, about schizophrenia was consistent with the finding of a Mendelian gene for this illness. I was in a state of shock for hours.

Information that had been available to the field for decades indicated that, although schizophrenia was strongly influenced by genetic factors, it was very unlikely to be due to the effects of one or even a small number of genes. Two sets of data are particularly relevant here. First, the risk of schizophrenia in various classes of relatives (siblings, parents, nieces, nephews, and so on) does not in any way resemble that which would be expected by genes that act in a Mendelian manner. If one member of a pair of MZ twins has schizophrenia, the co-twin has a risk of only about 50%. In all Mendelian disorders, that rate would be 100%. Second, and probably more important, a number of researchers had attempted to find subsets of families where schizophrenia would look like a Mendelian condition. No such families had ever been convincingly found, despite many efforts, some in quite obscure corners of the world.

The logical way to tell whether the finding was correct or not was to repeat the experiment on a different set of patients. No one was able to replicate the chromosome 5 findings.

Discarding the idea that common psychiatric illness could arise from a single mutation left a serious problem. How did genes operate in these conditions? Were we dealing with ten genes, 100, 1,000, or more? How did these mutations work? Were they really mutations at all or just common variants that in some unlucky individuals combined in a particularly malignant manner to produce illness? How would we recognize them

at a molecular level? Almost all of the subsequent problems arose because we didn't know the answers to these questions, and so we guessed.

For a while, the field continued to pursue linkage analysis, but this time with a method that did not assume the existence of a single Mendelian genetic effect. Instead, the assumption was relaxed and investigators examined pairs of siblings, looking for regions of the genome that the siblings shared more than expected by chance.

On average, siblings share half their genome. However, at a risk locus, two affected siblings share more genetic material than 50% (as they both have the disorder). This means it should be possible to detect the locus by identifying the segments where there is excess sharing. The method works as follows. Investigators collect a few hundred affected pairs of brothers and sisters (sibling pairs), and genotype markers across their genomes. The aim is to determine the parental origin of each part of every chromosome, and from that to work out which bits are shared. The ideal situation for the investigator is to have a molecular marker that can distinguish the four parental chromosomes—let's call them F1 and F2 (father 1, father 2) and M1 and M2 (mother 1, mother 2) —and to use the markers' alleles to determine which chromosome each child inherited. Ideally, the marker should have at least four alleles to make this possible, although by combining information from two or more closely linked markers it's possible to work out accurately where a child's chromosome is descended from, even when the markers have only two alleles.

Suppose at one marker you find that one child has inherited F1 and M1, and the other child has inherited F2 and M2—then they share nothing at that locus. Alternatively, both children could have inherited the same alleles (e.g. they both have F1 and M1) or, most likely, they share one chromosomal allele (e.g. one child is F1, M2, and the other is F2, M2).

The linkage literature usually refers to sharing as identity by descent (IBD) and the three states of sharing as 0, 1, and 2. Once the IBD state of each chromosomal segment in all of the sibling pairs has been worked out, a statistical test is applied to determine whether an IBD value of 2 occurs in a chromosomal region more often than expected by chance. For decades, the affected sibling-pair method was applied to many complex traits, but even though it did not make the strong (and incorrect) assumptions about Mendelian inheritance of a single mutation, it still failed to robustly identify risk loci.

One problem was that it wasn't clear what the most appropriate significance threshold should be. In a seminal paper in 1995, the mathematicians turned geneticists Eric Lander and Leonid Kruglyak derived significance thresholds for linkage mapping that were to become the standards thereafter. Table 5.1 shows their thresholds as both *P* values and logarithm of the odds (LOD) scores (Lander and Kruglyak 1995). An LOD score is the log-likelihood ratio of the data assuming that that the allele-sharing proportion has the observed value, compared with the hypothesis that there is no excess sharing. LOD scores were commonly used in the era of sibling-pair studies (their use comes up again in our discussion of mapping loci using inbred strains of mice: see Chapter 15, section 15.6.2). The values given in Table 5.1 are gold standards for mapping complex traits using linkage-based approaches—the only problem was that it was rare for anyone to obtain a result that exceeded the Lander and Kruglyak thresholds (and if they did, the result usually failed to replicate). Why?

The reason, explained in a highly influential paper in 1996 (Risch and Merikangas 1996), is that the linkage studies were underpowered. We describe this important paper in more detail in Chapter 6, section 6.1, but for now the important point to note is that very large samples, hundreds of thousands of families, orders of magnitude larger than anyone had considered necessary, were needed to find loci using the sib-pair linkage strategy. By contrast, genetic association studies might be successful with a few thousand individuals. In the next section, we introduce genetic association studies to our readers.

5.2 Genetic association analysis

Linkage analysis has the virtue of interrogating the entire genome and requires no prior knowledge of the genes that might be involved. This makes linkage attractive: our ignorance about the causes of psychiatric disorders is not an impediment to success. However, linkage analysis has two important drawbacks. First, as Risch and Merikangas (1996) pointed out, it has low power. It is only good at detecting relatively large genetic signals—loci that substantially alter risk for a disorder. Second, even when you find positive results, the linkage signals are very broad, spread over regions of the chromosome containing thousands of genes. By contrast, association analysis is more powerful, requiring smaller sample sizes, is able to detect smaller effects, and has higher mapping resolution: each marker gives information only over a small part of the genome, typically stretching over tens of thousands rather than tens of millions of DNA base pairs. The drawback to association is that it needs hundreds of thousands of markers to interrogate the genome.

The design of association analysis is simple (Figure 5.3). Suppose we want to know whether a genetic variant increases the risk of schizophrenia. We gather affected individuals, our cases, and a group of matched unaffected individuals, our controls, and we determine whether the frequency of the marker variants differs between the two groups. Association testing

Table 5.1 Significance thresholds for human linkage mapping studies. From Lander and Kruglyak (1995).

	Suggestive *P* value	LOD	Significant *P* value	LOD
Linkage mapping (classical LOD score analysis for Mendelian traits)	1.7×10^{-3}	1.9	4.9×10^{-5}	3.3
Allele sharing: sibling pairs	7.4×10^{-4}	2.2	2.2×10^{-5}	3.6

LOD, logarithm of the odds.

Figure 5.3 The ingredients of a genetic association test for a quantitative trait. A polymorphic locus has two alleles (*a* and *g*) that form three genotypes (the two homozygotes, *aa* and *gg*, and the heterozygote, *ag*). A statistical test compares the numbers of genotypes in the cases and controls to determine whether they are significantly different.

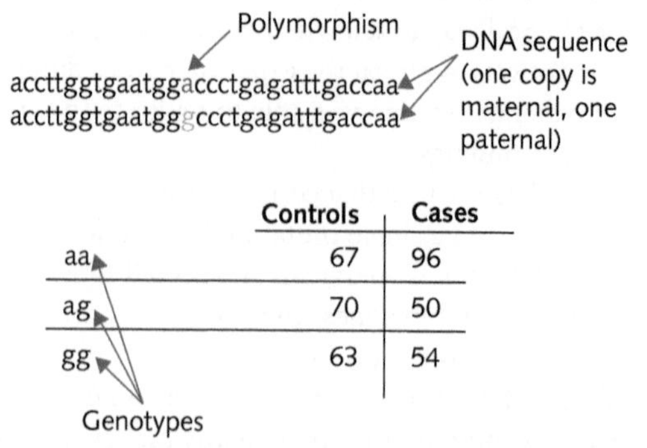

	Controls	Cases
aa	67	96
ag	70	50
gg	63	54

is also straightforward for a quantitative trait (scores from a personality questionnaire or cognitive test): in this design, the test is whether there are differences in the scores for the different genotypes: for example at a marker with two alleles, *a* and *g*, determine whether scores for those with the genotype *aa* differs from those with *ag* and *gg*. Subjects are given a test, for example a personality test, and the scores are recorded for each genotype. A statistical test is applied to determine whether the mean scores differ significantly (these tests are described in detail in Chapter 21, section 21.2).

5.2.1 Candidate gene association tests

The first application of association tests was in testing variants in candidate genes. The candidate gene approach itself comes in two flavors. In the more common approach, the gene is picked because you think it might have something to do with the physiology of the illness. Thus, these genes are called **physiological candidate genes**. In the second approach, genes are picked because of where they lie in the genome, in particular because they are under linkage peaks. Such genes are called positional candidates.

A good illustration of how physiological candidate genes were selected is through the hypothesis that alteration in the metabolism of dopamine was involved in psychosis. Dopamine is a neurotransmitter, a molecule that is released from a synapse of a neuron, and activates a receptor on a neighboring neuron.

Neurotransmitters have always fascinated psychiatrists. Many of the drugs used to treat psychiatric and neurological illnesses act by mimicking, blocking, prolonging the life of, or otherwise interfering with neurotransmitters. Many illicit psychoactive drugs act in the same way: the fact that you can make yourself excited, calm, happy, paranoid, depressed, transcendental, or traumatized by interfering with neurotransmitter function has spurred psychiatrists to investigate whether abnormalities of neurotransmitter function are a cause of psychiatric disease. The neurotransmitter dopamine has been a magnet for the attention of psychosis researchers in large part because, with stunning consistency, drugs that treat the symptoms of schizophrenia block one particular class of dopamine receptors in the brain. Furthermore, their potency at treating symptoms and blocking the receptor are uncannily well correlated. These two sets of observations—which are widely accepted—form the basis of the dopamine hypothesis of schizophrenia, which is far and away the most influential theory from its first proposal in the 1970s to today. This theory posits that, in individuals with schizophrenia, the dopamine system is somehow and somewhere hyperactive. Dopamine itself is not encoded by a gene: the body makes it from raw ingredients using a series of enzymes that are encoded by genes. Once released, the chemical can be degraded enzymatically, or by reuptake into neurons via a transporter in the cell membrane (Figure 5.4). Theoretically, alterations in **any** of these steps could lead to functional hyperactivity: excess production, excess release,

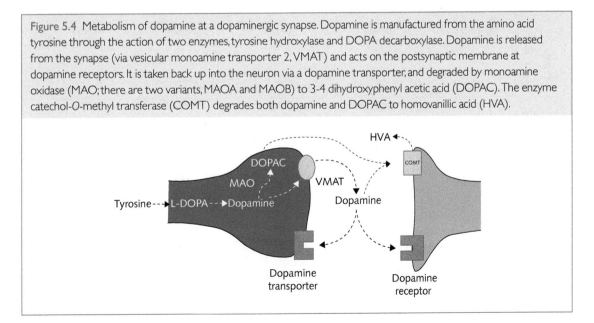

Figure 5.4 Metabolism of dopamine at a dopaminergic synapse. Dopamine is manufactured from the amino acid tyrosine through the action of two enzymes, tyrosine hydroxylase and DOPA decarboxylase. Dopamine is released from the synapse (via vesicular monoamine transporter 2, VMAT) and acts on the postsynaptic membrane at dopamine receptors. It is taken back up into the neuron via a dopamine transporter, and degraded by monoamine oxidase (MAO; there are two variants, MAOA and MAOB) to 3-4 dihydroxyphenyl acetic acid (DOPAC). The enzyme catechol-*O*-methyl transferase (COMT) degrades both dopamine and DOPAC to homovanillic acid (HVA).

excess stimulation, excess receptor number, or reduced removal. Despite the accumulated biochemical and neuroanatomical knowledge about dopamine, after 30 years of research no one has been able to prove whether the dopamine hypothesis is true or false.

There are literally thousands of papers reporting the results of physiological candidate gene association tests, but it's not too harsh to say simply that these studies have taught us nothing useful about the genetic basis of psychiatric disease (FARRELL et al. 2015). Figure 5.5 shows, dramatically, the numbers of genetic association papers published for 25 genes tested for association with schizophrenia from 1990 to

2015. All of the findings are now considered to be false positives. There are four reasons for this failure.

Perhaps most central to the failure of the physiological candidate gene approach is that you had to be able to pick good candidate genes, genes that had a plausible chance of being involved in the physiology of the disorder. To pick good physiological candidate genes, you needed to know something about what caused the disorder, which, of course, we did not when it came to schizophrenia or any of the other major psychiatric disorders. Just to be clear about what we mean here—two examples of good physiological candidate genes would be the insulin gene if you were

Figure 5.5 Candidate gene studies in psychiatry. For each candidate gene, the number of publications is indicated on the y-axis and the year on the x-axis.

Reproduced with permission from Farrell, M.S., et al. Evaluating historical candidate genes for schizophrenia. *Molecular Psychiatry*, 20: 555–562. Copyright © 2015, Springer Nature. https://doi.org/10.1038/mp.2015.16.

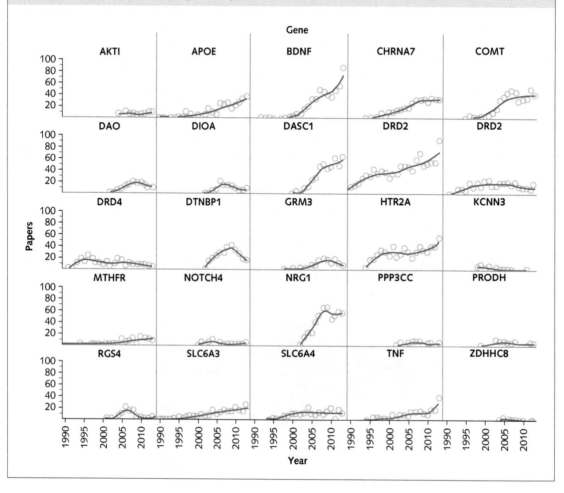

studying diabetes or a gene for a cholesterol receptor if you were studying heart disease. While there is no guarantee that variants in these genes would influence risk for the disease, at least you know that the products of these genes are directly involved in the disease process. We had good evidence that antipsychotics blocked dopamine receptors, which then reduced the symptoms of schizophrenia, but direct evidence that dopamine itself contributed to the illness has been elusive. Reasoning from treatment back to cause is likely to be an unreliable approach. Few would argue that common headaches result from a deficiency in endogenous aspirin!

The second reason for the poor performance of candidate gene studies lies with poor analysis. The single biggest contributor to the mountainous pile of false positives is the consequence of multiple testing: if you do enough statistical tests, a few of them will turn out to be positive by chance alone, even when there is nothing to find. Think about rolling three dice. It is, by chance, quite unlikely that you will roll all '1s' (if you have true dice, it should occur once out of every 6^3 times or once in 216 rolls). But if you sit there for a while and roll the dice enough times, you are guaranteed, if you are patient enough, to get three ones (see Chapter 21, section 21.3 for further discussion of multiple testing and how to deal with it). For the same reason, even if there is

no effect at a variant in the dopamine transporter gene, when 20 studies are carried out, one of them is likely to be significant (i.e. to find an association at $P < 0.05$). Even if you don't get a significant result after running 20 experiments, keep going, because eventually you will. The impact of multiple testing might not matter if all the studies were published so that we could see, for example, that 19 were negative and one was positive, consistent with no effect at the locus, but it turns out that scientific journals afford more importance to positive than to negative results. What is published does not represent what scientists have actually done. For example, Figure 5.6 shows the effect sizes (y-axis) and the year of publication (x-axis) for analyses of the association between a dopamine receptor (DRD2) with alcoholism (MUNAFÒ et al. 2007). Larger effect sizes represent more significant findings. The correlation is clear: highly significant studies with the largest effects are published first; negative studies (where the effect is close to 0) appear later (Figure 5.6). Again, this might not matter if there was really an effect, but as we now know, candidate gene studies were also underpowered: the genetic effects were just too small to be detected with the number of subjects that were typically used (a few hundred). No single study had enough subjects to detect an effect reliably, thus giving rise to many inconsistent results.

Figure 5.6 Association between year of publication and effect size estimate (log of odds ratio (OR)). The year of publication is negatively associated with individual study effect size ($P = 0.032$), reflecting a decrease in individual study effect size over time.

Reproduced with permission from Munafò, M.R., et al. Association of the DRD2 gene Taq I A polymorphism and alcoholism: a meta-analysis of case–control studies and evidence of publication bias. *Molecular Psychiatry*, 12:454–461. Copyright © 2007, Springer Nature. https://doi.org/10.1038/sj.mp.4001938.

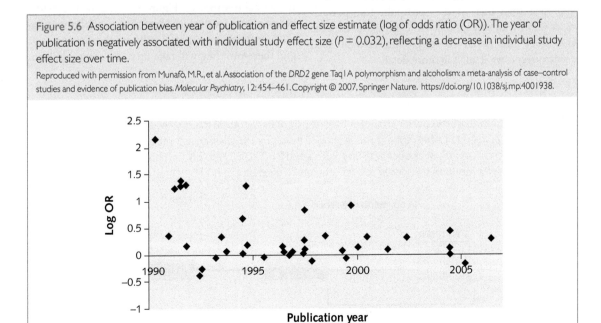

The third reason is that the association analysis may not adhere to a key requirement that the two groups, cases and controls, are equivalent in all respects other than the disease being tested. To appreciate this problem, consider what happens when the assumption is relaxed. For example, suppose you carried out a study of the genetic basis of religion and compared gene frequencies in groups who practice Buddhism with those who do not, without ensuring that the groups were matched in all other respects. You would be very likely to find many significant differences, solely due to the fact that Buddhism is commoner in East Asians than it is in Europeans, and there are many genetic differences between East Asians and Europeans. Even a slight degree of mismatching in the ethnicity of the cases and controls could introduce bias into the results. Adequately dealing with population structure remains a key issue for all genetic association studies. There are ways to deal with it, as we explain in the description of genome-wide association studies in Chapter 6, section 6.5. For now, bear in mind that it can cause false positives.

The fourth reason comes from the realm of the sociology of science. Put simply, '*It made such a good story; it just had to be true.*' Scientists have strong affiliations with their theories, not unlike many people's affiliation to their political parties. No better example can be found of the persistence of the candidate gene approach than that of the serotonin transporter's association with susceptibility to depression. Because this story has lasted so long—and created such controversy—we'll tell it in more detail.

5.2.2 A genetic variant in the promoter of the serotonin transporter

In the same way that the dopamine theory has dominated neurobiological theories of the cause of schizophrenia, 'something wrong with serotonin' has been a standard theory of the origin of depression for some time, again despite the lack of conclusive evidence. Serotonin, like dopamine, is a neurotransmitter in the brain, released by neurons and degraded or reused by reuptake via a transporter in the cell membrane, the serotonin transporter. There's evidence that the amount of serotonin available to neurons correlates with mood: low levels correlate with the presence and the lethality of suicide attempts, and can be used to predict future suicide attempts. Prozac and other drugs that block the reuptake of serotonin (selective serotonin reuptake inhibitors or SSRIs) are effective antidepressants and anxiolytics (i.e. reduce anxiety), presumably because of the increased availability of serotonin at the synapse. Finding any evidence that genetic variants in the serotonin system are associated with mood disorders could lead to a '*genetic test for vulnerability to depression and a way to predict which patients might respond best to serotonin-selective antidepressants*' (Tom Insel, Director, National Institute of Mental Health, 2005, personal communication). When Peter Lesch found a polymorphism at the start of the serotonin transporter gene that increased the amount of functional transporter (Figure 5.7) (LESCH et al. 1996), you can see how the idea must have struck home: people with the polymorphism will have more transporter and therefore they will take up more serotonin from

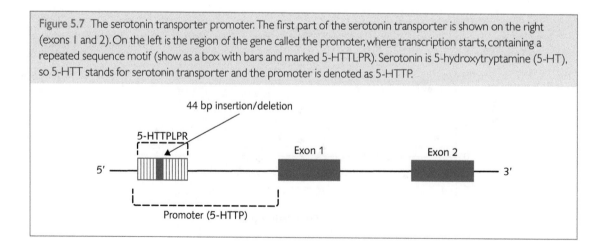

Figure 5.7 The serotonin transporter promoter. The first part of the serotonin transporter is shown on the right (exons 1 and 2). On the left is the region of the gene called the promoter, where transcription starts, containing a repeated sequence motif (show as a box with bars and marked 5-HTTLPR). Serotonin is 5-hydroxytryptamine (5-HT), so 5-HTT stands for serotonin transporter and the promoter is denoted as 5-HTTP.

the presynaptic cleft, leaving less available serotonin to act synaptically, causing them to be depressed.

Lesch found differences in the DNA sequence at the start of the serotonin transporter gene, differences that altered the amount of gene product. There were two alleles, which he called *s* (for short) and *l* (for long) at the transporter. Individuals with an *s* allele were more neurotic and more prone to anxiety and depression than individuals with an *l* allele. This ought to mean that the *s* allele increases the amount of transporter. However '*the basal activity of the* l *variant was more than twice that of the* s *form of the serotonin transporter gene promoter*' (LESCH et al. 1996). Lesch argued his way out of this unexpected finding on the following grounds: '*The therapeutic effects of the SSRIs have primarily been demonstrated in neuropsychiatric patients, who may have some primary serotonin or other neurotransmitter dysfunction that is ameliorated by the SSRIs, whereas our findings are in a sample of the general population*' (LESCH et al. 1996). In other words, serotonin behaves differently in patients than it does in you or me.

The subsequent decade saw literally hundreds of papers investigating the role of the serotonin transporter gene in psychiatric disorders. By June 2018, there were 5,318 citations of Lesch's paper (to put this in context, most scientists are happy if anyone cites their paper, and getting 100 citations is a pretty impressive achievement). Overall, however, as with all other candidate gene studies, there was little consistent, robust evidence for an association (for an assessment of the failure of candidate gene studies, see BORDER et al. 2019). But what makes the serotonin transporter example remarkable, and such a good example of how enthusiasm for the candidate gene approach has persisted, is the story of gene–environment interaction at the locus.

5.2.3 Gene–environment interaction at the serotonin receptor locus

Richie Poulton and colleagues in New Zealand have for many years been collecting information on the same 1,000 people, contacting them every year and flying them back home if necessary, to find out what has happened to them and assess how they are. The cohort were in their 20s when Poulton together with

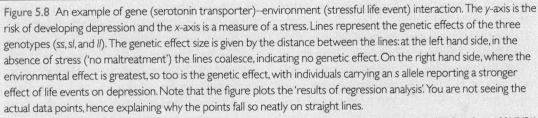

Figure 5.8 An example of gene (serotonin transporter)–environment (stressful life event) interaction. The *y*-axis is the risk of developing depression and the *x*-axis is a measure of a stress. Lines represent the genetic effects of the three genotypes (*ss*, *sl*, and *ll*). The genetic effect size is given by the distance between the lines: at the left hand side, in the absence of stress ('no maltreatment') the lines coalesce, indicating no genetic effect. On the right hand side, where the environmental effect is greatest, so too is the genetic effect, with individuals carrying an *s* allele reporting a stronger effect of life events on depression. Note that the figure plots the 'results of regression analysis'. You are not seeing the actual data points, hence explaining why the points fall so neatly on straight lines.

Reproduced with permission from Caspi, A. Influence of Life Stress on Depression: Moderation by a Polymorphism in the 5-HTT Gene. *Science*, 301 (5631): 386–389. Copyright © 2003, American Association for the Advancement of Science. DOI: 10.1126/science.1083968.

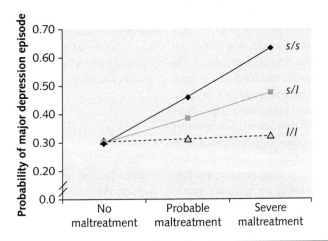

Terrie Moffitt at the Institute of Psychiatry in London, and her partner Avshalom Caspi, carried out a genetic analysis of the serotonin transporter gene. They reported in *Science*: 'the effect of life events on informant reports of depression was stronger among individuals carrying an s allele than among l/l homozygotes. These analyses attest that the serotonin transporter gene interacts with life events to predict depression symptoms, an increase in symptoms, depression diagnoses, new-onset diagnoses, suicidality, and an informant's report of depressed behavior' (CASPI et al. 2003). The interaction between gene and environment is discussed in more detail in Chapter 14, section 14.1. Here, we are interested only in a specific example, which is illustrated in Figure 5.8, taken from Caspi and Moffitt's *Science* paper (CASPI et al. 2003). This figure is one of the most influential in the entire human behavior genetics literature.

Psychiatrists and psychologists around the world have loved this piece of work; it's inventive and interesting and suits our belief that genes act in complicated ways, in combination with the environment, to work their effects. Genetic tests for this gene variant are even marketed on the internet. And there were, of course, attempts to replicate the result, with a mixture of positive and negative results, typical for candidate gene studies. By March 2019, the paper had been cited 8,519 times. Is the finding true?

The one study that came closest to replicating the original design, a longitudinal study of a birth cohort in New Zealand, failed to replicate the first report (FERGUSSON et al. 2011). Two meta-analyses (with sample sizes of 2,999 and 14,250 subjects) found no evidence for an interaction (MUNAFÒ et al. 2009; RISCH et al. 2009), while two concluded that there was an effect (sample sizes of 40,749 and 54,996) (KARG et al. 2011; SHARPLEY et al. 2014). The larger sample sizes might make you believe that the positive studies are more powerful and more trustworthy, but the authors of the negative studies claimed that the authors of the positive studies included results from studies with incompatible statistical and genetic models, and included outcomes other than depression. In turn, authors of the two positive studies criticized authors of the negative studies for being too restrictive in their inclusion of studies.

In an attempt to resolve the disagreement, a collaborative meta-analysis was set up to 'harmonize phenotypes across studies, to prioritize specific analyses a priori, and to apply identical de novo statistical analyses across all participating studies' (CULVERHOUSE et al. 2018).

Optimistically, the consortium claimed: 'With this approach and the large number of contributing samples, we are well positioned to clarify the relationship between 5-HTTLPR, stress and depression'. And what happened? First, the consortium found nothing ('We found no subgroups or variable definitions for which an interaction between stress and 5-HTTLPR genotype was statistically significant'), and second, the protocol of the consortium was accused of introducing bias against the detection of the interaction (MOFFITT AND CASPI 2014). The debate looks set to continue for a while (ANCELIN AND RYAN 2018), but if the interaction is present (and to be clear, the authors of this book haven't found evidence that it has been robustly detected), it is making a small contribution.

None of this is to say that gene–environment interactions do not occur. In fact, there is substantial evidence that genes and environment do interact, evidence that we review in Chapter 14, section 14.1. It's just that at the 5-HTT locus the evidence for interaction is unconvincing. But before leaving the subject, it's worth asking again why scientists have been so wedded to the idea that serotonin is involved in depression. One factor is the importance of the drugs used to treat depression—the efficacy of SSRIs, for example—which has led to the idea that mood disorders really must be due to a disorder of serotonin metabolism, or neurotransmission. If an association could be found between a genetic variant in the serotonin system and depression, it would prove what so many psychopharmacologists suspect: that there is a causal connection between serotonin dysfunction and depression. Attempts to use genetics to support the serotonin hypothesis, like similar attempts to use genetics to support the dopamine hypothesis as a cause of schizophrenia, are examples of the power that genetic analysis has to identify the causes of disease. Chapter 21 provides a detailed summary of the statistical issues involved, and we recommend that readers familiarize themselves with the arguments.

5.3 Endophenotypes

What could be done to improve success rates? Since neither linkage nor candidate gene studies worked, perhaps the problem was with the phenotype? After all, some psychiatric geneticists have, for some time, been frustrated with the disorders they study, feeling that they are too difficult to assess accurately (requiring lengthy careful interviews) and too hetero-geneous (many different subtypes have been proposed for disorders such as depression or schizophrenia). Furthermore, they suspect that the path from genetic variation to psychiatric disorder is too long and too tortuous a road.

This is not unreasonable. Think for a moment about the genes for eye color. If you lack both copies of a key enzyme needed to make a darker eye pigment, you get blue eyes. It's one simple step from gene to pheno-type. Now imagine how many biological steps might be needed to go from genetic variant to vulnerability to recurrent severe sad mood, to auditory hallucina-tions, to the propensity to develop false persecutory ideas about people around you. The number is likely to be very large.

It is from this background that the concept of 'endo-phenotype' emerged during the candidate gene era of research (GOTTESMAN AND GOULD 2003). The initial hope was that the adoption of this approach would solve the problem of negative findings and poor repli-cations (MEYER-LINDENBERG AND WEINBERGER 2006). The basic idea is simple. Was there a construct that could be measured with reasonable accuracy that sat between a candidate gene on the one hand and a given psychiatric disorder on the other, as illustrated in Figure 5.9a? That is, was there an endophenotype that sat closer to the genes than does the psychiatric disorder?

The endophenotype strategy sounds good in prin-ciple, but it was much less successful than many peo-ple anticipated when it was first proposed (FLINT AND MUNAFÒ 2007). We point out two problems (KENDLER AND NEALE 2010). First, since endophenotypes are 'closer' to the genes, the genetic influences should be stronger for endophenotypes than for psychiatric dis-orders. In the language of genetics, the genes should be more 'penetrant' for the endophenotype than for the disorder. But that often did not turn out to be true for endophenotypes used in psychiatric disorders such as schizophrenia and depression. For example, assess-ments of some endophenotypes were not very stable over time, introducing 'noise' into their measure-ment, which attenuated genetic influences. In other cases, the endophenotypes turned out to be equally, if not more, polygenic than the disease phenotypes (using the personality trait neuroticism as an endophe-notype for depression is an example).

Second, in many instances, endophenotypes were chosen because they correlated with the disorder, rather than because they lay on a causal pathway from

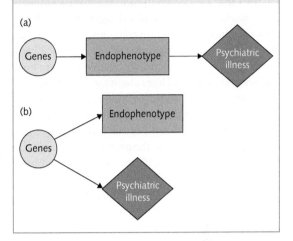

Figure 5.9 An illustration of relationships that might exist between genes, an endophenotype, and a psychiatric disorder. (a) The pathway from genes to disorder goes through the measured phenotype. (b) There is no direct pathway from the endophenotype to the disorder; they are correlated because they share genetic risk factors.

gene to illness. For example, individuals with schizo-phrenia perform more poorly on a range of cognitive tests than individuals with similar backgrounds with-out schizophrenia. Certain brain imaging findings reflecting emotional processing were proposed as an endophenotype for depression. But how could we tell whether endophenotypes just shared genetic or envi-ronmental risk factors with the disorders rather than that they sat in the causal pathway? This point can be best illustrated by comparing Figure 5.9a and b (where for simplicity we leave out environmental effects).

It is not as easy as you might think to differentiate the models depicted in Figures 5.9a and b. For exam-ple, both predict that unaffected siblings of individuals with the disorder will have levels of the endopheno-type somewhere in between those found in affected individuals and in the general population.

Why does it matter which of the two models is cor-rect? It actually matters a lot. For example, if I wanted to try to prevent the disorder from emerging in a vulnerable individual, the model in Figure 5.9a sug-gests that if I could lower the levels of the endophe-notype, it should help reduce risk. The second model (Figure 5.9b) makes no such prediction. If I wanted to understand disease pathways, the first model indicates that understanding the causes of the endophenotype would get me at least part way to understanding how the disease itself arises. Again, the second model makes no such claim.

In summary, the endophenotype concept is sound in principle, but it has been of limited utility in psychiatric genetics. This is in contrast to its use in genetic studies of a number of medical disorders. Here, biological tests that are known to sit in the disease pathway (e.g. measures of average blood sugar levels in diabetes or of antibody levels in autoimmune diseases) have been helpful in tracking down genetic effects. This is part of a story you will hear again: that the very poor understanding we still have about the biological pathways involved puts efforts to understand the genetics of psychiatric disorders at a disadvantage compared with such efforts in the rest of medicine.

Many of us working on the genetics of common disease, not just psychiatric illness, became increasingly frustrated. The job ought to be easy: we knew that the disorders had a heritable component and molecular genetics gave us the power to investigate them at a molecular level. Why could we not find robust results? During the time that the genetic association studies of psychiatric disorders were being carried out, other diseases whose origins were also obscure were subject to the same genetic analysis: cancer, diabetes,

hypertension, stroke, arthritis, asthma, and other common illnesses. Researchers in all of these areas were facing the same difficulties (IOANNIDIS 2006).

A few papers pointed to the importance of sample size. For example, an analysis of over 3,000 people in a study of a gene thought to be involved in type 2 diabetes (*PPARG*) gave a clear indication that the investigators had found the correct variant (ALTSHULER et al. 2000). Studies with similar sample sizes detected a signal of roughly the same effect size, and combined analyses improved the significance of the result (rather than weakening it as we found in meta-analyses of psychiatric association studies).

Studies that analyzed thousands of cases and controls were extremely rare in psychiatry. It was becoming clear that they were needed. But we also needed a way to interrogate something other than candidate genes. Given the problems of the endophenotype approach, the next suggestion was to try something radically different: an unbiased screen of all genes in the genome. But testing all genes was an idea that, at the heyday of the candidate gene studies in the 1990s and early 2000s, was technically impossible to realize. The next chapter explains the technological and methodological developments that gave birth to genome-wide association studies.

Summary

1. The construction of a genetic map of anonymous DNA markers across the human genome provides a resource for genetic linkage analysis.

2. Genetic linkage analysis searches for genetic markers that co-segregate with traits that follow one of the three patterns of Mendelian inheritance: dominant, recessive, or sex linked.

3. Genetic linkage analysis is a powerful method when applied to diseases that arise from Mendelian mutations segregating in families, such as Huntington's disease, but has less application in behavior genetics. One important exception

is speech and language disorders that arise from mutations in a gene called *FOXP2*.

4. Genetic association uses unrelated individuals (not families), and tests whether allele frequencies at a polymorphic genetic marker are correlated with a quantitative trait (such as personality) or, in the analysis of disease, whether they differ between cases and controls.

5. Genetic association analysis using variants at candidate genes has failed to demonstrate any robust, replicated results in analysis of behavior or psychiatric disease.

6

Genome-wide association studies

[On the use of variants from across the entire human genome, what they have told us about human DNA diversity, and the genetic architecture of common, complex disease]

6.1 Genome-wide association is a more powerful strategy than linkage

From the 1980s, human geneticists began asking for research funds to identify the genetic basis of common, complex diseases, such as diabetes, high blood pressure, stroke, and schizophrenia. They argued that the molecular techniques then available would soon find the genes involved, and from those discoveries it would be a short step to developing new and better treatments. By 2007, some of their promises had, finally, been fulfilled.

This chapter describes the advent of the genome-wide association study (more commonly known by the acronym GWAS, pronounced 'gee-wass'), the method that has successfully dissected the genetic basis of complex traits, that is, common diseases such as stroke, cancer, schizophrenia, and autism whose genetic origin cannot be explained by a single Mendelian mutation. The success of GWAS turned human genetic analysis of complex traits from a field littered with unreplicated, highly contentious findings to a

rapidly expanding area of research where there is little argument about the validity and importance of the findings. But, as we shall see later, while the difficulties of locating genetic signals on chromosomes have been solved, interpreting those signals and turning them into biological insights remains a challenge.

One simple, and unavoidable, interpretation of the relative failure of the linkage mapping described in the previous chapter is that the method was underpowered, a point made by, among others, Neil Risch and Kathleen Merikangas in 1996: *'The modest nature of the gene effects for these disorders likely explains the contradictory and inconclusive claims about their identification'* (RISCH AND MERIKANGAS 1996). From this follows the obvious question: 'How large does a gene effect need to be in order to be detectable by linkage analysis?' Risch and Merikangas answered this question as follows.

First, they defined the gene effect as the increase in risk of developing illness that a genotype confers, or in short, the genotype relative risk (GRR). A GRR of 2 means that if you have the genotype that confers susceptibility, your risk of becoming ill is twice that of someone who does not have the genotype. From working out how many families will be needed to detect different GRR values, they concluded that *'linkage analysis for loci conferring GRR of about 2 or less will never allow identification because the number of families required (more than ~2500) is not practically achievable'*. In fact more than 10,000 families are needed, and for smaller effects we're talking about millions of families.

An alternative strategy, genetic association, which compares the frequencies of genotypes in those with and without disease, is more powerful but had only been applied to testing candidate genes (with no success, as we described in Chapter 5; BORDER et al. 2019). Now, suppose we had polymorphisms in every gene in the genome, and could test them all. Surely then we'd be able to detect the variants that cause common disease? Risch and Merikangas noted a key problem with this idea. The large number of genes in the human genome means we'd have to carry out hundreds of thousands of individual tests. The more tests you carry out, the more likely you will be to find a positive result by chance alone. Suppose the positive outcome you are hoping to find is to score 6 when you roll dice. Roll a single die once and the chance of getting a six is 1 in 6; roll the die 100 times and you'd be extraordinarily unlucky not to get at least one 6 ($P = 1.4 \times 10^{-8}$). This is the multiple testing problem (see Chapter 21, section 21.3 for further discussion). A simple way to take into account multiple testing is divide the P value by the number of tests you carry out. If you accept as significant a probability of less than 1 in 20 ($P < 0.05$), then for 100 tests you accept a probability of less than 1 in 2,000 ($0.05/100$) as a threshold for rejecting the null hypothesis. Testing a million polymorphisms requires you to beat the odds of 1 in 20,000,000 ($0.05/1,000,000$). This is a ridiculously small number, but as Risch and Merikangas showed, it can be obtained by genetic association, with some reasonable assumptions about genetic effect size and action: *'... the required sample size for the association test, even allowing for the smaller significance level, is vastly less than for linkage, especially for affected sibling pair families when the value of p is small. Even for a GRR of 1.5, the sample sizes are generally less than 1,000, well within reason'* (RISCH AND MERIKANGAS 1996). As we'll see, the GRR of 1.5 and samples less than 1,000 turned out to be overly optimistic, but the numbers needed were still within reason.

At the time Risch and Merikangas wrote, genotyping a couple of polymorphisms was onerous. Genotyping a million was as impossible as sequencing the genome. The impediments to implementing genome-wide association were technical; how many markers were needed and how these were to be found and then genotyped emerged as the critical questions, which, if answered, could make possible genetic studies of common disease that had the capacity to interrogate the **entire** genome. Without a genome sequence, without knowing how many genes there were, let alone how many sequence variants, everything was obscure—it was all theoretical speculation and plain guesswork. Gradually, however, answers to these questions began to emerge, as we explain next.

6.2 Reading human history in our DNA

How much genetic diversity was there in human populations? The more there was, the more difficult it would be to carry out GWAS because capturing all that variation would need millions of markers. At about the same time as human genetic linkage studies were first attempted, molecular data emerged that provided empirical evidence about the extent of diversity within, and between, human populations.

Our species, *Homo sapiens*, dates back about 250,000 years from its origins in Africa. One hypothesis of how humans came to populate the rest of the world is that there was a migration out of Africa, between 70,000 and 50,000 years ago. There are various versions of how this might have happened, and the dates and the number of migrations vary considerably, but one piece of evidence could not be gainsaid: genetic diversity, from sequence and genotyping data, is greater in Africans than in other populations. This suggested that a small number of people (initial estimates even suggested a handful) had left Africa to populate the rest of the world.

The 'Out of Africa' hypothesis has some slightly worrying implications. It means that anyone of European or Asian descent is much more closely related than we might suspect, close enough for the relationship still to be discernible in our DNA. Go to the DNA Ancestry Project website where you will learn that *'Marie Antoinette was the vivacious Queen of France who was, and still is, associated with the extravagant lifestyles of the 18th century monarchy ... With a simple swab of your mouth, you can be compared with Marie Antoinette and discover your relation to her. You will be able to trace your heritage back to times of grandeur, to times of kings and queens and royal balls, and possibly even to Marie Antoinette.'* (https://www.dnaancestryproject.com/features/famous-dna/).

Many people want to know their descendants, and the following story is fairly typical. While doing some work on his family history, Tom Robinson, an Associate Professor of Accounting at the University of Miami, decided to try one of the DNA-based ancestry projects, Oxford Ancestors (http://www.oxfordancestors.com).

In April, I received a call from Oxford Ancestors letting me know that they had done additional research (not at my request) based on recent research on the descendants of Genghis Khan. While Genghis Khan's DNA has not been found, I understand from published research that it has been 'inferred' from living individuals. Subsequently I received a certificate (suitable for framing) that states: This is to certify that Thomas R Robinson carries a Y-chromosome which shows him to be of probable direct descent from Genghis Khan, First Emperor of the Mongols.

https://www.nytimes.com/2006/06/06/science/06genghis.html

Even if you are not as closely related to Marie Antoinette, Genghis Khan, and the Romanovs as you suspect, it's still true that, if you are of European or Asian descent, then the genetic differences between you and other Europeans and Asians is less than between Africans, and, in general, less than that which differentiates species. This has important implications for genetic studies of common disease.

Soon after Risch and Merikangas's argument in favor of genetic association, Eric Lander laid down ten goals for genomics (*'with the aim of stimulating ferment'*) (LANDER 1996). One goal was the discovery of all common sequence variants within genes (coding variants) because, Lander argued, these variants would be a major contributor to the causes of common disease. His argument ran as follows:

… human diversity is … quite limited in that most genes have only a handful of common variants in their coding regions, with the vast majority of alleles being exceedingly rare. The effective number of alleles … is rather small, often two or three. This limited diversity reflects the fact that modern humans are descended from a relatively small population that underwent exponential explosion in evolutionarily recent time. … The catalog of common variants will transform the search for susceptibility genes through the use of association studies.

LANDER (1996).

Cataloging and testing common variants in coding regions assumes that these variants are causative, enabling a direct association between the disease and one of the molecular changes that confers susceptibility. Given the arguments in favor of reduced diversity, a complete catalog might consist of a few hundred thousand common variants.

6.3 The nature and number of genetic variants needed for GWAS

While everyone accepted that not all of the functional variants would be in the coding regions, on the basis of what was then known, it seemed reasonable to think that noncoding variants would play a minor role only. Common variants had been found in apolipoprotein E that explained a substantial fraction of the risk for Alzheimer's disease (CORDER et al. 1993), as well as risk for cardiovascular disease. Other genes, for factor V, angiotensin-converting enzyme (ACE), and chemokine receptor 5 (CKR-5), had common variants associated with, respectively, the risk of thrombosis (blood clots), heart disease, and resistance to human immunodeficiency virus infection. Why not expect this pattern to continue? The common disease–common variant hypothesis was attractive and, if true, made common disease amenable to genetic dissection. All it needed was the list of common coding variants.

But suppose the susceptibility variants did **not** lie in the exons. Suppose the causative variants lay in introns, or worse in that vast unexplored territory lying between genes making up more than 95% of the genome. What then? How could the effects of these variants ever be detected? Surely we would need complete DNA sequence data on everyone—an unimaginable thought.

As it turned out, it proved possible to interrogate the entire genome, not just the coding section, because of the way that genetic variants are correlated. In 1997, Francis Collins, Mark Guyer, and Aravinda

Chakravarti argued for a dense map of genetic variants, based on theoretical speculation about how disease alleles might have arisen and be distributed throughout the population (COLLINS et al. 1997). They focused on one type of sequence variant, the single-nucleotide polymorphism (SNP), which typically has just two alleles (e.g. A or G). SNPs are common and relatively straightforward (and relatively cheap) to genotype. Here, in their words, is their justification for the approach that was in the end adopted:

This strategy is based on the hypothesis that each sequence variant that causes disease must have arisen in a particular individual at some time in the past, so the specific array of polymorphisms (haplotype) in the neighborhood of the altered gene in that individual must be inherited in all of his or her descendants [a haplotype refers to the combination of alleles at different loci, but on the same chromosome; see Figure 1.1 in Chapter 1]. The presence of a recognizable ancestral haplotype therefore becomes an indicator of the disease-associated polymorphism. The size of this region (in which the genetic markers are said to be in 'linkage disequilibrium') will vary with the age of the variant.

COLLINS et al. (1997).

What this boils down to is that a sufficiently high-density set of SNPs would capture all (or at least the great majority of) common sequence variation in the genome (where 'common' means with an allele frequency greater than 5%). The strength of correlation between SNPs is the major determinant of the number of SNPs that would be needed: if it were low (meaning that most SNPs were independent of their neighbors), then more markers would be required, perhaps millions. If it were high (meaning that most SNPs were highly correlated with their neighbors), then fewer would do the job. On this basis, Francis Collins and his co-authors concluded: 'Further improvements in the technologies for the discovery and detection of SNPs should be immediately and aggressively pursued. Two goals should be targeted simultaneously: (i) the development of a dense map of at least 100,000 SNPs and (ii) the identification of common cSNPs in as many genes as possible' (cSNPs is short hand for sequence variants in coding regions of the genome, the coding SNPs that we mentioned above) (COLLINS et al. 1997).

In December 1997, a meeting in the National Institutes of Health Genome Research Institute (NIHGRI) decided to collect at least 100,000 SNPs in human DNA donated by 100–500 people in four major population categories: African, Asian, European, and Native American. Costs for the project were put at between US$20 million and US$30 million over 3 years. That large investment of public money was intended to forestall big academic laboratories and companies from patenting SNPs. While the meeting participants agreed on the need to do the work, there was considerable debate about where the samples would come from and how to ensure that privacy was safeguarded. As the genetic data would be publicly available, it might be possible to identify donors from their DNA profiles (and determine whether they were descendants of Genghis Khan). These ethical, social, and legal issues became increasingly important in human genetics.

6.4 Linkage disequilibrium and the HapMap consortium

Francis Collins, then head of NIHGRI, emerged as chief architect of the new initiative, which came to be known as the HapMap (haplotype map of the human genome). By then, it was generally assumed that 100,000 SNPs would do the job. Even 100,000 SNPs was going to cost a vast amount of money, so when geneticist Leonid Kruglyak argued in 1999 that 500,000 SNPs were necessary (KRUGLYAK 1999), there was a degree of upset in the human genetics community. Based on what was then known about human evolution, Kruglyak estimated that a map of 500,000 SNPs with an average spacing of 6 kb across the genome, was necessary (KRUGLYAK 1999). The HapMap consortium later reported the discovery of more than 1 million SNPs, driven in part by Kruglyak's arguments and made possible by remarkable falls in the cost and methods of genotyping (ALTSHULER et al. 2005).

Figure 6.1 is a figure from the HapMap consortium that illustrates the data the consortium collected, and exemplifies the linkage disequilibrium structure across the genome that means a subset of SNPs is sufficient for GWAS. We introduced linkage disequilibrium in Chapter 5, section 5.1, where we told you that it is a measure of the correlation of genotypes in a population. Here is why linkage disequilibrium is the key concept in human genetics that allows us to genotype a few hundred thousand markers, rather than sequence the genome, in order to map the genetic basis of psychiatric disease and other traits.

Suppose we have two polymorphic loci, two SNPs that are close to each other (let's say separated by 10 kb, i.e. 10,000 bp). We find that every person who has the genotype CG at marker 1 has the genotype AT at marker 2; in fact, the correlation is so good that there is no point in genotyping both markers because we can completely predict the one from the other. In an association test, I would not need to genotype both markers if I had information from just one, as both would give the same result. I can relax the requirement that the information from one marker is completely predictive of a second, let's say to require a correlation of greater than 0.8. Instead of a complete catalog of variants, a set of variants distributed across the genome can be chosen that is in sufficient linkage disequilibrium to capture information from nearly every other polymorphic locus in the genome, coding or not (note that this strategy will not find rare

variants; it still assumes that the common disease–common variant hypothesis is true).

Figure 6.1 shows that the human genome is broken into segments of highly correlated genotypes, which are known as haplotype blocks; the haplotype blocks are chromosomal segments on which alleles at adjacent loci are inherited together. From information about the haplotype structure of the human genome, it is possible to select a subset of markers that will capture almost all of the genetic variation attributable to common variants. Markers on genotyping microarrays are selected on this basis (see Chapter 3.8 for a description of genotyping microarrays).

There is an important caveat. The degree to which genotyping arrays capture genomic information is population specific, because population history affects the extent of linkage disequilibrium. Importantly, consistent with the 'Out of Africa' hypothesis, African

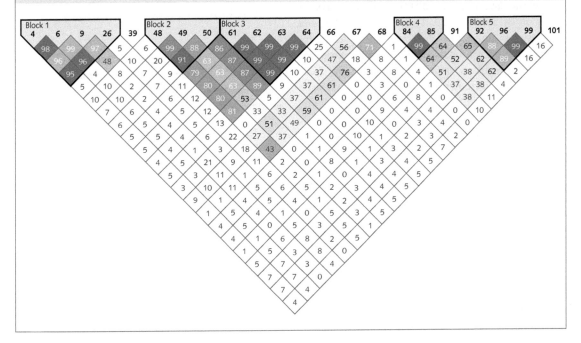

Figure 6.1 A haplotype map: linkage disequilibrium in the human genome. The panel shows the results for estimating the correlations between genotypes at 22 markers, covering a region of about 100 kb. The genetic correlation between each pair of markers is shown by the intensity of the shading in the diamonds, each diamond representing the result for a single pairwise correlation. The numbers in the triangles are the squared correlation coefficients between the genotypes (expressed as a percentage), when these are less than 100. The top row of diamonds shows the results between a marker and its neighbor, the next row between a marker and its next nearest neighbor and so on, with the apex of the pyramid representing the correlation between the furthest two markers. Where a set of markers is in very high linkage disequilibrium, a line has been drawn around the triangle to indicate a haplotype block.

populations show much lower levels of genetic correlation between markers than populations in the rest of the world. This means that if genotyping arrays designed for European populations are used in African populations, they won't capture as much common variation, and as a consequence, standard strategies for GWAS won't work nearly as well (TEO et al. 2010).

6.5 The first results of GWAS of common complex disease

In 2003, as the HapMap data began to emerge, the next question was: could these dense SNP maps find loci that altered the risk for common diseases such as diabetes, hypertension, and schizophrenia? Such a study was bound to be large, to involve many clinical investigators, to require access to high-throughput genotyping, molecular biologists, computer programmers, and perhaps most importantly statisticians; it would mean working in an interdisciplinary consortium, on a scale and at a cost common in physics but almost unheard of in biology.

At the time, the Sanger Institute, a large genomics center outside Cambridge (UK), was generating data for the HapMap project. The Sanger geneticists had focused on chromosome 20 (which at 60 Mb, i.e. 60,000,000 bases, is one of the smallest), and by 2004 had all the data from over 30,000 working assays for SNPs on chromosome 20. Together with colleagues at a sister institute, the Wellcome Trust Centre for Human Genetics in Oxford (UK), scientists proposed genotyping markers on chromosome 20 in 2,000 DNA samples, including cases with different diseases, at a cost estimated to be £5 million (US$7 million). Over the next year, these ideas matured and the use of a single set of controls emerged (rather than using a set for each disease), as did arguments about which diseases to analyze, but most importantly the cost of genotyping fell, so dramatically in fact that by 2006 a whole-genome association study was affordable with a projected budget of £8 million (US$11 million). By this stage, the Wellcome Trust Case Control Consortium (WTCCC) had decided to genotype 2,000 cases of each of seven diseases, and a common set of 3,000 controls (WELLCOME TRUST CASE CONTROL CONSORTIUM 2007). The results of analyzing 16,179 individuals with 469,557 SNPs were published in June 2007. Fortunately for all concerned, there were some positive results.

6.5.1 The Manhattan plot

Figure 6.2 displays the results from the WTCCC for all seven disorders. Plots of this form became known as 'Manhattan' plots, because of their resemblance to a city skyline. In each plot, the y-axis represents the significance of the association, which is plotted as the negative logarithm (base 10) of the P value. Thus a P value of 0.001 will be shown as 3, $P = 0.00001$ as 5, $P = 1 \times 10^{-8}$ as 8 and so on. The x-axis indicates the positions of genotyped variants, laid out along the genome from chromosome 1 to chromosome 22 Some studies will also report results for the X chromosome (but rarely are results shown for the small Y chromosome). Each dot on the graph is the result from a single SNP; the continuous blocks of color are the consequence of results from hundreds of thousands of markers merging together.

We are interested in places in the genome where the significance exceeds expectations. Regions of significance emerge, like skyscrapers, because linkage disequilibrium means that markers close to but not directly at the peak of association will have P values lower than the background rates (and hence higher $-\log_{10}(P)$ values). If you see a Manhattan plot with just one dot at high significance and nothing nearby showing any sign of association, then it's reasonable to suspect that the single result is a false positive. Adjacent markers on genotyping arrays almost always have some degree of linkage disequilibrium, so the results will be partly correlated.

6.5.2 A genome-wide significance threshold

The WTCCC used a significance threshold of $P < 5 \times 10^{-7}$ based in part on the number of markers tested, and found this threshold was exceeded at 24 loci. They summarized their results as follows:

Case-control comparisons identified 24 independent association signals at $P < 5 \times 10^{-7}$: 1 in bipolar disorder, 1 in coronary artery disease, 9 in Crohn's disease, 3 in rheumatoid arthritis, 7 in type 1 diabetes and 3 in type 2 diabetes. On the basis of prior findings and replication studies thus-far completed, almost all of these signals reflect genuine susceptibility effects. ... Our yield of novel, highly significant association findings is comparable to, or exceeds, the number of those hitherto-generated by candidate gene or positional cloning efforts. For many of the compelling signals, replication has already been obtained ... For others, replication is required to establish a definitive relationship with disease.

WELLCOME TRUST CASE CONTROL CONSORTIUM (2007).

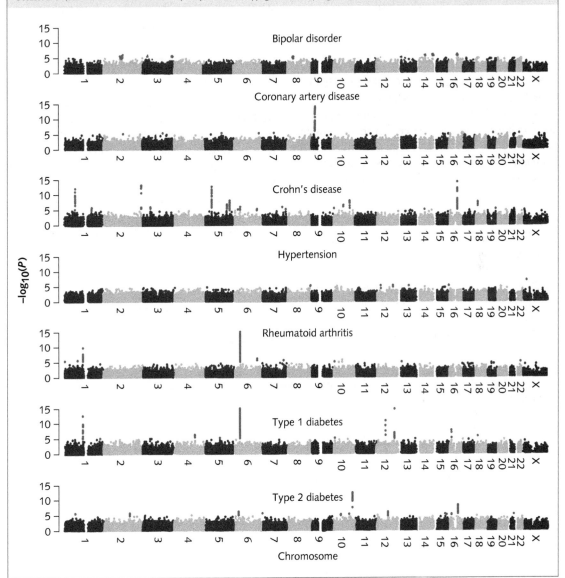

Figure 6.2 Manhattan plots for the seven disorders that were mapped by the Wellcome Trust Case Control Consortium (2007). The y-axis is the negative logarithm (base 10) of the association P value and the x-axis is the position on the genome of each marker for every chromosome tested. Blue dots represent SNPs whose significance is close to, or exceeds, the WTCCC significance threshold.

The fact that the WTCCC found anything at this threshold is because of the sample size they used. Had only 1,000 cases and 1,000 controls been collected and analyzed (rather than the 2,000 cases for each disease and the joint set of controls), just two significant signals would have been detected. The distribution of hits varied markedly across diseases: none was found for hypertension, one for coronary artery disease, but nine for Crohn's disease. The one psychiatric disorder analyzed, bipolar disorder, yielded only one significant hit, which later failed to replicate in an independent sample.

The genome-wide significance threshold applied by the WTCCC was not, however, to last long. Although it seems a ridiculously severe threshold (better than 1 in 50 million!), we will see in a moment that an even stricter threshold was soon adopted: $P < 5 \times 10^{-8}$ (Pe'er et al. 2008), a value that all complex trait geneticists now have tattooed on their hearts.

6.6 How to interpret a GWAS result

What does it mean to say a GWAS has identified genetic loci for a disease or a trait of interest? Can we say, for example, that we have found a gene for the condition? And can I be certain that the finding is robust? How do I know we are not still facing the problems of candidate gene studies? The four interpretations of a GWAS are laid out in Figure 6.3.

6.6.1 A false positive

The first interpretation is that the result is wrong, a false positive. We can control (but not exclude) false positives with an appropriately rigorous threshold for determining significance. The currently accepted standard is $P < 5 \times 10^{-8}$ (Pe'er et al. 2008) (as stated above, this is more rigorous than the threshold that the WTCCC used in 2007 of $P < 5 \times 10^{-7}$). While its theoretical justification has been debated, nearly everyone agrees that that the 5×10^{-8} threshold has controlled false positives, and it has now become standard in the field. There is a second criterion to determine robustness and rigor for GWAS, and that is replication. Nowadays, no one is going to believe your association result unless you have identified the variant at a $P < 5 \times 10^{-8}$ threshold and you have replicated the finding, preferably in a completely independent sample.

6.6.2 The genotyped marker alters disease risk

The second interpretation of a GWAS result (one that we hope is true) is that the genome-wide significant marker is the variant that alters risk for the disease: it is the causal variant. Unfortunately, because only a very small fraction of variants are genotyped (as we explained in section 6.3), the chances that the set includes the causal variant are very low. A causal associated marker is not the same as a causally associated gene, a point we return to in section 6.9.

6.6.3 The genotyped marker is in linkage disequilibrium with a causal variant

The third interpretation is that the significant marker is in linkage disequilibrium with the causal variant. This is an indirect association. Indirect and direct

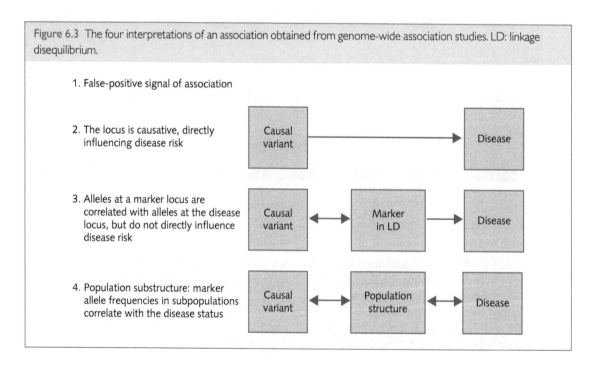

Figure 6.3 The four interpretations of an association obtained from genome-wide association studies. LD: linkage disequilibrium.

1. False-positive signal of association

2. The locus is causative, directly influencing disease risk

   Causal variant → Disease

3. Alleles at a marker locus are correlated with alleles at the disease locus, but do not directly influence disease risk

   Causal variant ↔ Marker in LD → Disease

4. Population substructure: marker allele frequencies in subpopulations correlate with the disease status

   Causal variant ↔ Population structure ↔ Disease

associations can't be distinguished on the basis of the data collected for the GWAS alone. Indeed, the problem of identifying the causal variant poses considerable challenges. We don't yet have an agreed method to identify causal variants.

6.6.4 **The result is due to population structure**

The fourth interpretation is that a confound explains the results, most likely population structure. In order to obtain cohorts of sufficient size for a GWAS, researchers almost always have to combine samples, collected in different ways, at different times, in different parts of the world. If cases and controls come from different ethnic groups whose allele frequencies differ, then genetic differences that a GWAS detects between cases and controls could be driven by ethnicity, not disease (i.e. population structure) (Figure 6.4).

There are two ways to deal with population structure. The first exploits the fact that correlations between genotypes reflect the shared descent of the population. In one of the most highly cited papers in genetics (23,000 citations by 2018), Jonathan Pritchard,

Matthew Stephens, and Peter Donnelly showed how a standard statistical approach called principal component analysis could identify a set of factors that reflected population structure. Including these factors in GWAS controlled for population structure (PRITCHARD et al. 2000). Principal component analysis works by identifying components that explain most of the correlation. Typically, the first few components explain most of the correlational structure, and a standard way of illustrating the analysis is to plot values of the first two principal components for all individuals in a GWAS. Figure 6.5 is an example (a nice illustration of the effectiveness of principal components in recognizing population structure comes from applying the method to identify ethnicity; John Novembre and colleagues reported the results of this analysis in a paper published in 2008. It's one of those papers whose succinct title almost means you don't have to read the text: 'Genes mirror geography within Europe'; NOVEMBRE et al. 2008).

In a second method, Eleazar Eskin and colleagues proposed using a genetic relationship matrix, which records pairwise relationships between individuals in the study, to model the degree to which each

Figure 6.4 Population structure. In the total sample collected for a genome-wide association study, there is a higher frequency of the 'risk' allele in cases than in controls. However, there are two populations within this sample. In population 1, the 'risk' allele is at a higher frequency than in population 2. Restricting attention to population 1 reveals that the 'risk' allele frequency is the same in cases and controls (2/3). The same is true for population 2 (1/6). Cases and controls have been sampled disproportionately from the two populations, resulting in a false-positive association for the 'risk' allele.

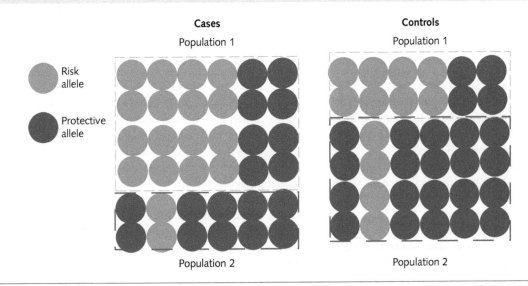

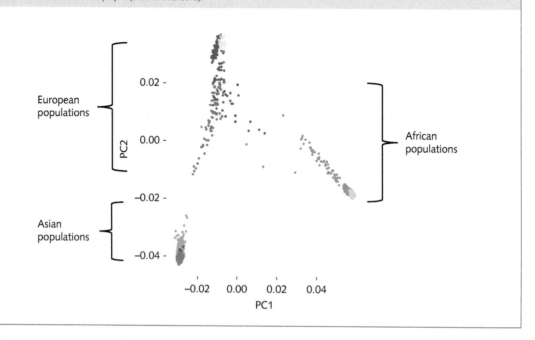

Figure 6.5 Principal component analysis of human populations. Principal component analysis of genetic relatedness, assessed by pairwise correlation coefficients at hundreds of thousands of genetic markers, identifies a large number of components, but the contribution of most is very small. Here only the first two (PC1 and PC2) are plotted. Each dot represents results from an individual.

Data are from the 1000 Genomes project (ABECASIS et al. 2012).

individual is related to every other individual (KANG et al. 2008). Again, as with the principal components method, shared relationships reflect descent from common ancestors; risk loci are expected to be present in all groups, regardless of their descent.

6.6.5 Quantile–quantile plots: a useful quality control step

The consequences of population structure, or indeed of other unknown confounds that affect the results, can be checked by comparing the observed distribution of the association statistics with that expected if there were no true signal. This is the purpose of the quantile–quantile (QQ) plot.

A QQ plot is obtained by ranking the negative logarithm (base 10) of the P values of the GWAS and plotting them against the negative logarithms for the same number of observations. For example, if you had 100 results, to get the expected values (assuming there is no true effect) calculate the negative logarithms (base 10) for 100 equally spaced numbers running between 0 and 1 (0.01, 0.02, 0.03 ... 0.99, 1.00). Plot these

against the negative logarithm of the GWAS results. Figure 6.6 is an example.

In Figure 6.6, the thin central line is the expected values (the negative logarithms (base 10) for 500,000 equally spaced numbers running between 0 and 1). The thick dotted line is the observed set of P values (converted into negative logarithms, so a P value of 10^{-6} is 6, and so on). The two lines overlap, which means that the P values we see are consistent with there being no effect (the outer lines are confidence intervals for the observed values).

In Figure 6.7, there are many points on the thick dotted line that are above the expected values (the central diagonal line). We'd like to find values that exceed the expected—after all, those are the signals that indicate we have a significant result—but we expect to see this for only a relatively small proportion of the results. In Figure 6.7, the two lines diverge across nearly the entire distribution. This is a bad sign and almost certainly means that something—perhaps population structure, perhaps some other artifact—is artificially inflating the results. A QQ plot like Figure 6.7 is evidence for P value inflation.

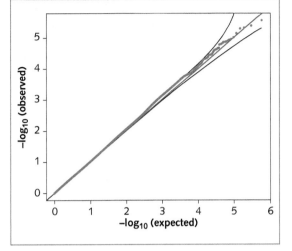

Figure 6.6 Example of a quantile–quantile plot. On the x-axis are the negative logarithm of the P values expected if there were no effect ($-\log_{10}$(expected)); on the y-axis are the observed values from a GWAS ($-\log_{10}$(observed)). The outer lines are the 95% confidence intervals.

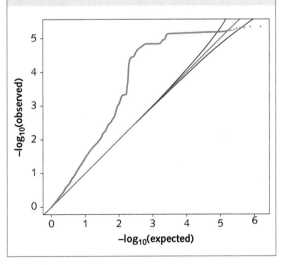

Figure 6.7 Example of a quantile–quantile plot where there are significant deviations from the expected values.

The WTCCC provided QQ plots for their findings, and while they plotted a chi (χ) statistic rather than a logarithm of a P value, the idea is exactly the same. For five of the seven diseases, the expected values lie above the observed. The WTCCC results show an upward deviation (outside the confidence intervals) **only at the right end of the curve. If the deviation** were found across the entire distribution (from 0 to 20), it could be attributed to inflation.

6.7 GWAS developments since 2008: more SNPs (both real and imaginary)

Since 2008, more and more SNPs have now been routinely genotyped, up from half a million in the 2007 WTCCC study to millions after 2015. At the same time as the number of genotyped markers has increased, a statistical innovation has also extended the scope of genotyping arrays to interrogate even more genetic variation: imputation (HOWIE et al. 2009).

Imputation relies upon the linkage disequilibrium structure of the human genome and does something similar to the following. I know the genotypes at marker A. I also know that marker A's genotypes are highly correlated with those at marker B (let's say, about 90%). This means I can estimate (some of my not-so-quantitatively-trained colleagues say 'guess') the probability of genotypes at marker B, **even though I have not genotyped marker B in my sample.** Using the very high density of variants available from the 1000 Genomes Project (GENOMES PROJECT CONSORTIUM et al. 2015), it's possible to interrogate millions of ungenotyped variants. Figure 6.8 illustrates the idea.

The figure ignores the fact that the linkage between SNPs in a haplotype block is not always 100%. As Figure 6.1 made clear, the correlations take a range of values. To take this into account, geneticists use a probabilistic estimate of the genotypes. We assess the probability of each of the three genotypes, given the information available (which will include the genotypes of the adjacent flanking SNPs, the reference haplotypes, and the linkage disequilibrium between the markers). The genotype probabilities must sum to 1, so for a SNP with G and C alleles, the imputation method might report GG = 0.1, GC = 0.2 and GG = 0.7 (i.e. the most likely genotype is GG).

Currently, almost all genotypes are scored as genotype 'dosages'. This is a way of expressing the probabilistic estimates of three genotypes as a single number. First, we score the genotypes as 0, 1, and 2 for the homozygote of one allele, the intermediate value of the heterozygote, and the homozygote of the other

Figure 6.8 Genotype imputation. At the top of the figure is shown a short stretch of sequence from two chromosomes, in which there are two SNPs, shown in larger bold. These two SNPs form part of a larger haplotype block. Below the sequence, four haplotypes are shown. This is not DNA sequence; it is the alleles at 11 loci that constitute the four haplotypes. Arrows indicate the position within the haplotype block of the two SNPs from the sequence. The haplotypes, generated by deep sequencing of large numbers of people, are called reference haplotypes. The third section of the figure shows genotypes obtained from an array. Not all of the loci that make up the haplotypes have been genotyped. The missing loci are shown as 'N'. Imputation uses data from the reference haplotypes to work out the most likely genotypes in the genotyped sample. You can see that for someone with a genotype of TA at the first position and GC at the fourth position, the genotypes at the second and third positions have to be GC and GA. The last lines of the figure show the imputed genotypes (shown as two haplotypes).

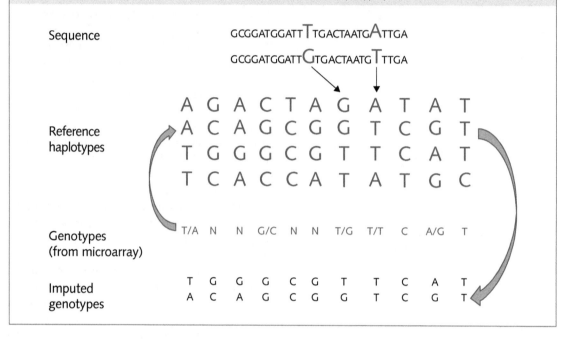

allele, respectively (the justification for this is that we use an additive model to carry out the association analysis, as explained in more detail than you may want to know in Chapter 21, section 21.2 and in the Appendix, sections A10 and A11). We then multiply the probabilities of the genotypes by their respective values, namely 0, 1, and 2, and sum to get a single number, the dosage. In the example above, the dosage is $(0 \times 0.1) + (1 \times 0.2) + (2 \times 0.7) = 1.6$. Dosage summarizes the information neatly: rather than three probabilities, the genotype is expressed as a single number. Just one warning: some people score the genotypes as 0, 0.5, and 1 (rather than 0, 1, and 2), in which case dosages extend from 0 to 1. If you analyze dosages, make sure you know which system is being applied!

As more and more whole-genome sequencing data accumulate, with better and better estimates of allele frequencies and of the haplotype structure of the human genome, it's becoming possible to impute millions and even tens of millions of variants. Why is this useful? There are two reasons. First, it's straightforward to combine data from studies that used different genotyping arrays—we simply impute genotypes to a common set of markers. Results can then be combined, using meta-analysis, thereby increasing sample size and hence the robustness of findings. Second, markers with lower allele frequencies can be imputed, so that the reach of GWAS to interrogate the genome is expanded. A corollary is that the accepted threshold of 5×10^{-8} may no longer be appropriate. However, this has not stood in the way of the GWAS community claiming that (with 20 million markers) they have found many loci that exceed the genome-wide significance threshold.

6.8 Three axioms that emerge from GWAS results

By the end of 2006, a dozen or so loci were widely accepted as susceptibility loci for complex disease. By the end of 2008, that number was approaching 500. By mid-2018, there were 75,977 genome-wide significant loci held in the database of GWAS findings (maintained at http://www.ebi.ac.uk/gwas/docs/file-downloads) for 2,762 different phenotypes. These results can be boiled down to three important observations that we will refer to here as the three axioms of a GWAS.

6.8.1 **A GWAS finds association with allelic variants at a locus**

Just so there is no misunderstanding, a locus here is a geneticist's term for a place—a place in the genome where the genetic variant is found (see Chapter 1, section 1.4). To put the same axiom negatively, a GWAS does not find association with a gene. It's hard to overemphasize the importance of this statement. If the variant found by a GWAS altered a coding region, as was initially hoped, then it would be straightforward to say which genes were involved in the trait under investigation. But GWAS hits turned out not to be coding SNPs. Rather they lie in noncoding regions of the genome, in introns or in regions between genes.

When GWAS was first proposed, one suggestion was to test all sequence variants in genes that would have functional consequences: changes that would turn off a gene, or would alter its protein product into something strange or even ghastly. Finding nonsynonymous variants (genetic variants that change proteins) associated with a disorder is good because it directly implicates the gene in which the variants occur. Knowing which genes are involved helps identify the relevant biology and pinpoint pathways to illness, thereby potentially leading to new treatments.

One arm of the WTCCC study specifically addressed this question: 14,436 nonsynonymous SNPs were tested in 1,000 cases of ankylosing spondylitis, autoimmune thyroid disease, multiple sclerosis, and breast cancer (BURTON et al. 2007). Two new loci related to ankylosing spondylitis were found, and two others were identified for autoimmune thyroid disease, although both had been reported previously—not a big yield.

The general rule is that changes to the protein-coding part of a gene are the exception. From the 2018 list of GWAS results, less than 5% are associated with a coding variant (and the causal relationship of these has yet to be established). Rather, the vast majority lie in regulatory regions of the genome (see Chapter 4, section 4.2.1), particularly in enhancers (MAURANO et al. 2012).

Figure 6.9 shows an example of a GWAS significant locus that alters the risk of major depressive disorder (CONVERGE CONSORTIUM 2015). You can see that the peak of the signal lies in an intron of a gene called *LHPP*. Unfortunately, this does not necessarily mean that the locus acts by altering the expression, or some other feature, of *LHPP*.

In 2012, it was discovered that regulatory regions of the genome (i.e. regions of the genome that control gene expression, described in Chapter 4, section 4.2.1) are enriched for genome-wide significant hits (MAURANO et al. 2012). It is often assumed that the regulatory region containing the GWAS hit regulates the closest gene. In fact, we know that the location of the hit is not a good guide to the gene affected, as was made clear from work on the genetics of obesity.

In one of the very first GWAS results, genome-wide significant variants associated with obesity (in short, an 'obesity locus') were found within an intron of a gene called *FTO* (FRAYLING et al. 2007). A series of papers were published that assumed that *FTO* was the gene (FISCHER et al. 2009). For example, analyses of genetically engineered knockouts of *FTO* in mice demonstrated that *FTO* affected mouse weight (CHURCH et al. 2009). However, no one tested whether a causative variant at the human obesity locus really acted on the *FTO* gene. When that experiment was finally carried out, it was discovered that the obesity locus didn't influence the expression of the *FTO* gene but instead acted on two other genes, *IRX3* and *IRX5*, that lay more than 1 MB from the obesity locus (SMEMO et al. 2014; CLAUSSNITZER et al. 2015).

6.8.2 **Genetic effects at GWAS loci are small**

Figure 6.10 plots the effects attributable to loci that affect the risk of disease, expressed as an odds ratio (an odds ratio of 2 means a doubling of the risk) (HINDORFF et al. 2009). The median odds ratio is 1.18, which is almost certainly an overestimate as studies are underpowered to detect smaller effects (this is implied from

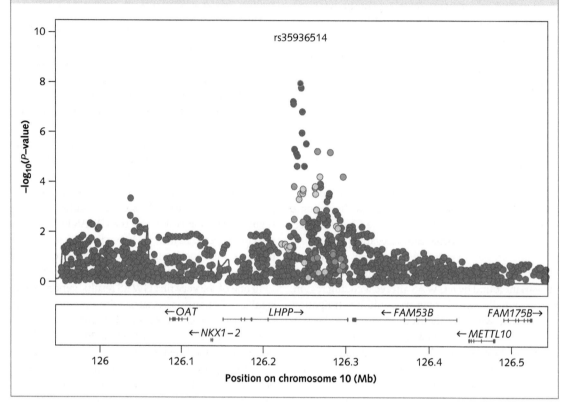

Figure 6.9 A genome-wide significant locus that alters the risk for major depressive disorder lies in an intron of the *LHPP* gene. The *x*-axis is the negative logarithm (base 10) of the association between marker genotypes and the risk of disease. Each dot represents the results for markers whose position on chromosome 10 is given by the scale on the *x*-axis (in megabases, Mb). Below the association plot is a schematic to show the position of genes, with arrows indicating the direction of transcription and vertical bars showing the position of exons. The peak of the association lies in the last intron of the *LHPP* gene.

the drop in frequency of the lowest category of odds ratio on the left of the figure, which is less than 500, whereas the next category contains more than 3,000 results). Recall the estimates of effect size put forward by Neil Risch (he thought that an odds ratio of 2.0 was reasonable): small as his estimates seemed at the time, they have turned out to be overly optimistic.

6.8.3 Genome-wide significant GWAS hits only explain a small fraction of the heritability of a trait

From the very start of the GWAS revolution, it became clear that the risk loci that were found did not explain all of the heritability. For example, twin and family studies have shown with great consistency that height is highly heritable: at least 80% of the variation

has a genetic basis. With the first studies of height, 54 SNPs were found to have genome-wide significance, a great advance on what we knew before the advent of GWAS, but the total heritability accounted for by all these SNPs was only about 5%. Each variant contributed much less than 0.5%. Another way of putting this is that if you inherit an allele that increases your height, you'll be about 0.4 cm taller as a consequence. And all the loci are of this size or smaller; there are no big effects, no exponential distribution with a few large ones and a tail of smaller effects. The same is true for disease genetics, with a few exceptions (such as the effect of the HLA locus on autoimmune diseases, including type 1 diabetes). The risk loci that have been discovered account for only a fraction of the heritability. This third GWAS axiom is important enough to deserve additional elucidation, particularly since

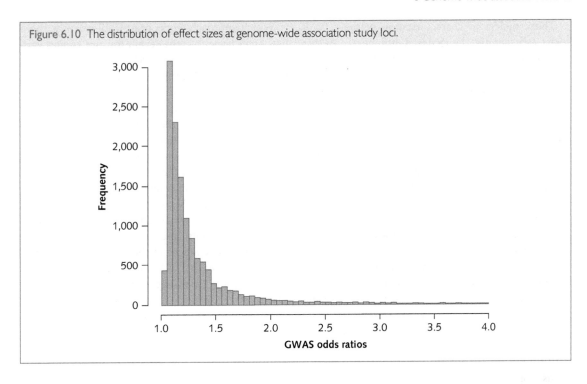

Figure 6.10 The distribution of effect sizes at genome-wide association study loci.

6.9 Turning genetic loci into genes

work on schizophrenia was to play an important part in explaining so-called 'missing heritability'. We discuss this issue in Chapter 7, section 7.5.

We stressed that GWAS results don't find genes. That prompts the obvious question: how do we find genes from a GWAS? Let's summarize the situation. The genetic architecture of complex traits (i.e. behavior, and psychiatric disease) consists of many thousands of small-effect loci that are located in the noncoding regions of the genome. The sequence variants at the loci alter the function of regulatory elements, primarily enhancers (although other less well-characterized elements are also affected). The consequences are changes in gene expression and gene splicing. Genetic mapping indicates the position of risk loci but does not unambiguously indicate which variant is causal at a locus. How do we turn knowledge of the location of sequence variants associated with the disease into information about which gene is involved?

One way is simply to assume that the identity of relevant genes is indicated by their proximity to a risk locus. This is implicit in the reports of many studies that claim gene identification (e.g. NETWORK AND PATHWAY ANALYSIS SUBGROUP OF PSYCHIATRIC GENOMICS CONSORTIUM 2015). However, as discussed in section 6.8.1, analysis of an obesity locus demonstrates that proximity alone does not prove gene identity (CLAUSSNITZER et al. 2015). Please don't follow the example of people who point at a locus where the GWAS identifies a genome-wide significant hit and confidently tell you, oh, this is the gene involved. If all they rely on is proximity, then they are likely to be correct at best 50% of the time (we'll explain below how we know this). Just remember: the closest gene is not guaranteed to be correct.

A second method is to sequence genes at a locus where there is a significant GWAS association to reveal rare variants that support the candidacy of a gene (NEJENTSEV et al. 2009). The assumption here is that large-effect, loss-of-function mutations can be found at causal genes for complex traits. The discovery of such mutations is difficult (MACARTHUR et al. 2014), and it is not certain that every GWAS locus will harbor both common and rare variants. An extension of this idea is simply to give up on GWAS and search for any rare coding variant that is associated with the trait. We discuss results from applying this approach to schizophrenia in Chapter 7, section 7.8 and to autism in Chapter 8, section 8.5.

Third, molecular characterization of the GWAS locus can identify the genes it influences. As we have pointed out, the vast majority of GWAS hits lie in regulatory regions of the genome (ERNST et al. 2011; MAURANO et al. 2012; SCHAUB et al. 2012; GUSEV et al. 2014). The role of these regulatory elements is described in Chapter 4, section 4.2.1. Briefly, the DNA containing regulatory elements, separated by up to hundreds of kilobases from the gene they regulate, combines with proximal elements to control gene transcription (DEKKER et al. 2002; KADAUKE AND BLOBEL 2009; SEXTON et al. 2009; SCHOENFELDER et al. 2010). The DNA is thought to form loops, bringing the distant regulatory regions (the enhancer) into contact with the promoter of a gene. Suppose we identify the set of regulatory elements that contain risk loci. We then isolate the base of the loops, where the regulatory elements touch each other (using chromosome conformation capture; see Chapter 4, section 4.7), determine which gene a regulatory element regulates, and thereby work out which gene the GWAS variant affects. This is how the *FTO* obesity locus was found to control the expression of *IRX3* and *IRX5* (CLAUSSNITZER et al. 2015). This idea is illustrated in Figure 6.11.

Which approach works best? Until recently, there was great hope that the search for rare coding variants would be the most efficient way to find genes involved in complex traits. The short answer is that rare variants have identified genes giving rise to intellectual disability and autism (see Chapter 8), but not schizophrenia, and certainly not depression or anxiety. Identifying rare causal variants for common psychiatric disease is turning out to be very hard.

As expectations for the success of rare variant analysis have decreased, more interest has turned to the use of 'omics data (see Chapter 4, section 4.7). In a systematic analysis of the relationship between GWAS hits, regulatory elements, and the genes they regulate, Dan Geschwind, a geneticist at the University of California, Los Angeles (UCLA), used chromosome conformation capture (see Chapter 4, section 4.7) to show that most regulatory elements (approximately 65%) don't interact with their adjacent genes (WON et al. 2016); only about half of GWAS hits for schizophrenia influence the closest gene. Characterizing interactions between genes and regulatory elements seems to be a good way to go! But there are two problems.

Jesse Engreitz, working with Eric Lander at the Massachusetts Institute of Technology and Harvard,

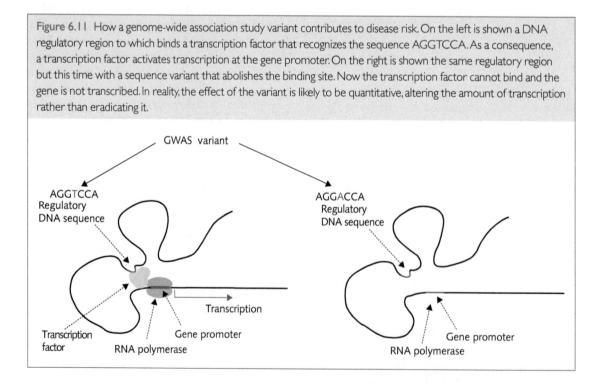

Figure 6.11 How a genome-wide association study variant contributes to disease risk. On the left is shown a DNA regulatory region to which binds a transcription factor that recognizes the sequence AGGTCCA. As a consequence, a transcription factor activates transcription at the gene promoter. On the right is shown the same regulatory region but this time with a sequence variant that abolishes the binding site. Now the transcription factor cannot bind and the gene is not transcribed. In reality, the effect of the variant is likely to be quantitative, altering the amount of transcription rather than eradicating it.

developed a method to systematically interrogate the relationship between thousands of regulatory elements (enhancers) and promoters. He found that the relationship is more complex than expected (Fulco et al. 2016, 2019). While most enhancers are located within 100 kb of their target genes, they often regulate more than one gene and, importantly, they have a wide range of quantitative effects on gene expression. It seems that rather than the interactions being all or nothing, with some enhancers dedicated to serving the regulation of specific genes, they are more quantitative than qualitative. So it is currently probably not possible to say 'this variant lies in this enhancer, which means this gene must be involved', because an enhancer's attachment to a single gene is fickle.

A second problem with characterizing enhancers and their targets is that the effects are specific to tissues and cells. This means the experiments require extracting relevant tissue (not just blood) from many people (probably thousands). That's expensive, time-consuming, and pretty well impossible if you want to sample brain tissue.

6.10 Transcriptome-wide association studies

There is an inventive solution to the problem of determining the identity of genes involved in psychiatric disease, which avoids the need to sample brain tissue in patients (Pasaniuc and Price 2017). Suppose we obtain the tissues we need from dead people, and then map the effects of genetic variants on the transcriptome, the amount of RNA expressed, and splicing (the data are publicly available; GTEx Consortium 2017). You might ask what is the use of that? We need to know the effects of the variants on transcripts in our GWAS sample, on our patients with schizophrenia, not in dead people!

The answer is that we can use the information from the tissues to predict (or, more colloquially, 'guess') the transcript levels in our GWAS sample based on the genotypes we have obtained. How is that possible? Well, the risk variants are common, so many people without schizophrenia, or whichever disease we are interested in, will have them. With frequencies of greater than 5%, we can be sure to find those variants in samples taken from the tissues of a few hundred people. A few hundred is enough, because genetic loci that contribute to transcriptional variation (expression quantitative trait loci, or eQTLs) have much larger effect sizes than loci that contribute to disease risk or other traits. eQTLs often explain 20% or more of the variation (GTEx Consortium 2017). We can then calculate the effect of the variants in the sample of tissue. Now that we know what the variants do in the tissues of interest to us, the trick is to use that information to 'guess' the consequences of possessing the variants for the subjects in our GWAS sample.

The way we guess is conceptually similar to the genotype imputation approach we discussed in section 6.7 (in which information from a reference sample is used to learn how to impute genotypes at SNPs that were not tested). For example, we know from the reference data set (the data from tissue) that a particular genotype increases a certain gene transcript in the brain by a certain value. We then infer that anyone in our GWAS sample who has that genotype has that same amount of transcript in their brain (or whichever tissue we assayed). We can do that for every genotype that affects gene expression, and hence end up with an estimate for the transcript level (or splicing isoform) of every gene in the genome (Gamazon et al. 2015). In fact, as my colleague at UCLA, Bogdan Pasaniuc, showed, you don't even need the genotypes from the GWAS; you can do the same guessing game using the summary statistics from a GWAS (Gusev et al. 2016). This makes it possible to rapidly estimate transcript abundance for all the genes in the genome, and then test for association between the predicted transcript abundance and disease. This is the method that Pasaniuc calls transcriptome-wide association analysis (TWAS). The significance of the association between disease state and each predicted transcript level can be plotted to produce a TWAS Manhattan plot (a significance threshold of 2.2×10^{-6} can be used, which is lower than for a genotype Manhattan plot, as only 23,000 genes are tested rather than the million or so genetic markers).

There is one twist. It is easy to imagine that the genes found through TWAS are causally associated with disease, but this is not necessarily the case. Let's say we have a locus that we know is associated with the trait; in this case, we know that we have a causal relationship between genotype and trait. If the

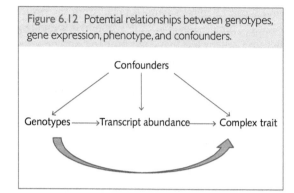

Figure 6.12 Potential relationships between genotypes, gene expression, phenotype, and confounders.

expression level of the gene has an effect on a trait, different genotypes will have different phenotypes, that is, the genetic variant will also show an effect on the trait. Does that mean the gene is also causal? This question is similar to the one answered by Mendelian randomization, a method we discuss in Chapter 13, section 13.5. In principle, Mendelian randomization analysis could be used to identify functionally relevant genes at the loci identified in GWAS for complex traits (Zhu et al. 2016).

Figure 6.12 shows the potential relationships between the elements of TWAS and GWAS experiments. The relationship we want to find runs horizontally from the genotypes via changes in transcript abundance (or splicing variation) to the disease. But there are other possible ways in which the correlation between trait, genotype, and gene expression could occur, as indicated by the arrows in the diagram. We will come across variations of this diagram elsewhere (see Chapter 5, Figure 5.9 and Chapter 13, Figure 13.2). It's worth noting its implications, as they come up elsewhere in this book.

In brief, these are the three rules that must be obeyed if we are to conclude from the analysis that the genotypes act through the gene expression variable to alter disease risk: (i) genotypes must be associated with the transcript abundance; (ii) genotypes must not be associated with the disease, except through the transcript abundance; and (iii) genotypes cannot be related to measured or unmeasured confounds. Ensuring that the three rules are obeyed is very hard (especially the last!). TWAS, and other methods using 'omics data can suggest candidates, but to prove which candidate a GWAS variant acts on to alter behavior requires an experiment, such as

those described for the *IRX3* and *IRX5* genes at the *FTO* locus (Claussnitzer et al. 2015).

6.11 A conversation about genetic architecture

We have had results from GWAS for more than a decade and a consistent pattern of the polygenic architecture of behavior is emerging. We'll describe these findings in subsequent chapters, which, while familiar to those working in the field, are still a surprise to those who are not. I (J.F.) had lunch recently with Alcino Silva, a neuroscientist whose work on the molecular basis of memory is described in Chapter 16, section 16.4. Alcino, by temperament, is enthusiastic and as we sat down to our kale salad was excitedly talking through with me some ideas about the causes of depression. We were discussing what we might learn from giving interferon-α to mice (there is a long history of this drug causing depression as a side effect of its use as a treatment for hepatitis C, a viral liver infection).

'We should run a parallel human experiment,' Alcino was urging me. 'You have found genes involved in depression. Let's test those to see if there is a difference between those who do and those who don't get depressed on interferon therapy.'

'Alcino, that won't work, we'd need thousands and thousands of patients.'

'Surely not. We'll be looking at such a specific group, there's bound to be a large genetic effect, due to, what would you say, maybe half a dozen genes?'

'No, it doesn't work like that!'

'Well tell me, how does it work?'

'It's like this: either a disease or a trait is Mendelian, with a single mutation, or it's polygenic and then there will be thousands of loci. There's no evidence that the response to interferon is Mendelian, so it's going to be polygenic.'

'But how do you know that? You might be wrong!'

'Well, of course I might be wrong, but human geneticists have been carrying out GWAS experiments for a while now, and for behavior we really don't see anything other than deeply polygenic traits.'

'Really?'

'Really.' There was a pause.

'You know I didn't really get that point till now. Wow, now you have made me depressed.'

Summary

1. Genotyping microarrays capture information from almost all common variants in the genome in European and Asian populations but are less comprehensive for African populations.

2. Genome-wide association studies (GWAS) of common variants, using sample sizes of thousands of individuals, can detect the genetic signals that are responsible for complex traits, including psychiatric disease and normal behavioral variation.

3. GWAS findings have earned their reputation as robust for three reasons: (i) a stringent significance threshold is applied ($P < 5 \times 10^{-8}$); (ii) replication is (usually) provided; and (iii) every attempt is made to mitigate the effects of confounds (particularly population structure).

4. A GWAS does not find association with a gene (please don't forget this!). A GWAS finds association with allelic variants at a locus, not with genes. Molecular characterization of a GWAS locus can identify the genes it influences, but the approach is time-consuming and difficult, especially for behavior, because it is not clear which part of the brain is involved, which cell type, or which developmental stage.

5. Genetic effects at GWAS loci are small. The median odds ratio, estimated over thousands of genome-wide significant loci for hundreds of diseases, is 1.18.

6. Significant GWAS hits only explain a small fraction of the heritability of a trait. This is known as the missing heritability problem and is discussed in more detail in Chapter 7, section 7.5.

7. Analysis of 'omics data can be used to identify genes involved with complex traits.

7

Molecular genetics of schizophrenia

[On how molecular genetic approaches to schizophrenia discovered risk loci and uncovered the disorder's highly polygenic architecture]

7.1 Genome-wide association studies of schizophrenia

When the Wellcome Trust Case Control Consortium (WTCCC) demonstrated that a genome-wide association study (GWAS) with 3,000 cases was sufficient to identify risk loci for common biomedical disorders, but not for bipolar disorder (WELLCOME TRUST CASE CONTROL CONSORTIUM 2007), did that mean psychiatric disorders just weren't tractable to molecular genetic approaches? Or did it mean that the sample size employed was too small? For those diseases where GWAS had worked, it was clear that the effect of individual risk loci was much smaller than first thought (thus explaining why linkage studies had failed), and, furthermore, that the effects varied among diseases (we still don't know why risk loci for inflammatory bowel disease are, on average, larger than those for obesity).

To detect small effects, a large sample size is needed, so it was reasonable to assume that GWAS hadn't found risk loci for psychiatric diseases because

not enough cases had been recruited. You might think that if the genetic effect at a locus is half what you thought it was, that would mean you would need to double the sample size. Unfortunately, the relationship isn't like that. Power to detect an effect is related to the square root of the sample size (see Chapter 21, section 21.5). Suppose the effect size we want to detect increases the risk of disease twofold; then the genetic effect at that locus has an odds ratio of 2. Figure 7.1 shows that a sample size of over 1,000 (that's 500 cases and 500 controls) has 80% power of finding the effect in a GWAS study. But if the genetic effect has an odds ratio of 1.5, a sample size of more than 3,000 is needed to have the same chance of detecting the locus, and if the odds ratio is 1.1, the sample size required is about 30,000.

Obtaining large sample sizes—thousands or even tens of thousands of samples—seemed prohibitive and indeed would have been impossible for any single research group. But the WTCCC not only revealed for the first time the genetic architecture of common disease, it also showed the value of consortia efforts.

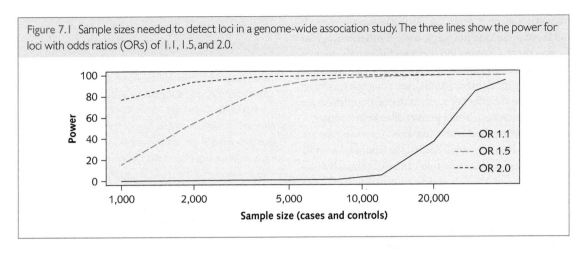

Figure 7.1 Sample sizes needed to detect loci in a genome-wide association study. The three lines show the power for loci with odds ratios (ORs) of 1.1, 1.5, and 2.0.

It was not a change in the methods of science that led to the first successful genetic dissection of a psychiatric disease but a change in the sociology of the science.

As individual groups accumulated more negative schizophrenia GWAS findings on their own, they started to form into consortia pooling their samples. This meant a move from the individualistic ethos common in academic research to a more communal approach. Part of this was motivated by genuine altruism: to try to get to the answers more quickly to help those suffering from schizophrenia. In part, it was also the result of pragmatic calculations. It was better to be a small part of a big important study than to have a small unimportant study all to yourself.

The first noteworthy group was the International Schizophrenia Consortium (ISC). In 2009, they published their results using 2,601 cases and 3,345 controls (PURCELL et al. 2009), a sample size a little larger than that used by the WTCCC. They found nothing significant. But the ISC was not the only group undertaking a schizophrenia GWAS. Two other groups had independently collected similar-sized samples: the SGENE-plus Consortium had amassed 2,663 cases and 13,498 controls from eight European locations, and the Molecular Genetics of Schizophrenia Consortium had genotyped 2,681 cases and 2,653 controls (SHI et al. 2009; STEFANSSON et al. 2009). None of these groups found loci that exceeded genome-wide significance, but when the three groups shared their data, now using combined sample sizes of 8,008 cases and 19,077 controls, they all found a genome-wide significant locus on chromosome 6, in a region of the genome called the major histocompatibility complex, a 3 Mb stretch of chromosome 6p21 that contains

genes involved in immune function such as human leukocyte antigen (HLA) genes. Its discovery marked a turning point in the application of genetic approaches to solving the mystery of what causes schizophrenia.

7.2 Polygenic risk scores

The GWAS results were not the only development in our understanding of the genetic basis of schizophrenia. One of the members of the ISC was a geneticist with a very different background from those involved in GWAS, a training that meant he looked at the findings from a different angle from those whose interest was finding genome-wide significant hits. Peter Visscher had trained in agricultural genetics, where a main goal is to use molecular approaches to make predictions from genetic data about the expected crop yield, milk production, or size and quality of meat in an animal. Agricultural geneticists don't care about genome-wide significance. They want genotypes to predict phenotypes to help farmers pick which cows to breed or which corn seeds to propagate. Genome-wide data on a large human sample allowed similar questions of prediction to be addressed for human traits, including psychiatric disease.

Visscher proposed creating a polygenic risk score (usually abbreviated to PRS) for schizophrenia, which exploits the fact that many variants failing to meet genome-wide significance could still have an effect on the trait. For instance, a locus that did not meet genome-wide significance is not by that evidence alone excluded as a risk locus for schizophrenia; it might just happen to have too small an effect to be robustly detected with the available sample size. If schizophrenia has a polygenic basis of inheritance, that is, that there are many loci, perhaps thousands, each of small effect contributing to disease risk (as predicted by Fisher's infinitesimal model; see Chapter 1, section 1.3.3), then the effect of variants in one population (or group of patients) should predict disease in another.

The PRS method works like this. Two samples are needed, which are usually called the 'training' sample and the 'test' sample (the two can be obtained by randomly dividing the full sample into two halves). The PRS uses markers from one sample to predict disease risk (or a trait) in the second sample. You don't want to use all genetic variants, because some are very highly correlated due to linkage disequilibrium (see Chapter 6, section 6.4). There are various ways to pick the best set of variants for analysis. The number of genetic variants that can usefully predict a trait can't be known in advance, so researchers typically use a variety of cut-offs, starting by including markers with highly significant P values and then using increasingly less stringent P values (the P values are those from a GWAS). As the threshold becomes less stringent, the number of markers included increases, until, at a value of $P < 0.5$, almost half the markers are included. The assumption is that, for highly polygenic traits, even genetic variants with very slight evidence for association will contribute, so that increasing the number of markers used will lead to an increase in the prediction accuracy of the PRS. But if large numbers of the added genetic variants are unrelated to the disease, this could lead to a dilution of the predictive power.

To generate the PRS requires two key bits of information about every genetic marker from the training sample: (i) the magnitude of the association with the disorder; and (ii) the identity of the variant that increases risk. Then, sum the effects of every one of the variants you are considering, where the effect is a product of the direction of effect and the effect size of that variant. Finally, ask how well the score predicts disease in the test sample.

What the authors found surprised many people. Using the ISC data as the training sample, the PRS predicted about 3% of the variance in risk for schizophrenia in the test sample, the European–American sample from the Molecular Genetics of Schizophrenia cohort (PURCELL et al. 2009). The result was highly significant: $P = 2 \times 10^{-28}$ (this is shown as the SCZ column in Figure 7.2). As a control, they used the PRS to predict

Figure 7.2 Polygenic risk scores derived from schizophrenia (PURCELL et al. 2009). This is the figure that astonished psychiatrists as it demonstrated the unexpectedly high genetic correlation between bipolar disorder and schizophrenia, and showed that schizophrenia was a highly polygenic disorder. The variance explained in the target samples on the basis of scores derived in the entire sample for five significance P value thresholds (P_T) of <0.1, <0.2, <0.3, <0.4, and <0.5 are plotted left to right in each study. The y-axis indicates Nagelkerke's pseudo R^2, a measure of the variance explained by the polygenic risk score. SCZ, schizophrenia; BPD, bipolar disorder; CAD, coronary artery disease; CD, Crohn's disease; HT, hypertension; RA, rheumatoid arthritis; T1D, type 1 diabetes; T2D, type 2 diabetes.

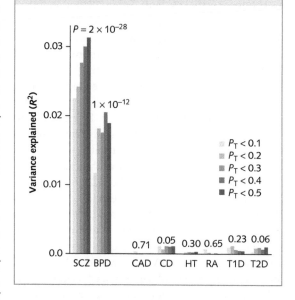

risk in six other disorders from the WTCCC: none of the results was significant (Figure 7.2). So, to the question of whether the variants from their GWAS for schizophrenia that were not genome-wide significant were still an index risk for schizophrenia, they got a resounding answer: yes.

However, they learned more than that. For any particular PRS, you set a P value threshold for the genetic markers to include. The predictive power of the PRS kept getting larger as the criteria for selecting genetic markers was loosened. The maximal predictive power was seen from including any genetic marker with a value of $P < 0.5$; that is, the best score was obtained when using half of the total independent genetic markers in the genome (Figure 7.2). This was a jaw-dropper, as it suggested that a substantial proportion of all the common genetic markers in the human genome contained some small aggregate risk for schizophrenia.

The PRS analysis produced two more important results. First, when the ISC group used an African-American subsample, the predictive power of the PRS was much smaller: the percentage variation explained was less than 0.5%, compared with 3% in the European comparison, suggesting that results could differ substantially across major ethnicities. Finally, they tested their PRS on a case–control sample of bipolar cases (Figure 7.2). Remarkably, the PRS explained 2% of the variance in the WTCCC cases with bipolar disorder (PURCELL et al. 2009).

Polygenic risk scoring has become a standard tool in behavior genetics. Indeed, to some extent, one of the most important outcomes of a GWAS is that it allows the construction of a robust PRS. Significant effects attributable to the use of PRSs can be obtained in relatively small sample sizes (less than 100 is often sufficient), making this a tool useful for individual research groups, not just international consortia. PRSs can be used to answer important questions not just in behavioral genetics, but in the social sciences more generally—we give an example in our discussion of the PRSs for educational attainment, discussed in Chapter 12, section 12.1.7. The ability to index genetic predisposition gives social science researchers an extraordinary tool whose application is only beginning to be explored. It also raises ethical issues, which we return to in Chapter 20, section 20.9.

The routine use of PRSs might suggest that it is a robust tool, but some caution is required. Accurate and unbiased estimation of the effects of all single-nucleotide polymorphisms (SNPs) are difficult to obtain in a structured population when environmental differences cannot be controlled, as is the case with data collected in large consortia. Methods that deal with population structure in GWAS can avoid false-positive genetic associations, but these methods are not so well adapted for dealing with PRSs, which may lead to bias in results from the scoring (BERG et al. 2019; SOHAIL et al. 2019). A useful summary of the problem is given in a short review by BARTON et al. (2019).

7.3 The Psychiatric Genomics Consortium

By late 2010, a total of 17 studies had been collected, making up a total of 9,394 schizophrenia cases, to which were matched 12,462 controls. The results, published in September of 2011 (RIPKE et al. 2011), increased the evidence for the chromosome 6 locus, and in total identified seven loci with genome-wide significance. This paper convinced many groups working in schizophrenia genetics, who had up until then been sitting on the sidelines, to join the collaboration, which by then had become had become known as the Psychiatric Genomics Consortium (PGC).

The PGC grew out of a National Institutes of Health initiative (Genetic Association Information Network or GAIN) initiated in 2006. GAIN was a public–private partnership established to investigate the genetic basis of common diseases through a series of collaborative GWAS. GAIN proposed an approach that included data deposition and distribution, and collaborative analysis with the hope of fostering a commitment to shared scientific knowledge that would facilitate rapid advances in understanding the genetics of complex diseases. Four of the GAIN projects involved psychiatric disorders: major depression, attention deficit hyperactivity disorder, schizophrenia, and bipolar disorder.

The PGC emerged in 2007 from discussions of the leaders of the psychiatric projects within GAIN. The group, whose first publications appeared in 2009, organized themselves into six working groups. Five of these were focused on individual disorders, adding autism to the four psychiatric disorders funded by GAIN. The sixth or Cross-Diagnostic Group was to examine patterns of genetic risk across the five disorders, trying to determine which factors were diagnosis specific rather than shared across multiple conditions. One of us (K.S.K.) has been deeply involved with the PGC, co-directing the Cross-Diagnostic Group and serving as a working member of the schizophrenia group from its inception.

The PGC established three core principles. First, they were to practice 'open, inclusive, participatory, and democratic science'. Second, they sought to use a pre-specified '*rigorous methodology [and a] commitment to producing robust, replicable, and secure findings*'. Third, they were going to use a '*mega-analysis framework*', which would require that '*PGC members share raw genotype data so that all samples can be processed using a uniform quality control, imputation, and analysis pipeline*'.

What this meant in practice was that raw GWAS data had to be submitted to a single secure server and analyzed using a common methodology. This permitted a much more rigorous approach to the many subtle issues in GWAS analyses than would have been feasible if only summary data were supplied.

The PGC has had a major positive impact on the field of the molecular genetics of psychiatric illness. It changed the sociology of this field, moving research groups away from an individualistic focus common in science to a more mixed model, where collaborations with PGC represent a major part of many groups' research agenda.

The PGC has not, however, been without its limitations. One of the most critical problems is that the collaboration was established after its component studies were established. Thus, the ascertainment and clinical assessments of cases and controls vary widely. We cannot be certain that the assessments were done in similar ways or that the diagnostic criteria were always applied in the same manner.

Many of these limitations were unavoidable, given the design of the PGC. Collecting the large sample sizes that have been so critical for the breakthroughs we document would not otherwise have been feasible. But the field is now confronting the question of whether, building on the basis of the PGC, it is time to prospectively plan large enough studies that would include very careful (or 'deep') phenotyping and assessment of environmental risk factors. The PGC has worked relatively well at the detection of molecular risk variants, but the wide variability in sample design makes it less ideal for understanding how those variants relate to environmental exposures and to subtler and more carefully assessed phenotypic variables moving forward.

7.4 GWAS for schizophrenia identifies hundreds of risk loci

The PGC had found some loci explaining a small fraction of the heritability of schizophrenia. More samples were needed to find more genome-wide significant loci, with the expectation doing so would deepen our understanding of the biology of schizophrenia. As the consortium grew and the data analysis became more complex, we decided to introduce data freezes, that is, we wouldn't peek at the data because it was so much work to standardize all the samples. So we would wait, sometimes for more than a year, as the sample size increased.

One of the tasks of the PGC schizophrenia group was to develop a quick way to do a 'quality check' on the results. We didn't expect the molecular genetic findings to agree exactly across groups. Diagnostic approaches might differ and, perhaps more importantly, there were sometimes systematic differences in the ascertainment of cases that groups collected: some centers sampled older and more chronic patients, while others studied more acute patients, often in the early stages of their illness.

During one of the many PGC conference calls, I (K.S.K.) clearly recall looking at a large spreadsheet of molecular findings from a number of research groups. Each row of the spreadsheet showed information for a variant with low P values. There were about 30 variants reported. The columns were different samples. Each entry gave the odds ratio for a given variant in a given sample in schizophrenia cases versus controls. Given the complexities in the way samples were collected, and differences in ethnic composition, one thing we wanted to know was how consistent the association findings were for these variants. Our conference-call discussion was ranging over a variety of rather complex correlational statistics, when my eye was caught by an obvious pattern in the data. As I scanned down adjacent columns, I noticed that the direction of effect of a variant appeared to be the same in each group more often than would be expected by chance. The direction of effect means whether the odds ratio was greater than 1 (the allele at the variant is associated with an increase in risk for schizophrenia) or less than 1 (the allele is associated with a decrease in risk for schizophrenia). I stopped listening to the discussion and went to work out the P value for one of the oldest and most simple of statistical tests: the sign test. Think of a sign test as a way to tell if you have a true coin. Flip the coin 50 times. If you get 28 heads, the P value of the sign test is 0.48. You are likely to have a true coin. But if you flip 35 heads …

The first pair of columns agreed for 24 of the 30 variants. I thought that was likely to be very significant. The P value was 6.3×10^{-5}. I went back to the call and at the next pause said, '*Guys, I think we are looking in the wrong place here. This may be a simpler problem than we are making out. The sign test could be a good method.*' I explained my observations. Everyone was intrigued, and the sign test was adopted as a way of getting a quick and easy measure of the agreement at individual markers across samples. For the schizophrenia PGC, sign tests typically show between 70 and 80% agreement across moderately significant sets of markers from different centers in Europe and North America.

We decided to make the next freeze on the accumulating data 2 years later in 2013. By this point, the sample size had grown dramatically with the final numbers equaling 34,241 cases and 45,604 controls from 49 different case–control samples, including three from East Asia, the first time the PGC group had included non-European samples. I remember the call when we saw the first draft of these results. There was a lot of excitement. While we had expected some increase in

the number of genome-wide significant findings, this was more than many of us had expected: a total of 108 independent loci (SCHIZOPHRENIA WORKING GROUP OF THE PSYCHIATRIC GENOMICS CONSORTIUM 2014). When Stephan Ripke first presented these results publicly at a key meeting of the World Congress of Psychiatric Genetics, the audience broke into applause at the sight of the figure showing the results (Figure 7.3). It was a big moment for the field: the sense that we had finally arrived.

It's worth pausing a moment to consider the Manhattan plot. As we explained in Chapter 6, section 6.5.1, each dot represents the result of an association test from a single variant. The y-axis is the P value of the association statistic, expressed as the negative logarithm (base 10), so the higher the value, the less likely the null hypothesis of no association. What is most striking is the peak, the Manhattan skyscraper, on chromosome 6. This is the locus that was first

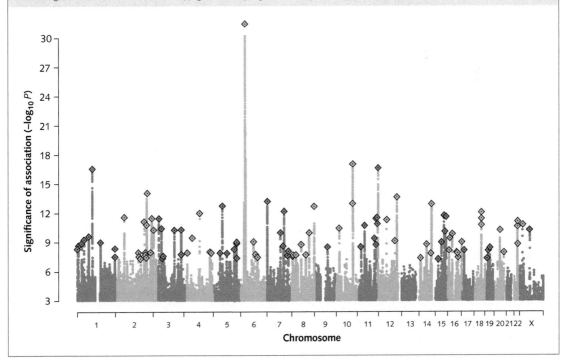

Figure 7.3 The Manhattan plot for a schizophrenia genome-wide association study from 2014. The *x*-axis represents the whole human genome laid end to end from chromosome 1 on the far left to chromosome 22 on the far right, plus the X chromosome. The *y*-axis represents statistical significance. Each dot represents the result from a single association test. Markers at each peak are shown as a diamond.

Reproduced with permission from Schizophrenia Working Group of the Psychiatric Genomics Consortium. Biological insights from 108 schizophrenia-associated genetic loci. *Nature*, 511: 421–427. Copyright © 2014, Springer Nature. https://doi.org/10.1038/nature13595.

found, in 2009, and has been found in every study since of European samples. The association statistic is twice as significant, on the logarithmic scale, than any other. Does this mean that the effect attributable to the HLA region is the most important? We should be careful here—the locus is **not** found in GWAS of East Asians (LAM et al. 2019). Steve McCarroll, a geneticist in Harvard, has determined that the chromosome 6 locus contains more than one genetic effect, and that one of the sources of the signal lies in a small rearrangement (not a SNP) that affects a gene called *C4* (SEKAR et al. 2016). The absence of this locus in East Asians indicates that, despite the prominence of the locus in Europeans, the genetic basis of schizophrenia may not be entirely conserved among different populations. Note that this is consistent with a Fisherian, highly polygenic model: of the many potential risk loci, only some fraction may be needed to cause psychosis. Which fraction will depend on the frequency of the alleles, which will vary from population to population.

The success in 2014 has not stopped the collection and genotyping of more samples. By the fall of 2018, analysis of the third freeze of the schizophrenia PGC was well underway. With data from over 65,000 cases—nearly double the number of cases compared with the previous freeze—the number of genome-wide significant loci increased to well over 230 (PARDINAS et al. 2018). And these loci still do not account for all the heritability of schizophrenia. It seems there may be hundreds more still to discover.

What about the genes involved? What are they? What do they do to cause schizophrenia? Just to reiterate, **a GWAS does not find genes**. Of the methods discussed in Chapter 6, section 6.9 to turn GWAS hits into genes, two have been applied to identify genes involved in schizophrenia. In the first, Dan Geschwind applied chromosome conformation capture analysis of fetal brain tissue to relate regulatory elements to gene promoters (WON et al. 2016). He looked for SNPs associated with schizophrenia that were positioned within regulatory regions of developing cortex, and used the chromosome conformation capture data to find genes physically coupled to these regulatory regions. This identified approximately 500 genes that were neither adjacent to the associated SNPs nor in linkage disequilibrium with them. In the second approach, Bogdan Pasaniuc applied transcriptome-wide association analysis (described in Chapter 6, section 6.10) to identify 157 unique genes

(GUSEV et al. 2018). Forty-six were attributed to variation in splicing and 42 were significantly associated with nearby chromatin phenotypes. Of the 108 loci from the PGC GWAS, 47 were located near at least one of these genes (SCHIZOPHRENIA WORKING GROUP OF THE PSYCHIATRIC GENOMICS CONSORTIUM 2014). Finally, Pasaniuc's data showed that both transcriptome and chromosome conformation data supported the candidacy of genes at 59 loci. This set of genes stands as our current best guide to the biology of schizophrenia.

Neither approach definitively proves that a gene is causally related to schizophrenia. The transcriptome method, in common with other tests of causal relationships (see our discussion of Mendelian randomization in Chapter 13, section 13.5), can't distinguish between pleiotropy (in which a SNP or linked SNPs influence schizophrenia and expression independently) and truly causal susceptibility genes. The two signals are statistically indistinguishable. But let's not carp! Here's a better set of genes than anything else has generated! What do these genes tell us about the origins of the disease? Do they point to any mechanism? Do they support any of the prior ideas about schizophrenia's biological basis (there are quite a few)? Well, to be honest, the gene set does not tell us a lot we did not already know. From the expression pattern, we can say that schizophrenia is a disorder of the brain (no news there), and as almost a third of the genes were present only in the developing brain, we can say that schizophrenia is a disorder of the developing brain, consistent with other evidence that it is a neurodevelopmental condition. Intriguingly, Geschwind's analysis indicated that one SNP lay in a regulatory region that physically interacted with the promoter of a dopamine receptor, DRD2, the target of antipsychotic drugs (WON et al. 2016). So the dopamine theory of the origin of schizophrenia (see CHAPTER 5, SECTION 5.2.1) gets some genetic support at last. Other than that, there are genes involved in synapse function, in gene regulation, and in developmental processes. No one has yet used the gene information to propose a theory of the biological origins of schizophrenia.

Overall, then, there is mounting evidence for the identification of dozens of genes that are involved in schizophrenia, although definitive proof requires experimental verification. We turn next to consider to what extent the variants found in GWAS explain the heritability of schizophrenia. How much more is there to discover, how many more loci are there, and what do we now know about the genetic architecture of the disease?

7.5 SNP-based heritability

The discovery of 108 genome-wide significant loci settled discussions about the genetic architecture of schizophrenia (in brief, it's polygenic), but the 108 loci explain only 3% of the heritability of schizophrenia. And increasing the number of loci to 240 doesn't do much better: they account for less than 10% of the heritability (PARDINAS et al. 2018). Where's the remaining 90%? This is the problem of 'missing heritability' that has turned up with every other GWAS for every other trait mapped. For example, the heritability of human height is about 80% (SILVENTOINEN et al. 2003), of which the 40 genome-wide significant loci that had been found by 2008 only explained about 5% (VISSCHER 2008), and the 180 found by 2010 explained 10% (LANGO ALLEN et al. 2010). Answers to the missing heritability problem arise from developments that weld quantitative and molecular genetic techniques together in a novel way. This section describes the background and theoretical justification for assessing heritability from genotypes, or, more simply, SNP heritability.

We will present two methods to estimate SNP heritability, one that Peter Visscher introduced that uses genotypes (VISSCHER et al. 2010; YANG et al. 2010), and another, linkage disequilibrium score regression from Ben Neale and colleagues (BULIK-SULLIVAN et al. 2015), that uses summary statistics from GWAS data. These methods have had a profound impact on behavioral genetics, and we recommend you make the effort to understand how they work, as so doing is really the only way to also understand what they reveal. However, if you only want to know the answer to the missing heritability problem in schizophrenia (and other complex diseases), then skip the next two sections.

7.5.1 GREML: the first method to estimate heritability in unrelated individuals

Peter Visscher's strategy to estimate heritability from genotypes has been called 'the first new quantitative genetic method in a century' (PLOMIN AND DEARY 2015)

and has become routine in the genetic analysis of all complex traits. How does his method work? The approach derives from agricultural genetics. Yes, I understand a discussion of dairy cows might seem irrelevant, but let me explain it this way. Before dairy farmers spend a lot of money on a bull (to be precise, on the bull's sperm), they want to be sure that the cows they get will produce lots of milk. Milk production is a heritable trait, and one way to estimate the milk production is to calculate the breeding value of the sire (the breeding value is defined as twice the deviation from the population average of the mean milk yield of his progeny, but think of it as an estimate of how much milk his daughters will produce). Charles Henderson showed how it was possible to predict phenotypes from genetic information (then derived solely from pedigrees), but his publications in such magazines as the *Journal of Dairy Science* are rarely recognized as the starting point for the major advance in psychiatry that they were to become. Here, in Henderson's own words, is the way to make predictions:

The following method for maximum likelihood estimation of fixed elements of mixed linear models has been derived by one of us (H.).

Let the mixed linear model be

$$y = X\beta + Zu + e$$

where β is a vector of fixed effects, while u and e are independent vectors of variables that are normally distributed with zero means and variance-covariance matrices $D\sigma^2$ and σ^2 respectively.

HENDERSON et al. (1959).

Henderson introduces two important terms in this introduction. The first is 'mixed linear models'. This turns up a lot in current genetic analyses, and most people who use it will have forgotten, if they ever knew, that Henderson introduced mixed models into genetics. 'Mixed models' means that there are both fixed effects and random effects in a linear model (see our description of linear models in the Appendix, sections A10 and A11). Fixed effects are unknown constants, while random effects have values that are drawn from some underlying distribution. In the equation above, X and Z are two matrices associated with the fixed and random effects, respectively.

The second term to note, one that Henderson introduced, is BLUP, which means best linear unbiased predictor. In this case, BLUP is the predictor of the vector of random effects (Z in the above equation). If you find this terminology confusing (Why best? Why linear? What is an unbiased predictor?), you are not alone (see, for example, https://andrewgelman.com/2015/06/10/best-linear-unbiased-prediction-is-exactly-like-the-holy-roman-empire/). The important thing here is the following: Henderson was able to relate the prediction (the BLUP) to the heritability. Technically, $BLUP(u) = H(y - X\beta)$ where $H = GV^{-1}$. The last term is a sort of heritability matrix, a representation of Fisher's infinitesimal model (the method is analogous to regression to the mean, described in the Appendix, section A4). In this famous equation, u can be a vector of genetic (additive) effects, which at Henderson's time was estimated from pedigrees, but we can now obtain the same information from genotypes. In other words, heritability can be directly estimated from the genotypes. This is precisely what Peter Visscher did (VISSCHER et al. 2010; YANG et al. 2010) (the GCTA software that implements his method is available here: http://cnsgenomics.com/software/gcta/#Overview). His method is called genomic relatedness matrix restricted maximum likelihood, or GREML for short.

You might ask, how can this possibly work. How can genotype data from unrelated individuals ever provide information about heritability? It sounds like a contradiction; unrelated surely means individuals without genetic relatedness, while heritability is a measure of a common genetic effect (due to relatedness). The answer is that, at some level, everybody is related to everybody else; there are genetic variants we share because of common descent from some distant ancestor (we came across this idea in our description of the 'Out of Africa' hypothesis in Chapter 6, section 6.2). It follows that a cohort of people who share a disease that is heritable, say schizophrenia, will share more of their genome than a cohort of people without the disease. By detecting and estimating the extent of that shared component, we can calculate the extent to which the disease is heritable. The amount of genetic relationship between unrelated individuals is very small, but by examining a large number of individuals (usually at least 3,000 are needed for this method to work), at hundreds of thousands of markers, there is enough information to obtain a reasonable estimate (tens of millions of results are generated from calculating the pairwise relationships for thousands of people). Finally, in a quote from Peter Visscher, here is a shorthand way to understand his method: '*Another*

way to think about what GREML estimates, is to do the following thought experiment: if we had a GWAS with a sample size of 10 million, then we could perform a multiple regression analysis where we fit all, say, 500,000 genetic markers together in the model. The regression R-squared from that model is the same as the SNP-heritability.'

Note that in our explanation, and in Henderson's equation, there has been no mention of 'significance', no mention of the number of variants needed. The heritability calculations use all of the genetic markers, regardless of their significance. This is a radically different way of approaching the genetics of schizophrenia, and its importance took some time to be fully recognized.

Heritability calculated from the genotypes is referred to as SNP heritability and makes different assumptions from family-based methods. Some of these assumptions have led to discussion (at times heated) about the best method for estimating SNP heritability (see YANG et al. 2017).

One further important application of GREML is in assessing genetic correlations (LEE et al. 2012). The method is dependent on the same assumptions about SNP heritability and has revealed an unexpected amount of genetic correlation between psychiatric disorders (CROSS-DISORDER GROUP OF THE PSYCHIATRIC GENOMICS et al. 2013), a point we elaborate in Chapter 13, section 3.4. Also you should remember the acronym GRM for genetic relationship matrix. This acronym turns up a lot in the literature, and later too in this book.

7.5.2 Linkage disequilibrium score regression

The second method to estimate heritability from genetic markers is linkage disequilibrium score regression (usually shortened to LD score regression or LDSC) (BULIK-SULLIVAN et al. 2015). The idea is very neat and can be explained in the following way. We explained in Chapter 6, section 6.6.5 that a standard test of the quality of a GWAS uses a quantile–quantile (QQ) plot to compare the distribution of the observed test statistic with the expected values (the expected values are calculated assuming there is no genetic effect). The numbers plotted in QQ plots are often the negative logarithm of the P value of the statistic, or the statistic itself, such as χ^2, from testing the association. When the observed statistic deviates from the expected over the entire distribution of values, from very small to very big, that's an indication that something is wrong, that perhaps population structure was not adequately dealt with. Almost all large GWAS involve consortia, with samples collected under different conditions and from different parts of the world, so the QQ plot is one way to find out whether the differences between the collecting centers are, artifactually, contributing to the genetic signal. Inflation due to population structure and cryptic relatedness are always concerns in GWAS. Ben Neale and colleagues at Harvard reasoned that it might also mean that the trait is highly polygenic—that there are so many variants contributing to variation in the trait that their effects are seen even at loci across the entire distribution of P values. How could the two explanations for the QQ plot inflation be distinguished?

Neale's solution was as follows. At a locus that influences the phenotype (a risk locus for a disease) the amount of inflation (deviation from the expected, as shown for example in QQ plots; see Chapter 6, section 6.6.5) depends on the extent to which a marker is in linkage disequilibrium with the causal variant. That is not true for inflation due to population structure and cryptic relatedness. You can see this from the following argument.

Genetic variants lie in linkage disequilibrium blocks (see Chapter 6, section 6.4), regions of the genome that recombine rarely if at all (recombination occurs at specific positions in the genome, rather than occurring randomly). So the more variants you genotype in a block of linkage disequilibrium, the more likely you are to have genotyped the causal variant. Suppose, as is the case for a large GWAS, your samples come from various populations, among which allele frequencies of variants differ. For a variant that has no effect on the disease, its linkage disequilibrium with other variants is uncorrelated to the magnitude of allele frequency differences between the populations (they all just vary randomly) but it is correlated when the genetic marker is in linkage disequilibrium with a causal variant.

Neale and colleagues estimate a linkage disequilibrium score (LD score) for each genetic marker, which measures the extent to which a genetic marker is in linkage disequilibrium with its neighbors (where 0 means no correlation). This can be calculated from public databases of genetic markers (provided by the Thousand Genomes project; GENOMES PROJECT CONSORTIUM et al. 2015) The LD scores for each genetic marker are then compared with the association statistics of that genetic marker in a GWAS, by regression. Figure 7.4 plots the association statistics (χ^2 from the association analysis) against the LD scores for genetic markers in a schizophrenia GWAS (SCHIZOPHRENIA WORKING GROUP OF THE PSYCHIATRIC GENOMICS CONSORTIUM 2014).

In Figure 7.4, the intercept (where the black line crosses the y-axis) of the LD score regression should be at 1.0 (for the χ^2 statistic). Any departure is due to some artifact, such as confounding, poor quality control, or overlapping samples in the meta-analysis. Here, the value is 1.02, which is close enough that we can conclude that there are no significant confounds. The slope of the LD score regression (the slope of the black line in the figure) provides an estimate of heritability (BULIK-SULLIVAN et al. 2015) (the estimate may be biased due to some effects of selection that we won't discuss here, but see GAZAL et al. 2017).

Neale had thus found a way to distinguish true genetic association results from results that are due to spurious correlation, **and** he had a method to estimate heritability from the summary statistics, rather than the genotypes, of a GWAS (BULIK-SULLIVAN et al. 2015).

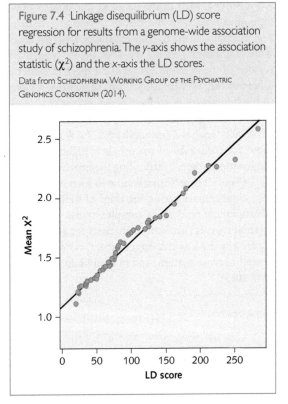

Figure 7.4 Linkage disequilibrium (LD) score regression for results from a genome-wide association study of schizophrenia. The y-axis shows the association statistic (χ^2) and the x-axis the LD scores.

Data from SCHIZOPHRENIA WORKING GROUP OF THE PSYCHIATRIC GENOMICS CONSORTIUM (2014).

The latter is important because summary statistics are easily obtainable (unlike genotypes, they don't contain any personal information so can be released without prejudicing the privacy of subjects in a GWAS), and there are far fewer of them than genotypes, making LD score regression a very fast way to calculate heritability (for a study of 10,000 subjects with 500,000 markers, there will be 500,000 summary statistics, but 5,000,000,000 genotypes, all of which are needed to compute the SNP heritability using GREML).

7.5.3 Genetic correlations and regional heritability

There are two important extensions to the SNP-heritability methods. The first extension is to estimate genetic correlations (which, as noted earlier, is also possible with GREML). This method has been key to determining the genetic relationships between psychiatric disorders (see Chapter 13, section 13.4).

The second extension is estimating heritability attributable to regions of the genome, rather than the entire genome (FINUCANE et al. 2015). This is particularly important because it becomes possible to answer questions about the relative contribution of different functional parts of the genome, and different cell types and tissues, to the risk of developing disease. For example, from the observations that genome-wide significant genetic markers lie in regulatory regions of the genome, we would expect to detect a relatively large contribution from regulatory regions to the heritability of schizophrenia. Is that true of all genetic markers that contribute to disease? Regional heritability analysis (FINUCANE et al. 2015) shows that sequence conserved in mammals is highly enriched for genetic markers contributing to heritability. Overall, 2.6% of all genetic markers lie in such regions and they contribute to about one-third of heritability for schizophrenia (and other complex traits). The same method reveals that genetic markers for schizophrenia are enriched, as you might expect, in tissues in the central nervous system, and in fetal brain (FINUCANE et al. 2015).

7.5.4 PRSs and SNP heritabilities are not the same

Finally, it's worth emphasizing the difference between PRSs and SNP heritability. The former is used for making predictions, for example between two diseases or between two populations with the same disease, to ask how much of the disease risk can be predicted from known risk factors. Typically, this is low, in the order of a few percent, but it improves as we obtain more complete catalogs of genetic markers associated with a trait, and more accurate assessments of their effect size. PRSs can be used to make a prediction in an individual. Genetic relationship matrices can't be used to make predictions in individuals; instead, they are used to assess the contribution of heritability (or genetic correlation) of a disease or a quantitative trait within a group (e.g. cases versus controls).

7.6 Schizophrenia and the problem of missing heritability

We can turn now to the issue that in part prompted the development of SNP heritability, namely 'missing heritability'. Where could that missing heritability possibly be hidden? Is it in rare variants invisible to the genotyping arrays? Is it in interactions between loci (epistasis)? Is there an epigenetic contribution? (Epigenetics is always a good guess when you have no idea what else might be to blame; see Chapter 4, section 4.6). Are there errors in the family- and twin-based heritability estimates (MANOLIO et al. 2009)?

What are the SNP-heritability estimates for schizophrenia? The SNP-heritability estimates are always lower than the twin- and family-based estimates, sometimes substantially so (perhaps half as much). Estimates for schizophrenia are about 25–30% (BULIK-SULLIVAN et al. 2015; LOH et al. 2015; BRAINSTORM CONSORTIUM et al. 2018), much less than the twin-based estimate of about 70% we discussed in Chapter 2, section 2.6.5. That looks bad for the missing heritability question. Why the discrepancy?

SNP heritability quantifies the contribution of the genotypes they are given, from markers on the genotyping arrays. These are, of course, the common variants, so SNP heritability is the phenotypic variance captured by common SNPs. Family- and twin-based methods capture all contributions to heritability, regardless of the allele frequency. Assuming these estimates of heritability of schizophrenia are correct, the SNP-based assessments tell us that less than half of the heritability is explained by common variants.

Where does the 'remaining missing heritability' hide? As we'll see (to save you reading the next couple of sections), we don't have any evidence that rare mutations of large effect make a contribution (as was once hoped; McCLELLAN AND KING 2010). The most parsimonious explanation is that the missing heritability is hidden in lower frequency variants, of relatively small effect. When LD score regression was first applied to schizophrenia data, the authors concluded that 'the mean χ^2 statistic of 1.613 results mostly from polygenicity' (BULIK-SULLIVAN et al. 2015). Schizophrenia, along with most other psychiatric diseases and quantitative traits, is highly polygenic, but we can now add an important addendum to this conclusion: the genetic architecture consists of risk loci with allele frequencies across the entire spectrum, from very common to very rare. This is an important addendum, as you may easily come to the conclusion from the literature that there are contributions from 'common' and 'rare' variants, as if there was a dichotomy. Nothing could be further from the truth. Genetic variants range across all allele frequencies, and all of them contribute to the risk of developing schizophrenia (VISSCHER et al. 2012). The remaining question, still of considerable interest, is to define precisely the frequency distribution of allelic variants that cause psychiatric disease, and to understand what forces contribute to this distribution.

7.7 Copy-number variants

Maria Karayiorgou is a US-trained psychiatrist of Greek origin now working in New York. She has spent her career trying to identify the genes involved in schizophrenia. In particular, Karayiorgou explored one of those rare but potentially informative clinical observations that are often overlooked. Clinical geneticists had noted that deletions on chromosome 22 were not only associated with cleft palate, heart anomalies, and unusual facial features but also with psychiatric problems, including psychosis. There were also hints from the linkage-mapping approaches that there was a susceptibility locus on chromosome 22 (PULVER et al. 1994). Karayiorgou set out to characterize patients with schizophrenia for the presence of the 22q deletion and found that it was present in 2% (she found no deletions in the control group) (KARAY-IORGOU et al. 1995). This observation, now one of the most robustly replicated findings in all genetic analysis of schizophrenia, laid the basis for the idea that structural variation in the human genome can cause psychosis.

The discovery of the 22q11.21–q11.23 deletion as a cause of schizophrenia, made before the success of GWAS and before genome sequencing, is a key advance in the biological understanding of psychosis. It was the first time that an unambiguous association was established between a biological marker, a small chromosomal deletion, and schizophrenia.

The chromosome 22q deletion is about 2 Mb in size and contains between 30 and 40 genes, many of which have still not been characterized functionally. We now know that the reason why the deletion is relatively common is because it is flanked by large repetitive regions that occasionally confuse the recombinational machinery, so that when two copies of the chromosome pair up, they misalign (the high degree of similarity between the repetitive regions results in the occasional mismatch, as region A mistakenly pairs with region B). This type of variation in the genome is an example of structural variation and is often called a copy-number variation (see Table 3.3 in Chapter 3).

The discovery that a structural variant on chromosome 22q contributed to the risk of schizophrenia suggested that copy-number variants (CNVs) in other parts of the genome might also be important causal factors, but in the 1990s there was no way to test this hypothesis; we did not have a complete human genome sequence and we did not have the tools to interrogate structural variation. By the early 2000s, the developments in genomic technology made genome-wide screening possible.

In 2008, a year before the first successful GWAS finding in schizophrenia, a group led by Jonathan Sebat tested for the presence of structural variation in 150 individuals with schizophrenia and 268 controls (WALSH et al. 2008). They found deletions and duplications that contained genes in 5% of controls versus 15% of cases, a highly significant difference. The researchers showed an enrichment of CNVs in schizophrenia but were unable to identify a significant enrichment of any single structural variant.

We will not give you a blow-by-blow description of the subsequent developments in the CNV literature but rather will turn to the latest report, which came from the PGC CNV group led by Sebat and published in 2017. It was by far the largest sample yet, including 21,094 cases and 20,227 controls (MARSHALL et al. 2017).

This report had four important findings. First, in a replication of the 2008 report on a far larger sample, a striking excess of CNVs was found in the schizophrenia cases versus controls. The effect size of CNVs is harder to express than the simple odds ratio for a single genetic marker. The unit used is per kilobase of sequence contained in CNVs that were either insertions or deletions. In this study, for every extra kilobase of CNV, the risk for schizophrenia increased 11% (i.e. the OR was 1.11).

Second, the effects of deletions on schizophrenia risk were much stronger than duplications. With respect to an individual's risk for schizophrenia, it is more harmful to lose parts of your genome than to have multiple copies of the same genomic region.

Third, eight individual CNVs were identified at a genome-wide significant threshold that increased the risk of developing schizophrenia, at 1q21.1, 2p16.3, 3q29, 7q11.2, 15q13.3, distal 16p11.2, proximal 16p11.2, and 22q11.2. The researchers were able to calculate ORs for the individual CNVs in the same way as was done for genetic variants. Focusing on those that were frequent enough to yield reasonably accurate estimates, the range was from 3.8 for 1q21.1 to 67.7 for 22q11.2.

Fourth, all the CNVs were rare. For example, both the 22q11.1 and 1q21.1 CNVs were seen in less than 0.3% of cases. The vast majority of cases of schizophrenia do not carry any pathogenic CNVs.

Which clinical features are seen in individuals with schizophrenia with a CNV compared with those cases who do not have one? Examining the traditional key symptoms of schizophrenia (e.g. hallucinations, delusions, thought disorder), no appreciable differences were found. However, much higher rates of both learning disabilities and seizures were seen in the carriers of CNVs compared with noncarriers. Some authors have suggested that many individuals with CNVs have multiple neurological problems of which schizophrenia is only one. They question whether the nature of schizophrenia in those with CNVs is the same as that seen in those without these structural variants.

Several attempts have been made to estimate the total impact of common SNPs on risk for schizophrenia in the population versus CNVs. All the results suggest that, at a population level, common genetic variants as a class have a much stronger effect on schizophrenia risk than all the rare CNVs considered as a group. The early hope that CNVs would reflect the 'royal road' to understanding molecular genetic effects on schizophrenia has been disappointing.

We have recently gained important insights into how PRSs for schizophrenia (i.e. the aggregate effects of common genetic variants) and CNVs together contribute to risk for schizophrenia. The way this question was tackled was to compare PRSs for schizophrenia in individuals with schizophrenia who did versus those who did not also have a pathogenic CNV. If these two kinds of risk factors roughly added together to produce a total risk for schizophrenia, then we would expect that carriers of CNVs affected with schizophrenia would have lower PRSs than noncarriers. That is exactly what was found. This finding was taken a step further by looking at PRSs in carriers of specific CNVs. These results showed that the average PRS was lowest in those who were carriers of CNVs with the largest effect (the 22q11.1 deletion), intermediate in cases who were carriers of more modest-effect CNVs and strongest in those who carried no CNVs. These results suggest that, to a first approximation, the risk for schizophrenia can be understood as the sum of that derived from common polygenes and rare CNVs (BERGEN et al. 2019).

7.8 Exomes and rare variants as a cause of schizophrenia

The discovery that CNVs contributed to disease risk in schizophrenia demonstrated that there were other sources of sequence variation important for understanding the genetic underpinnings of common disease. Moreover, although rare, each CNV has a relatively large effect size. Figure 7.5 plots the effect sizes (vertical scale) of SNPs and of CNVs against their frequency in the population (horizontal scale). You can see that CNVs are rare, and have large effects, while SNPs are common and have small effects.

The problem of missing heritability and the discovery of large-effect rare CNVs suggested that there might be other rare large-effect genetic variants that so far had escaped detection. Such variants might have a large biological effect if they disrupted gene function, for example by abolishing transcription, or truncating or altering the structure of the protein, all things achieved by a variant lying in the coding region of the genome (referred to as the exome, the summation of the exons of all genes).

Even before the GWAS discoveries, there was evidence that rare coding mutations could contribute to common disease. In 2004, Helen Hobbs and Jonathan

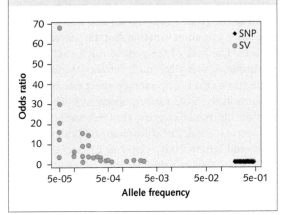

Figure 7.5 The effect sizes of risk variants for schizophrenia plotted against their frequency in the population. SV, structural variant; SNP, single-nucleotide polymorphism associated with schizophrenia.

Data from Marshall et al. (2017) and Schizophrenia Working Group of the Psychiatric Genomics Consortium (2014).

Cohen of the University of Texas Southwestern Medical Center in Dallas sequenced genes known to be involved in the control of cholesterol levels (Cohen et al. 2004). They were able to detect an enrichment of coding variants in people at the extremes of the normal distribution. Richard Lifton at Yale University School of Medicine found the same thing for blood pressure variation, again sequencing candidate genes (Ji et al. 2008). Could the same be true for schizophrenia? The difficulty was that Cohen, Hobbs, and Lifton sequenced genes known to be involved in the disorder, and for schizophrenia there were no such genes. That was why genetics was being used in the first place—to find genes. However, with the development of next-generation sequencing, it became possible to screen the entire exome for such variants.

In 2011, Maria Karayiorgou reported the results of sequencing the exomes (just 1.2% of the genome) of 53 family trios, where a trio is an affected individual and both parents (Xu et al. 2011). The importance of this design is that it makes it possible to determine whether a mutation is inherited or not. A mutation that is found only in the affected individual is a new mutation (a *de novo* mutation), most likely having arisen in the germ line of one of the parents. As the parents of trios that Karayiorgou sequenced were all unaffected, *de novo* mutations found in the offspring (but not the parents) might be causative of the disease. Karayiorgou found that 27/53 cases (51%) carried at

least one *de novo* mutational event, while there were only seven exonic *de novo* variants in 7/22 control subjects. Another group, sequencing a much smaller cohort (just 14 trios), found the same thing, an enrichment of *de novo* mutations in subjects with schizophrenia (Girard et al. 2011).

Two points should be emphasized. First, these findings did not prove the involvement of any specific gene; rather, all that could be said was that there was an overall enrichment of mutations in the exomes. This is the same observation made for CNVs, where the first studies reported an enrichment of CNVs in cases. Second, these findings could not explain the missing heritability of schizophrenia (by definition, *de novo* mutations are not inherited). But the findings electrified the field because they implied that by screening exomes additional mutations would be found to prove the role of specific genes, and that inherited variants might exist. Sadly, it proved to be much harder than anticipated to find genes using exome sequencing, and the contribution of rare SNPs to heritability has turned out, so far, to be modest.

There are two problems here. The first, and perhaps unexpected, is the realization that deleterious mutations occur in the normal population much more frequently than expected. Everyone carries about 100 deleterious mutations, and both copies of about 20 genes are inactivated (MacArthur et al. 2012). So contrary to many people's preconceptions, just finding an inactivating mutation in someone with a disease isn't sufficient to prove that the mutation is the cause of that disease. It could be, and most likely is, just an innocent bystander.

The second problem is that the mutations in the candidate genes turned out to be much rarer than people expected, requiring the screening of tens of thousands of individuals. This is not surprising on evolutionary principles. Variants that have an impact on the function of important genes are often selected against because they are associated with reduced evolutionary fitness. Selection does not permit them to become common in the population.

In retrospect, the rarity of these harmful mutations should have been obvious, the highly polygenic nature of psychiatric disease implied that many genes were involved, as did the fact that sequencing the exomes of just a few people was sufficient to find enrichment (this could only be true if the number of potential causative genes is large). With such a large set of targets, there would be few instances of the same gene suffering multiple hits—thus explaining why so many people need to be screened to accumulate sufficient information to prove that a gene is indeed involved in the disease.

Exome sequencing of 5,079 individuals (2,536 schizophrenia cases and 2,543 controls) confirmed the enrichment of mutations in exomes, but wasn't able to unequivocally identify the involvement any single gene (Purcell et al. 2014b). In 2016, whole-exome sequences of a sample twice as large implicated one gene: three *de novo* mutations and seven loss-of-function variants (but none in controls) were found in *SETD1* (Singh et al. 2016). *SETD1* encodes a methyl-transferase that catalyzes the methylation of lysine residues in a histone protein (for a description of histones and the importance of their methylation, see Chapter 4, section 4.2.5). It remains unclear what a mutation in that gene does to cause schizophrenia. But we have many more genes to identify. We are still at an early stage in the rare-variant story on schizophrenia. Very large samples are likely to be needed, and it may be several years before the overall picture is clear.

The final frontier in schizophrenia molecular genetics is whole-genome sequencing. Modest sample sizes are beginning to emerge. Two major forces limit developments here. The first is the expense of whole-genome sequencing. The second is the substantial challenge of the analysis of whole-genome sequences. While we have a good idea of what exomes do and how to interpret the potential harmfulness of variants, we understand much less about the effects of much of the noncoding genome. Currently, it is still too early to try to summarize meaningfully any results.

We have related, in some detail, the story of the striking success of the GWAS method applied to schizophrenia. This has now become the poster child for psychiatric genomics, and with good reason. But let us take a step back and see what we have learned.

First, we now know much more about the genetic architecture of schizophrenia, that is, how nature distributed the genetic risk. GWAS assesses one part of that space, common variation. But the general picture is clear. The bulk of the genetic risk for schizophrenia is due to many common variants of individually quite small effect. The average effect size of genetic variants in the PGC study is about 1.08. This means that an individual carrying that risk variant has an 8% increased risk for schizophrenia. Recall from our twin- and family-study review in Chapter 2, section 2.6.5 that if you are a monozygotic twin and your co-twin has schizophrenia, your risk for schizophrenia is about 50%, or about 50–80 times the general risk in the population. This translates into a 5,000–8,000% increase in risk. Each individual variant contributes to only a very small part of that total effect.

Second, as the sample sizes for schizophrenia grow ever larger, so do the number of genome-wide significant findings. It has to stop somewhere—but where? Figure 7.6 shows the number of genome-wide significant findings as a function of sample size for four traits. The *y*-axis is the number of genome-wide specific hits and the *x*-axis is the sample size. Focus first on the line that depicts the results for schizophrenia. For every 1,000 new subjects, three to four new genome-wide significant loci have been found. When will adding more cases stop yielding new loci? While a variety of

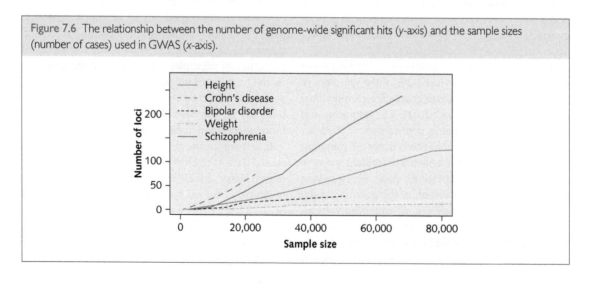

Figure 7.6 The relationship between the number of genome-wide significant hits (*y*-axis) and the sample sizes (number of cases) used in GWAS (*x*-axis).

different approaches have been taken to try to answer this question, we really do not know. It is also interesting to note that the slope of the curve is much steeper for some other complex disorders. Crohn's disease is an inflammatory bowel disorder. On average, the effect sizes of the individual risk variants for this disorder were larger than for schizophrenia, so more hits were found for the same sample size. Also note that the curve for schizophrenia is similar to that seen for human height. Importantly, this curve is only for GWAS. For exome or whole-genome sequencing, the curve is very likely to be flatter because the variants are so much rarer.

Third, one of the promises of molecular genetic studies of schizophrenia was clarification of etiologic pathways and the identification of new drug targets. The PGC GWAS findings have provoked a wide variety of network and expression analyses trying to find biological meaning behind the detected statistical signals. It is still early days as the complexity of these problems is daunting.

While some in the field continue to argue that molecular genetic results for complex conditions such as schizophrenia will be incoherent and will never result in meaningful biological understandings, the authors of this book do not share this opinion. There are plenty of hints in the studies to date and a few particular findings that render us hopeful that key biological insights will emerge from GWAS and sequencing analyses. These insights, we suggest, will both eventually increase substantially our knowledge of the etiology of schizophrenia and, hopefully, provide the opportunity for the development of new biological treatments.

Summary

1. Success in identifying genetic loci that increased the risk for developing schizophrenia required sample sizes in excess of 10,000 cases. To collect a sample of the necessary size involved a change in the sociology of the science, rather than a change in the methods of science. The creation of large collaborative groups, such as the Psychiatric Genomics Consortium, was key to the successful genetic dissection of schizophrenia.

2. Genome-wide association studies (GWAS) of schizophrenia have identified hundreds of risk loci, but even 240 loci only explain about 10% of the heritability of schizophrenia.

3. Polygenic risk scores (PRSs) applied to the genetic analysis of schizophrenia revealed that the disease is polygenic and that it shares genetic risk factors with bipolar disorder. PRSs use the results from a GWAS to assign a genetic effect to each allele, and in so doing can be used to predict the outcome of an individual's genotype.

4. SNP heritability can be estimated from genetic relationship matrices, which use the genotypes obtained in a GWAS, or from linkage disequilibrium score regression, which uses the summary statistics from a GWAS.

5. SNP-heritability methods can also be used to estimate genetic correlations and estimate heritability attributable to regions of the genome, rather than the entire genome.

6. Schizophrenia is highly polygenic, and its genetic architecture consists of risk loci with allele frequencies across the entire spectrum, from very common to very rare.

7. Cases of schizophrenia have more copy-number variants (CNVs) than controls. Individual CNVs are being discovered that are risk alleles, among which a locus on chromosome 22q is among the most common and has the largest effect.

8. Cases of schizophrenia have more de novo deleterious coding variants than controls. Coding variants are rare, and they make only a small contribution to the risk of schizophrenia in the population, but they can make a relatively large contribution in an individual. The identification of individual risk genes from the analysis of rare coding variation has identified the gene SETD1 as being involved in schizophrenia.

9. The bulk of the genetic risk for schizophrenia is due to common variants of quite small effect. The average effect size of genetic variants in the PGC study was about 1.08. This means that an individual carrying this risk variant has an 8% increased risk for schizophrenia.

10. The total number of risk loci for schizophrenia is not known. It is likely to number in the thousands.

8

Autism spectrum disorder

[On the phenomenology and genetic and molecular epidemiology of autism spectrum disorder]

8.1 What is autism?

Leo Kanner's careful description of autism in 1943, in a paper that first brought the disorder to the attention of the medical community, identified in 11 children abnormal social interaction, abnormal speech, and repetitive and stereotyped play (KANNER 1943). For many years, the diagnosis of autism depended on recognizing these three features:

1. Deficits in social reciprocity: social behavior was characterized by a lack of cooperative group play, a lack of empathy and a failure to perceive other people's feelings and responses, and failure to make personal friendships.

2. Deficits in communication: for example, a child might say 'You want apple' when they mean 'I want apple.' Autistic children might repeat stereotyped phrases and repeat back phrases spoken to them (echolalia). Kanner reported a child's response to the question 'If I were to buy 4c of candy and give the store keeper 10c, how much would I get back?' as 'I would draw a hexagon.' As another example, 'Gary T. at 5 years designated a bread basket as "home bakery", and after this, every basket to him became a "home bakery".'

3. Presence of restricted, repetitive behaviors and interests. Autistic children's attachments to objects are sometimes so strong that they need to have the objects with them at all times. As they grow older, they manifest unusual preoccupations (e.g. with bus routes and train timetables), interests that they pursue to the exclusion of almost anything else. They develop obsessional behaviors, touching things repeatedly for example, and become extremely distressed if things around them change: for instance, pieces of furniture must remain in the same place. Kanner wrote that autistic children manifest a 'preservation of sameness'.

However diagnostic systems changed in 2012, and the diagnosis of autism now depends on features recognized in only two domains: (i) social communication/ social interaction; and (ii) restricted, repetitive behaviors and interests (see http://www.DSM5.org). This is because the decision to classify a behavior as social or as communication is arbitrary: facial expressions, eye contact, and hand gestures represent forms of communication and social interaction (GOTHAM et al. 2007). Language development is currently treated as separate, which means a diagnosis of autism can be made with or without a language disorder.

8.2 The heritability of autism, and autism spectrum disorder

Autism, like schizophrenia discussed in Chapter 2.5.2, is part of a spectrum of conditions, and the evidence for the spectrum comes in part from genetic studies. Kanner noted that parents had some features in common with their autistic children: '*Most of the parents declare outright that they are not comfortable in the company of people; they prefer reading, writing, painting, making music, or just "thinking." ... Matrimonial life is a rather cold and formal affair*' (KANNER 1949). Kanner reported a similar lack of emotion in the parents' interactions with their children, which he observed when parents brought their child to his clinic. '*As they come up the stairs, the child trails forlornly behind the mother, who does not bother to look back. The mother accepts the invitation to sit down in the waiting room, while the child sits, stands, or wanders about at a distance. Neither makes a move toward the other. Later, in the office, when the mother is asked under some pretext to take the child on her lap, she usually does so in a dutiful, stilted manner, holding the child upright and using her arms solely for the mechanical purpose of maintaining him in his position.*'

The presence of these autistic-like features in first-degree relatives had two interpretations. One possibility was that shared etiology, probably genetic, gave rise to autism in the children and a similar, related phenotype, perhaps less severe, in the relatives. At that time, there was no evidence for a genetic basis to autism, so Kanner preferred a second interpretation: that the impersonalized mechanization he observed in the interactions between parents and children could be the cause of autism: '*Most of the patients were exposed from the beginning to parental coldness, obsessiveness, and a mechanical type of attention to material needs only....They were kept neatly in refrigerators which did not defrost. Their withdrawal seems to be an act of turning away from such a situation to seek comfort in solitude.*' The concept of the refrigerator mother, popularized by Bruno Bettelheim at the University of Chicago (BETTELHEIM 1967), gained ground at the expense of the evidence

for a biological cause, and at the cost of laying the blame on parents for their inadequate parenting style as the cause of their children's condition (see POLLAK 1998 for how Bettelheim's reputation has since fared).

Michael Rutter, a child psychiatrist in London, carried out the experiment that showed, contrary to expectations, that heredity was an important contributor to the etiology of autism. While a history of autism in parents or grandparents is almost unknown, the rate of autism in siblings is between 2 and 6% (SMALLEY et al. 1988). Autism, as then defined by Rutter, was rare, with a frequency in the population of about 4 per 10,000, so the increased rates in siblings could be an indicator of a genetic origin. But it was equally compatible with the refrigerator parent hypothesis.

One way to decide between the two hypotheses was to carry out a twin study. Rutter identified 11 monozygotic and ten dizygotic twin pairs (FOLSTEIN AND RUTTER 1977). None of the dizygotic pairs was concordant for autism, whereas four monozygotic pairs were, strongly supporting an important role of heredity. Subsequent studies, with few exceptions, have confirmed the view that autism is highly heritable. A meta-analysis of all twin studies carried out up to 2015 gives heritability estimates of between 64 and 91% (TICK et al. 2016). Genetic effects on autism are substantial.

Rutter's paper not only established the substantial genetic contribution to the etiology of autism but also indicated that autism is one extreme of a continuously distributed characteristic, in which the nonautistic co-twins display features similar to those seen in autism. Indeed, six of the unaffected children showed a cognitive abnormality. As is true for schizophrenia (see Chapter 2, section 2.5.2), there exists a spectrum of conditions related to autism, found at higher frequency in relatives, which reflect a shared genetic etiology, rather than a shared environment. Between 12 and 20% of siblings are affected with a 'broad phenotype', the variability in the estimate depending on how the spectrum is defined (BOLTON et al. 1994). Heritability for parent- and teacher-rated autistic traits in middle childhood and older varies between 60 and 90% (RONALD AND HOEKSTRA 2011; TICK et al. 2016). To take into account the diversity of the phenotype, autism is now known as autism spectrum disorder (commonly abbreviated to ASD).

It is important to differentiate three different meanings of 'spectrum' in ASD. First, spectrum refers to differences in the severity and presentation of symptoms among those with a diagnosis of ASD. Second, spectrum refers to the fact that autism is heterogeneous, arising from different roots, and that there exist subgroups. Genetic findings have developed and justified this idea, as we will see later. Perhaps it is better to talk about 'autisms', rather than autism (there are similar issues surrounding the diagnosis of depression, as we discuss in Chapter 10, section 10.1.8). Third, we can recognize, using questionnaires, the presence of autistic traits in populations. For example, everyone can be ranked along a continuum, a spectrum, of social responsiveness. Simon Baron-Cohen, a psychologist at Cambridge University, defines an autism spectrum quotient, determined by questions such as 'I tend to notice details that others do not.' The autism spectrum quotient measures the presence of a broader autism phenotype in the population (WHEELWRIGHT et al. 2010) and turns out to be genetically correlated with ASD.

We know the following about the epidemiology of ASD. About one in 59 children meets the diagnostic criteria for ASD (BAIO et al. 2018). However, autism, strictly defined, is much rarer, with estimates of the prevalence around one in 500 (FOMBONNE 2009). ASD is found in all racial, ethnic, and socioeconomic groups, with complex variation in prevalence (BAIO et al. 2018). Among children meeting the criteria for ASD, the male:female ratio is between 4:1 and 3:1 (LOOMES et al. 2017; BAIO et al. 2018). These features, you should note, are similar to those for schizophrenia, and you won't be surprised that both conditions are considered to be neurodevelopmental disorders. The epidemiology of autism and schizophrenia contrasts with that of anxiety and depression, which are primarily adult-onset diseases with high prevalence rates, are more common in women, and for which the environment plays a large part in their causation, as we discuss in Chapter 10, sections 10.1.1 and 10.3.1.

8.3 Molecular genetics of ASD

At a small meeting at Warwick University in the 1990s, Michael Rutter proposed undertaking a molecular genetic study of autism. At that time, a new center established in Oxford was embarking on exploring the genetic basis of complex traits by analyzing the segregation of markers in hundreds of pairs of siblings, a method that John Todd's group in Oxford had used to identify risk loci for type 1 (insulin-dependent) diabetes

(DAVIES et al. 1994). What, Rutter wanted to know, was necessary to make this work for autism? At that time, the answer was thought to be 'a lot of affected siblings' (around 200), which is hard to find for autism as, strictly defined, the condition is rare. But Rutter was internationally known for his work on autism and thought he could assemble a large enough cohort. The other thing needed was a molecular geneticist to head the project, a role taken on by Tony Monaco, famous at that time for identifying the mutation and cloning the gene underlying Duchenne's muscular dystrophy (a remarkable achievement, given the lack of genomic resources that we now take for granted) (MONACO et al. 1985). Monaco had immense expertise in the molecular genetics of Mendelian disorders. Tackling autism was his first foray into the molecular basis of complex traits, and the methodology he was handed was based on the assumption that the genetic effects were small and additive, rather than Mendelian. Ironically, it was the discovery of Mendelian versions of autism that was to yield key insights into the biology of the disorder.

By 1998, Rutter had assembled an international consortium that identified 99 families with multiple affected individuals. In most families, ASD does not follow a simple mode of inheritance, so the genetic-mapping approach assumed that an unknown number of loci contribute to disease risk, and that the relevant loci could be identified as regions in the genome where affected siblings share more of their genetic material than expected by chance (on average, siblings share 50%). This is the sibling-pair linkage method described in Chapter 5, section 5.1.2. The first genetic analysis found nothing (INTERNATIONAL MOLECULAR GENETIC STUDY OF AUTISM CONSORTIUM 1998), and by the end of the century, three more screens had been completed with still no positive results (PHILIPPE et al. 1999; RISCH et al. 1999; LIU et al. 2001). By 2003, nine independent studies had been complete, of which the largest, an analysis of 345 families, yielded only suggestive evidence of linkage (YONAN et al. 2003). One of the conclusions from these failed studies was that 'a large number of loci (perhaps ≥15)' were likely to contribute to disease risk (RISCH et al. 1999). Given what we know now, 15 was a huge underestimate, and the idea that sample sizes of 345 families would be enough was incredibly optimistic. As happened with other studies of psychiatric disorders, success did not come from family studies that relied on genetic linkage. But neither did it come about with the application of genome-wide association studies (GWAS), as was the case with schizophrenia. Instead, two breakthroughs occurred in autism genetic research, both driven by technological innovations, and both dependent on earlier work indicating that chromosomal rearrangements and single gene mutations were a cause of ASD.

8.4 Copy-number variants

The first breakthrough came from the application of methods that screened the whole genome for small chromosomal rearrangements. Chromosomal abnormalities had occasionally been reported in association with autism, such as a duplication of a region on chromosome 15q (15q11–13; Baker et al. 1994; Bundey et al. 1994), suggesting that it would be worth screening the genome of children with ASD for rearrangements. But doing so remained unimaginable until the human genome project gave us the necessary reagents to develop tools that screened for deletions or extra copies of DNA in any region of the genome. The method was called comparative genomic hybridization (CGH) (Figure 8.1).

CGH works as follows: DNA from a patient is labeled with a fluorescent dye, so that, for example the DNA is colored green. DNA from someone whose genome is known to be intact is labeled with another color, such as red. The DNA is denatured and allowed to hybridize to DNA sequences on a microarray that represent the entire genome (for a description of hybridization and how it works, see Chapter 3 sections 3.4 and 3.8). The strength of the hybridization signal reflects the relative amount of DNA, so the method detects deletions (loss of one or both copies of a DNA region) and extra copies of DNA (duplications, triplications, or other multiples of the normal diploid state). Because of what the method is able to detect, the rearrangements are often referred to as copy-number variants (CNVs), a form of structural variation that we have seen plays a part in the etiology of schizophrenia (see Chapter 7, section 7.7).

When patients with ASD were screened with these new tools, cases were found to have more CNVs than controls (Jacquemont et al. 2006; Sebat et al. 2007), but in addition, Jonathan Sebat, a year before he reported evidence for CNVs in schizophrenia, found that a proportion of the CNVs were not present in

Figure 8.1 Comparative genomic hybridization. On the left is shown the principle of the method. The patient's DNA and a reference (normal) DNA are labeled; here one is labeled in blue and one in gray, but in practice red and green fluorescent molecules are used. Equal amounts of the two samples are mixed and allowed to hybridize to a microarray, shown here as a series of dots. Each microarray dot contains DNA from a known location in the genome. The intensity of the two colors hybridizing to each dot is measured. If there are equal amounts of DNA at a dot, then the signal of the each color should be the same, indicating no abnormality. On the right is the result for a family with an autistic child. The mother is shown in black, the father in blue, and the child in gray. There is a *de novo* deletion in a region containing the neurexin gene (*NRXN1*). For both graphs, the *y*-axis represents the amount of DNA detected relative to normal, where the normal amount is 1.0. Right panel: from Levy et al. (2011).

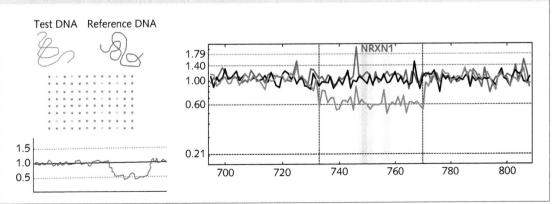

either of the parents (14/195 cases compared with 2/196 controls). This meant that the mutations had arisen as new mutations in the sperm or eggs of the parents: the mutations had arisen *de novo*. The *de novo* CNVs were relatively large (with a mean size of 2.3 Mb) and were rare (<1% frequency) in the general population.

The discovery that *de novo* rearrangements cause ASD meant that genetic causes were likely to be more important than previously thought. The impact of *de novo* chromosomal rearrangements can't be detected from twin or family studies, as the CNVs are not usually inherited; individuals with severe ASD rarely reproduce. Cases due to *de novo* CNVs would thus be considered to have an environmental origin. Molecular approaches thus identified a new source of genetic variation, one that had not previously been expected to play a role as a cause of ASD.

There was considerable interest in confirming and extending observations about CNVs, which led to four findings:

1. **CNVs are enriched in children with ASD:** 6–7% of cases of ASD are now thought to be attributable to CNVs (Glessner et al. 2009; Pinto et al. 2010; Gilman et al. 2011; Levy et al. 2011; Sanders et al. 2011).

2. **Some *de novo* CNVs that predispose to ASD are found in the same region of the genome in different individuals.** Thus, some CNVs are recurrent, an observation that shows that a chromosomal rearrangement causes ASD (i.e. the same CNV occurs in enough people with ASD to demonstrate a significant genetic association). The recurrent CNVs were not necessarily inherited, because the genome, like the earth's crust near plate tectonic boundaries, has unstable regions where rearrangements arise spontaneously. Although CNVs were found in many hundreds of independent regions, recurrent *de novo* events have so far been found at seven loci: chromosomes 1 (at 1q21), 2 (at 2p15), 7 (at 7q11.23), 15 (at 15q11.2–13.1 and 15q13.2–13.3), 16 (at 16p11.2), and 22 (22q11.2) (Sanders et al. 2011).

3. **The association between some CNVs and ASD suggested a more complex relationship than at first suspected.** This point needs some elaboration. Sometimes, one of the unstable CNV regions might harbor deletions in some people and duplications in others. You might imagine that the deletion and duplication would necessarily have different consequences and thus manifest with different phenotypes, but one region of chromosome 16p11.2 influences susceptibility to ASD when it is **either** deleted **or** duplicated (Weiss et al. 2008b). By contrast, rearrangements at 7q11.23 do show contrasting patterns: ASD is associated with duplications, while deletions are associated with a rare form of syndromal intellectual disability called Williams–Beuren syndrome, which is characterized by a social personality, apparently the opposite of an autistic phenotype (Sanders et al. 2011). It also became clear that sometimes people with rearrangements found in autism had other conditions. CNVs have pleiotropic consequences: the same rearrangement can manifest with different clinical features. CNVs at 1q21.1 and 15q13.2–13.3 are found in schizophrenia (International Schizophrenia Consortium 2008; Stefansson et al. 2008) and idiopathic epilepsy (Helbig et al. 2009), while microduplications of 16p11.2 are associated with schizophrenia (McCarthy et al. 2009).

4. **Genetic analysis identifies individuals with different disease severity.** Inherited susceptibility to ASD, the type of polygenic effect we have seen underlying schizophrenia, and the large-effect *de novo* mutations can be used to distinguish groups of individuals with different probabilities of disease severity. Familial risk of ASD (due to the presence of heritable, polygenic variants) is present in cases with high intelligence quotient (IQ; >120), while cases with very low IQ (<50) have a familial risk similar to that found in cases of severe intellectual disability (i.e. their risk is due to large-effect *de novo* mutations) (Robinson et al. 2014). This result underscores the close relationship in people with ASD between the underlying genetic architecture (*de novo* structural variants versus transmitted polygenes) and the level of intellectual functioning.

8.5 *De novo* mutations in genes that cause ASD

The second breakthrough was the discovery of a few families with mutations in single genes associated with ASD. It had been known for many years that ASD is not a single condition. ASD is associated with some

Mendelian disorders, where a single mutation segregates as either recessive, dominant, or X-linked. Examples include tuberous sclerosis, fragile X syndrome, Rett syndrome, and Timothy syndrome. Collectively these conditions may account for 10–20% of all cases (BERG AND GESCHWIND 2012). This observation suggested that there might be other mutations in single genes that caused ASD—but how to find them? One clue was to use chromosomal rearrangements as a guide.

Thomas Bourgeron's group gained fame for identifying mutations in two neuroligin genes (*NLGN3* and *NLGN4*) (JAMAIN et al. 2003) and in *SHANK3* (DURAND et al. 2007) as causes of autism. In each case, Bourgeron's discovery relied on evidence from chromosomal rearrangements that strongly implicated candidate genes. Molecular characterization of a chromosome 22q translocation in a patient with autism identified a breakpoint in a gene called *SHANK3* (BONAGLIA et al. 2001). Bourgeron decided to screen 227 families for abnormalities in *SHANK3* and found three cases with mutations in the gene (DURAND et al. 2007). These mutations are the same single base pair mutations familiar to geneticists as the cause of Mendelian conditions such as Huntington's disease (discussed in Chapter 5, section 5.1). Perhaps there were more cases of autism that could be explained by such mutations, but how could they be found in the absence of a strong clue, such as the chromosomal rearrangements? New sequencing strategies provided an answer.

The arrival of next-generation sequencing technologies (Chapter 3, section 3.5) made it possible to look at the entire coding region of the genome at a reasonable cost. Four papers, three in *Nature* and one in the journal *Neuron*, all published in 2012, demonstrated the power of exome sequencing to identify likely causal sequence variants in ASD (IOSSIFOV et al. 2012; NEALE et al. 2012; O'ROAK et al. 2012b; SANDERS et al. 2012). These studies used the same design: they compared the coding sequences of affected children with their parents, looking for the presence of changes that were present only in the autistic children. They were looking for new mutations—mutations that arise in the sperm or eggs of the parents. The remarkable thing is that exome sequencing of a few hundred subjects was sufficient to find these rare events (compare this with the failure to find anything with the sib-pair studies).

De novo mutations are sufficiently rare that you are lucky to find one new mutation per generation. So are there more new mutations in children with ASD than in controls? In fact, the **total** number of mutations is the same. When you look at the distribution of all mutations, regardless of where they occur in the gene, there is no difference between children with and without ASD (approximately one *de novo* point mutation is found per exome). The observation from the ASD sequencing studies was that if you count mutations that disrupt gene function in autistic patients and controls, then there **is** a difference: all four studies reported an accumulation of the rare variants that damage genes. This enrichment is most marked in genes known to be intolerant of mutations (such genes are found by examining the exomes of very large numbers of people and ranking the genes by the number of mutations observed; LEK et al. 2016). We return to this issue in section 8.6.

The year 2012 was a good year for autism genetics. These four papers showed that, at last, the field had matured so that consistent answers were emerging. Everyone agreed that cases of ASD have more deleterious *de novo* mutations than controls (Table 8.1). By 2015, exome sequencing had been deployed to thousands of families with a child with ASD (IOSSIFOV et al. 2014).

Table 8.1 *De novo* mutations that disrupt genes are enriched in ASD

Mutation class	Mutations per person (mean (variance))		P value
	Probands	Siblings	
All	1.10 (1.23)	0.97 (1.00)	0.05
Silent	0.35 (0.37)	0.36 (0.41)	0.59
Missense	0.61 (0.64)	0.54 (0.53)	0.21
Gene disrupting*	0.15 (0.14)	0.06 (0.07)	2.63×10^{-5}

* Gene-disrupting mutations include frameshift, stop gain and loss, and splice-site mutations (see Chapter 3, Table 3.2 for explanations).

8.6 Rare variant analysis: how do we know that a mutation causes ASD?

How can we be certain that mutations in a specific gene cause ASD? The simplest solution is to find multiple cases with mutations in the same gene, and to show that the recurrent observation is unlikely to have arisen by chance. But how many cases do we need? And what significance threshold is appropriate?

Looking across the four studies from 2012 (Iossifov et al. 2012; Neale et al. 2012; O'Roak et al. 2012b; Sanders et al. 2012), there was evidence for recurrent de novo variants in seven genes (*DYRK1A, POGZ, CHD8, NTNG1, GRIN2B, KATNAL2,* and *SCN2A*). But was this evidence good enough? Investigators set out to resequence these (and other) candidate genes in larger groups of patients (Michaelson et al. 2012; O'Roak et al. 2012a). Evan Eichler, for example, sequenced one gene, *CHD8*, in 3,730 children with developmental delay or ASD. His group identified a total of 15 independent mutations and found no deleterious mutations in 8,792 controls (Bernier et al. 2014). They estimated the significance of association with *CHD8* to be $P = 1.01 \times 10^{-5}$. That sounds impressively significant, but is it sufficient to declare that mutations in *CHD8* cause ASD?

During the gold rush of exome sequencing, some scientists complained that '*investigators should not simply assume that the presence of two or more independently occurring* de novo *mutations in the same gene within a sequenced cohort is definitive evidence of a causal role for that gene; such a threshold results in ever increasing numbers of false positives as the number of sequenced cases increases*' (MacArthur et al. 2014). The recommendation was made to adopt a conservative genome-wide significance threshold: '*a Bonferroni-corrected P value of 1.7×10^{-6} (that is, 0.05 out of 30,000)*'. That's a lot lower than the *P* value for *CHD8*. Does that mean we should reject the *CHD8* finding as a false positive? However, this is not just a problem of statistical significance.

First, just because a family study reveals that a variant is *de novo* (arising in the child as a new mutation) does not necessarily mean that the variant causes disease. The ones we are seeking—highly deleterious, pathogenic *de novo* mutations—could be confused with *de novo* mutations that are much more benign. How can we tell the difference?

One argument runs as follows. The current size of the human population is about 7 billion. A *de novo* variant occurs about once in every exome, which works out to mean that every coding variant is present as a *de novo* mutation at least once in the human population. *De novo* variants that are relatively benign (meaning that those people who carry them can reproduce at nearly normal rates) will be passed on to the next generation, and will thus join all the other variants present in the population as standing variation. In general, because of their severity, deleterious *de novo* mutations won't be passed on, so a simple test of whether a *de novo* mutation is deleterious or not is to see whether the variant is found in surveys of unaffected populations (Kosmicki et al. 2017). Information on the distribution of variants from exome sequencing of unaffected subjects can be found in databases such as The Exome Aggregation Consortium (ExAC), a repository of exomes from 60,706 adult individuals without severe developmental abnormalities (Lek et al. 2016). Analysis of this database shows that the *de novo* variants enriched in ASD are not found in unaffected individuals and are thus likely to be pathogenic (Kosmicki et al. 2017). However, failure to find a variant in ExAC could be because ExAC is too small (for the time being, at least—the databases are continuously expanding); by contrast, if the variant is found in the database, we can be confident that it is not pathogenic.

There is a second problem. The idea that a deleterious mutation has to have a phenotype isn't true. This means that just finding an enrichment of deleterious mutations in a gene might not be sufficient to prove the gene is causative. A healthy genome contains about 100 deleterious mutations, and every one of us carries mutations that inactivate both copies of approximately 20 genes (MacArthur et al. 2012). (The explanation for why these mutations are found in people who, to all intents and purposes, are perfectly healthy is almost certainly the presence of other genetic variants that compensate for the loss of gene function.) Even with the application of rigorous criteria to define a highly deleterious mutation, supposedly pathogenic mutations still occur in healthy people.

In order to deal with this second issue, we need to know how well genes tolerate functional genetic variation on a genome-wide scale. If we sequence lots of exomes in unaffected people, will we find that mutations are randomly distributed across all genes or are there some that are less likely to contain mutations? Again, exome databases such as ExAC can answer this question. And there are indeed sets of genes that are constrained, genes that are less likely to carry mutations (PETROVSKI et al. 2013; SAMOCHA et al. 2014). Information about constraint can predict whether mutations in a gene are likely to cause disease: genes responsible for Mendelian diseases are significantly more intolerant to functional genetic variation than genes that do not cause disease (PETROVSKI et al. 2013). A list of 1,003 genes defined as constrained was discovered to be enriched for genes known to cause human disease, including ASD (SAMOCHA et al. 2014).

Returning to the *CHD8* example, we can now estimate the likely chance that the recurrent mutations found in this gene are pathogenic. At the time of writing, the estimate is indeed highly significant ($P = 8.38 \times 10^{-20}$) (SAMOCHA et al. 2014). By concentrating on recurrent mutations not found in databases, and that occur in constrained genes, the mutational excess in ASD is found to occur in 18% of genes—those that are most functionally intolerant of mutations (KOSMICKI et al. 2017).

8.7 Mosaicism

Sequencing revealed another, unexpected, source of genetic variation as a cause of ASD. Next-generation sequencing obtains sequence from multiple copies of the target DNA, partly in order to ensure that a sequencing error is not mistaken as a variant (the method has an error rate of approximately 1%) but mostly because the sequencing obtains random fragments (called 'sequencing reads') from the target: large numbers of fragments are required to ensure that there are enough sequencing reads to cover the entire target. For exome sequencing, the target is often covered by as many as 100 reads. There is some redundancy with this level of coverage, which has been exploited to demonstrate the presence of sequence mosaicism.

As the genome exists in two copies and every cell has the same two copies, in theory there can only be two alleles at a locus. But every time a cell divides, as must happen during development, there is a small possibility that a mistake is made when the cell's DNA is copied, introducing sequence differences into the cell's progeny. Suppose this error happens early on in development, when the embryo consists of just four cells. If one of those four cells has a mutation that arose during cell division, then that mutation will be passed on to all the cell's descendants, so that a proportion of the cells of the adult organism will carry the mutation (if some of these cells populate the germline, the mutation can even be passed on to the next generation). These mutations are called postzygotic mutations, and the state of having some cells with a different sequence from others is called mosaicism. Because the cells that give rise to the actual embryo come from the inner cell mass, which does not become segregated from the rest of the embryonic cells until there are many more than four cells, the proportion of mosaic cells will vary (e.g. it's unlikely to be a quarter).

Chris Walsh, a neurogeneticist at Harvard, decided to see whether there was evidence of mosaicism in the exome sequences of autistic children (LIM et al. 2017). He found that mosaicism is indeed likely to play a part in the etiology of the disease. His group estimates that 7.5% of *de novo* mutations are somatic mutations, and that children with ASD have more deleterious somatic mutations than their unaffected siblings. Although the mosaic mutations were found in blood, because they arise early enough in development to be present in high enough frequencies for detection, they are certainly present in brain tissue as well. This intriguing observation extends, again, our understanding of how genetic effects can give rise to psychiatric disease.

8.8 Common variants that contribute to the risk of ASD

You could be forgiven, reading the frankly spectacular findings from molecular studies of single variants and CNVs, which are transforming our knowledge of ASD, for believing that the genetic basis of ASD arises primarily from rare variants. So just how many cases of ASD are due to these types of mutations? Figure 8.2 provides an answer to this question. Based on a Swedish epidemiological sample, it is estimated that only 3% of cases are due to *de novo* mutations, while almost 50% are due to common, inherited genetic variants (the sort we have seen as important in other psychiatric disorders; GAUGLER et al. 2014).

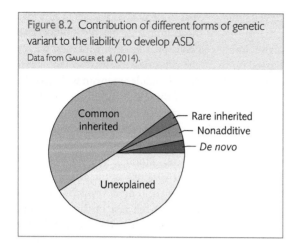

Figure 8.2 Contribution of different forms of genetic variant to the liability to develop ASD.
Data from Gaugler et al. (2014).

Progress with understanding the common inherited component has so far lagged well behind the analysis of rare variation; we have, to date (as of March 2019) one GWAS that has identified five genome-wide significant loci from analysis of 18,381 individuals with ASD and 27,969 controls (Grove et al. 2019). As with other common disorders, ASD risk arises from a polygenic burden of thousands of common variants in a dose-dependent manner.

There is an intriguing finding about common variants: the effect of common risk alleles for ASD is positively associated with intelligence and educational attainment (Clarke et al. 2016; Hagenaars et al. 2016). Educational attainment simply means the number of years of education. We discuss the genetics of this trait in Chapter 12, section 12.1.6. The excess of alleles associated with higher intelligence and educational attainment appears to occur only in ASD individuals from higher-functioning categories (so, for example, the excess is not seen in individuals who have comorbid intellectual disability) (Grove et al. 2019). Apparently, if you carry alleles that increase your risk of developing ASD, you also carry alleles that increase your IQ!

The result was obtained from an analysis of the inheritance of polygenic risk scores (PRSs) for educational attainment, as we describe next. The polygenic risk of a trait can be approximated by the sum of the effects of risk alleles. For example, an individual might have 50 loci that increase risk by a certain amount and 60 that decrease risk; the PRS is the sum of effects at all risk loci (see Chapter 7, section 7.2 for a fuller description of PRSs). If the parents are unaffected by ASD (as is usually the case), then we can test whether

the cause in an affected child is due to polygenic risk (i.e. it is due to the sum of lots of risk alleles) by determining whether the child has more than the expected 50% sharing of polygenic risk alleles (50% is the expected quantity as, on average, each child gets half of the risk from one parent and half from the other). Furthermore, we can examine whether that excess is associated with other polygenic risk, such as educational attainment. Remarkably, polygenic risk for ASD, educational attainment, and schizophrenia are significantly overtransmitted in affected children ($P < 1 \times 10^{-6}$) but not in unaffected siblings (Weiner et al. 2017). On average, individuals with ASD inherit more IQ-increasing alleles than their typically developing siblings.

The same method can be used to assign individuals a risk attributable to the rare, *de novo* CNVs and then compare that risk with the polygenic risk, thereby providing a way to assess the relative contributions for both and to see how much they overlap. In fact, overtransmission of the ASD PRS provided evidence that common and rare risk variants for ASD act additively, implying that they probably confer risk for ASD in different ways (Weiner et al. 2017). This is similar to the way in which PRSs and CNVs contribute to schizophrenia risk. We can see this too from the associated phenotypes. From the epidemiological studies, we know that a strongly acting, *de novo* variant risk factor has an impact on a limited subset of cases. These types of variants are associated with intellectual disability, seizures, and global neurodevelopmental impact (Sanders et al. 2015). By contrast, polygenic risk for ASD appears similarly relevant to individuals with high and low IQs, and those with and without a strongly acting *de novo* mutation. As noted above, the common polygenic influence is not associated with severe disability (Robinson et al. 2014). Instead, it is associated with better educational and cognitive outcomes.

In summary, common genetic variation is the major route by which genetic effects alter the risk of developing autism, and these variants are associated with a higher IQ. Recall that the diagnosis of autism is enriched in individuals with intellectual disability, a counterintuitive finding, unless we assume that the rare variants—all those coding and structural variants we have discussed—primarily (or indeed solely) act on cognitive function and **not** on autism. Consistent with this view, analysis of the common variant GWAS hits shows minimal overlap with rare variants: there was no significant overlap with 71 genes in one study (SANDERS et al. 2015), although some overlap was identified with genes identified from a larger study of 50,000 families (SPARK CONSORTIUM 2018). There is still work needed to determine exactly what the mutations described in this chapter are doing. It is our view that we still haven't found any mutations that specifically alter the risk of developing ASD, only intellectual disability.

8.9 Altering the consequences of mutations

Our last comment on ASD takes us to a far-reaching implication of genetic analysis, and that is to suggest that the consequences of some genetic mutations are not fixed. In the last few years, an increasing number of studies have suggested that phenotypes associated with some mutations that cause autism might be reversible. The findings come from the following experiments.

First, a gene known to cause Rett syndrome (see Chapter 9, section 9.5.2), which may manifest with ASD, was inactivated before birth in a mouse but turned back on in the adult, with the remarkable effect that the disease phenotype resolved (GUY et al. 2007); the technology and its applications are described in Chapter 16. A similar observation came for analysis of neuroligin-3 knockout mice (a model for nonsyndromic ASD). Again, the phenotype could be rescued by re-expression of neuroligin-3 in juvenile mice (BAUDOUIN et al. 2012).

Second, a drug has been used to reverse autistic symptoms in children with Dravet syndrome. Affected children have recurrent intractable seizures, cognitive deficit, and ASD, due to a mutation in a channel in neurons that allows the passage of sodium, known as SCN1A (voltage-gated sodium channel $Na_V1.1$). Mice that were created with a mutation in SCN1A (see Chapter 16 for a description of how this is done) recapitulated the human behavioral phenotype, and treatment with clonazepam, an antiepileptic drug that acts on receptors for the inhibitory neurotransmitter GABA (γ-aminobutyric acid), completely reversed the abnormal social behaviors (HAN et al. 2012).

Third, dietary manipulations can reverse autistic phenotypes in mice with mutations in two genes associated with ASD. Mutations in the gene encoding BCKDK (branched-chain ketoacid dehydrogenase kinase) in families with ASD, epilepsy, and intellectual disability result in lower-than-normal levels of a class of amino acids calledbranched-chain amino acids. When mice with the *Bckdk* mutation were put on a branched-chain amino acid-enriched diet, they became phenotypically normal (NOVARINO et al. 2012). Similarly, mutations in SLC7A5 (solute carrier transporter 7a5), a large neutral amino acid transporter localized at the blood–brain barrier, disrupt branched-chain amino metabolism and cause ASD. Intracerebroventricular administration of branched-chain amino acids ameliorates abnormal behaviors in adult mutant mice (TARLUNGEANU et al. 2016).

These remarkable results indicate that there is more plasticity than might be expected in ASD due to a single gene mutation. Correction of the underlying defect allows the adult brain to compensate for, or even correct, a severe developmental disorder. These investigations encourage a revision of the view that ASD is a developmental disorder, incurable from birth, and suggest that, as our knowledge of the genes and mechanisms underlying ASD advances, new treatments may become available.

8.10 Causes of autism

The complexity of the genetic causes of ASD, from Mendelian to polygenic, and the presence of so many different types of molecular variation, from structural variant to single-nucleotide polymorphism, make ASD a good place to review the different ways in which genetic variants contribute to psychiatric disease. Therefore, we provide in Figure 8.3 (adapted from DE LA TORRE-UBIETA et al. 2016) an overview of the contents of this chapter. Many of the illustrations of how genetic effects operate apply to disorders described elsewhere in this book.

At the top left are shown the Mendelian patterns of inheritance: dominant, recessive, and X-linked. *De novo* inheritance is typically dominant. The top right shows common variants contributing additively to disease risk (this is the polygenic model that underlies most of the risk of autism and other psychiatric disorders). Underneath are shown different types of genetic variation (left) and the developmental disorders (right) associated with autism. Genes that have been associated with ASD are also indicated.

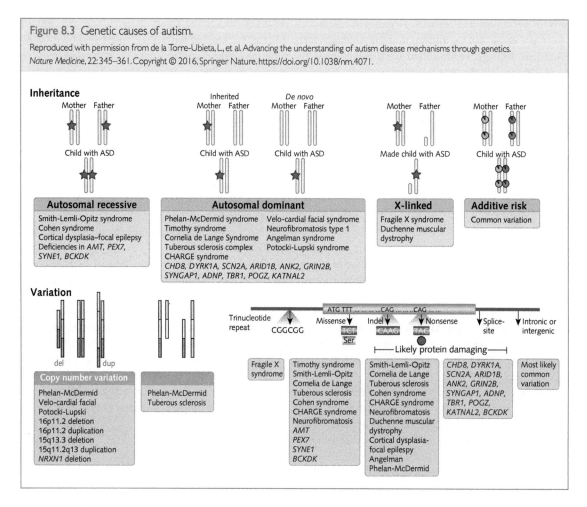

Figure 8.3 Genetic causes of autism.

Reproduced with permission from de la Torre-Ubieta, L, et al. Advancing the understanding of autism disease mechanisms through genetics. *Nature Medicine*, 22: 345–361. Copyright © 2016, Springer Nature. https://doi.org/10.1038/nm.4071.

Summary

1. Autism is characterized by abnormalities in social development and stereotyped, repetitive behaviors. There exists a set of conditions related to autism found at higher frequency in relatives, reflecting a shared genetic etiology. To take into account the diversity of the phenotype, the condition is known as autism spectrum disorder (ASD).

2. About one in 59 children meets the diagnostic criteria for ASD, while autism, strictly defined, has a prevalence around one in 500. The male:female ratio is between 4:1 and 3:1. The heritability of ASD lies between 64 and 91%.

3. Copy-number variants (CNVs), particularly *de novo* deletions, are enriched in 6% of cases of ASD.

4. Some *de novo* CNVs that predispose to ASD are recurrent, such as that on chromosome 16p, which increases the risk for ASD when it is either deleted or duplicated.

5. Exome sequencing has identified *de novo* deleterious mutations in genes as a cause of ASD. The mutational excess in ASD is found to occur in 18% of genes: those that are most functionally intolerant of mutations. However, proving the involvement of any one gene is hard. One gene for which there is good evidence is *CHD8*.

6. Current observations indicate that 3–4% of ASD is due to *de novo* mutations, but it is predicted that the figure may be as high as 15%. Almost 50% is due to common, inherited genetic variants. While no single common variant has been identified, the effect of common risk alleles for ASD has been positively associated with intelligence and educational attainment.

7. A number of molecular genetic studies have suggested that autistic phenotypes might be reversible.

9

Intellectual disability and developmental disorders

[On the genetic basis of intellectual disability, the types of mutation that are responsible, the genes identified, and the lessons they teach about the biology of the condition]

9.1 Intellectual disability, developmental delay, or neurodevelopmental disorder?

The first issue to tackle in discussing intellectual disability is terminology. Anyone reading the literature will meet a multitude of similar-sounding terms (developmental disorder, neurodevelopmental disorder, cognitive disorder), some obsolete ('pervasive development disorder', replaced in 2013 by autism spectrum disorder (ASD)), and some apparently identical (learning disability, intellectual disability, and mental retardation). Are all these the same thing? If not, how do they relate?

Let's begin with the simplest distinction: acceptable and unacceptable nomenclature. In October 2010, US President Obama signed into law a bill that requires the US federal government to replace the term 'mental retardation' with 'intellectual disability'. This was at the instigation of parents and patients who argued that the term 'mental retardation' was derogatory.

For the rest, well, things are not so easily settled. 'Developmental delay', widely used by pediatricians, refers to the slower acquisition of psychomotor milestones. This doesn't necessarily (but almost always does) include impairment in cognition. The (apparently) more specific terms 'intellectual developmental disorder', 'neurodevelopmental disorder', or 'developmental cognitive impairment' stress the failure to develop intellectually but are usually subsumed under the blanket term 'intellectual disability'. It would be useful to distinguish between global cognitive impairment and specific impairments of cognitive function, such as language or memory, although there is little consensus on how to do so. Some researchers use 'learning disability' to mean specific impairments, such as dyslexia (reading difficulty) and dyscalculia (lower-than-expected mathematical ability), associated with academic underperformance but not with a lower overall cognitive ability.

In this chapter, we discuss the genetics of a term that has the broadest category: 'intellectual disability', diagnosed on the basis of deviation from intellectual abilities expected of a given age. Intellectual disability is usually divided into mild and severe, based on assessments of intelligence quotient (IQ): mild intellectual disability has an IQ of 50–70 and severe has an IQ of less than 50. The distinction is important etiologically because the causes of severe intellectual disability are largely due to single insults (both genetic, such as a chromosomal rearrangement, and environmental, such as infection), while mild intellectual disability has a complex etiology, often attributable at least in part to polygenic variation (see Chapter 12; section 12.1.6 presents data on genetic effects on intellectual functioning within the normal range).

By any measure, intellectual disability is common. Assessment measures vary considerably, yet even so, between 1 and 3% of the population has mild intellectual disability, and severe forms occur in about 0.4% (McLaren and Bryson 1987; Roeleveld et al. 1997). Intellectual disability is commoner in men, with a 30–50% excess of males over females.

The causes and clinical features of severe intellectual disability are multifarious, united only in the fact that they (almost) all occur prenatally (head injury after birth being an obvious exception). Genetic causes fall into three categories: polygenic, chromosome abnormalities, and single gene mutations. Here, we discuss chromosome abnormalities and single mutations. Chapter 12 discusses the polygenic basis of IQ.

9.2 Chromosomal abnormalities as a cause of intellectual disability: Down syndrome ·

The commonest single cause of intellectual disability is Down syndrome, due to the triplication of the whole (or less commonly a part) of chromosome 21. Trisomy 21 is found in 0.3% of infants (it also occurs in about a quarter of spontaneous abortions). Cognitive impairment is always present, but otherwise Down syndrome has a highly variable clinical presentation. Some or all of the following features occur in individuals with trisomy 21: a characteristic facial dysmorphology, a small brain, and early-onset Alzheimer's disease. The disorder was first described in 1866 (Down 1866) and the cause, trisomy 21, was reported a century later (LeJeune et al. 1959).

The extra chromosome 21 in Down syndrome arises because of errors in meiosis (the cell division that produces gametes: eggs and sperm). Instead of replicating and separating neatly into haploid cells with a single complete set of chromosomes, an extra copy of chromosome 21 is retained. Nondisjunction can affect other chromosomes, but almost all embryos with other trisomies do not survive. Almost all trisomies are due to errors in maternal meiosis, thought to arise because meiosis in oocytes is paused for many years, in contrast to the production of male gametes that quickly pass through the two meiotic divisions required to produce gametes. This would also explain why trisomy 21 is related to maternal age (Morris et al. 2002).

Not all Down syndrome is due to nondisjunction. About 2.5% of cases are due to a translocation, another meiotic error, in this case occurring when chromosomes incorrectly pair and swap material, for example involving chromosome 21 pairing with chromosome 15. Translocations swap parts of a chromosome and so show that Down syndrome cannot be attributable to trisomy of the entire chromosome 21 (Nadal et al. 1997). Chromosome 21 is 45 Mb in size, the smallest of the human chromosomes, containing 670 genes, while trisomy of the smallest region that results in Down syndrome was found to be about 5.4 Mb with about 80 genes.

One simple hypothesis to explain the pathogenesis of Down syndrome is to assume that a few genes in the critical region are dosage sensitive, so their over-expression could explain all the phenotypes of Down syndrome, including the behavioral manifestations. This hope drove considerable efforts to reduce the critical interval further and identify the genes, efforts that were largely unsuccessful (Shapiro 1997), but you'll find in the literature claims for single candidate genes such as *SIM2* and *DYRK1A*.

An extra chromosome means extra copies of genes, resulting in changes in gene expression, so an obvious task is to identify those genes whose expression changes due to the trisomy. Among this set of genes should be those responsible for the disease. Given the ease with which we can now interrogate the expression of every gene in the entire genome using RNA sequencing (see Chapter 4, section 4.7), it's surprising how hard it proved to find these genes. When gene expression was compared between trisomy 21 cells and normal cells, the distribution of expression overlapped for more than half of the transcripts (Prandini et al. 2007). Normal variation in gene expression obscures the effect of the trisomy. With such a large amount of normal variation in gene expression, how is it possible to attribute a causal role to any gene? The solution came from finding a very rare event: a pair of monozygotic twins discordant for Down syndrome. As, apart from chromosome 21, the genomes are identical, gene expression alteration must be due to the additional chromosomal material and not due to differences at other genetic loci.

The existence of such a discordant pair seems extraordinary: monozygotic twins are genetically identical, and Down syndrome is due to a genetic imbalance, so how can monozygotic twins differ with respect to a whole chromosome? The origin of an extra chromosome in just one twin is indeed remarkably complex (at least four processes were involved), involving events both before conception and during early development (DAHOUN et al. 2008).

Stylianos Antonarakis, a human geneticist who has devoted his career to the study of Down syndrome, analyzed gene expression in the discordant monozygotic twins and found 182 genes whose expression was different between the twins (LETOURNEAU et al. 2014). Remarkably, gene expression differences were **not** restricted to chromosome 21, but occurred across the genome, and the differences were not just increases—there were decreases too. Most intriguingly, the differential expression was not spread randomly across the genome but followed a pattern. They found that the differences between trisomic and normal cells were organized into chromosomal domains, sharing differential expression profiles (LETOURNEAU et al. 2014). There was also an epigenetic change that accompanied the domains, a marker indicating the presence of active genes (H3K4me3, see Chapter 4, section 4.2.5).

This finding, the biggest advance in our thinking about the cause of Down syndrome since the discovery of the additional chromosome 21, indicates that the phenotypes we see in Down syndrome have (at least) two origins: they are in part due to overexpression of genes on chromosome 21, but they are also due to the dysregulation of the expression of genes that are not on chromosome 21 at all.

9.3 Small chromosomal abnormalities: a question of dosage

A careful search for chromosomal abnormalities in cases of intellectual disability, before the advent of molecular techniques, led to the identification of a small number of syndromes associated with the deletion of a specific chromosome region, each of which has a characteristic phenotype. Clinical geneticists diagnose these conditions on the basis of their physical features. Each condition includes some variable degree of intellectual disability. The surprise was that when the behavior of children with the same disorder was examined, the chromosomal deletions were associated with (relatively) specific psychological abnormalities, not just simply undifferentiated intellectual disability. Psychologists identified characteristic patterns of behavior and cognitive profiles, including the overly social behavior of Williams syndrome children (BARAK AND FENG 2016), the obsessional behavior of Prader–Willi sufferers (GRIGGS et al. 2015), and the happy, sociable disposition of children with Angelman syndrome (BUITING et al. 2016). The 22q11.2 deletion, found in children with intellectual disabilities, heart defects, and characteristic facial features, is associated with psychosis (KARAYIORGOU et al. 1995) (see Chapter 7, section 7.7 for a fuller description). The recognition of 'behavioral phenotypes' associated with a chromosomal abnormality implied that each had a different molecular pathogenesis, and suggested that molecular characterization would lead to the identification of genes with relatively specific behavioral functions.

The causative lesion of each syndrome is a deletion, rendering the possessor haploid at the locus (i.e. having only one copy), so molecular pathogenesis involves dysregulation of gene expression. Most researchers assumed that there would be a critical dosage-sensitive gene but, as with the assumption that Down syndrome was due to the overexpression of genes on chromosome 21, this explanation has not always held up. The most egregious exceptions are Prader–Willi and Angelman syndromes, caused by identical deletions on chromosome 15 but with very different phenotypes. As we pointed out earlier (see Chapter 4, section 4.4), the different syndromes arise because of the parental origin of the deletion: Prader–Willi syndrome is due to a deletion on paternally inherited chromosome 15, whereas in Angelman syndrome the deletion is maternally inherited (NICHOLLS et al. 1989).

At some level, the dosage hypothesis must account for the phenotypes, but we still don't have a thorough understanding of which genes are critical for the intellectual disability, and we are even further from understanding how the molecular lesion gives rise to a behavioral change. Despite extensive molecular characterization of each syndrome teaching us a lot about the genetic mechanisms involved, to date we haven't learnt much about the biological pathways that give rise to the behavioral phenotypes.

There has been much greater success discovering the genetic causes of other forms of intellectual disability. If microscopically observable changes in chromosome structure gave rise to intellectual disability, then it was a reasonable bet that smaller changes, invisible to the microscope, would also be responsible. Screening the entire genome for small chromosomal rearrangements became possible when, as a consequence of the human genome project, arrays were developed that detected alterations in the amount of DNA at any locus (this is described in Chapter 8, section 8.4 and Figure 8.1). The method is called comparative genomic hybridization, and current arrays detect deletions and duplications as small as 10 kb. Three important findings have emerged from the analysis of copy-number variants (CNVs):

1. **CNVs (deletions and insertions) are more common in children with intellectual disability than in controls.** The arrays detect CNVs in up to 20% of cases of intellectual disability compared with about 10% in controls (COOPER et al. 2011). About 10% of children have a *de novo* arrangement, make it likely that these rearrangements are the cause of the disability (DE VRIES et al. 2005). For comparison, in ASD the diagnostic yield is around 6–7% (mostly *de novo* CNVs) and about 5% for adults with schizophrenia and epilepsy.

2. **Recurrent rearrangements are a cause of intellectual disability** (e.g. on 17q21.31, 16p12.1, and 7q11.23). The discovery of recurrent CNVs not only added to the evidence for the cause of the associated intellectual disability but also allowed comparisons of phenotypes. This, in turn, revealed that the newly identified CNVs transcended phenotypic boundaries: a 1.5 Mb deletion on 15q13.3 identified in individuals with intellectual disability is also present in some cases of ASD and schizophrenia, and is found in approximately 1% of cases with epilepsy (GIRIRAJAN et al. 2011b). Similarly, a 1.6 Mb deletion on 1q21.1 occurs in intellectual disability, ASD, and schizophrenia, and in children with cardiac defects. CNVs, as we also noted in the discussion of autism in Chapter 8, section 8.4, are

pleiotropic (different clinical features can arise from the same genetic lesion).

3. **Phenotypic variability is explained by the presence of additional CNVs,** as suggested by analysis of a 16p12.1 rearrangement in which it was found that about a quarter of individuals with the deletion had a second chromosomal abnormality (GIRIRAJAN et al. 2010). The severity of the phenotype is influenced by the presence of additional variants (GIRIRAJAN et al. 2011a). The current best estimate is that, among children with intellectual disability, about 10% carry a second CNV; those with two CNVs are eight times as likely to have intellectual disability as controls (GIRIRAJAN et al. 2012). An important implication of this finding is that the presence of a single CNV, although not in itself sufficient to cause intellectual disability, acts as a predisposing factor, and that it may be associated with subtler changes in cognition. One example that supports such a view is the analysis of phenotypes associated with a chromosome 15q11.2 deletion: carriers have a history of dyslexia and dyscalculia, and the deletion has modest effects on other cognitive traits (STEFANSSON et al. 2014).

9.4 X chromosome abnormalities

There is a long history of an interest in the relationship between sex chromosomes and cognitive abilities (LEHRKE 1972). Sex chromosome abnormalities are found in two to three children per 1,000 (NIELSEN AND WOHLERT 1991), of which Klinefelter syndrome (XXY), XYY, XXX, and Turner syndrome (45,X: a single X chromosome) are the commonest. All of these are associated with some degree of cognitive impairment. There is a male excess of intellectual disability (the sex ratio is 1.3–1.4:1 for male:females), and the X chromosome bears more than its fair share of mutations that give rise to intellectual disability—a finding inferred from the existence of families in which intellectual disability segregates on the X chromosome (X-linked intellectual disability is referred to in the older literature as XLMR for X-linked mental retardation). What, if anything, does this indicate about the genetic basis of behavior?

The studies of entire X chromosome abnormalities have not told us much. Differences in whether the single X chromosome of Turner syndrome came from the mother or father at one point seemed to relate to alterations in social behavior (SKUSE et al. 1997), but this finding hasn't stood up well (LOESCH et al. 2005). There has been little attempt, and less success, in identifying any critical regions that contribute to cognition.

Far more has been learnt from the study of X-linked intellectual disability due to mutations in single genes. By 2017, more than 100 genes had been identified, accounting for up to 10% of intellectual disability in males. With the exception of fragile X syndrome, almost all of the X-linked intellectual disability genes are vanishingly rare. We turn next to a discussion of fragile X syndrome, and then discuss other genes, both X-linked and autosomal.

9.4.1 Fragile X syndrome

Fragile X syndrome is one of the commonest causes of intellectual disability, called 'fragile' because a cytogenetic abnormality on the X chromosome, in which the X chromosome breaks when chromosomes are prepared from cultured cells, is found in affected males. Fragile X syndrome has a frequency of one in 5,000 males and accounts for about 0.5% of all cases of intellectual disability (a small percentage but a large number of individuals) (COFFEE 2009). Clinical features are complex, and variable, including learning disabilities, perseverative behaviors, hyperactivity, language deficits, and disrupted sleep. Fragile X syndrome has long been recognized as a mutation that presents with some features of ASD, including social anxiety, gaze aversion, and difficulties forming peer relationships. Children also exhibit a group of sensory alterations, also noted in ASD, that include hypersensitivity to sensory stimuli and hyperarousal to seizures. The latter are perhaps the only features that come close to explanation from the molecular pathogenesis.

Fragile X syndrome was among the first behavioral disorders to be characterized at a molecular level, and the discovery of the nature of its molecular basis came as a great surprise: rather than being mutations in the coding regions of a gene, or deletions or insertions that removed or altered key components of the coding regions, the mutation turned out to be a trinucleotide repeat expansion of CGG in the promoter region of a gene called *FMR1* (so-called because it was the first 'fragile X mental retardation' gene to be found—there are some other fragile sites on the X chromosome, also associated with intellectual disability). The gene was named before the ban on using the term 'mental retardation'). Affected individuals have more than 200 of the CGG repeats, while unaffected individuals have less than 50 and people with between 55 and 200 carry a premutation, an unstable array that can expand in offspring to generate the pathogenic sequence. Despite the unorthodoxy of the mutation, the expansion behaves phenotypically just like any mutation that disrupts a coding frame; the expansion shuts off expression of the *FMR1* gene.

What does the *FMR1* gene do? Despite being discovered in 1991 (OBERLE et al. 1991; PIERETTI et al. 1991; VERKERK et al. 1991; YU et al. 1991), after more than 25 years of work, we still don't know for sure. What we do know is that the protein that *FMR1* encodes (called FMRP, for fragile X mental retardation protein) binds RNA. FMRP functions in every cell in the body, although at particularly high levels in neurons and in the testes (large testicles, called macroorchidism, is one of the clinical features of fragile X syndrome). It has two paralogs (genes with similar functions) called *FXR1P* and *FXR2P*. The best-understood function of FMRP is in translational control, although other functions might be equally, or more, important for explaining clinical features.

One line of research has been to characterize FMRP's targets, which include a wide variety of genes involved in regulating neuronal excitability. FMRP binds to as much as 4% of the mRNA in the mammalian brain. More than 400 putative mRNAs associate with FMRP (but only 14 have been validated by a biochemical assay, so we can't talk with certainty about the FMRP targets). Intriguingly, and for unknown reasons, the list is enriched in risk genes for schizophrenia.

For some time, the leading theory on how FMRP exerted its effect in the brain was through its action on a specific receptor subtype, metabotropic glutamate receptor type 5 (mGlur5). In the presence of the fragile X mutation, it was believed that mGlur5 stimulated the internalization into the cell of a related receptor, the AMPA receptor, which altered neuronal synaptic plasticity; that is, mGlur5 affected the ability of neurons to alter neurotransmission, which is believed to be a key component of how learning and memory occur in the brain (see Chapter 16, section 16.5 for a fuller description of this phenomenon). If true, this meant that blocking the mGlur5 receptor would rescue the phenotype—it would cure fragile X syndrome (MICHALON et al. 2012). A large-scale drug trial was set up to test this hypothesis but failed to find an effect (BERRY-KRAVIS et al. 2017). The hypothesis was based on the drug's effect on synaptic physiology in the mouse, not on the human behavioral phenotype, so the likely explanation is that different outcomes were measured in mice and human studies (BERRY-KRAVIS et al. 2018). But more fundamentally, the theory of how FMRP works may have been flawed.

Rather than enquiring about the targets of FMRP, another line of investigation led Jennifer Darnell to ask: what does FMRP do to its targets? The answer is that FMRP stalls or stabilizes ribosomes during the elongation phase of translation (DARNELL et al. 2011). FMRP seems to be associated with 'traffic jams' on ribosomes. In fact, FMRP may be binding to the ribosomes rather than the mRNA (CHEN et al. 2014). It forms part of a ribonucleoprotein (RNP) complex, a mixture of RNA and RNA-binding proteins, that takes up inefficiently translated mRNAs. Larger mRNAs preferentially associate with these granules and, as a consequence, their translation is repressed. One of the functions of FMRP is to counteract the tendency of large mRNAs to be segregated into inactive RNP complexes (GREENBLATT AND SPRADLING 2018). The specific effect of FMRP arises because it affects the translation of mRNAs in a length-dependent manner. Genes expressed in the brain are on average longer than genes expressed elsewhere, which might explain why the loss of FMRP affects brain function more than other organs. Furthermore, maintaining translation of large mRNAs at thousands of synapses is a challenge for neurons that other cells don't face (most cells don't have to deal with the problem of multiple, distributed sites of protein synthesis). This would also explain why the targets of FMRP are risk genes for other psychiatric diseases, such as schizophrenia. We don't yet have a full understanding of how the mutation leads to the behavioral disorder, but we are getting closer. Perhaps the main lesson is just how complex the biology is turning out to be, and how hard it is, even in a Mendelian disorder, to relate mutation to disease.

9.5 Single-nucleotide variants as a cause of intellectual disability

In this section, we consider what has been discovered from the molecular characterization of other mutations in individuals with learning disabilities. To do so, we introduce a distinction between syndromal and nonsyndromal conditions. Fragile X is an example of a syndromal condition: learning disability is just one of a number of features that characterize fragile X syndrome, including anatomical and physiological abnormalities. In nonsyndromal conditions, intellectual disability may be the only recognizable feature.

9.5.1 Nonsyndromal intellectual disability

Identifying the genes that cause nonsyndromal forms of intellectual disability is potentially a powerful way to find genes involved in behavior. Nonsyndromal intellectual disability means intellectual disability in which the major or often sole phenotypic manifestation is a cognitive or behavioral trait. The intellectual disability can almost always be attributed to a single causative mutation (compare this with the genetic effects on other complex traits in this book), and identifying the mutation (almost) always leads directly and quickly to finding a gene (this is because large-effect mutations typically disrupt coding regions in very obvious ways, such as deletions or disrupting reading frames (classes of mutations are described in Chapter 3, section 3.9.1 and Table 3.3)).

Identifying the mutations that cause nonsyndromal intellectual disability is not straightforward. People with severe intellectual disability rarely have offspring, so genetic linkage mapping approaches are hard to implement. Furthermore, intellectual disability is very heterogeneous, with many hundreds of genes involved, making association approaches impractical: it's hard to find enough cases of people with the same condition. Finally, even when the complete sequence is available for an individual with intellectual disability, finding a mutation in a gene is not proof that the mutation is causative. As noted earlier, every one of us carries mutations that inactivate both copies of approximately 20 genes (MACARTHUR et al. 2012).

Three approaches have successfully found genes involved in nonsyndromal intellectual disability. The first examined families with X-linked intellectual disability. In X-linked intellectual disability, the disorder is only manifest in men, but the mutation can be passed on to the next generation through women, so pedigrees can, and do, exist. Genetic linkage mapping (see Chapter 5, section 5.1) can be used to identify the region of the X chromosome that contains the gene, and sequencing can then be used to identify the mutation in the gene (this is how the fragile X syndrome gene was identified). In 1998, the first three nonsyndromal X-linked genes were identified using this approach: *GDI1* (D'ADAMO et al. 1998), *PAK3* (ALLEN et al. 1998), and *OPHN1* (BILLUART et al. 1998). To date, more than 100 genes for X-linked intellectual disability have been identified, accounting for about 10% of intellectual disability in males (LUBS et al. 2012; VISSERS et al. 2016).

Second, for autosomal mutations, it is possible to map the location of causative mutations in families where individuals have married relatives (consanguineous populations). For example, in some cultures it is common practice to marry first cousins, with the result that there are higher-than-expected rates of recessive disease, as the parents share more of their genome, including pathogenic mutations, than unrelated individuals (STOLL et al. 1994). The shared regions of the genome in individuals from consanguineous populations are homozygous; there are often many megabases of identical sequence, which provides a guide to the location of recessive mutations. By screening a population for homozygous regions, it is then possible to restrict the search for causative mutations to the regions of homozygosity and find the relevant genes by sequence analysis. The largest single application of this strategy, which was applied to families from Iran and elsewhere, identified 78 causative mutations in 50 novel genes (NAJMABADI et al. 2011).

Third, researchers have looked for mutations that have a *de novo* origin. Comparison of the genomes of parents with those of their children can identify new mutations, present only in the child, that are most likely the cause of intellectual disability (VISSERS et al. 2010). Figure 8.3 in Chapter 8 illustrates a *de novo* inheritance pattern. You don't need a large sample to find things using this approach and you don't need the entire genome, just the coding sequence, or exome. The study that demonstrated the efficacy of the approach sequenced the exomes of just ten patients with unexplained severe intellectual disability, as well as their unaffected parents (VISSERS et al. 2010). Larger studies have confirmed the viability of the method (DE LIGT et al. 2012; RAUCH et al. 2012). The largest study has now sequenced the exomes of 4,293 families, a sufficiently large number that the investigators could find multiple *de novo* mutations in the same gene, providing highly convincing evidence for the gene's roles in intellectual disability (DECIPHERING DEVELOPMENTAL DISORDERS STUDY 2017). However, we should emphasize that finding a *de novo* mutation is not proof that the mutation is causal, as we discussed in the case of *de novo* mutations found in ASD (see Chapter 8, section 8.5).

9.5.2 Syndromal intellectual disability

There are 1,177 traits in the catalog at Online Mendelian Inheritance in Man (http://omim.org) associated with intellectual disability, and the genetic basis of many of them is now known (Figure 8.3 in Chapter 8 provides a simple review of Mendelian inheritance patterns). The great majority are syndromal forms of intellectual disability; that is, the intellectual disability is one of many features, including anatomical and physiological abnormalities in many systems. Molecular characterization of syndromal intellectual disability is less likely to give insights into the genetic basis of behavior, partly because it is not clear to what extent the intellectual disability is a secondary symptom of the genetic lesion. Mendelian disorders of lipid storage often include intellectual disability as a phenotype; cranial malformation syndromes that involve abnormal growth of the skull impair brain development and thus are a cause of intellectual disability. Characterizing genes affecting lipid metabolism and skull growth has not taught us much about behavior. However, there are some exceptions. Rett syndrome deserves mention for one remarkable lesson that it teaches.

Rett syndrome is a devastating illness. Children initially show normal development, but by the second year of life suffer an unusual pattern of loss of motor skills, in which stereotypical movements replace purposeful hand movements. About half cannot walk; those who walk do so with a wide-based and unsteady gait. Children with Rett syndrome lose speech and develop abnormalities of breathing with periods of hyperventilation and breath-holding. In addition, they have gastrointestinal problems, including severe constipation, and cardiac electrical problems. Seizures, contractures, and fractures are common.

The syndrome is unusual in being an X-linked dominant condition, so almost all sufferers are girls. This is because affected boys have a much more severe illness and most either do not survive to term, or die soon after. Rett syndrome is due to a single mutation in an X-linked gene encoding methyl-CpG-binding protein 2 (MeCP2) (AMIR et al. 1999). MeCP2, which is highly expressed in the brain, binds to methylated DNA, and its function is to repress transcription (the role of methylated DNA is described in Chapter 4, section 4.2.4).

Rett syndrome is an example of a neurodevelopmental disorder, a category that would include ASD and indeed most genetic forms of intellectual disability.

One of the assumptions that most people hold about severe childhood-onset neurodevelopmental disorders, particularly when they are due to genetic mutations, is that they are irreversible. Most of us imagine that the damage must have been done very early on, well before birth, so that the chances of fixing anything are about zero. Yet we would be wrong to believe this.

In a remarkable experiment, Adrian Bird, a geneticist from Edinburgh University, was able to reverse the consequences of the Rett mutation, at least in mice (GUY et al. 2007). His team engineered the genome of a mouse so that they had control over the expression of MeCP2, allowing them to turn expression off during development (as happens in the human mutation) and also turn expression back on again after birth (how this was done is described in Chapter 16). They showed, convincingly, that the phenotype was reversible: when the gene's expression is turned on in the adult animal, there is restoration of neuronal function. This experiment clearly showed that developmental absence of MeCP2 does not irreversibly damage neurons. It raises the intriguing possibility that other disorders that we consider 'neurodevelopmental' may also be reversible (a point we discussed in Chapter 8, section 8.9, in the context of ASD).

9.6 The biology of intellectual disability

When the very first nonsyndromal intellectual disability genes were found in 1998—*GDI1* (D'ADAMO et al. 1998), *PAK3* (ALLEN et al. 1998), and *OPHN1* (BILLUART et al. 1998)—there was widespread enthusiasm for the idea that these genes would serve as a foothold for attempts to understand how the brain works. The fact that all three genes acted through the same biochemical pathway (involving a protein called Rho GTPase) was incredibly exciting. But, consistent with other stories in this book, there has been more promise than delivery.

Currently, more than 700 genes for intellectual disability have been found. From the perspective of 1998, that is an unimaginable achievement, and is testimony to the transformational power of next-generation DNA sequencing (see Chapter 3, section 3.5). Mutations in about 60% follow autosomal-recessive inheritance, mutations in 20% are autosomal dominant (mostly *de novo*), and the remainder (20%) are X-linked

(Figure 8.3 in Chapter 8 provides a simple review of Mendelian inheritance patterns). Screens for X-linked genes are now returning fewer and fewer novel genes, suggesting that the catalog is close to completion. It is difficult to reliably predict how many more genes are still to be discovered, but perhaps something over 1,000 will eventually be found.

While the story that emerges from the gene discovery program reveals some patterns, the overall picture is one of heterogeneity, of multiple processes leading to intellectual disability. One simple way to see this is to look for similarities in the known biological functions of the 700 or so genes (KOCHINKE et al. 2016). You can obtain annotations about possible, suspected, or known biological functions from databases, such as The Gene Ontology Resource (http://geneontology.org/), otherwise referred to as 'GO annotations'. A single gene is usually assigned multiple functions, which are hierarchically based, so, for instance, a gene could be annotated as being involved in development (a high-level classification) as well as glycosylation (a lower-level category). It's perhaps not surprising that 200 intellectual disability genes are associated with metabolism, the single commonest function. The next largest categories, in terms of numbers, are transporters (proteins that transport things across cell membranes), nervous system development, RNA metabolism, and transcription.

A useful way to show function is to calculate the amount by which a group of genes is enriched in each functional category, compared with a random set of genes. Figure 9.1 shows the enrichment for 20 functional categories (GO annotations) for genes known to cause intellectual disability. The most enriched terms are signaling pathways, peroxisomes (small, membrane-enclosed organelles that contain enzymes involved in metabolic processes), glycosylation, and cilia (hair-like structures that stick out of cell surfaces). Frequently discussed functions of intellectual disability genes, such as synaptic and chromatin-related processes, are much less highly enriched. Figure 9.1 illustrates how an unbiased assessment of function, compared with reviews of the detailed studies that explore function of individual genes, can generate rather different views of what genes do.

There are four functions that deserve mention;

1. **A number of genes are involved in development:** defects in the production of new neurons (neurogenesis), neuronal differentiation, and migration are often associated

Figure 9.1 Enrichment of gene functional annotations (listed on the *x*-axis) found for mutations that cause intellectual disability. The *y*-axis is the fold enrichment for each class of annotation. MAPK, mitogen-activated protein kinase; TOR, target of rapamycin.

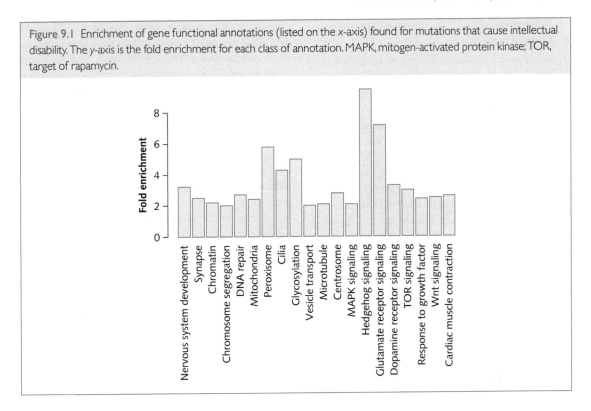

with intellectual disability. The disorders are frequently severe and include several with marked abnormalities of the brain, such as the lissencephalies, where the brain loses its normal folds and appears almost smooth (KEAYS et al. 2007).

2. **A number of mutations compromise synaptic function.** Synapses are the places where neurons communicate, releasing neurotransmitters from one side of the synapse and activating (or inhibiting) neurons at neurotransmitter receptors on the other (see Chapter 16, section 16.5). Synapses are found on dendrites, which are thread-like extensions to neurons. The dendrites of most neurons are covered with small protrusions known as dendritic spines, the main sites of excitatory synaptic input. The shape and structure of synapses and spines can change in response to stimuli, conferring synaptic plasticity on the neuron, a property thought to be the cellular correlate for learning and memory. All these aspects of synaptic function are altered by mutations that cause intellectual disability (examples include the Rho GTPases, such as

Cdc42, RhoA, and Rac). Furthermore, synapses are part of another piece of cellular machinery, called the postsynaptic density because of its appearance under an electron microscope. Of 1,500 proteins assigned to the postsynaptic density, 133 have been associated with nervous system disorders, most of which include intellectual disability (BAYES et al. 2011). This is significantly more than would be expected by chance.

3. **Intellectual disability genes are involved in cellular signaling processes.** After a neurotransmitter binds to its receptor, changes occur downstream, within the cell body, sometimes leading to profound changes of cellular metabolism. There are numerous signaling pathways that convey information from the cell surface to other parts of the neuron, and genes involved in this process are mutated in intellectual disability. Examples include the 'RASopathies', mutations in the RAS–MAPK (mitogen-activating protein kinase) pathway, which regulates growth factors and embryological development. The RASopathies include intellectual disability

syndromes, such as Noonan syndrome and Costello syndrome.

4. **Some intellectual disability genes regulate gene expression through epigenetic mechanisms,** as discussed in Chapter 4, section 4.2. A good example is the *ATRX* gene, which encodes a chromatin-remodeling protein, causing changes in gene expression and thus acting as a transcription factor (Gibbons et al. 1995). Mutations in transcription factors typically result in a large number of phenotypes, some apparently unrelated to intellectual disability (*ATRX* mutations cause a red blood cell disorder), presumably because of the disregulation of many different physiological processes.

9.7 The nonspecificity of genetic action

One other lesson to emerge from the genetics of intellectual disability is the relative lack of specificity of genetic action. There is considerable overlap between the genes found underlying intellectual disability and ASD: about 17% of all genes (Krumm et al. 2014). However, as about 70% of individuals with ASD also present with intellectual disability, this finding is not in itself surprising. Nor is it surprising that there is an overlap with epilepsy. However, four genes have been found that overlap with schizophrenia, a more surprising observation.

Figure 9.2 summarizes the current state of our knowledge about the overlap between intellectual disability, epilepsy, ASD, and schizophrenia (Vissers et al. 2016).

Nonspecificity of genetic action is a theme that recurs throughout this book. Genetic action appears never to adhere to psychiatric nosology; genes that are involved in intellectual disability turn out to be important in ASD and schizophrenia. These three conditions have in common that they are all neurodevelopmental disorders. We turn in the following chapters to examine what is known about disorders of emotion (anxiety and depression), drug and alcohol abuse, and personality. There are similarities in the genetic architecture of these conditions but, as we will see, also some profound differences.

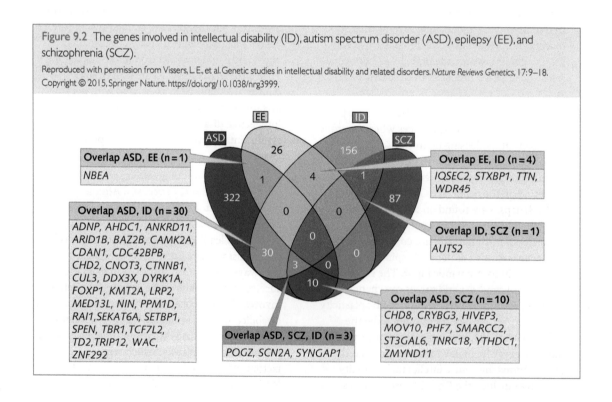

Figure 9.2 The genes involved in intellectual disability (ID), autism spectrum disorder (ASD), epilepsy (EE), and schizophrenia (SCZ).

Reproduced with permission from Vissers, L. E., et al. Genetic studies in intellectual disability and related disorders. *Nature Reviews Genetics*, 17:9–18. Copyright © 2015, Springer Nature. https://doi.org/10.1038/nrg3999.

Summary

1. Between 1 and 3% of the population has mild intellectual disability; severe forms occur in about 0.4%. Intellectual disability is commoner in men, with a 30–50% excess of males over females.

2. Genetic causes fall into three categories: polygenic, chromosome abnormalities, and single gene mutations. Polygenic causes are the commonest but least well understood.

3. Down syndrome, due to trisomy of chromosome 21, is the commonest cause of intellectual disability. Almost all trisomies are due to errors in maternal meiosis, which likely explains why trisomy 21 is related to maternal age.

4. The phenotypes of Down syndrome have (at least) two origins: overexpression of genes on chromosome 21, and dysregulation of the expression of genes that are not on chromosome 21.

5. Copy-number variants (CNVs; deletions and insertions), some of which are recurrent, are more common in children with intellectual disability than in controls. CNVs are pleiotropic: for example, a deletion on chromosome 15 that causes intellectual disability is also found in cases of ASD and schizophrenia, and in approximately 1% of cases of epilepsy.

6. Phenotypic variability is explained by the presence of additional CNVs: among children with intellectual disability, about 10% carry a second CNV, while those with two CNVs are eight times as likely to have intellectual disability as controls.

7. Fragile X syndrome, due to a cytogenetic abnormality on the X chromosome, accounts for 0.5% of all cases of intellectual disability. It is due to a trinucleotide repeat expansion of CGG in the promoter region of a gene called *FMR1*, which encodes a protein called FMRP. FMRP binds to RNA. Its function is still not fully understood

8. Genetic linkage mapping identified regions of the X chromosome that contain mutations, and led to the identification of a series of X-linked genes that cause intellectual disability.

9. Exome sequencing has shown that deleterious mutations, particularly *de novo* mutations, are enriched in cases of intellectual disability.

10. More than 700 genes for intellectual disability have been found, and it is predicted that something over 1,000 will eventually be found. The genes are involved in development, synapse function, and cellular signaling.

11. There is considerable overlap between the genes found underlying intellectual disability and autism spectrum disorder, schizophrenia, and epilepsy.

10

Anxiety, depression, and eating disorders

[On the phenomenology, genetic epidemiology, and molecular basis of three common psychiatric disorders]

The disorders described in the previous three chapters—schizophrenia, autism, and intellectual disability—have their roots very early in life. Some forms of intellectual disability are entirely genetically determined, but even for those that are not, the causative factors lie in childhood or before birth. In this chapter, we turn to disorders that are very different. Anxiety and depressive disorders are much more common than schizophrenia and autism. For major depression and anxiety disorders, the environment plays a large role, providing an additional complication for our attempts to explain how genes relate to behavior. This chapter also discusses bipolar disorder, which is grouped with major depression as an affective disorder but often has some clinical features more akin to schizophrenia (delusions and hallucinations). Some claim that it has features of a neurodevelopmental disorder. The last part of the chapter turns to eating disorders, which have characteristics that put them closer to anxiety and depression.

10.1 Major depression

It was the worst experience of my life. More terrible even than watching my wife die of cancer. I am ashamed to admit that my depression felt worse than her death but it is true. I was in a state that bears no resemblance to anything I had experienced before. It was not just feeling very low, depressed in the commonly used sense of the word. I was seriously ill. I was totally self involved, negative and thought about suicide most of the time. I could not think properly, let alone work, and wanted to remain curled up in bed all day.

WOLPERT (2006).

Like many terms in psychiatry, 'depression' is slippery, having multiple meanings that can easily result in misunderstanding. Depression can mean just a transient sad or gloomy mood that at one time or another occurs to all of us. That is not what psychiatrists mean when they use this term. Rather, they refer to a syndrome that starts with a mood that is persistently low, sad, blue or 'down in the dumps', but is also accompanied by a range of other symptoms including altered appetite, disturbed sleep, difficulty with concentration, impaired ability to experience pleasure, persistent guilt, hopelessness, and, in more severe cases, suicidal thoughts or behaviors. To discriminate the psychiatric concept of depression from sad mood, psychiatry has for many years used the term **major** depression, or as we will call it, MD.

MD is diagnosed when depressed mood, or a loss of interest or pleasure in daily activities, is present for more than 2 weeks, and if five or more out of nine symptoms (including low mood and loss of interest) occur nearly every day. The diagnostic criteria are not claimed to represent a coherent biological construct (and we will see evidence in section 10.1.4 from genetics that MD is heterogeneous); the justification for the diagnostic criteria is that they are reliable, sensitive, and relatively specific. Psychiatrists are rightly proud of this achievement (REGIER et al. 2013), but it would be a mistake to believe that because we have reliable criteria for MD that we understand the pathology of the disorder (KENDLER 2016). We don't, and our knowledge of the disease is still primarily reliant on patients' descriptions of the experience of what it is like to be depressed.

Lewis Wolpert's description of his depression captures the idea that low mood and MD are different things. For Wolpert, severe depression is not just being lower than usual; it is not just being 'down'. William Styron, another sufferer, explains that depression is a word '*that has slithered through the language like a slug, leaving little trace of its intrinsic malevolence and preventing by its very insipidity a general awareness of the horrible intensity of the disease when out of control*' (STYRON 1992). To make concrete what it means to say the disease is 'out of control', about half of all suicides can be directly attributed to depression. Depression isn't just low mood—it's a condition that takes lives.

But is MD, the disease that psychiatrists treat, qualitatively different from low mood, or is it the extreme end of what we all experience, from sadness and low mood to misery? And once we resolve that question, how do we know that depression in one person is really the same as in another? Is it always the same disease, with the same underlying pathology, or does it represent the final common outcome of many different pathways (AKISKAL AND MCKINNEY 1973)?

10.1.1 Genetic epidemiology of MD

MD is disturbingly common. About one out of five women and about one out of eight men will, at some point in their lives, meet the criteria for this disorder (KESSLER et al. 2003). MD is common everywhere in the world (although the frequency is lower in some places, notably East Asia; PHILLIPS et al. 2009), and it imposes a heavy burden on each country's health system and economy, as well as directly impacting sufferers and their families (USTUN et al. 2004; TURNER et al. 2008). Depression imposes such a burden that in 2017, the World Health Organization ranked depression as the leading cause of disability in the world (WORLD HEALTH ORGANIZATION 2017).

What do we know about the genetics of MD? Many studies have shown that depression 'runs in families'. Our best investigations show that the risk for MD in the first-degree relatives of patients with MD is about threefold higher than that seen in the general population (Sullivan et al. 2000). What is the source of this familial aggregation? We can answer this question with reference to adoption and to twin studies.

10.1.2 MD: adoption studies

Until recently, only three major adoption studies had examined depression, and each has one or more major limitations (such as reliance only on hospital or work-leave records, never directly interviewing the relatives, or having quite a small sample size.) Of these three studies, two produced results suggesting that genetic risk factors were important for the etiology of MD (SULLIVAN et al. 2000).

In 2018, I (K.S.K.) and Swedish colleagues used a large Swedish population register to examine the resemblance for MD in parents and children from four family types: intact (2,041,816 offspring), adoptive (14,104 offspring), not-lived-with father (116,601 offspring), and stepfather (67,826 offspring) (KENDLER et al. 2018c). A few of these terms need explanation. A not-lived-with father had a child but then never lived with the child (or lived nearby). A stepfather, by contrast, lived in a family with no biological father for at least ten of the first 15 years of the child's life. Using these four family types, we quantified the parent–offspring resemblance for MD into 'genes plus rearing', 'genes only', and 'rearing only'.

The results for the different family types were combined to obtain a weighted estimate (e.g. combining results from adoptive fathers and stepfathers who both provide only rearing to their offspring). For fathers, we found a modest resemblance for liability to MD in the 'genes only' category: a correlation of +0.09. Surprisingly, the estimate for rearing only was exactly the same: +0.09. In addition, the estimate from intact families, where parents contribute both genes and rearing for their children was +0.15, very close to the simple addition of the estimated 'genes only' and 'rearing only' relationships. The pattern was relatively similar in mothers, although the 'genes only' relationship was lower (+0.05). However, this was known imprecisely (there is a very large confidence interval). So this large sample study ended up with a simple finding: parents contribute to the risk for MD in their children and the contributions from genes and from rearing are similar in magnitude; think of it as roughly equal doses of nature and nurture.

10.1.3 MD: twin studies

When a careful meta-analysis of twin studies of MD was completed in 2000, six studies were of sufficient quality to be included (Sullivan et al. 2000). These six studies all agreed well with one another, but they were all done in predominantly European populations in Europe, the USA, and Australia. So we do not know whether similar results would be obtained in other ethnic groups. Across all the studies, the estimated heritability for MD was 37%. This figure is much lower than the heritability estimates for schizophrenia and alcohol dependence. So we can draw an important conclusion: there are significant differences in heritability among psychiatric disorders.

Since this meta-analysis was published, one additional very large-scale twin and sibling study of MD has been completed (Kendler et al. 2018b). We used several population-based registries in Sweden, including a newly created primary care registry and, in addition to standard twin modeling approaches, we employed a different way of estimating heritability—comparing resemblance between full siblings (who on average share half of their genetic variants identical by descent) and half-siblings (who on average share one-quarter). The sample sizes were impressive: twin and sibling/half-sibling analyses including 26,892 and 1,691,971 pairs, respectively. Unlike most previous studies, however, our investigation did not rely on personal structured interviews but rather on diagnoses recorded by diverse clinicians. This approach has

some advantages—for example, you do not need to rely on accurate recall from subjects about depressive episodes that might have occurred decades ago or to worry about their unwillingness to report episodes they do remember because of embarrassment. The disadvantage is that different clinicians may use different approaches in diagnosing MD. As we have said before, there is no perfect study in human genetics. In general, you should have faith in results that are reached by a variety of different methods as the biases of each are unlikely to be the same.

Men and women were analyzed separately in this large study. Using the standard twin method, heritability of MD in males was estimated at 41% but with quite large confidence intervals. In the much more precise estimates using the full and half-siblings, heritability in males was estimated at 36% (which was statistically no different from the twin estimate). In women, heritability estimates were higher, and were very similar in the twin and sibling samples: 49 and 51%, respectively.

There are three major take-home messages from this large study. First, heritability estimates for MD obtained from twin and sibling samples were very similar. This suggests that many of the concerns about twin studies we reviewed in Chapter 2, section 2.6 (e.g. that the striking physical resemblance of monozygotic twins might cause them to be more similar than dizygotic twins for social–environmental reasons) do not meaningfully impact on risk for MD. Second, the heritability estimates obtained (36–51%) are in the range of the previous meta-analysis. Third, there is a sex difference: MD has a higher heritability in women, a point that we elaborate on next.

10.1.4 MD is heterogeneous

Those unfamiliar with the literature debating the division of MD into subtypes may be surprised not only at the diversity of the proposed classificatory systems employed (e.g. dimensional, hierarchical, or categorical), but also at the vehemence with which each position has been defended or, more usually, attacked (Eysenck 1970; Parker 2000). The importance of this acrimonious debate is the extent to which genetic research strategies might resolve it, and potentially guide interpretation of the underlying pathogenic mechanisms. Genetic data do, in fact, indicate heterogeneity (Flint and Kendler 2014). Most striking is the effect of sex.

Epidemiological studies of MD have consistently shown a higher prevalence rate for women

(WEISSMAN et al. 1993, 1996). Therefore, twin researchers have also been interested in asking whether the heritability of MD differs across sexes and whether the same genetic factors impact on risk for MD in men and women. The two major studies that have addressed this question found similar answers (KENDLER et al. 2001a, 2006c). In both studies, MD was appreciably more heritable in women than in men (40 versus 30% and 42 versus 29%, respectively), and both studies reported evidence for sex-specific genetic effects with genetic correlations between men and women estimated at +0.55 and +0.63. So while a substantial proportion of genetic risk factors for MD are shared in men and women, an appreciable proportion are relatively sex specific in their effect. However, the new and much larger Swedish twin and sibling study we reviewed above (KENDLER et al. 2018b) found a considerably higher genetic correlation for MD between the sexes: +0.89. While we know the genetic effects for MD are not entirely the same in men and women, the degree to which they differ is still somewhat unclear.

Heterogeneity is also evident at a phenotypic level. Do the nine symptomatic criteria for MD given in the *Diagnostic and Statistical Manual of Mental Disorders* reflect a single underlying genetic factor? Apparently not: three genetic factors best explain MD concordance in 7,500 adult twin pairs, reflecting the psychomotor/cognitive, mood, and neurovegetative features of MD (KENDLER et al. 2013).

Is there a more genetic form of MD? An old distinction between 'endogenous' and 'reactive' MD (GILLESPIE 1929) is based on the presumed occurrence of depressive episodes that are independent of environmental adversity (e.g. stressful life events), compared with episodes that are a reaction to adversity. Is it possible that the endogenous form of MD is more genetically determined than others? The short answer to that question is: no. In fact, the opposite is true: those reporting more stressful life events are more likely to have a family history of MD (KENDLER AND KARKOWSKI-SHUMAN 1997). However, this finding does indicate, as the large amount of literature on familial MD confirms, that clinical differences exist between those with and those without a family history of MD (reviewed in RUTTER et al. 1999; SULLIVAN et al. 2000). Distinguishing features are relatively nonspecific: those with a family history of MD have a more severe illness, tend to present at an earlier age, and suffer higher rates of recurrence (KENDLER et al. 1994b; 1999a; LIEB et al. 2002; WEISSMAN et al. 2006).

Environmental influences are also likely to stratify MD. Evidence from twin studies (KENDLER et al. 1995a, 2004; SILBERG et al. 2001) indicates that genetic risk factors for MD not only alter average risk but also alter the extent to which environmental adversity, particularly various forms of childhood maltreatment and recent stressful life events, increase the risk of depression. It seems that some people have an inherited tendency to react badly to stressful life events, while others inherit a resilience to adversity. This is often called a gene–environment interaction and we discuss it in more detail in Chapter 14, section 14.1. Here, we want only to point out the possibility of subdividing MD on the basis of the response to the environment.

10.1.5 MD: candidate gene studies

As is true of other psychiatric diseases, candidate gene studies of MD have generated many publications but few robust findings (Flint and Kendler 2014). Almost 200 genes have been subject to testing, many by multiple groups (LOPEZ-LEON et al. 2008; BOSKER et al. 2011). A thorough review of 18 candidate genes found little support for any (BORDER et al. 2019).

The difficulty, common in candidate gene research, is that few groups agree with each other, and although resolution of conflicting results can be attempted through meta-analysis, the results don't necessarily satisfy everyone (see Chapter 5, section 5.2.1). This problem is particularly egregious for the factions warring over the role of a variant in the serotonin transporter gene and its interaction with stressful life events, as explained in Chapter 5, section 5.2.3.

10.1.6 MD: genome-wide association studies

Three genome-wide association studies (GWAS) of MD have produced robust, replicated results, and all have very different designs. The first, which two authors of this book ran (K.S.K. and J.F.), was based on our view that MD was likely to be genetically heterogeneous. We pointed out above that genetic risk factors probably differ between the sexes, so that studying just one sex should provide a genetically more homogeneous sample, with greater power to detect loci. There are additional hints in the literature about likely sources of heterogeneity; for example, people who have suffered recurrent depression are also likely to have more and possibly different genetic risk factors than those who suffer just one episode. For these reasons, our study of the genetics of depression focused on women with recurrent depression.

We decided to carry out our study in China, and we are often asked, why work in China? In brief, the size of China's population means we could identify and recruit women to our study relatively quickly. We established a consortium, which we named, after an afternoon arguing in a Shanghai restaurant, CON-VERGE, for China, Oxford and Virginia Common-wealth University Experimental Research on Genetic Epidemiology. In China, we trained Chinese psychia-trists to interview cases for about 90 minutes to collect a wealth of information to ensure the diagnosis was accurate, and to obtain information about other rel-evant risk factors (e.g. stressful life events that might cause depression). Very few of the Chinese women whom we identified smoked, almost none of them drank alcohol, and there was no indication that any of them even knew of the existence of nonprescription drugs, thus providing us with a population almost free of the confounds that exist in most Western popula-tions. It took 6 years to collect a sample of almost 6,000 cases and 6,000 controls, genotype them all, and then identify two loci that contributed to the risk of MD (CONVERGE CONSORTIUM 2015).

The second GWAS used a sample collection approach that could not have been more divergent (HYDE et al. 2016). The study was run by a consumer genetics company called 23&Me, which makes money by selling genotypes to the public: you send in a sample of your spit and they send back data from a genotyping array, annotated to provide you with information on your ancestry and on risk factors for disease. 23&Me's data can tell you something about where your grand-parents were born, the parts of the world your ances-tors came from, and who you might be related to (you can imagine the surprises that might provide). When you send 23&Me your spit, they ask you to fill out a questionnaire, much of it about your health. One of the questions is: 'Have you ever been diagnosed by a doctor with any of the following conditions: Depression? (yes, no, I don't know)'. If you answered 'yes', you became a case in the GWAS study. The study analyzed 45,773 cases and 106,354 controls (23&Me runs a popular ser-vice, so they have a lot of data) and robustly identi-fied 15 regions that increased the risk of MD. None of them replicated the findings from the Chinese study and, conversely, the Chinese study supported none of the results from 23&Me.

Finally, an international consortium, the depression arm of the Psychiatric Genomics Consortium (PGC), used a range of methods for assessing MD: direct inter-views, national treatment registers, self-reported MD symptoms, and self-report of treatment for MD by a medical professional. They obtained 130,664 cases and 330,470 controls (most of the sample came from the 23&Me subjects) and found 44 risk variants, of which 12 overlapped with the 23&Me findings and none over-lapped with the Chinese sample (WRAY et al. 2018).

We don't yet know to what extent the discrepan-cies reflect differences in ancestry (genetic risk factors are likely to differ among ethnic groups because of variation in allele frequencies) or arise because of dif-ferences in the way the disease is diagnosed.

10.1.7 The genetic architecture of MD

We know the following about the genetic architec-ture of MD (Flint and Kendler 2014). First, it is highly polygenic, as revealed by single-nucleotide polymor-phism (SNP) heritability (i.e. the heritability estimated from genotype data; see Chapter 7, section 7.5). SNP heritability estimates for MD range from 8.7% (WRAY et al. 2018) to 21% (LEE et al. 2013) to 30% (LUBKE et al. 2012; PETERSON et al. 2016). SNP heritability is lower than family-based heritability estimates (38%), because SNP heritability does not yet access all the causal variants (Chapter 7, section 7.6 explains why this might be) (YANG et al. 2010). Because SNP-based heritability is a lower bound on the contribution to heritability from common variants, the fact that it is as big as it is indicates that common variants of small effect are a major contributor to the genetic suscep-tibility to MD. There is no support for the idea that there are rare variants of large effect.

Second, a component of genetic risk for MD is shared with other psychiatric disorders. We describe in Chapter 13, section 13.4 how GWAS data can be used to calculate the correlations in genetic risk and determine whether the same genetic loci that increase susceptibility to MD also increase susceptibility to other disorders. Two disorders that most frequently overlap diagnostically with depressive illness are anxi-ety disorders and bipolar illness.

Although the methods are quite different, genotype data agree with twin data that risk factors for MD over-lap considerably with those for anxiety (MIDDELDORP et al. 2005; CERDA et al. 2010; DEMIRKAN et al. 2011). The comorbidity can be attributed, in part, to a common genetic basis. Indeed, at a genetic level, generalized anxiety disorder and MD are the same: the correlation is not significantly different from unity (we discuss this finding in more detail in Chapter 13, section 13.2).

The story for the relationship between MD and bipolar disorder is more complex (Flint and Kendler 2014). For many years, genetic data have been employed to support a separation of MD from bipolar affective illnesses: relatives of those with bipolar disorder are more likely to develop bipolar disorder, and, conversely, relatives of MD probands are more likely to develop MD (Perris 1966). With few exceptions, subsequent studies have confirmed this observation: bipolar illness aggregates in the families of bipolar probands much more than in families of MD probands (Weissman et al. 1984). Using SNP-heritability approaches (So et al. 2011; Yang et al. 2011), there are now estimates of the genetic correlation between MD and bipolar disorder (Lee et al. 2013). The genetic correlation with bipolar disorder is 0.47 (with a standard error of 0.06), comparable to a twin-study genetic correlation of 0.64 (McGuffin et al. 2003). This finding suggests an overlap between MD and bipolar illnesses; many loci contribute to both conditions. Further consideration of the genetic effects shared between MD and other disorders is given in Chapter 13.

Third, genetic and environmental effects on MD interact to cause disease. There is evidence for a genetic susceptibility to environmental adversity, which includes stressful life events and childhood abuse. We review the evidence for this in detail in Chapter 14, section 14.1.

10.1.8 MD: outstanding issues

Genetic analysis of MD has been among the greatest challenges facing health researchers (Collins et al. 2011), but success in finding risk loci demonstrates that genetic dissection can often raise more questions than it answers. What actually has been mapped? Is the disorder one or many? We are still grappling with these questions. While there is considerable agreement in the literature that MD has heterogeneous causes, there is much less agreement about its homogeneity as a clinical disease (Parker 2000). The picture is consistent with a fairly undifferentiated phenotype emerging as the final common outcome of diverse processes, a process called 'equifinality' in the developmental literature. What are these processes? What are the pathways? The list of possibilities

is large: in addition to long-running favorites such as abnormalities of monoamine metabolism (including postreceptor components of the downstream cyclic AMP signaling pathway; Duman et al. 1997) and impaired corticosteroid receptor signaling (Holsboer 2000), more recent hypotheses include the involvement of neurotrophins (Samuels and Hen 2011), fibroblast growth factors (both ligands and receptors; Turner et al. 2012), GABAergic deficits (Luscher et al. 2011), and epigenetic changes, specifically alterations in methylation and acetylation profiles at the promoters of glucocorticoid receptors and brain-derived neurotrophic factor (McGowan et al. 2009). Genetics still does not support the primacy of one theory over another; indeed, to date, genetics does not support any of the biological theories put forward.

10.2 Bipolar disorder

The following is a description of a patient with bipolar disorder, seen by K.S.K.

Margaret was an unmarried 54-year-old school librarian from Vermont, described by her relatives as quite proper and highly religious when well. She had begun to suffer from bipolar illness in her late 30s and due to a poor treatment response was admitted to our ward for careful study. Upon admission, she barely moved, spending most of her day curled up in a fetal position on her bed, sometimes sobbing quietly. She had no appetite and often took a full minute to respond, in a very low voice, to a simple question. Ten days later, she got up and began to move about the ward, at first slowly with a sad downcast demeanor. But then her pace started to quicken. She began to smile and then put on subtle make-up. Soon she was occasionally laughing, talking loudly, and flirting with the male patients and physicians. A few days later, her make-up changed dramatically to include bold red lipstick and lots of powder and rouge. She became progressively less restrained, talking a mile a minute, laughing, singing, and skipping up and down the hallway, clearly euphoric. She tried several times to disrobe in public and started grabbing the private parts of male patients, openly inviting them to have sex.

Bipolar illness, along with schizophrenia, was first described by the German psychiatrist Emil Kraepelin in 1899. It is characterized by a life-long history of intermittent episodes of depression and mania. We have already talked about the clinical features of depression.

Mania is in many ways its opposite, with euphoria, hyperactivity, increased quantity and rate of speech, and the dissolution of social constraints. Patients often lack judgment about the potentially painful effects of their own often outrageous behavior. Margaret had a form of rapidly cycling bipolar illness in which depressive episodes moved straight into a manic episode with only a few days of normality in between. Bipolar illness is relatively uncommon, typically affecting about 1% of the population (WEISSMAN et al. 1996).

10.2.1 **Bipolar disorder: genetic epidemiology**

There is a large literature on the genetics of bipolar, second only to that on schizophrenia. At least 18 major family studies have been published, with average results suggesting a risk of bipolar illness in first-degree relatives of 8.7% compared with relatives of controls of 0.7% (SMOLLER AND FINN 2003). Similar to schizophrenia, close relatives of individuals with bipolar illness have a risk that is about ten times greater than that expected in the general population.

Studies of families with bipolar illness reveal an interesting wrinkle. Not only is the risk for bipolar illness substantially increased, but so is the risk for MD: it is about threefold higher. In fact, the risk for MD in relatives of bipolar patients is very similar to that seen in the relatives of people with MD itself. Now you might think, 'Oh, that is just because they failed to discover the manic episodes or they just haven't happened yet in the relatives.' But that is not true. Careful studies have followed up these relatives. They nearly all have classical recurrent MD without a hint of manic episodes (which are usually pretty dramatic and hard to miss). What this means is that in ways we only modestly understand, the familial risk factors for bipolar illness and MD are partly intermingled.

We have both twin and adoption studies for bipolar illness to help us disentangle the sources of resemblance in relatives. For bipolar illness, the answer is clear. Looking at the three best studies (reviewed in SMOLLER AND FINN 2003), if one monozygotic twin has bipolar illness, rates of bipolar illness in the co-twins are very close to 50% (i.e. at least 50 times greater than in the general population). Concordance rates in dizygotic twins are much lower, with an average estimate from these three recent studies of 8%. The best estimates for heritability range from 60 to 88%, with a mean of 75%. The results from twin studies of bipolar

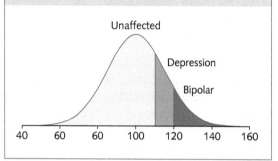

Figure 10.1 A multi-threshold model of the relationship between bipolar disorder and depression. A hypothetical score is shown on the x-axis, where a threshold above 110 triggers a diagnosis of depression and a score above 120 qualifies as a diagnosis of bipolar disorder.

illness are clear: it is highly heritable and runs in families, mostly, or entirely, as a result of genetic effects.

One of these studies (KENDLER et al. 1995b) tried to understand, from a genetic perspective, the relationship between bipolar illness and MD. The approach was to use a multiple-threshold model. This model assumes that there is a liability to 'mood disorders' that has two thresholds (Figure 10.1).

If you have a moderately high liability, you are prone to get only MD. If you have a really high liability on that dimension, you are more likely to get bipolar illness. The model fits the twin data well. So maybe (but this is still debated, as discussed in section 10.2.2) the two disorders are closely related and differ only as a function of the severity of underlying liability.

Only one rather small high-quality adoption study has ever been done of bipolar illness (MENDLEWICZ AND RAINER 1977). The sample was collected in Belgium and included 29 individuals who had both been adopted at a quite young age and had developed bipolar illness. Interestingly, two comparison groups were available. The first was a sample of 22 normal, adopted individuals. The second was 31 individuals with bipolar illness who came from nonadopted homes. Compared with the families of the control adoptees, rates of bipolar illness were elevated in the biological parents of the bipolar adoptees but not in their adoptive parents. Furthermore, the rates of bipolar illness were as high in the biological parents of the adoptees as they were in the natural parents from the intact families with a bipolar individual. So, we have evidence that bipolar illness is transmitted from parents to children genetically but not as a result of the rearing environment. To put that more crudely, parents don't 'teach' their children to have bipolar illness in the way they teach them a language, an ethnic identity, or political and cultural beliefs. While we might wish for more adoption studies, nonetheless these results are reassuring in that, using a quite different method, they agree with the twin findings. Bipolar illness runs in families for genetic and not, in any major way, environmental reasons.

10.2.2 **Bipolar disorder: molecular genetics**

Like other psychiatric disorders, a substantial number of linkage and candidate gene association studies were performed for bipolar illness. In fact, of the three very high-profile early linkage findings for all psychiatric disorders, two were on bipolar illness. Both were published in the high-profile journal *Nature* in 1987. One reported a locus on chromosome 10 based on pedigrees found in the Old Order Amish in Pennsylvania in 1987 (Egeland et al. 1987) and the other reported one on the X chromosome from pedigrees studied in Israel (Baron et al. 1987). Although they created a lot of excitement in the field, neither was replicated and both eventually were seen as false-positive findings.

Let us turn to the results of GWAS studies from the PGC Bipolar Working Group. As of May 2019, 20,352 bipolar cases and 31,358 controls had been genotyped (all of European descent). The GWAS revealed 19 genome-wide significant loci, which increased to 30 when a replication sample was added (Stahl et al. 2019).

There are two sets of findings that deserve comment. First, the genetic correlations (r_g) calculated from this GWAS for bipolar disorder were much more highly correlated with results from studies of schizophrenia $(r_g = 0.70)$ than with depression $(r_g = 0.35)$. These results, which suggest that, genetically, bipolar disorder is much more closely related to schizophrenia than to MD, are at odds with what Emil Kraepelin, the originator of our current diagnostic system, would have expected. They also differ from what family studies have predicted. We noted above how much MD is concentrated in the close relatives of individuals with bipolar illness. Most of the field assumes that the molecular results are correct and that there must be something wrong with the earlier family studies. The authors of this book are not so sure and are rather troubled by these discrepancies.

A second set of interesting findings depends on knowing that bipolar disorder can be divided into two forms. The first, bipolar I disorder, includes full manic episodes, as we described above for Margaret. The second type, bipolar II, involves only 'hypomanic episodes', which are more subdued than those of classic mania. Typical symptoms would be a reduction in sleep, greater energy levels, often more impulsivity, and sometimes poorly thought-out behaviors. Hypomania will rarely get anyone admitted to a hospital. Often, the episodes are only noticed by those close to the individual, especially if they have been warned to be on the lookout. The GWAS validated the clinical distinction between these two types. The genetic correlation with schizophrenia calculated from all the relevant SNP markers was moderately stronger from the bipolar I cases $(r_g = 0.71)$ than that seen with the bipolar II cases $(r_g = 0.51)$. The reverse and even stronger effect was seen in the genetic correlation with MD, which was more than twice as great with bipolar II cases $(r_g = 0.69)$ than in bipolar I cases $(r_g = 0.30)$. This is one of a small number of research findings from modern molecular genetics that has validated prior clinical intuitions.

10.3 Anxiety disorders

As a group, anxiety disorders are the most common psychiatric disorders. One reason is because some forms of anxiety are likely to be adaptive. There are, after all, a lot of dangers in the world that should either be avoided or confronted with caution. Because of this, evolution has probably tuned us to be vigilant of our environment. A modest amount of anxiety in new situations is common, and can improve performance (think of test taking or athletic competitions). But some individuals are prone to experience anxiety in ways that are disproportionate to the environment with which they are confronted. Often, such individuals suffer substantially, and their anxiety interferes with their well-being and functioning. These are the people in whom psychiatrists diagnose an anxiety disorder.

Of the many varieties of anxiety disorders, we focus on only two: panic disorder and phobias. We meet a third, generalized anxiety disorder, in Chapter 13, section 13.1. Most patients with panic disorder, which occurs in 2–4% of the general population (Kessler et al. 2006), are not normally anxious but then experience brief, rapid-onset periods of terror. Here is a typical case:

Jane was 18 and a freshman in college when she visited a large shopping mall the day after the Thanksgiving holiday. She was in a long line at a check-out counter with dozens of people crowding about her. All of a sudden, she felt her face flush, her heart start pounding and her hands shake uncontrollably. She could not catch her breath and she felt she was being smothered. She had intense anxiety and the fear that either she was going crazy or having a heart attack. An ambulance was called. She was rushed to the emergency room. Tests were made and all were normal. The doctor reassured her that nothing medical was wrong and said that most likely she was having a panic attack. If these recurred, he suggested she consider seeking psychiatric help as there were effective treatments available. In fact, the panic attacks came back regularly several times per week. She started avoiding crowded places as they seemed to bring on the attacks. Her friends and family, realizing something was wrong, finally convinced her to seek help.

Panic attacks are not innately abnormal. In a life-threatening situation, many of the symptoms of a panic attack are adaptive and commonly experienced. One of the authors (K.S.K.) was, in his youth, climbing in the Alps and made the foolish judgment to try to cross a wide scree slope some 10 yards above a cliff edge with a long drop down to a glacier.

As I started across the scree, it began sliding down. Spreading eagle, I helped it to slow down and eventually stop. I very carefully picked my way back to solid ground but not before I realized I was having nearly all the symptoms of a panic attack. I had just been in mortal danger and evolutionary pressures programmed me to respond in this way.

The abnormality in panic disorder is thus not having panic attacks per se but rather having them at the wrong time, in the wrong place, when there is no real danger, such as while shopping in a crowded mall.

10.3.1 Anxiety disorders: genetic epidemiology

The genetic epidemiology of panic disorder is well understood. It runs in families. First-degree relatives of panic-disorder patients have a fivefold increased risk of panic disorder, with rates of about 10% in relatives of affected individuals versus 2% in the relatives

of matched controls (HETTEMA et al. 2001). A meta-analysis of high-quality twin studies estimated the heritability of panic disorder at 48%, moderately higher than that seen for depression (HETTEMA et al. 2001).

Phobias are much more common than panic disorder, with lifetime prevalences for the individual subtypes ranging from 7 to 13% (MAGEE et al. 1996). They are defined by two key clinical features: (i) an irrational fear of a situation or object; and (ii) a fear, or the associated avoidance of the situation, that is significantly impairing. Most of us have one or more irrational fears, some of which can be adaptive. Consider how you feel about snakes, rats, thunder, spiders, closed-in places, flying, public speaking, heights, or needles. Such fears do not alone qualify as a psychiatric disorder. The fear or avoidance has to be life changing. Think of an individual terrified of speaking in public from an early age. They could not give reports in school, which impacted negatively on their grades. They liked debating but couldn't join the debating club. They wanted to be a lawyer but had to rule that out because of the need to speak in public. They avoided a promotion in their job because it would have involved having to present to the board. This is someone with a phobic disorder.

We can classify phobias into five subtypes: agoraphobia, social phobia, animal phobia, situational phobia, and blood-injury phobia. The meanings of several of these categories are clear from the name. Agoraphobia is a fear of open but also crowded places. Situational phobia includes two main forms: acrophobia (fear of heights) and claustrophobia (fear of closed-in places)

We know a fair bit about the genetic epidemiology of phobias, although they have not been studied nearly as extensively as some psychiatric disorders. Like virtually every other disorder we have examined, they run in families. First-degree relatives of probands with a phobic disorder have a risk for phobia 4.1 times greater than expected (HETTEMA et al. 2001). The one high-quality twin study of phobias estimated heritability ranging from 15 to 40% (HETTEMA et al. 2001).

These results don't address what is the most interesting question about the genetics of phobias: how do genes predisposing to individual phobias relate to each other? Do we inherit a general liability to be fearful of things in the world around us, a general liability that takes a specific form due to chance, to a childhood experience, such as being knocked down by a big scary dog, or falling out of a tree? Or, alternatively, do we inherit vulnerabilities to develop specific fears? One especially interesting question arises from an important difference between blood-injury phobia and all the other phobias. Individuals suffering from agoraphobia, social phobia, animal phobia, or situational phobia who are exposed to their phobic stimulus experience something very close to a panic attack. Their heart rate and blood pressure rise, they sweat, their hands shake, and so on. However, individuals with a blood-injury phobia have the opposite effect. At the sight of blood, in affected individuals, their pulse slows, their blood pressure drops, they become pale and light headed, and they often faint. Would we expect the genes that predispose to blood-injury phobia to be different, given the different physiological effects of the fear?

A multivariate twin model was used to address this question (KENDLER et al. 2001b). The key results are shown in Figure 10.2. We can first see that the total heritabilities of the phobia subtypes are different. Even with the large sample size of twins studied here (1198 male–male twin pairs), these estimates are known imprecisely, so we are only modestly confident that animal phobia is more heritable than agoraphobia or social phobia. More importantly for our current purposes, the figure also shows results from the model that allows us to distinguish a proportion of the genetic risk reflecting a 'general genetic fear' factor common to all phobias, and a proportion indexing genetic

risks specific to that phobia subtype. The ratios differ widely across subtypes. Thus, we have an answer to our first question: the genetic risks for individual phobias are partly shared with a general propensity to fears, and are partly fear specific. For our second question, our initial hypothesis was completely wrong. Blood-injury phobia is not at all different genetically from other phobias and in fact is highly similar to agoraphobia for having the highest proportion of its genetic risk originating from the general genetic fear factor. These results provide us with an interesting lesson. The genetic risks for the types of phobia seem unrelated to the specific physiological effects of confronting the phobic object. Although someone having an acute response to a blood-injury phobia (pale, slow pulse, ready to faint) looks very different from someone with agoraphobia stuck in a crowded restaurant (sweaty, flushed, pounding heart), the genetic basis of their fear propensity is actually quite similar.

10.3.2 Anxiety disorders: molecular genetics

We will pass over the large candidate gene literature for these disorders, as, consistent with what was seen with other psychiatric disorders, no robust replicated findings emerged. There have been only a few GWAS analyses published. Here, we focus on one published in 2016 (OTOWA et al. 2016). The authors began by noting that anxiety disorders as a group appeared to share some genetic effects. We saw this was true for different forms of phobias, but this is also true across a broader range of anxiety disorders. They studied seven different samples from the USA, Europe, and Australia with a total of around 18,000 subjects.

Their first approach was to count, as an affected case, any individual with any anxiety disorder. Second, using a more sophisticated measurement approach, they conducted a factor analysis of all forms of anxiety disorder in that study, which allowed them to assign every individual a liability score for the common anxiety disorder factor. Each approach yielded one genome-wide significant signal, but not the same one. However, when they calculated SNP heritability from their sample, the estimate using the factor analytic approach (0.095) was moderately higher that using the simpler case–control design (0.072). This hints that the factor analytic approach is a more effective way to tap molecular genetic risk for anxiety disorders. One other interesting result was evidence for a substantial

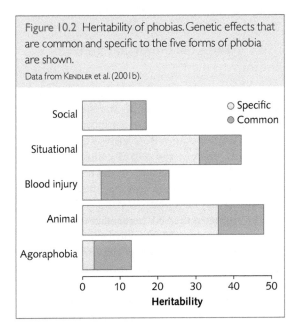

Figure 10.2 Heritability of phobias. Genetic effects that are common and specific to the five forms of phobia are shown.

Data from KENDLER et al. (2001b).

SNP genetic correlation (+0.68) between anxiety disorder variants found in this sample and those found in the PGC studies of MD (WRAY et al. 2018). We will take up the relationship between genetic risks for anxiety and depression in more detail in Chapter 13, section 13.1.

10.4 Eating disorders

Although 16 years old, Laura looked more like she was 12, standing in front of a mirror in her underwear on the psychiatric ward. She was 5' 4" and that morning had weighed 52 pounds. She resembled pictures of prisoners rescued from concentration camps or children dying of starvation. Most remarkably, as I walked past, she was pulling at her skin in her abdomen, muttering about how fat she was.

Laura was my (K.S.K.) introduction to the extraordinary psychiatric disorder we call anorexia nervosa. She had all the classical symptoms: extremely low weight, a refusal to eat anything but the most restricted diet—in her case, mostly small quantities of lettuce and occasionally a swallow of milk—and a markedly distorted perception of her body shape. Anorexia is the most severe of the class of what we call eating disorders and the one we will examine in this book. While likely to be an old disorder, modern genetic studies of anorexia date back no further than the last 35 years. It is a rare disorder, seen in less than 0.5% of women, with much lower rates in males.

10.4.1 Anorexia: genetic epidemiology

One of the best and most carefully conducted family studies was reported in 2000 (STROBER et al. 2000). It found that 3.4% of the close female relatives of women with anorexia (their mother and sisters) also had anorexia compared with 0.3% of matched controls—an 11-fold greater risk. This study also examined rates of bulimia nervosa, the other major eating disorder. While both anorexia and bulimia share a central feature of preoccupation with body size and shape, other features are quite different. Bulimia is characterized by binges of eating large quantities of high-calorie foods and subsequent efforts to lose the weight gained, most often by self-induced vomiting (purging). Interestingly, the rates of anorexia in the relatives of bulimics and the rates of bulimia in the relatives of anorexics were substantially elevated in this study. These

findings suggest that, whatever the factors that cause these two disorders to run in families, the two eating disorders share many of them.

No adoption study has ever been reported for anorexia, so our ability to disentangle the effects of nature and nurture for this disorder using traditional genetic designs has to rely on twin studies. Two different kinds of twin studies have been conducted for anorexia, and we will describe here one of each. The first approach identified twins with anorexia in hospitals. The classical study using this method was reported by HOLLAND et al. (1984). They found, after much searching, 30 informative female–female twin pairs. Among the 16 monozygotic pairs, nine (56%) were concordant for anorexia, while the parallel figure for the 14 dizygotic pairs was one (7%)—an impressive difference. Years later, researchers applied standard twin modeling methods to these data (BULIK et al. 2000). Heritability was estimated to be very high (88%), with the remaining effects due entirely to individual specific environment factors (12%).

The other approach to twin studies of anorexia utilized general population twin samples where a history of anorexia was detected by personal interview and/or medical records. The best example is a report from Sweden based on more than 30,000 twin pairs (BULIK et al. 2006). The heritability of anorexia in this sample was 56%. The heritability figures from these two twin studies were not significantly different from each other because they were not very precisely known due to the small sample of affected twins.

10.4.2 Anorexia: molecular genetic findings

While there have been a few candidate gene studies of anorexia (e.g. TRACE et al. 2013), results have, as in other psychiatric disorders, been inconclusive. The largest GWAS to date, a PGC analysis of 16,992 cases of anorexia nervosa and 55,525 controls, identified eight significant loci (the authors did not have access to a replication sample, so bear in in mind that not all of these loci may be robust) (WATSON et al. 2019). The authors estimated the SNP heritability to be between 0.11 and 0.17, with a highly polygenic genetic architecture, in other words, findings consistent with other psychiatric disorders we have reviewed. Again, they observed highly significant and remarkably large genetic correlations with obsessive compulsive disorder (0.45; $P = 4.97 \times 10^{-9}$), major depressive disorder

$(0.28; P = 8.95 \times 10^{-5})$, anxiety disorders $(0.25; P = 8.90 \times 10^{-8})$, and schizophrenia $(0.25; P = 4.61 \times 10^{-18})$. Intriguingly, they also noted significant negative correlations with metabolic traits: including body mass index (a measure of weight corrected for a person's height) $(-0.32; P = 8.93 \times 10^{-25})$, obesity $(-0.22; P = 2.96 \times 10^{-11})$, and type 2 diabetes $(-0.22; P = 3.82 \times 10^{-5})$. The genetic correlations were not due to the effects of common variants associated with body mass index. This finding throws into question the assumption that low body mass index is a consequence of the drive for thinness and dissatisfaction with body shape. Instead, it raises the interesting possibility that metabolic dysregulation may contribute to anorexia nervosa. It's a good example of how genetic studies cast new light on the etiology of psychiatric disorders.

Summary

1. Major depression (MD) is common: lifetime prevalence is about one in five for women and about one in eight for men.

2. Heritability estimates for MD obtained from twin and sibling samples are very similar, with estimates of about 40%. While we know the genetic effects for MD are not entirely the same in men and women (they are higher in women), the degree to which they differ is still unclear.

3. MD is a heterogeneous condition arising from a number of distinct biological and environmental pathways.

4. Genetic risk factors for MD alter the extent to which environmental adversity, particularly various forms of childhood maltreatment and recent stressful life events, increases the risk of depression. This is an example of gene–environment interaction.

5. Genome-wide association studies (GWAS) have successfully identified risk loci for MD. Common variants of small effect are a major, and possibly sole, contributor to the genetic susceptibility to MD. Genotype data agree with twin data that risk factors for MD overlap considerably with those for anxiety, and suggest an overlap between unipolar and bipolar illnesses in which some loci contribute to both conditions.

6. Bipolar disorder is characterized by a life-long history of intermittent episodes of depression and mania. It is relatively uncommon, typically affecting about 1% of the population.

7. The best estimates for heritability of bipolar disorder range from 60 to 88%, with a mean of 75%.

8. GWAS have identified a number of risk loci for bipolar illness. Genetic correlations calculated from genotype reveal higher genetic correlations with schizophrenia than with depression, contradicting the classical view dating back to Kraepelin.

9. Anxiety disorders are the most common form of psychiatric disorders. Panic disorder has a heritability of 48%. Phobias are much more common than panic disorder. They include five conditions: agoraphobia, social phobia, animal phobia, situational phobia, and blood-injury phobia, with lifetime prevalences for the individual subtypes ranging from 7 to 13%. Heritabilities range from 15 to 40%. The genetic risks for individual phobias are partly shared with a general propensity to fears, and are partly fear specific. No robust molecular findings have yet emerged.

10. The heritability of anorexia is 56%. A GWAS has identified eight genome-wide variants and suggests that metabolic dysregulation may contribute to anorexia nervosa.

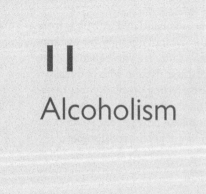

11
Alcoholism

[On the genetic epidemiology and molecular genetic basis of alcohol-use disorder]

11.1 Alcohol-dependence syndrome

As an introduction to the human face of alcoholism, this quote from James Joyce's *Dubliners* describes the sequelae of an evening's heavy drinking by an office worker in Dublin around 1900:

A very sullen faced man stood at the corner of O'Connell Bridge waiting for the Sandymount train to take him home. He was full of smouldering anger and revengefulness. He felt humiliated and disorientated; he did not even feel drunk; and he had only twopence in his pockets. He cursed everything. He had done for himself in the office, pawned his watch, spent all his money and he had not even got drunk. He began to feel thirsty again and he longed to be back again in the hot, reeking public-house. He had lost his reputation as a strong man, having been defeated twice by a mere boy. His heart swelled with fury and when he thought of the woman in the big hat who had bumped against him and said *Pardon*! his fury nearly choked him.

His tram let him down at Shelbourne Road and he steered his great body along in the shadow of the wall of the barracks. He loathed returning to his home. When he went in by the side-door he found the kitchen empty and the kitchen fire nearly out. He bawled upstairs:

—Ada! Ada!

… A little boy came running down the stairs … When the lamp was lit, he banged his fist on the table and shouted:

—What's for my dinner?

—I'm going to cook it Pa, said the little boy.

The man jumped up furiously and pointed to the fire.

—On that fire! You let the fire out! By God, I'll teach you to do that again!

He took a step to the door and seized the walking stick which was standing behind it.

—I'll teach you to let the fire out! he said, rolling up his sleeve in order to give his arm free play.

The little boy said *Oh Pa!* and ran whimpering round the table, but the man followed him and caught him by the coat. The little boy looked about him wildly but seeing no escape fell on his knees.

—Now you'll not let the fire out the next time! said the man striking him vigorously with the stick. Take that you little whelp!

JOYCE (1914).

Alcoholism is a very different disorder from schizophrenia or major depression, and sorting out what causes it is even more complicated, partly because of the multifarious nature of the disorder—its mix of physiological and psychological components—and partly because of the importance of cultural factors (drinking patterns vary widely among Italians, English, Japanese, and Jews, for example), developmental differences, and gender effects. For one thing, alcoholism is much commoner than psychosis. In the most recent large-scale representative study of the US population (GRANT et al. 2015), the lifetime risk for alcohol-use disorder—the new and rather broad definition of alcoholism adopted by the *Diagnostic and Statistical Manual of Mental Disorders: Fifth Edition* (AMERICAN PSYCHIATRIC ASSOCIATION 2013)—was 36.0% for males and 22.7% for females. Many in the field would favour the narrower definition of severe alcohol-use disorder, but even this disorder is far from rare, with estimated lifetime prevalences of 18.3% in men and 9.7% in women.

Then there are issues of diagnosis. For one thing, alcohol-use disorder can't happen when there's no alcohol (alcoholism is uncommon in Islamic countries and other cultures that ban the use of alcohol). Furthermore, human volition is clearly involved. You have to do something to develop alcohol-use disorder (at a minimum, acquire and consume large amounts of alcohol-containing beverages). While addiction can greatly increase the motivation for individuals to seek out and use psychoactive substances, they cannot literally cause you to pour yourself 'a strong one' and drink it. Some have argued that the whole problem of alcohol-use disorder arises because some people simply don't have the strength of will to say no when offered a drink. They argue that alcohol-use disorder is largely a disorder of willpower.

Because its effects are so corrosive and so destructive of relationships, the social consequences of alcoholism are different from those of schizophrenia. It's difficult to study the families of alcoholics, because often they have none. While for many families, having several members with schizophrenia brings the family together and increases family cohesion, alcoholism has the opposite effect. It tears families apart. This makes it difficult to study the genetics of alcoholism because it is difficult to obtain the family data.

After completing the large-scale study of families with a high density of schizophrenia in Ireland described in Chapter 2, section 2.2.1, one of us (K.S.K.) undertook a similar study of alcoholism. When I

started, I felt it would be a comparably easy study to conduct, but it proved much harder. Ask a sibling about how to contact their alcoholic brother and we would frequently get responses like:

I haven't seen him in 15 years. I don't know where he is and I don't want to know. He has ruined his own life, and for many years, he just about ruined mine. I am sorry for him, but he has brought it on himself. I have just had enough of it. I won't have anything to do with him ever again.

One way to avoid traveling around winding Irish lanes, staying in rural boarding houses with no heat after 8 p.m., is to obtain access to data someone else has already collected. Taking advantage of the records kept by the Swedish medical system, this proved possible for an extraordinary twin sample of alcoholism. About the time that prohibition was gathering force in the USA, Sweden took a different approach to the problems of alcohol abuse by developing a system of local temperance boards. Physicians, police, and public prosecutors were legally enjoined to report all information on alcohol-related problems to these boards, which then had to investigate each individual case. An individual's record would follow them throughout their life. While initially existing only on paper, the Swedes eventually computerized the records of all the local temperance boards into one large national registry. However, in the early 1970s, at a time when concern in Sweden was particularly high about privacy, the government made a decision to destroy the temperance board records.

11.2 A twin study of alcoholism in Sweden

Sweden also maintained a twin registry, which has not been destroyed. We've already mentioned that the heritability estimates provided by most twin studies have very broad confidence intervals: even though the mean estimate of heritability for many conditions may look impressive, often the study is not powerful enough to rule out quite different values. So you can imagine the attraction of a national twin registry for people like me (K.S.K.), who have been frustrated by years of working on smaller samples:

The Swedish Twin Registry had accumulated various accretions and appendages over the years; it has been collected by numerous people and worked over by numerous researchers, each with their own interests and own ways of working. I had worked with Nancy Pedersen at the Swedish Twin Registry on the heritability of depression in a small subset of the registry, just 486 pairs [Kendler et al. 1993e]. I was interested to see what else might have been collected that was relevant to depression, so I requested a list of the whole database. It arrived and was quite faint, having passed through several generations of photocopies. Furthermore, it was in Swedish, a language with which I was only faintly familiar. I wasn't expecting much, as I sat in my office, flipping through the multipaged document. Generally, these exercises turn up very little. But as I leafed through, a heading caught my eye. Even without knowledge of Swedish, I knew that *registrerings-kontor* might mean the temperance board registration data. If so, here, in front me, was evidence of the largest twin data set ever collected for the study of alcoholism. This was the early days of email. I wrote to Nancy Pedersen at once.

Two days later, I had an answer. Shortly prior to the formal destruction of the temperance board data, somebody had matched it, using the Swedish ID system, which uniquely identifies each citizen, to the Swedish Twin Registry. For some reason, this was only done on male–male twin pairs. The data had been sitting in a data file unrecognized and unanalyzed for over two decades. We had complete data on alcohol-related problems in over 8,900 twin pairs born from 1902 to 1949. Registration was not a rare thing. Fully 14% of the sample, 2,516 twins to be precise, had presented to medical or legal authorities with alcohol-related problems [Kendler et al. 1997]. What was so remarkable about this data set is that the information in it was extraordinarily unbiased.

If you had set out to collect data on alcoholism, the more typical approach would be to interview twins or adoptees and ask about their past experiences: how much they drank, how it interfered with their lives, whether they had withdrawal symptoms, and so on. Inevitably such data (we call it 'self-reporting') will have its biases: some people won't want to divulge this information, and even if they do, their memory may not be very reliable (particularly if they have been alcohol dependent). By contrast, the information from the temperance board records was contributed by third parties (e.g. doctors, police) at the time that the alcohol-related problems occurred. It's likely that a large majority of individuals with true alcohol problems had come into contact with the temperance board, as alcoholism is typically a public

phenomenon. While some psychiatric disorders, such as major depression, often go untreated as the affected individuals live what Thoreau has called 'lives of quiet desperation', few who are severely affected by alcoholism escape coming to the attention of legal or medical personnel. Thus, the data were of high quality, and were available in a large quantity.

We found that the concordance rates for temperance board registration were substantially higher (47.9%) in monozygotic than in dizygotic (32.8%) twins. Model fitting using the kinds of statistical models that we explain in the Appendix, section A5 produces a heritability estimate of 54% for alcohol abuse. The results of this quite large Swedish study were consistent with findings from population-based twin studies in the USA and Australia, which involved personal interviews. Nearly all of these studies have found estimates of the heritability of alcoholism in the range of 50–60% (Pickens et al. 1991; Kendler et al. 1992a; Heath et al. 1997; Prescott and Kendler 1999). Indeed, a formal meta-analysis of all twin and adoption studies of alcoholism found the best-fit estimate of the heritability of alcohol-use disorder was 0.49 (95% confidence intervals: 0.43–0.53) (Verhulst et al. 2015).

To complement the Swedish twin study of alcoholism described above, we recently completed an adoption study of alcohol-use disorder in Sweden (Kendler et al. 2015a). While adoption studies were quite commonly used in psychiatric genetics in the 1960s and 1970s, for the rest of the twentieth century they were nearly eclipsed by the surge of twin studies, which are much easier to accomplish. But, as a result of the increasing availability of national registries from the Scandinavian countries, mostly Denmark and Sweden, adoption studies are making a comeback.

Like twin studies, the point of adoption studies is to disentangle the sources of familial resemblance for traits or disorders we care about. The two sets of causes we are most interested in discriminating between are, at first glance, the same ones we examine in twin studies—genes and family environment (nature and nurture). But there is an important difference. Twins are always the same age. Parents and children are always a generation apart. Twin studies look at environmental effects that make siblings similar as they are members of the same generation. Adoption studies look at environmental sources of parent–offspring resemblance. So, the familial environment we examine in twin and adoption studies are not the same.

Thinking about alcohol consumption, teenage twins might resemble each other because they share friends, they attend the same school, or they both have an older brother who encourages their drinking and even buys them a six-pack now and again. However, these kinds of environmental effects won't impact much on the resemblance between parents and offspring.

For adoption studies, we typically think about two kinds of cross-generational environmental processes: direct and indirect. In direct transmission, the parent teaches the child a particular behavior or attitude. Psychologists have called this process social learning. Children might end up resembling their parents in their drinking habits because they learned them from watching their parents. We are, after all, pretty sure that parents influence lots of attitudes that their children have about political and social values, and so it makes sense that a parent directly influences their children's attitude toward alcohol.

Indirect transmission is more complex. A parent might literally teach their child to develop alcoholism—that heavy alcohol is a good way of dealing with problems, that heavy drinking and partying is an attractive (and, for men, a macho) life style, and so on. But, they could transmit a risk for alcoholism to their children in a quite different way. It turns out that bad parent–child relationships predispose to risk for alcohol and drug abuse. Maybe alcoholism in parents increases the risk for alcoholism in their kids because alcoholic parents are often not good parents, paying more attention to their drinking than to their children, and cause conflict in the home that is not good for the children growing up in it.

We now turn to our Swedish adoption study of alcoholism (KENDLER et al. 2015a). Like the Swedish twin study we reviewed above, we did not use personal interviews with the relatives. In this study, we assessed alcohol-use disorder from three sets of administrative records: complete national medical records, complete criminal records, and complete pharmacy records (receiving drug treatment for alcohol-use disorder).

Using a standard adoption design (eliminating individuals adopted by biological relatives and those who were above the age of 5 years when placed into their adoptive home), we found that the risk for alcohol-use disorder in the adoptee was increased by 40% if one of their adoptive parents had alcohol-use disorder, and 46% if one of their biological parents had alcohol-use disorder. These figures do not statistically differ from

one another. In other words, transmission of risk for alcohol-use disorder from parent to child was approximately equal when the parent raised the child but had no genetic relationship, and when they had a genetic relationship but played no role in rearing.

The validity of adoption studies depends on low levels of selective placement, meaning that children with a high genetic risk are not placed into homes with high environmental risk at above-chance levels. We could directly test this in our Swedish adoption sample. We found that the correlation between genetic and environmental risk (from, respectively, their biological and adoptive parents), while positive, was too small to produce any substantial bias: +0.07.

Another concern with adoption studies is the unrepresentativeness of the adoptive parents. They are screened by adoption agencies, and in Sweden, this is a particularly rigorous process. Individuals with a history of severe psychiatric and drug-abuse problems are much less likely to be selected to adopt a child. Biological parents must also be willing to go through the rigors of a legal adoption. We developed a different adoption-like design for which we used the informative but rather awkward phrase: triparental families (KENDLER et al. 2015b). We identified 41,360 individuals in Sweden born between 1960 and 1990 who were raised in a triparental family, which consisted of (i) a biological mother who raised the child; (ii) a not-lived-with biological father who sired the child but never subsequently lived with him or her, or even in the same town; and (iii) a stepfather who lived with the child for at least ten of their first 15 years.

The elegance of this design is that in a single family we have one parent (the biological mother) who provided for the child both genes and rearing, one parent (the not-lived-with father) who provided genes but no rearing, and one parent (the stepfather) who provided rearing but no genes (Figure 11.1). As should be clear, to a first approximation, the not-lived-with father is like the biological father from a classic adoption design and the stepfather is like the adoptive father.

In these triparental families, the risk for alcohol-use disorder in the offspring if the biological mother, not-lived-with father or stepfather had alcohol-use disorder was increased to 123, 84, and 27%, respectively. Like the classic adoption design, we found that offspring risk for alcohol-use disorder could be influenced both by parental genes and by rearing. However, while the adoption design suggested that they were of similar impact, our triparental study,

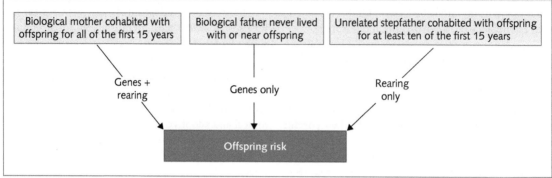

Figure 11.1 Triparental families. The biological mother who raises the child contributes both genes and rearing environment to the offspring. The biological or not-lived-with father never resides with or near the offspring and therefore contributes, as a first approximation, only genes to the offspring. The stepfather, who is not biologically related to the offspring, contributes only the rearing environment.

for reasons we do not entirely understand, provided a somewhat different answer: overall, genes are more important than rearing.

The triparental study also provides a couple of additional features. Another way to estimate the importance of rearing would be to compare the increased risk from the biological mother (genes + rearing) with the not-lived-with father (genes only). This gives an estimate of a 39% increased risk, close to what we find for the direct test of rearing effects that we get with the stepfathers. We obtain another estimate for the role of genes in the transmission of alcohol-use disorder by comparing the increase in risk from the biological mother (rearing + genes) and the stepfather (rearing only). Here, we estimated a 96% increase in risk, again reassuringly close to what we find in our direct test with not-lived-with fathers.

Our Swedish study was not the first adoption study of alcoholism. Prior studies had been done in Denmark, Sweden (using a sample that did not overlap with ours), and Iowa, all of which found that risk for alcohol-use disorder was increased in the adopted-away children of biological parents with alcohol-use disorder. So genes mattered. However, no prior study had had a very good look at adoptive or step-parents. Our two studies, the classic adoption and the triparental design, therefore were the first to clearly show that rearing also matters.

Twin and adoption studies together provide consistent results. Alcoholism is not merely a weakness of will or a personality quirk. Rather, it is a quite heritable condition, although somewhat less so than

schizophrenia. But, importantly, its transmission within families is not only a result of genes—environmental factors also play a role.

11.3 The molecular genetics of alcoholism

One way in which molecular genetic studies of alcoholism differ from most other psychiatric disorders is that we have good physiological candidate genes for alcohol dependence. We have candidates that are every bit as good as the insulin gene is for diabetes or a cholesterol receptor is for heart disease. The deep difference between alcoholism and other psychiatric disorders can easily be summed up: the ethanol molecule (ethanol is the chemical name for the kind of alcohol that humans consume). We know a lot about how ethanol is metabolized in the body and how it acts on the brain. It does not take a physiological genius to figure out that differences in how the ethanol molecule is handled in the body might make a difference to the risk for alcoholism. And, indeed, that is the case. So, in contrast to the situation with the rather dismal results with candidate genes for psychiatric disorders described elsewhere, the physiological candidate gene approach has worked for alcoholism.

The ethanol molecule is almost entirely broken down in the human body in two steps. As shown in Figure 11.2, the first is by a group of enzymes called alcohol dehydrogenases, or ADHs, for short. There are seven known variants of this enzyme, which

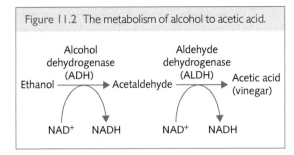

Figure 11.2 The metabolism of alcohol to acetic acid.

converts alcohol into a compound called acetaldehyde. Acetaldehyde is toxic. Even moderate concentrations induce dysphoria, and higher concentrations consistently produce bad headaches, nausea, and flushing. The second family of enzymes, called acetaldehyde dehydrogenases or ALDHs, convert acetaldehyde into acetic acid (vinegar), which is either excreted in urine or converted to a metabolic intermediary (called acetyl coenzyme A). The level of acetaldehyde is critical to the subjective effect we have after drinking alcohol. If the level is kept low, few acute side effects are experienced. If the level rises, the drinker often doesn't feel good, and if the level becomes quite high, the drinker feels ill.

Figure 11.2 shows that high levels of acetaldehyde arise in two ways. Either you make it too quickly or break it down too slowly. It turns out that both paths are important for influencing the risk for alcoholism, although the latter is more dramatic than the former. In East Asian populations, a substantial proportion of people (43% in one study; CHEN et al. 1996) have a form of one key ALDH enzyme (ALDH2) that barely works at all. Individuals who possess even one copy of this form (called ALDH2*2) do such a poor job of breaking down acetaldehyde that even after only one good drink, the concentration of acetaldehyde rises to high levels. The results are predictable: the individual turns as red as a beet, develops a splitting headache, and feels generally awful. This is sometimes called the "flushing reaction". Many studies have shown that individuals with this variant of ALDH are unlikely to develop alcoholism (AGARWAL AND GOEDDE 1992). Although it might be considered yet another example of science proving the obvious, we now know with some confidence that it is hard (but not impossible) to develop alcoholism when drinking substantial amounts of alcohol makes you ill.

This is not the end of this story: the molecular genetics revolution arrived for alcohol-use disorder.

In the last few years, a genome-wide association study (GWAS) successfully identified risk loci for alcohol-use disorder (JORGENSON et al. 2017). Gelernter and colleagues used a GWAS to examine alcohol dependence, an older diagnosis roughly equivalent to severe alcohol-use disorder (GELERNTER et al. 2014). Samples for this study came from multiple sources, including both European and African-American subjects, yielding a total of over 16,000 individuals (including cases and controls, and initial and replication samples), so power was available to detect variants of moderate to large effect.

The most significant findings were in the ADH gene complex on chromosome 4. This is reassuring. Unlike GWAS for schizophrenia, autism, or depression, studies of alcoholism have a kind of built-in expected signal. We expect to find a signal at the ADH locus. Intriguingly, while variants in the ADH gene complex were found in both the African-American and European-American samples, they were not the same variants; when you look at different human populations, you can't always expect the same specific variants to show up. In general, however, you would expect there to be more similarity at the level of the gene, as we have seen here.

Only one other genomic region, in a gene desert on chromosome 2 (an area of the genome containing no gene sequences), was replicated across both ethnic subsamples. Three other genome-wide significant findings emerged that were restricted to one of the two groups. The general picture is that, aside from the alcohol-metabolizing genes, the results look tentative, as one might expect for an early study of this complex syndrome.

Another indirect approach to finding genetic variants that impact on the risk for alcohol-use disorder is to study alcohol consumption. While a proper diagnosis of alcohol-use disorder requires an interview with a trained clinician, alcohol consumption can be easily measured by a few questions on a questionnaire that might also be examining dozens of other traits. So it is much easier and cheaper to find large samples of subjects with information about alcohol consumption than alcohol-use disorder.

We highlight one large study that examined drinker/nondrinker status and the regular quantity of alcoholic drinks consumed per week over the last year in over 86,000 individuals of mixed ancestry (JORGENSON et al. 2017). As expected, the strongest effect was with the well-known variant in the *ALDH2* gene in East Asians. The authors also found a similar impact for the variant in *ADH1B*, which was seen in non-Hispanic and Hispanic subjects. Combining all samples, they found evidence for two other variants, one upstream of a gene called β-Klotho (*KLB*) and the other an intergenic variant near a gene called gluco-kinase regulator (*GCKR*). These findings are consistent with a large meta-analysis of European samples, mapping daily alcohol intake (SCHUMANN et al. 2016).

Would we expect genetic variants that impact on the quantity of alcohol consumption also to impact the risk for alcohol-use disorder? If a genetic variant made you slightly sick when you drink (like the ALDH and ADH variants apparently do), then it seems very likely that it would both lower the average amount of alcohol consumed and also reduce the chances of developing alcohol-use disorder. We have presented some preliminary evidence that this supposition is true, but the correlation is not likely to be perfect. This is because asking about current drinking does a poor job of assessing alcohol-use disorder. For example, some individuals might have had major alcohol problems earlier in their life, and when they were assessed for the genetic study were either abstainers or light social drinkers. Furthermore, every clinician knows people who consume large amounts of alcohol for years, and sometimes a lifetime, yet never experience major adverse social, medical, or legal problems. They handle their liquor, and their drinking, although often heavy, never gets out of control. Indeed, twin studies suggest that asking people about how much they consumed when they drank the most in their life provides a reasonable proxy for alcohol-use disorder. Finally, when asking about the amount of alcohol consumed in general population samples, most of the difference is between people who drink very little and those who drink moderately. It remains an open question whether this variation has anything to do with the differences between people who drink moderately and those who develop severe alcohol-use disorder.

Summary

1. Alcohol-use disorder is common. The lifetime risk is 36.0% for males and 22.7% for females.

2. The heritability of alcohol-use disorder is about 50%.

3. Using a classic adoption and a novel triparental design, it has been shown that rearing also matters for the development of alcohol-use disorder.

4. Knowledge about the metabolism of ethanol (the chemical name for alcohol) provides robust candidate genes to study.

5. A key enzyme in ethanol metabolism is acetaldehyde dehydrogenase (ALDH). Mutations in the *ALDH2* gene are common in East Asia and produce a flushing reaction when the carrier drinks alcohol. Individuals with this variant are unlikely to develop alcoholism.

6. Genome-wide association studies (GWAS) of alcohol-use disorder have identified the ALDH locus and one locus on chromosome 2. GWAS of alcohol consumption also identified ALDH as well as two other loci that did not overlap with that from the GWAS of alcohol-use disorder.

12

The genetics of intelligence, personality, and personality disorders

[On the genetic basis of quantitative behavioral traits in human populations, focusing on intelligence and personality, and a discussion of the difference between personality types and personality disorders]

12.1 Intelligence

This chapter deals with science that has run into trouble, namely research into the genetic basis of intelligence. Genetic analysis of intelligence has been a battlefield because it has been used to explain differences between populations and between the sexes. Worse, genetics has been used as justification for the differential treatment of races; we're in the territory of eugenics here, and some very unpleasant history. Unfortunately, the science is so tied up with this history that it's disingenuous to say we can present the science and ignore how it has been used. Later in the chapter, we cover somewhat less controversial topics, particularly the genetics of personality and personality disorders.

12.1.1 A difficult history

As we introduced in Chapter 2, section 2.6.3, there is a long historical relationship between the genetics of behavior and eugenics. While the worst manifestations of this relationship emerged in Nazi Germany,

eugenics has not been the preserve of any one political ideology or country (KEVLES 1985). In England in the nineteenth century, Francis Galton saw genetics as a means of improving the human race through selective breeding, a philosophy that gave rise to the eugenics movement. In the USA, Charles Davenport—the same individual who studied the wandering impulse as a Mendelian trait—studied hereditary 'feeblemindedness', motivated at least in part by his wish to eliminate what he saw as a genetic plague. Through his efforts and those of others, the movement eventually succeeded in reinforcing the prejudices of US public policymakers in several infamous cases. The first person to be sterilized under Virginia's 1924 eugenic sterilization law, intended to prevent the conception of 'genetically inferior' children, was Charlottesville native Carrie Buck. On 2 May 1927, the US Supreme Court in *Buck v. Bell* affirmed the Virginia law. After Buck, 8,000 other Virginians were sterilized. Laws mandating the compulsory sterilization of mental patients were passed in many states. By 1929, in California alone 6,255 operations had been carried out, and from 1932 to 1941, sterilization was practiced in an increasingly large number of states (KEVLES 1985). Adolf Hitler's government in Germany promulgated a Eugenic Sterilization Law in 1933, making sterilization compulsory for all people who suffered from feeblemindedness, schizophrenia, epilepsy, blindness, and severe drug or alcohol addiction. From September 1939 until the end of the war, killing replaced sterilization (KEVLES 1985). It is worth noting that eugenic ideas did not disappear after 1945; Virginia's eugenics act was not fully repealed until 1974. Carrie Buck died in 1983.

While I (J.F.) was struggling with how to present the material in this chapter (I don't want to hurt anyone's feelings, but then again I don't want to gloss over too much), a friend of mine, David Reich, a geneticist at Harvard published an opinion piece in the New York Times that addressed the issue head on (https://www.nytimes.com/2018/03/23/opinion/sunday/genetics-race.html). Reich's piece is about the genetics of race, and it attracted a lot of attention. He began by stating the orthodox position that geneticists have maintained for many decades, exemplified by Richard Lewontin's paper in 1972 showing that 85% of variation in protein (an index of DNA variation) is explained by variation within 'races', and only 15% by variation across them (LEWONTIN 1972). In other words, although there are genetic differences between ethnic groups (Lewontin's analysis included West Eurasians, Africans, East Asians, South Asians, Native Americans, Oceanians, and Australians), these differences are far smaller than the differences within a group, so it made little sense to say that genetics define race. It simply couldn't. In Reich's words:

The orthodoxy maintains that the average genetic differences among people grouped according to today's racial terms are so trivial when it comes to any meaningful biological traits that those differences can be ignored. The orthodoxy goes further, holding that we should be anxious about any research into genetic differences among populations. The concern is that such research, no matter how well-intentioned, is located on a slippery slope that leads to the kinds of pseudoscientific arguments about biological difference that were used in the past to try to justify the slave trade, the eugenics movement and the Nazis' murder of six million Jews.

There is a lot we have learned since 1972 about the extent of genetic variation, and not all of it supports the orthodox view. Genetic differences between ethnic groups turn out to be nontrivial, in that they are related to biological differences. Reich quotes his own work showing that the higher rate of prostate cancer in African-Americans than in European-Americans can be entirely explained by genetic differences. He goes on to argue that ignoring evidence for biological, genetically driven differences between ethnic groups opens the door to worse things; his piece is a powerful defense of the view that scientists shouldn't avoid difficult topics. In his words, not speaking up will create a vacuum filled by pseudoscience.

As Reich points out, science is always producing surprises about our origins and about our genetic relationships. Reich's sequencing of ancient humans has recently revealed that 'whites' represent a mixture of four ancient populations that lived 10,000 years ago, each as different as Europeans and East Asians are today. And perhaps most unexpected of all is the discovery from Svante Pääbo that 1–6% of our genetic material originated from interbreeding with archaic humans about 50,000 years ago (Neanderthals and Denisovans, which are extinct species of archaic humans) (PÄÄBO 2014).

However, it is not only the difficult history of the subject but also the questions that genetic research raise about our biological nature that make the subject matter of this chapter so important. There are investigations into whether intelligence and personality are heritable, and if so how heritable. As you will see, to some extent the answers we have obtained are those that we learned from analysis of the genetics of psychiatric disorders, revealing the polygenic contribution to behavioral traits. But you will also see that genetic research on personality and IQ reveals that environmental effects, such as parenting, operate in unexpected ways. To pick up on David Reich's theme, what we learn from science often overturns our assumptions, and we should not shy away from assessing critically, and as fully as we can, the implications of these findings.

12.1.2 The genetic basis of intelligence

In 2009, the Behavior Genetics Association held its annual meeting in Minneapolis. They took this opportunity to celebrate the achievements of Thomas Bouchard, who for many years worked in Minneapolis running a study of monozygotic (MZ) twins who had been separated at or soon after birth. During the meeting, Bouchard gave an interview to *Science* magazine about his past and current work, and I give below a couple of quotes. Bouchard is famous for collecting data from twins who were reared apart to investigate the heritability of intelligence (and of other traits).

In the middle of the twentieth century, anyone who published heritability estimates for intelligence was in for a rough ride. Bouchard is not someone to shrink from a fight (BOUCHARD 2009):

In the '70s, when I was teaching research by [IQ researcher Arthur] Jensen and [twin researcher Francis] Galton, people picketed me, called me a racist, tried to get me fired. The progressive student association sent members in to ask hostile questions. ... So I put a tape recorder on the podium and said: 'I'm going to tape my lectures.' I never heard from them again. They knew what they were saying was nonsense and I would be able to prove it.

Bouchard states clearly what he thinks:

Nowadays, I'm sure there are people who are not publishing stuff on sex differences. Look what happened to Larry Summers [who resigned as president of Harvard University after suggesting that discrimination alone doesn't account for women's lower representation in math-based disciplines]. I talk about those things in my class all the time—that males and females have different interests; ... in a sense, females have a broader and richer view of life. There are a lot of people who simply won't talk about those things. Academics, like teenagers, sometimes don't have any sense regarding the degree to which they are conformists.

Perhaps a forthright approach was necessary to tolerate the hostile reception his work engendered. Perhaps you needed to behave like that to survive. But it would be a mistake to confuse being forthright with being unthinking. As we'll see in this section, the twin design reveals more than just heritability. But before we get to that, I'll review what is known about the heritability of intelligence. To start with, you need to know something about how intelligence is assessed.

12.1.3 IQ testing

The standard measure of intelligence is obtained using the Wechsler Adult Intelligence Scale (or the Wechsler IQ Test, or just the WAIS). There is also a version for children, the WISC (Wechsler Intelligence Scale for Children). I (J.F.) first came across the WAIS when I worked on a neuropsychiatry unit, a ward that admitted patients who had the types of problems described in Oliver Sacks' tales of people with unusual disturbances of cognitive function (*The Man Who Mistook His Wife for a Hat*). I had no idea what a WAIS was, but I was pretty certain I could administer one to our patient, a man in his 40s who had woken up after surgery for epilepsy to discover two people sitting at his bedside who were complete strangers (they claimed to be his wife and his 5-year-old son). But the WAIS can't be administered by anyone—you have to be trained in its use.

The test has a number of components, measuring different cognitive functions. There are memory tests, vocabulary tests, and tests of arithmetic. The results are reported as two forms of intelligence: verbal and performance IQ (VIQ and PIQ), or, roughly speaking, verbal and nonverbal skills. Just to give some idea of what is involved, consider the digit-span test. Take a series of random numbers and write down a selection of increasing length, such as 290, 1,745, 91,903, and 744,123. Then say to a friend, '*I'm going to say some numbers to you and I want you to repeat them back to me. For example if I say 8, 1, 4, then repeat back 8, 1, 4.*' The digit-span test consists of giving the numbers and finding the longest set that your subject can remember (a digit span of six is pretty good). The test can be made harder by asking your friend to repeat the series in reverse order (reverse digit span). You'll find that you can substantially alter their performance by the way you say the numbers: it's much easier to remember 8,946,743 if you give it as a phone number (894, 6743), than if you slowly state each number at 1-second intervals (as you are supposed to do). Training does make a difference. So obtaining a WAIS is not straightforward, and interpreting the results is complicated too. There are subscales, some of which can be used to assess the function of different parts of the brain. In my patient's case, we wanted to know if his memory loss was consistent with the pattern expected from damage to his temporal lobe and hippocampus, where the surgery was performed (we found that it was). However, in genetic studies, in most cases only a single measure is used: the overall test score.

The WAIS isn't the only way to assess cognitive ability (e.g. you will come across Raven's progressive matrices), but genetic studies usually assume that, whatever the test used, it measures the same thing. You will often see the score referred to as 'g', for general cognitive ability, on the basis that there exists a common factor present in all subtests, so that 'g' is roughly equivalent to the average of each subtest's correlation with every other subtest (the mean correlation among individual cognitive tests is about 0.3). Exactly how to interpret 'g' isn't obvious—does it represent a single biological process, something perhaps like processing speed (equivalent to the speed of the Intel Core i7 processor on my laptop: 2.2 Gigahertz)? Or is it measuring the joint effect of different processes? There is a literature arguing that 'g' is pretty meaningless (GOULD 1981), that 'g' is one of the best predictors of education and occupational status (STRENZE 2007), and that people with higher intelligence have better mental and physical health and live longer (DEARY et al. 2010). One final point about IQ scores: they are normalized to a population reference, where 100 is the population mean and 15 is the standard deviation (SD). This means that an IQ score of 130 is 2 SD beyond the mean, so only 2.2% of the population will have a higher score (see Appendix, section A8, for an introduction to the properties of normal distributions).

12.1.4 Twin and family studies of IQ

Figure 12.1 is from a review that Bouchard published in 1981 of all family and twin studies of intelligence (BOUCHARD AND MCGUE 1981). The figure summarizes 111 studies, revealing that the inheritance of IQ has been a focus of interest for behavioral geneticists for a long time.

The average correlation for 4,672 MZ twin pairs was 0.86, while for the dizygotic twin pairs the correlation was 0.6, a large difference of 0.26, but Bouchard was careful not to give a figure for heritability. Tests of heterogeneity were carried out within each group and many were significant, indicating that combining individual studies should be done with caution. Bouchard highlighted this, as well as the pattern of averaged correlations: highest for MZ twins, less for siblings, and less still for cousins. This pattern is consistent with polygenic effects (many loci of small effect). Bouchard stated: 'That the data support the inference of partial genetic determination for IQ is indisputable; that they are informative about the precise strength of this effect is dubious' (BOUCHARD AND MCGUE 1981).

What is the heritability of IQ? Sixteen years after Bouchard's review, a 1997 analysis of 212 studies of 50,470 pairs concluded that the total effect of genetic variants on IQ is 48% (DEVLIN et al. 1997). Adoption studies give similar estimates. Conscripts in the Danish military (almost all males in Denmark complete an IQ assessment) provide a remarkably representative sample of siblings brought up together and reared apart. The IQ correlations were 0.47 for full siblings reared apart, 0.52 for full siblings reared together, 0.22 for half-siblings reared apart, and 0.02 for adoptive siblings reared together (TEASDALE AND OWEN 1984). So adoption and twin studies indicate a substantial heritability for IQ.

This result shouldn't be much of a surprise; after all, by 2000, Eric Turkheimer had argued that one of the three laws of behavior genetics is that all human traits are heritable (TURKHEIMER 2000), but the history of IQ genetics has been anything but straightforward. Just to remind you how contentious this question has been, you might re-read the quote from Leon Kamin in Chapter 2, section 2.6.3. The arguments for and against the interpretation of the twin results, and the adoption results, take us back to arguments around the twin design, which we reviewed in Chapter 2, section 2.6.4.

12.1.5 Four (nongenetic) lessons from twin and family studies of IQ

As we've seen in other chapters, establishing that a trait or disease is heritable is just a starting point. What more can we say about the genetics of intelligence? There are four observations to note.

First, environmental effects are important, even when genetic effects are large. One striking example is the observation that IQ changes over time, that is, people born in the 1990s have higher IQs than people born in the 1940s. This was first noticed in a paper in 1982, which reported that the mean IQ among Japanese children was rising relative to American children (LYNN 1982). However, the person most associated with asserting that IQ has been increasing with time is James Flynn (FLYNN 1984). You can find a YouTube video about his work here: https://www.youtube.com/watch?v=9vpqilhW9uI and his TED talk here: https://www.ted.com/talks/james_flynn_why_our_iq_levels_are_higher_than_our_grandparents/transcript#t-59 303.

Second, heritability estimates depend on age (HAWORTH et al. 2010), the genetic contribution peaking between 60 and 70 years (PEDERSEN et al. 1992; DEARY et al. 2012; BRILEY AND TUCKER-DROB 2013), and then falling by 80 (REYNOLDS et al. 2005). Why? Surely the effect of genetics should go down with age? As we age, we experience more, we learn more; in short,

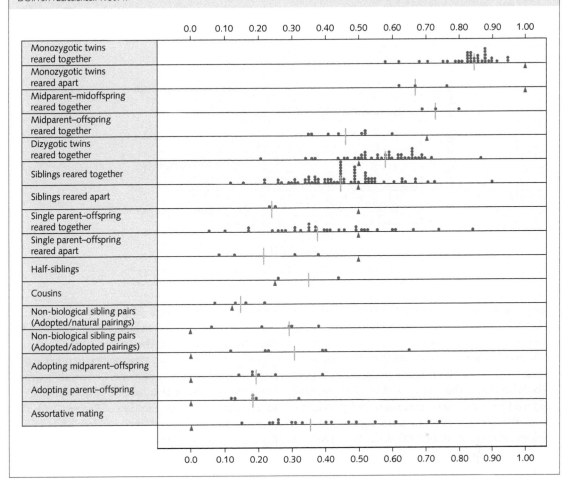

Figure 12.1 Familial correlations for intelligence quotient. The vertical bar in each distribution indicates the median correlation; the arrowhead is the correlation predicted by a simple polygenic model.

Reproduced with permission from TJ Bouchard Jr, and M McGue. Familial studies of intelligence: a review. *Science*, 212(4498): 1055–59. Copyright © 1981. DOI: 10.1126/science.7195071.

we are exposed to more environmental effects. It's counterintuitive that genetic effects should increase as we are exposed to more things. But apparently life is not quite as we imagine.

The most likely explanation is that as we age, we restrict environmental exposures, choosing our own environments, and these choices are themselves partly genetically determined. A shy, inward-looking person is going to avoid social events, while their extravert sibling will likely seek them out. This is an example of active gene–environment correlation. Bright individuals are more likely to select themselves into intellectually stimulating environments working as professors, reading more stimulating books, going

to more museums, and so on, which feeds back to augment their IQ.

Third, the environmental effects on IQ are largely limited to making the IQ of twins different, rather than making them similar. To explain this point, Thomas Bouchard demonstrated the results of studies of MZ twins reared apart. Figure 12.2 is from Bouchard's 1990 analysis of 100 twins (BOUCHARD et al. 1990).

The vertical scale of Figure 12.2 is the IQ difference between individuals and the results in the figure can be used to answer the question: how much difference can the environment make to an IQ score? Suppose parents sent one twin to an expensive private school and sent the other twin to a much cheaper, less

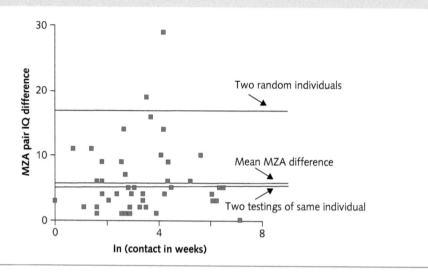

Figure 12.2 Within-twin pair intelligence quotient (IQ) differences for monozygotic twins reared apart (y-axis), plotted against the natural logarithm (ln) of pair contact in weeks (x-axis). MZA, monozygotic twins reared apart.

Reproduced with permission from Bouchard, T.J., et al. Sources of human psychological differences: the Minnesota Study of Twins Reared Apart. Science, 250(4978): 223–8. Copyright © 1990. DOI: 10.1126/science.2218526.

prestigious school. How big a difference would that make? Most of us assume quite a big difference, but Figure 12.2 argues otherwise.

The top horizontal line shows that the average IQ difference between two randomly selected individual is about 18 units, which is a result of all the genetic and environmental effects that make two people different. We are interested in assessing the environmental contribution to this difference. How big is it? The next horizontal line in Figure 12.2 answers this question because it is the IQ difference of MZ twins reared apart (MZA in the figure). Assume for the moment that they really were separated at birth and lived in very different places, with families from different backgrounds, so that we can say that they shared no environmental effects. Whatever makes their IQ scores different has to be the environment, and whatever makes them the same has to be genetic (there are assumed genetic differences between MZ twins and we argued that they shared no environmental effects). How big is the environmental effect? According to the vertical scale of Figure 12.2, the effect of the environment (in this case, all nonshared) is limited to a 5-point difference. The lowest horizontal line on Figure 12.2 is another measure of the environment. It represents the results if you were to take an IQ test now, and repeat the test next week. It's hard to argue that differences

between your two scores (about 5 IQ points, according to the vertical scale) can be anything other than the contribution of things in the environment that are not shared between the two time points, in other words it is the nonshared environment. The result is almost the same as same as that for MZ twins reared apart. In other words, the implication of Figure 12.2 is that parents' attempts to boost their children's IQ by sending them to better schools, or otherwise altering their environment, won't make a big difference.

On the face of it, Bouchard's result flies in the face of common sense—it suggests it really doesn't much matter, at least in terms of its effect on your IQ, whether you send your children to a really expensive school, whether you read to them every night, or take them to extracurricular mathematics and music lessons. You might reasonably object that this conclusion is likely to be biased, because even though reared apart, the twins shared a lot of environmental effects, at the very least they shared a womb, and this might contribute to the similarity in scores. Indeed, other results do show that the contribution of the shared environment is not as negligible as Bouchard's work implies. For example, children adopted into more prosperous homes show an increase in IQ (CAPON AND DUYME 1989). There is considerable evidence that the family environment contributes to similarity in the IQ of twins and siblings, but this is really only important while they are living at home. The effect attenuates after they leave home and in most studies is gone by the age of 30. In fact, no one has found a really big contribution from the shared environment.

One of us (K.S.K.) undertook a study with Swedish collaborators that used the records of all of Sweden to obtain information on hundreds of adoptions, and related genetic effects to IQ (which was measured when individuals were conscripted to the Swedish military service). Adoption conferred a significant (but relatively small) change in IQ: children adopted away into improved socioeconomic circumstances manifested an advantage of 2–5 IQ points (KENDLER et al. 2015c), which is not much different from the value shown in Figure 12.2.

Given the importance of education in our society, and the lengths parents will go to to ensure that their children obtain the best education possible, it is not surprising that there is some effect on IQ, even if it is less than most of us assume. We don't enforce such pressure on personality, and as we will see later, there is even less evidence for the contribution of the shared

environment to personality—presumably if a culture rewarded a particular personality type, the effect of a shared environment would emerge more clearly than it does at present. We'll return to the importance of the nonshared environment in our discussion of personality.

The fourth observation is that the genetic effects on intelligence are correlated across different tests for IQ. Genetic correlations among cognitive subtests are typically greater than 0.6, indicating that the same genes are responsible for the heritabilities of these tests (CALVIN et al. 2012); genetic correlations extend to other measures of intellectual ability (DAVIS et al. 2009).

Figure 12.3 shows additive genetic correlations assessed from twin studies (similar to the way we show correlations between different psychiatric disorders, in Chapter 13, section 13.2). There are large genetic correlations between 'g' and the three educational traits, reading, mathematics, and language skills (shown by the thick lines in the figure), with a mean value of 0.85. This is remarkable. It implies that the great majority of the genetic variation in intelligence is shared with variation in the three educational skills.

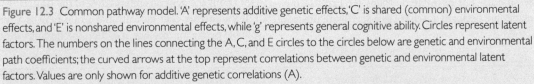

Figure 12.3 Common pathway model. 'A' represents additive genetic effects, 'C' is shared (common) environmental effects, and 'E' is nonshared environmental effects, while 'g' represents general cognitive ability. Circles represent latent factors. The numbers on the lines connecting the A, C, and E circles to the circles below are genetic and environmental path coefficients; the curved arrows at the top represent correlations between genetic and environmental latent factors. Values are only shown for additive genetic correlations (A).

Reproduced with permission from Davis, O.S.P., et al. Learning abilities and disabilities: Generalist genes in early adolescence. *Cognitive Neuropsychiatry*, 14 (4–5):312–31. Copyright © 2009, Rights managed by Taylor & Francis. https://doi.org/10.1080/13546800902797106.

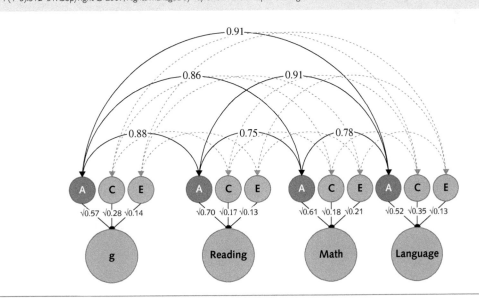

Why are genetic effects shared like this? It implies that there is a common set of genes involved in what we think of as different abilities, acting at different time points. It argues for the existence of a pleiotropic effect of genetic effects on intelligence. What are those genes? We hope to have clues from genome-wide association results.

12.1.6 Molecular genetic analysis of IQ

As we have seen, successful genome-wide association studies (GWAS) of psychiatric disorders and behavior require data from tens of thousands of individuals. Given the difficulties of measuring IQ accurately, is it possible to obtain a sufficiently well-powered sample for genetic mapping of intelligence? One answer is to take the results of the genetic correlation at face value and assume that variation in school achievement is likely to be attributable to many of the same genetic variants as contribute to IQ. Recall from what we described above that IQ has substantial genetic correlations with other, simpler, measures of intelligence. Rather than the time-consuming WAIS assessment, perhaps we can use one of these simpler and cheaper assessments? For example, perhaps we can use how well people perform at school as a surrogate measure for IQ?

A very simple index of educational attainment is the number of years of education you receive, a question that is frequently asked in genetic association studies to ensure that cases and controls have similar educational backgrounds. The idea that the answer to the question 'How many years were you in school?' could act as a proxy for intelligence led the economist Daniel Benjamin and colleagues to use educational attainment in a GWAS. They amassed data from 101,069 individuals, an extraordinarily large number at the time, and found three genetic loci that exceeded genome-wide significance (RIETVELD et al. 2013).

The Social Genomics Consortium that Benjamin and colleagues established has gone on to make remarkable contributions to behavioral genetics. They obtained a sample size of over 1 million people, and identified 1,271 independent loci that contribute to variation in educational attainment (Figure 12.4) (LEE et al. 2018). This, at this point in the book, is the single largest genetic study, and the largest number of genetic loci for a single trait (there's an even bigger study coming later in this chapter, so keep reading). Finding significant loci is, of course, only one of many things obtained from a GWAS. The results also tell us a lot about the genetic architecture of the trait.

If you accept that educational attainment is a reasonable proxy for intelligence, then the findings show

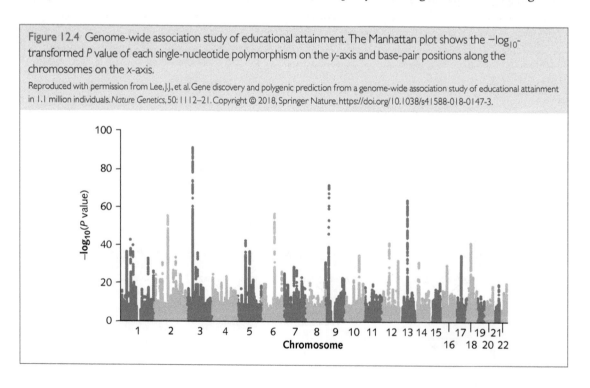

Figure 12.4 Genome-wide association study of educational attainment. The Manhattan plot shows the $-\log_{10}$-transformed P value of each single-nucleotide polymorphism on the y-axis and base-pair positions along the chromosomes on the x-axis.

Reproduced with permission from Lee, J.J., et al. Gene discovery and polygenic prediction from a genome-wide association study of educational attainment in 1.1 million individuals. *Nature Genetics*, 50: 1112–21. Copyright © 2018, Springer Nature. https://doi.org/10.1038/s41588-018-0147-3.

that intelligence is highly polygenic and that each locus makes only a small contribution to the phenotype (LEE et al. 2018). Heritability estimated from GWAS data gives a figure around 30% (MARIONI et al. 2014; HILL et al. 2018), consistent with the twin, family, and adoption studies we reviewed above (recall that single-nucleotide polymorphism-based heritabilities are usually about half the estimates from twin data).

What does the GWAS tell us about the biology of intelligence? With the usual caveats that genetic loci from GWAS are not genes, it's still good to learn that genes near the peaks in Figure 12.4 are overwhelmingly enriched for expression in the central nervous system (intelligence really is related to brain function). There's nothing from the gene list that's changed the way we think about the biology of intelligence, except perhaps that we can rule out the role of nonneuronal cells. One idea that the data make less likely is that superior intelligence requires a 'faster' brain, due, for example, to increased speed of transmission: the researchers found that there was no enrichment for genes involved in cells that produce and maintain myelin, a substance that determines the ability of axons to transmit fast electrical signals. Indeed, they found no enrichment of glial cells, the nonneuronal cells that make up about half the cells in the brain. The enrichments they **did** find were for genes involved in neurotransmitter secretion, the activation of ion channels and metabotropic pathways, and synaptic plasticity. None of this comes as news to neuroscientists.

The genotype data are informative about other aspects of intelligence. First, they point to associations with several neuropsychiatric disorders (HILL et al. 2018; SAVAGE et al. 2018). There are negative genetic correlations (r_g) with attention deficit hyperactivity disorder ($r_g = -0.36$), depressive symptoms ($r_g = -0.27$), Alzheimer's disease ($r_g = -0.26$), and schizophrenia ($r_g = -0.22$), and those with more educational attainment live longer: there is a positive correlation with longevity ($r_g = 0.43$). One interpretation is that being smart reduces your chances of smoking, of getting better healthcare, and so on. So educational attainment genes will show up in a GWAS for longevity in the same way that genes influencing how much people smoke emerge from studies of lung cancer (FURBERG et al. 2010).

Second, IQ genetic data have been used to relate intelligence to brain anatomy. It shouldn't come as a surprise to learn that brain shape varies, and that

variation has a genetic basis: the overall volume of the brain is heritable (TRAMO et al. 1998), and some brain structures (OPPENHEIM et al. 1989), but not all (BARTLEY et al. 1997), have heritable variation. Genetic control of brain structure mirrors the brain's functional organization: frontal, linguistic, and sensorimotor cortices have the highest heritability (THOMPSON et al. 2001).

This variation is correlated with IQ, and the correlation is due to shared genetic factors (as twin studies show) (POSTHUMA et al. 2002). There should be loci affecting brain structure and IQ. Does the mapping support this idea? Genetic mapping of brain structure is onerous because brain scans are slow and expensive, and thousands are needed for genetic mapping. Nevertheless, a consortium, called ENIGMA (for Enhancing Neuroimaging Genetics through Meta-Analysis), managed to amass sufficient scans to map brain features. With a sample size of 7,700 subjects, the consortium identified the first loci contributing to variation in hippocampal, total brain, and intracranial volumes (BIS et al. 2012; STEIN et al. 2012). One of the important negative findings was the lack of evidence for association with candidate genes (*BDNF*, *TOMM40*, *CLU*, *PICALM*, *ZNF804A*, *COMT*, *DISC1*, *NRG1*, and *DTNBP1*), mirroring the findings in studies of behavior and complex disease. Increasing the sample size to 30,000 individuals identified more loci (HIBAR et al. 2015), confirming the highly polygenic nature of the underlying genetic architecture, with each locus again making a small contribution: each locus explains between 0.17 and 0.52% of the phenotypic variance (0.51–1.40% differences in volume per risk allele). But so far, there has been no genome-wide significant locus found that contributes to both IQ and to brain volume. This does not contradict the genetic correlation results; it just means we haven't yet unequivocally found the shared loci.

12.1.7 Predicting phenotypes from educational attainment polygenic scores

The history of GWAS for educational attainment follows a familiar pattern: the detection of a small number of loci in a high-profile publication as a proof of concept that mapping works is followed by papers reporting increasingly large sample sizes, with no apparent end in sight to the number of loci detected (two is good, 50 is better, but do we really need 1,271?). Bigger is, however, better in one very important way:

the improvement in predictive power the results give us, from a polygenic score.

For psychiatric diseases, we explained how polygenic risk scores (PRSs) are created by summing up the effects estimated for each genotype that a person has (Chapter 7, section 7.2). For educational attainment, we can't really talk of PRSs as there is no known risk to greater intelligence, so it makes sense to call it a polygenic score. Larger sample sizes mean that the accuracy of the score improves because not only do we have large numbers of genome-wide significant loci, but we also have better estimates of the effect sizes attributable to each locus. The score from the largest educational attainment GWAS was able to explain about 10–12% of educational attainment (LEE et al. 2018); in other words, an increase of 1 SD in the score is associated with about 0.8 extra years of schooling. The power of the polygenic scores for educational attainment, which is only going to increase as ever more massive GWAS are completed, raises some interesting ethical issues (I can now, in theory, choose to have children who have higher scores for educational attainment), which we discuss in Chapter 20, section 20.9.

The scoring does more than predict educational attainment in cohorts. It provides a remarkably powerful tool to answer questions about the relationship between genetic effects and the environment, and about interventions: *'Just as a radiologist might administer a radioactive tracer to track the flow of blood within the body, researchers can use genetics as a molecular tracer to get a clearer image of how students progress through the twists and turns of a cumulative educational system'* (HARDEN et al. 2019).

Although it needs enormous sample sizes to collect the information to construct a polygenic score for educational attainment, it's possible to detect a significant effect in samples of less than 100. One example has been in the Dunedin study, a sample of about 1,000 people in New Zealand followed from the age of 3 years (this is the same study that was used to identify the serotonin transporter locus in a gene–environment interaction, discussed in Chapter 5, section 5.2.3). The analyses demonstrated that polygenic scores predict adult outcomes even after taking into account social-class origins, and that they remain predictive of behavior across a person's lifetime: high scores predicted an earlier age of starting to talk, of better reading skills, and even financial planning for retirement (BELSKY et al. 2016).

PRSs are also powerful tools to pick apart environmental effects. We discuss this in more detail in Chapter 14, but here point to one important application from the educational attainment polygenic scores, that is, the exploration of genetic effects that are dependent on the environmental context (gene–environment interactions). Early-life experiences, such as education, and genetic factors are independently associated with health in adulthood, and it's been thought that the two factors likely interact, although no one has robustly detected the interaction. For example, it's possible that additional education makes people more likely to adopt healthier lifestyles. This is the social determinants hypothesis—that interventions that increase education, income, or socioeconomic status can improve health. However, it could be that the effect of education on health is genetically mediated: those with higher IQs follow healthier lifestyles, regardless of the time they spend in school.

This question has been investigated in what is fast becoming one of the most useful resources for genetics, the UK Biobank, an enormous study of half a million people in the UK, aged 40–69 years at recruitment, all of whom have been genotyped and are currently (in 2019) being sequenced (BYCROFT et al. 2018). The UK Biobank is an experiment in open science: anyone can apply for the enormous data set of multiple phenotypes and genotypes. The large number of traits measured and the inclusion of medical records make the UK Biobank a treasure trove for genetic researchers. Polygenic scores applied to the UK Biobank made it possible to show that the effect of education on health varies with genetic constitution. It was possible to do this because of an unusual natural experiment in the UK: in 1972, the minimum age at which students could drop out of school increased from 15 to 16 years. The reform affected only students born on or after 1 September 1957, making it possible to compare the health outcomes of people born just before and just after 1 September 1957.

The investigators found no evidence that the effects of education on health depend on the educational attainment polygenic score. Instead, they depended on a polygenic score constructed from a large GWAS of weight (or more precisely from body mass index, which corrects weight for the height of the individual). Their conclusion was that *'the effects of education on health were not uniform across genetic backgrounds, benefitting those with greater genetic risk for obesity more. In other words, education not only affected health, corroborating the* social determinants hypothesis, but it also reduced the role played by genetic factors: the association between genetic predisposition to obesity and unhealthy body size was reduced among cohorts who were forced to stay in school longer' (BARCELLOS et al. 2018). These examples show the power of the polygenic scores, argue against the importance of genetic determinism, and demonstrate the complex ways that genetic predisposition acts with the environment to alter behavior.

12.2 Personality

When psychologists and psychiatrists talk about 'personality', they mean the stable part of an individual's attitudes, behavioral patterns, and emotional responses. Personality is something we cannot imagine being without, yet while we all have a general intuition of what it is, we all find it hard to describe it accurately. So it might seem strange that psychologists use short, self-report questionnaires to define personality.

Most people imagine that the truthfulness with which each question is answered is central to the interpretation of a personality questionnaire. But it is easy to see the problem with this interpretation. Here's an example from the psychologist Hans Eysenck: 'One might see some unfortunate individual sitting down with the questionnaire, his hands trembling and sweating with excitement, his face getting pale and flushed alternately, and his tongue licking his lips, his whole body in a tremor of nervousness; on going over to reassure him, one would find after the question, "Are you generally a nervous sort of person?" he had boldly put the answer "No"' (EYSENCK 1957).

In order to make sense of personality questionnaires, we have to abandon the idea that the answer a person gives should be interpreted as a truthful self-revelation. Instead, the answers must be treated as a pattern of responses to a set of stimuli (the questions). Psychologists exploit instead the correlations between responses, which are used to generate personality factors. Consider these two questions: 'Are your feelings hurt?' and 'Do ideas run through your head so that you cannot sleep?' There's no logical reason why you should not answer yes to the first and no to the second. But it so happens that people who answer yes to the first tend to answer yes to the second, and vice versa. And, as this is true for other questions, it is possible to pull out a set of correlated answers. The correlations

allow us to recover a relatively small number (usually five) of correlated responses, or factors, using the statistical technique of factor analysis. We describe below what is known about the factor structure of personality and its genetic determinants.

12.2.1 How to measure personality

Over the last 30 years, psychologists have come up with a bewildering array of different scales to measure human personality. From this tower of Babel, something approaching a consensus has emerged. A substantial proportion of personality researchers now agree that human personality can be described by five factors, often called the big five. For illustrative purposes, we focus on the 'Big Five Inventory' (JOHN AND SRIVASTAVA, 1999). Figure 12.5 shows part of the questionnaire:

An advantage of the big five system is the memorable acronym available for the five scales: OCEAN. Here is how the acronym works out, along with the selection of items sorted into the five dimensions (where '(−)' means the item is reverse coded):

O—Openness: items 5, 10, and 15
C—Conscientiousness: items 3, 8(−), and 13
E—Extraversion: items 1, 6(−), and 11
A—Agreeableness: items 2(−), 7, and 12(−)
N—Neuroticism: items 4, 9(−), and 14

While assessing a lifetime history of a psychiatric or drug-use disorder is a time-consuming and expensive business, personality is far more simply (and cheaply) measured. All you need to do is convince people to fill out the questionnaire.

A number of adoption and twin studies, some with enormous sample sizes, have examined the heritability of personality. A thorough review of all these studies found that the heritabilities for all five major dimensions of personality were in the range of 30–45%, with good agreement across studies (LOEHLIN, 1992). This means that between one-third and one-half of the differences between people in their personality come from differences in their genes.

One large twin study examined the sources of individual differences in a personality trait that we will hear more about: neuroticism. Neuroticism reflects the propensity of individuals to experience negative or unpleasant emotions such as anxiety, worry, sadness, anger, and irritability. People with low levels of neuroticism are typically easy-going and

Figure 12.5 Part of the personality questionnaire for the Big Five Inventory.
© John and Srivastava, 1999.

I see myself as someone who...

1....Is talkative

Disagree   1 ○   2 ○   3 ○   4 ○   5 ○   Agree

2....Tends to find fault with others

Disagree   1 ●   2 ●   3 ●   4 ●   5 ●   Agree

3....Does a thorough job

Disagree   1 ○   2 ○   3 ○   4 ○   5 ○   Agree

4....Is depressed, blue

Disagree   1 ○   2 ○   3 ○   4 ○   5 ○   Agree

5....Is original, comes up with new ideas

Disagree   1 ●   2 ●   3 ●   4 ●   5 ●   Agree

6....Is reserved

Disagree   1 ○   2 ○   3 ○   4 ○   5 ○   Agree

7....Is helpful and unselfish with others

Disagree   1 ○   2 ○   3 ○   4 ○   5 ○   Agree

8....Can be somewhat careless

Disagree   1 ●   2 ●   3 ●   4 ●   5 ●   Agree

9....Is relaxed, handles stress well

Disagree   1 ○   2 ○   3 ○   4 ○   5 ○   Agree

10. ...Is curious about many different things

Disagree   1 ○   2 ○   3 ○   4 ○   5 ○   Agree

11. ...Is full of energy

Disagree   1 ●   2 ●   3 ●   4 ●   5 ●   Agree

12. ...Starts quarrels with others

Disagree   1 ○   2 ○   3 ○   4 ○   5 ○   Agree

13. ...Is a reliable worker

Disagree   1 ○   2 ○   3 ○   4 ○   5 ○   Agree

14. ...Can be tense

Disagree   1 ●   2 ●   3 ●   4 ●   5 ●   Agree

15. ...Is ingenious, a deep thinker

Disagree   1 ○   2 ○   3 ○   4 ○   5 ○   Agree

imperturbable. It really takes a lot to get them distressed. By contrast, individuals with high levels of neuroticism are easily upset, and once upset, take a long time to calm back down. The study took data from over 45,000 twins and their relatives (LAKE et al. 2000). Reassuringly, the investigators calculated the heritability for neuroticism to be 41% in females and 35% in males, well within the range of results from the smaller studies.

12.2.2 The nonshared environment is a major determinant of personality

The fact that personality is heritable adds support to Turkheimer's observation that almost all human behavioral traits are heritable (TURKHEIMER 2000). But studies of the genetics of personality are important for another reason: they emphasize the importance of the role of the nonshared environment, an effect that turns out to be important for all sorts of traits, not just behavior, as we discussed in the context of intelligence. The paper that first highlighted the importance of the nonshared environment is one by Robert Plomin. It is now a classic in the field, and it asks the simple question, *'Why are children in the same family so different from one another?'* (PLOMIN AND DANIELS 1987).

Anyone with siblings is so familiar with the differences that exist between you and your brothers and sisters that it seems pointless to ask why your brother has worked all his life as a highly paid company executive, while you grew your hair into dreadlocks and spent years on a soul-searching trip in northern India, because that's just how life is. Anyone with children knows that it sometimes seems as if your offspring were selected at random from a telephone directory, so little do they have in common. Why are they so different?

Here is the key observation from twin studies, expressed in the words of John Loehlin, who has worked for more than 40 years on the genetic effects on personality *'Environment carries substantial weight in determining personality – it appears to account for at least half the variance – but that environment is one for which twin pairs are correlated close to zero'* (LOEHLIN AND NICHOLS 1976). We can see this from the twin correlations.

Let's start with the rough and ready way to estimate heritability: twice the difference in correlations of MZ and dizygotic (DZ) twins (while this is not

quite correct, it's good enough for our purposes here; see Appendix, section A5). Suppose you find correlations of 0.75 for MZ and 0.5 for DZ twins. The difference is $0.75 - 0.5 = 0.25$, and twice that is 0.50, or a heritability of 50%. Now suppose the correlations are 0.50 for MZ and 0.25 for DZ. Again, the heritability is 50% ($2 \times (0.5 - 0.25)$). Although the heritabilities are the same, in the second case the environmental effects that contribute to the correlations are much lower; in fact, the shared environmental contribution is negligible. We can work this out by doubling the DZ correlation and subtracting the MZ correlation (see Appendix, section A5 for an explanation of the formula). In the first case, the shared environmental effect is 25% ($2 \times 0.5 - 0.75$), and in the second case it is zero ($2 \times 0.25 - 0.5$).

The point that Loehlin noted, and that Plomin and Daniels expanded, is that the second situation is the one that is most commonly found, particularly for personality. In a review of ten twin studies of personality (GOLDSMITH 1983), the average twin correlations were 0.47 for identical twins and 0.23 for fraternal twins, so heredity accounts for 50% of the phenotypic variance, and nonshared environment and error of measurement explain the rest. Shared environment made no detectable contribution.

Most of us accept that our environment contributes to shaping our personality, and if pressed for examples, we would reply, 'It's something to do with how I was brought up, the schools I attended, the way my parents treated me when I was young, the habits they instilled in me.' These features—the school, the neighborhood, the family—are what twins and siblings have in common. If these factors are important, then we should see, from the twin studies, that the shared environment has a big effect on determining personality differences between families. Look at Table 12.1, from a 1990 paper published in *Science*, which reports the results of analyzing the correlations between twins reared apart and reared together (BOUCHARD et al. 1990). The correlation coefficients for the two personality scales that were used are almost the same, regardless of whether the twins were raised apart or raised together.

The sample size, by current standards, is very small (twins separated at birth are hard to find), but the message is clear and is consistent with the results from elsewhere: the things that you share with a sibling—the family you were brought up in, the school you went to, the neighborhood, the social standing of your

Table 12.1 Correlations in personality scores for twins raised together and raised apart

	Monozygotic twins reared apart	Number of twin pairs	Monozygotic twins reared together	Number of twin pairs
Multidimensional personality questionnaire	0.50	44	0.49	217
California psychological inventory	0.48	38	0.49	99

family, and the income your family enjoyed—these features make a negligible contribution to personality.

What are the nonshared environmental factors? Plomin introduced the 'gloomy' prospect that they might be entirely serendipitous, and quotes one example from, of all people, Charles Darwin:

The voyage of the Beagle has been by far the most important event in my life, and has determined my whole career; yet it depended on so small a circumstance as my uncle offering to drive me thirty miles to Shrewsbury, which few uncles would have done, and on such a trifle as the shape of my nose.

Darwin's comment about his nose refers to the quixotic captain of the *Beagle*, Captain Fitz-Roy, who nearly rejected Darwin for the trip because the shape of his nose indicated to Fitz-Roy that Darwin would not possess sufficient energy and determination for the voyage. (Darwin wrote that, during the voyage, Fitz-Roy became convinced that *'my nose had spoken falsely'*.)

The less gloomy prospect is that the nonshared environment consists of things, events, or stuff in the environment whose contribution to personality and other phenotypes could be assessed just like anything else: birth order, season of birth, interactions between brothers and sisters, differences in the way parents treat one child compared with the other, factors that systematically differ between siblings and can be measured and studied. This hope hasn't been so easily realized. A survey of studies attempting to identify elements of the nonshared environment concluded that the effects of nonshared environmental variables that had been measured explained only about 3% of the variation in behavioral outcomes (TURKHEIMER AND WALDRON 2000). This doesn't mean that the nonshared effects aren't there, rather that they are *'damnably hard to find'* (TURKHEIMER 2011).

12.2.3 The molecular genetic analysis of personality

The first successful mapping of the variants that contribute to personality came when GWAS were sufficiently well powered to detect the small genetic effects involved. In 2012, an analysis of 17,375 adults identified and replicated one locus for the personality factor conscientiousness (DE MOOR et al. 2012). Over 17,000 people might seem a large sample, but the fact that only one locus was found indicates that the sample was still underpowered. Personality data collected by the consumer genomics company 23&Me on 59,225 individuals identified six loci (Lo et al. 2017), but the real breakthrough came with the use of the UK Biobank, which had been assessed for neuroticism.

The UK Biobank (the study of half a million people that we introduced in section 12.1.7; BYCROFT et al. 2018) agreed to include Eysenck's measure of neuroticism (EAVES et al. 1989) after two of us (J.F. and K.S.K.) argued that it would be a valuable addition; the Biobank was set up as an epidemiological cohort, to find risk factors for cancer and heart disease (it was never designed to be a genetics study). Dominated by specialists in cancer and heart disease, the scientific steering committee had little interest in measuring a psychological trait. However, grudgingly, they allowed the short version of the Eysenck neuroticism scale to be administered to the participants. At that time, we had no idea how important this decision would be for successful GWAS of personality.

Two groups used the UK Biobank data and, with the first tranche of data (100,000 subjects), identified 11 variants associated with neuroticism (OKBAY et al. 2016; SMITH et al. 2016; TURLEY et al. 2018). With a larger sample of 390,000 participants, 116 loci were identified (LUCIANO et al. 2018) (the UK Biobank released its genotypes in tranches, which explains why there are publications reporting results from 100,000 individuals and later results from up to 500,000 participants). Intriguingly, there were two loci on chromosomes 8 and 17 that lay within regions of the genome where the DNA is inverted in some people (inversion polymorphisms). The significance of this finding is still not clear, but see our discussion of the impact of inversion polymorphisms on the behavior of animals in Chapter 19, section 19.5, where we introduce the idea of 'supergenes', regions of DNA containing a number of genes related to a phenotype and inherited as a unit because they occur together on an inversion.

The GWAS findings make the following points: personality is a highly polygenic trait, with hundreds, probably thousands of independent loci contributing. Each locus has only a tiny effect: in the range of explaining between 0.02 and 0.04% of the variance. With larger sample sizes, even more loci will be detected. At the time of writing, the first samples of more than 1 million participants are being analyzed for another personality-like trait: risk tolerance (defined as the willingness to take risks, typically to obtain some reward) has been assessed in both the UK Biobank and 23&Me, affording a sample of 1,448,690 subjects and leading to the identification of 611 loci (LINNÉR 2018).

Genotype data for personality has made it possible to explore in more detail the relationships between personality traits and psychiatric disorders. Figure 12.6 shows these relationships by plotting two principal components that extract features of the correlations between the personality and psychiatric disease (Lo et al. 2017). In this plot, correlation is shown by the angle between two arrows: an angle of less than 90° is a positive correlation, while an angle greater than 90° is a negative correlation (and traits are not correlated when two arrows are at right angles). The five personality traits are genetically correlated, with neuroticism negatively correlated with the other traits. Openness, bipolar disorder, and schizophrenia cluster in the first quadrant of the figure; three personality traits (conscientiousness, agreeableness, and extraversion) cluster in the second quadrant, and neuroticism and depression cluster in the fourth quadrant. Overall, the picture provides some support for the hypothesis that personality traits and certain psychiatric disorders exist on a continuum (TRULL AND WIDIGER 2013).

The molecular findings are thus similar to the discoveries made for other complex behaviors: a high

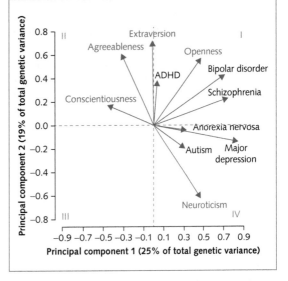

Figure 12.6 Loading plot of personality traits and psychiatric disorders on the first two principal components derived from a genetic correlation matrix. A small angle between arrows indicates a high correlation between variables, and arrows pointing in opposite directions indicate a negative correlation.

Reproduced with permission from Lo, M.T., et al. Genome-wide analyses for personality traits identify six genomic loci and show correlations with psychiatric disorders. *Nature Genetics*, 49: 152–6. Copyright © 2016, Springer Nature. https://doi.org/10.1038/ng.3736.

degree of polygenicity, and the presence of loci that contribute to other behavioral traits and psychiatric illnesses. This is a common thread throughout our book.

12.3 Personality disorders

Personality disorders are different from the kinds of psychiatric disorders we have considered up until now. To put it in its simplest terms, other psychiatric conditions such as depression, anxiety disorder, or alcoholism are something you can 'have'. Personality disorders, however, are more about what you 'are'. They represent stable and typically dysfunctional features of a person, often reflecting problems in the structure of the 'self' and/or in patterns of interpersonal relationships.

There is a long historical tradition of describing personality disorders in psychiatry. One of the best known was written by the mid-twentieth-century German psychiatrist Kurt Schneider (SCHNEIDER 1958).

He described ten forms of what he called 'psycho-pathic personalities':

hyperthymic, depressive, insecure (with sensitive and anan-kastic or obsessive subtypes), fanatic, attention-seeking, labile, explosive, affectionless, weak-willed, and asthenic (debilitated, enervated).

The current most influential typology for person-ality disorders is that proposed in the *Diagnostic and Statistical Manual of Mental Disorders, Third edition* (DSM-III) (AMERICAN PSYCHIATRIC ASSOCIATION 1987). It also includes ten forms, organized into three clus-ters. The 'odd' cluster includes paranoid, schizoid, and schizotypal personality disorder. The 'dramatic' cluster includes antisocial, borderline, histrionic, and narcissistic personality disorder. The 'anxious' cluster includes avoidant, dependent, and obsessive-compulsive personality disorder (note this is different from obsessive-compulsive disorder, a condition that you 'have'). The amount known about personality disorders, especially from a genetic perspective, dif-fers widely.

12.3.1 Schizotypal personality disorder

Eugen Bleuler, the Swiss psychiatrist who first pro-posed the name and concept of schizophrenia, used to insist that his psychiatrists in training spend some Sundays in the asylum so they could interview the relatives who came to visit their parents or siblings with schizophrenia. He thought this was important because he believed that a number of these relatives suffered from something he called 'latent schizophre-nia'. They were not psychotic; that is, they never demonstrated delusions, hallucinations, or a clinically severe thought disorder. However, in their mode of thought and their interpersonal relationships, they presented hints of the kinds of behaviors seen in their much more severely affected relatives; that is, they were often odd, suspicious, socially awkward, and prone to unusual ideas (that people were looking at them or making fun of them in public places), and they sometimes talked in ways that were difficult to follow.

Through several historical twists and turns, Bleul-er's concept of latent schizophrenia was renamed schizotypal personality disorder (SPD). We met this diagnostic category in Chapter 2, section 2.5.3

when we introduced the concept of a 'schizophrenia spectrum'.

Nearly all the research on SPD has used traditional family and twin designs. The most important ques-tion, especially in the early studies, was to test whether Bleuler was right: are close relatives of individuals with schizophrenia at increased risk for SPD? Several large-scale family studies have addressed this question and confirmed Bleuler's prediction. We will review two of them briefly. One study was performed in New York (BARON et al. 1985) and examined 750 first-degree rela-tives and found a rate of SPD that was much higher in the relatives of patients with schizophrenia (14.6%) than in the relatives of controls (2.1%). A somewhat larger study, part of the Roscommon Family Study described in Chapter 2, section 2.2.1 and carried out in the west of Ireland, which interviewed 1,753 first-degree relatives, had slightly less dramatic, but still sig-nificant, findings (KENDLER et al. 1993a). Rates of SPD were estimated at 6.9% in relatives of patients with schizophrenia versus 1.4% in relatives of controls.

As we have pointed out before, family studies can-not inform us why disorders run in families. For SPD and schizophrenia, however, we have results from the famous Danish Adoption Study of Schizophre-nia (Chapter 2, section 2.5.2) to answer this question. Using the National Adoptive sample (KENDLER et al. 1994b), the prevalence of SPD in biological relatives of schizophrenia and control adoptees was, respectively, 7.3 and 2.3%, a significant difference. It is notewor-thy how similar these findings are to the Irish family study. The rates of SPD were very low in the adoptive relatives of both the schizophrenic and control adop-tees (less than 1%) and were not significantly different. Thus, this provides strong evidence that the elevated rates of SPD in relatives of schizophrenia patients are a result of genetic and not familial environmental factors.

SPD has also been investigated in a population-based sample of twins, with the most rigorous study, based on personal interviews with 1,386 pairs of young adult twins, carried out in Norway (KENDLER et al. 2006a). But studying SPD in the general popula-tion has a problem. While many people have a few of the symptoms or signs required for SPD, having five or more (required to make the diagnosis) is rare. So what this study did—and something similar is used in many studies of personality disorders in population samples—was to study the number of criteria met for SPD. Using standard twin methods (see Appendix, section A5), SPD had a modest heritability of 26% with no evidence for shared environmental effects. So, these results are broadly consistent with the findings in the Danish adoption studies that the traits of SPD run in families largely or entirely for genetic reasons.

The results assume that the schizotypal traits found in the general population are closely related etiologically to SPD observed in close relatives of schizophrenic patients. PRSs for schizophrenia provide a way to evaluate this hypothesis empirically. PRSs use genetic markers from one sample to predict how much these markers predict disease risk (or a trait) in a second sample (see Chapter 7, section 7.2). While the matter is not settled, preliminary results suggest, somewhat surprisingly, little relationship between schizophrenia PRSs and schizotypy assessed in community samples (HATZIMANOLIS et al. 2018).

Given all the interest in the application of molecular genetic strategies such as GWAS to schizophrenia, one might imagine there would be large-scale molecular genetic studies of SPD as well. There are not. We review one study using molecular methods to try to validate the schizophrenia spectrum (BIGDELI et al. 2014). This study was performed in high-density pedigrees containing at least two well-validated cases of schizophrenia. Consistent with Bleuler's observation, such pedigrees contained multiple individuals who had other schizophrenia-like syndromes, including particularly SPD. Using PRSs, the investigators tried to determine where the overall genetic risk for schizophrenia could be differentiated in four groups of these subjects in these pedigrees in comparison with control subjects. The results are seen in Figure 12.7. The PRS was highest in the narrowly defined cases that met diagnostic criteria for classical schizophrenia. The intermediate and broad-spectrum cases (most cases of SPD were in the intermediate group) had significantly lower PRSs than the narrow group, but both were appreciably higher than was seen in unaffected members of these pedigrees. However, interestingly, the unaffected members of these high-density pedigrees had a much higher PRS for schizophrenia than did Irish control individuals. The molecular data are thus broadly consistent with the family and adoption findings, and support the schizophrenia spectrum concept. Individuals with disorders such as SPD appear, from a genetic perspective, to have appreciable genetic risk for schizophrenia, but on average this risk

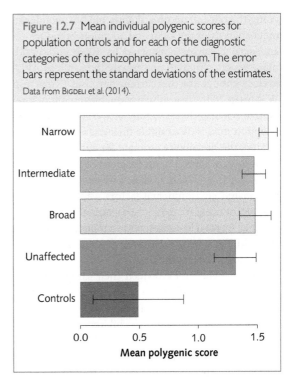

Figure 12.7 Mean individual polygenic scores for population controls and for each of the diagnostic categories of the schizophrenia spectrum. The error bars represent the standard deviations of the estimates. Data from BIGDELI et al. (2014).

is less severe than that seen with individuals with the classical severe psychotic disorder.

12.3.2 Antisocial personality disorder

The essential feature of antisocial personality disorder (APD) is a pervasive pattern of disregard for, and violation of, the rights of others; deceit and manipulation are central features (AMERICAN PSYCHIATRIC ASSOCIATION 2013). APD is related to but conceptually distinct from two other behavioral syndromes: criminality and psychopathy. Briefly, while APD is much more common among criminals than among noncriminals, many criminals are not antisocial (they can be good husbands and loyal to members of their gangs), and many individuals with APD never commit crimes. The relationship between APD and psychopathy is more complex. Most importantly, APD is defined by behaviors that individuals do, while psychopathy is defined by enduring personality-like traits, especially low empathy and callousness. In the general population, individuals with APD tend to be high on psychopathic traits, but the correlation is far from complete.

As with SPD, no definitive work has yet been done on the molecular genetics of APD. Here, we are going to concentrate on a slightly old but wonderful meta-analysis of the twin and adoption literature on antisocial behavior conducted by Rhee and Waldman, which, although not exactly the same as APD, is close enough for our current purposes (RHEE AND WALDMAN 2002). In a substantial achievement of integrative science, they carefully analyzed 51 twin and adoption studies. We outline their major findings.

First, the heritability estimate for antisocial behavior is 41%, a figure typical for personality measures. Unlike in many twin studies, however, they found that shared environmental factors were also important, accounting for 16% of the variance in liability. In other words, close relatives resemble each other in their risk for antisocial behavior for both genetic and environmental reasons.

Second, when they divided their sample by age, they found little difference in the impact of genetic factors. However, a shared environment was strongest in its effect on APD in children (20% of variance), intermediate in adolescents (16% of variance), and least important in adults (9% of variance). Why might this pattern of results occur? Most likely, living in the same home environment, as most twins do until they are in their late teenage years, impacts on their resemblance for APD above and beyond their genetic effects. 'Home environment' could include the community, school, and peer groups that twins share. But as twins grow up and move away from home, the impact of the rearing environment tends to dissipate, so the environmental sources of their resemblance for APD decline.

Third, APD is much more common in males than in females. This might lead us to wonder whether genetic effects might be stronger in affected females. Rhee and Waldman looked at this in detail. Examining studies that included both sexes, the estimated heritability was virtually identical, so genetic differences are not responsible for the differences in frequency of APD in males and females.

Finally, they compared transmission of APD in twin and sibling studies (members of the same generation) and in parents and offspring using adoption samples. It was stronger within than across generations (45 versus 31% heritability, respectively). This has been seen for a range of other disorders and behavioral traits. There are several plausible explanations, two of which merit description.

First, parents and children grow up at different times and in somewhat dissimilar cultures. Think about your childhood and that of your parents. Some things were likely pretty different, such as exposure to the internet and video games. Maybe the social environments were sufficiently different, in ways that would encourage or discourage APD, that genes expressed in the two generations are not entirely the same. In genetic-epidemiological jargon, this is called a 'gene by cohort interaction'.

Second, maybe the genetic influences of antisocial behavior in a 15-year-old adolescent and a 35-year-old adult are not entirely the same. They live in different life circumstances, often with different responsibilities and interpersonal relationships, and an adolescent versus a middle-aged brain and body. This is also plausible, and geneticists would call this (you can probably guess) a 'gene by age interaction'.

12.4 The three laws of behavior genetics

The story that has emerged of the genetic basis of IQ and personality is one that is repeated throughout this book: the genetic architecture is highly polygenic, consisting of multiple loci of small effect. It has been possible, with sufficiently large sample sizes, to robustly identify the genomic location of variants that contribute to variation in personality and in IQ, but this information has not yet delivered insights about the biological bases of the traits. Disputes about whether IQ or personality is heritable have now (almost) disappeared, and the research questions now concern the interpretation of genetic mapping results, which genes are involved, which biological processes are involved, and the genetic relationships with other traits, with psychiatric disease, and with brain structure.

The main message of this chapter is summarized in Eric Turkheimer's three laws of behavioral genetics (TURKHEIMER 2000):

1. All human behavioral traits are heritable.

2. The effect of being raised in the same family is smaller than the effect of genes.

3. A substantial portion of the variation in complex human behavioral traits is not accounted for by the effects of genes or families.

These are laws in the sense that they capture the main findings on the genetic basis of behavioral variation. In conjunction with the success of GWAS, the first

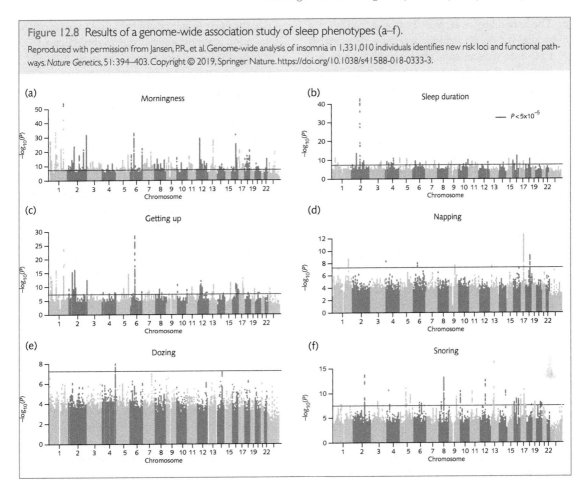

Figure 12.8 Results of a genome-wide association study of sleep phenotypes (a–f).

Reproduced with permission from Jansen, P.R., et al. Genome-wide analysis of insomnia in 1,331,010 individuals identifies new risk loci and functional pathways. *Nature Genetics*, 51: 394–403. Copyright © 2019, Springer Nature. https://doi.org/10.1038/s41588-018-0333-3.

law means that, with large enough sample sizes, any behavioral trait can be mapped. This is turning out to be true. Figure 12.8 shows the identification of loci that contribute to variation in various sleep-related phenotypes, from an enormous study of 1.3 million people (JANSEN et al. 2018). From this, you can see that there is a genetic component to napping, with one relatively large effect on chromosome 17. The information here can be used to work out if you have a propensity to dozing, or snoring, or if you are a morning person.

There's another paper telling you whether you have alleles that alter your coffee consumption (COFFEE CAFFEINE GENETICS CONSORTIUM et al. 2015), and no doubt by the time you read this book, someone will have mapped the genetic basis of what foods you prefer to eat, whether you like to keep pets, your tastes in music and art, and more that I can't imagine. What any of this means is beyond the scope of this book …

Summary

1. **Intelligence** is measured using intelligence scales that include memory tests, vocabulary tests, and tests of arithmetic. The results are reported as two forms of intelligence: verbal and performance IQ (VIQ and PIQ), or verbal and nonverbal skills. IQ scores are normalized to a population reference, where 100 is the population mean, and 15 is the standard deviation.

2. The heritability of IQ from twin and family studies is 48%. IQ changes over time, depends on age, and environmental effects on IQ are largely limited to those that are not shared (things that make IQs different rather than the same). The genetic effects on intelligence are correlated across different tests for IQ.

3. Genome-wide association studies (GWAS) have identified over 200 distinct genomic loci that contribute to educational attainment, a proxy phenotype for IQ. Genotype-based correlations identify negative genetic correlations between educational attainment and hyperactivity disorder, depressive symptoms, Alzheimer's disease, and schizophrenia; there is a positive correlation with longevity.

4. Genetic control of brain structure mirrors the brain's functional organization (the frontal, linguistic, and sensorimotor cortex have the highest heritability) and this variation is genetically correlated with IQ.

5. **Personality** refers to the stable part of an individual's attitudes, behavioral patterns, and emotional responses. Human personality can be described by five factors, often called the 'big five'.

6. Environment accounts for at least half the variance in personality, but this environment is one for which twin pairs are correlated close to zero. The shared environment makes almost no contribution to variation in personality (i.e. the environmental contribution to personality variation is almost entirely due to the effects of the nonshared environment). The heritability of personality traits is about 40%.

7. GWAS show that personality is a highly polygenic trait, with hundreds, probably thousands of independent loci contributing. Each locus has a tiny effect, contributing to between 0.02 and 0.04% of the variance.

8. **Personality disorders** are stable and typically dysfunctional features of a person, often reflecting problems in the structure of the 'self' and/or in patterns of interpersonal relationships. The elevated rates of schizotypal personality disorder in relatives of patients with schizophrenia are a result of genetic and not familial environmental factors, but there have been no large-scale genetic studies to dissect the molecular basis of the disorder. Similarly, no definitive work has yet been done on the molecular genetics of antisocial personality disorder.

9. Eric Turkheimer proposed three laws of behavior genetics:
   1. All human behavioral traits are heritable.
   2. The effect of being raised in the same family is smaller than the effect of genes.
   3. A substantial portion of the variation in complex human behavioral traits is not accounted for by the effects of genes or families.

13

Genes for what?

[On how diagnostic boundaries often do not coincide with genetic action, and on the interaction of gender and development with genetic variation]

In previous chapters, we considered one disorder at a time, and (mostly) asked: how important are genes in explaining the variability among people in their risk for developing these disorders? An implicit assumption behind our question is that genetic risk factors have a unique and privileged relationship with the individual diagnostic categories we have been using—schizophrenia (SCZ), major depression (MD), alcohol dependence, and so on. But what happens if genes influence risk for more than one disorder or trait?

In fact, as we saw, genetic factors are shared with other conditions. Polygenic risk scoring demonstrates that variants conferring risk to SCZ also confer risk to bipolar disorder (BPD; see Chapter 7, section 7.2), analysis of structural variants (in particular, of copy-number variants) revealed cases where the same mutation alters risk for intellectual disability, autism, and epilepsy (see Chapter 8, section 8.4 and Chapter 9, section 9.3), and twin data indicated that the genetic risks for individual phobias are only partly fear specific (see Chapter 10, section 10.3.1).

In this chapter, we explore in more detail how the definitions that psychiatrists apply to different disorders correspond to the structure of their genetic risk factor. We first look at genetic correlations using twin designs and then explore the same question using genotypes. Last, we look at how genetic effects can be modified by sex and by development.

13.1 Depression and anxiety

If asked, most of us would be able to distinguish between the subjective states of being anxious, tense, or jittery versus feeling sad, gloomy, or depressed. However, epidemiological and clinical studies in psychiatry have long indicated that if you meet the criteria for MD, you are far more likely to also have an anxiety disorder than would be expected by chance (this kind of association between diseases is called comorbidity).

One of us (K.S.K.) set out in the early 1990s, using over 1,000 female–female twin pairs from the Virginia

Twin Study of Psychiatric and Substance Use Disorders, to try to understand how genes contributed to this comorbidity. Specifically, we studied MD and the most closely related anxiety disorder, known as generalized anxiety disorder. The term generalized here means that the affected individual experiences relatively prolonged periods of excessive anxiety and worry that are hard to control, cause distress or impairment, and have a range of specific symptoms such as feeling restless, keyed up, or on edge, irritability, muscle tension, and sleep problems. The worry is typically about a range of concerns, not just one; hence the term 'generalized.' So this syndrome is quite different from those where anxiety arises only in one specific circumstance, such as before public speaking, getting on an airplane, or confronting a particularly nasty looking spider.

This first study published in 1992 of MD and generalized anxiety disorder reached a remarkable conclusion: from a genetic perspective, the two syndromes were for all intents and purposes the same disorder (KENDLER et al. 1992b). In this first study, the genetic correlation (which reflects the degree to which these disorders are influenced by the same genetic risk factors) between MD and generalized anxiety disorder was estimated at 1.00. Correlations can't get any higher than that.

This seemed at first a rather implausible result. Yes, these two disorders might share some risk factors—but to be perfectly correlated ... We repeated the study in 1996 with a follow-up of the same twin sample and found the same result, a genetic correlation of 1.00. Maybe there was something unusual about female twins from Virginia, although we had no idea what that might be. So we repeated the study in 1,484 male and female twins ascertained from the Swedish Twin Registry (ROY et al. 1995). The estimated genetic correlation between generalized anxiety disorder and MD in the sample was, again, 1.00. Being a persistent sort, we tried one last time—with a far bigger (and more statistically powerful) sample—37,296 twins reflecting an entire national sample from the Swedish Twin Registry. In the women in this sample, estimated genetic correlation between generalized anxiety disorder and MD in the sample was, again, 1.00 (KENDLER et al. 2007a). However, we found that in males the genetic correlation was less: 0.74—still quite high but not unity.

What do these results mean? In humans, our genes appear to put us at low or high risk for a general negative emotional state, which we sometimes call 'dysphoria', a term for a range of negative mood states. However, whether that state tends to be sad, depressive, and gloomy versus anxious, worried, and tense may actually largely be a result of our environmental experiences. If you have a genetic liability and are exposed to losses (e.g. a bad romantic break-up), you are likely to develop depression. If, by contrast, you are exposed to chronic danger (e.g. your factory might close and you will lose your job), you are more prone to develop generalized anxiety disorder. The preconception that the gene–behavior relationships come in nice neat packages, where one set of genetic risk factors impacts on only one behavioral disorder or trait, is probably no truer than that one gene governs one phenotype.

13.2 Seven common disorders

Let's expand our playing field. Instead of looking at two disorders at a time, let's look at seven disorders, again in the Virginia twin sample (KENDLER et al. 2003). The seven disorders are all relatively common. In addition to the two we have just met, MD and generalized anxiety disorder, we also studied phobias, alcohol dependence, drug abuse or dependence, adult antisocial behavior, and conduct disorder (phobias reflect irrational and disabling fears of a range of objects or situations, and conduct disorder describes children who are aggressive and have little empathy for the suffering of others). We looked at the pattern of genetic risk factors for these seven disorders and obtained the results shown in Figure 13.1.

Before looking at these results in detail, let's review what we might expect. One possibility would be that the people who designed our psychiatric diagnosis really got it right. There was one specific and unique set of genetic risk factors for every single disorder (or you might see it the other way—that the genes are very respectful of psychiatric categories). Alternatively, it could be that there is no specificity whatsoever with genetic risk factors for these common psychiatric illnesses; that is, there is just one big clump of genes that impact on your overall vulnerability to any psychiatric condition.

Let's now turn to what we found. Factor A_{C2} at the top left of Figure 13.1 is a set of genetic factors that impact strongly on the risk for MD, generalized anxiety disorder, and phobias. We have a name for these

Figure 13.1 Genetic risk factors for seven common psychiatric disorders. Dep, dependencies; GAD, generalized anxiety disorder. The highlighted lines and path estimates reflect paths that explain at least 10% of the variance in risk for the disorder—in other words the "important" paths.

Reproduced with permission from Kendler, K.S., et al. The structure of genetic and environmental risk factors for common psychiatric and substance use disorders in men and women. *Arch Gen Psychiatry*, 60(9):929–937. Copyright © 2003, American Medical Association. doi:10.1001/archpsyc.60.9.929.

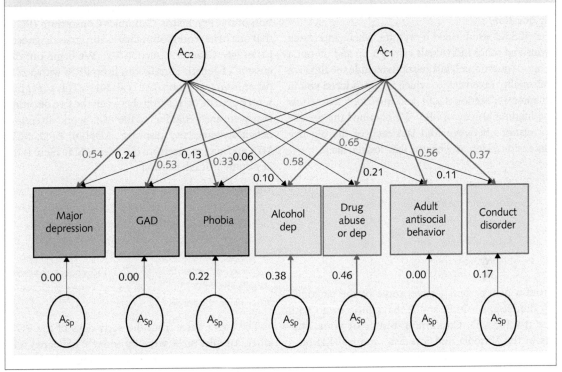

kinds of conditions in psychiatry: internalizing disorders. We call them 'internalizing' because they involve people feeling dysphoric in their internal life in ways that impact on how they feel about themselves and the world around them. Factor A_{C1} at the top right is a set of genetic factors that impact strongly on risk for alcohol dependence, adult antisocial behavior, and conduct disorder. We also have a name for these kinds of conditions in psychiatry: externalizing disorders. We call them 'externalizing' because they involve people acting out their distress, often in impulsive or aggressive ways, rather than experiencing it 'on the inside'.

In addition to these two general factors, there are disorder-specific genes (A_{sp}). Two of these are large, influencing alcohol dependence and drug abuse/dependence. These disorder-specific genes are expected to include **alcohol dehydrogenase**, which we discussed in Chapter 11, section 11.3, and other similar genes that have an impact **only** on the risk for alcohol dependence because they specifically affect the metabolism of alcohol.

The truth is probably somewhere in between our two prior hypotheses. Mostly, genetic risk factors (at least for these common disorders) impact on groups of disorders, not individual diseases. To put this another way, most genes that act on risk for psychiatric illness are not disease specific in their effect. But there is not just one large dimension of genetic risk for all disorders. Rather, there are at least two dimensions for the very common psychiatric disorders (and there would certainly be more if our model included other disorders such as SCZ, BPD, autism, and anorexia nervosa).

13.3 Major depression and personality

In our quest to understand the specificity versus non-specificity of genes with respect to psychiatric disorders, we have now explored two examples using genetic epidemiological designs. Before we leave this area, let's briefly examine a related question. What is the relationship between genes that impact on our personality and those that predispose to MD? The idea of a depressive (or melancholic) temperament goes back to Greek times. Most of us know people who, by nature, seem gloomy, worried, and depression prone. What does the data say about this intriguing question?

I (K.S.K.) established a website where interested twins and other individuals can sign up and fill out a range of questions. The questions include the Big Five Personality Inventory to which we introduced you in Chapter 12, section 12.2.1 and questions that assess the lifetime history of MD. We obtained the genetic correlations between MD and each of the big five dimensions. Here is what the data looked like:

O—Openness:	0.17
C—Conscientiousness:	−0.36
E—Extraversion:	−0.06
A—Agreeableness:	−0.18
N—Neuroticism:	0.43

From a genetic perspective, three of the personality dimensions (O, E, and A) have little to do with the risk for MD. Confirming results of prior studies in the Virginia and Swedish samples (KENDLER et al. 1993e; KENDLER et al. 2006b), the genes that predispose to high levels of neuroticism are moderately correlated with those that predispose to MD. More surprising was the inverse genetic relationship between conscientiousness and risk for MD. Readers who are consistently careful, reliable, and planful might take some solace in this result—you may have a bit of genetic protection against developing depression.

13.4 Polygenic risk scores and genetic correlations

We described in Chapter 7 section 7.2 the methods for calculating polygenic risk scores (PRSs) from genome-wide association data and mentioned the surprise when a PRS developed for SCZ also predicted risk for BPD. Now we want to explore this area further. In our prior discussion on MD and generalized anxiety disorder, we introduced the concept of genetic correlations and reported such results obtained from twin studies. But genetic correlations can also be predicted from genome-wide association data.

We described methods in Chapter 7, section 7.5 to estimate single-nucleotide polymorphism (SNP)-based heritability, and explained that the methods can be extended to estimate genetic correlations (see Chapter 7, section 7.5.3) (VISSCHER et al. 2014; BULIK-SULLIVAN et al. 2015). You may recall from our description of the Psychiatric Genomics Consortium (PGC) that one of its workgroups focused on cross-diagnostic issues (see Chapter 7, section 7.3). We begin our discussion of genetic correlations from PRSs with a 2013 paper from that group (LEE et al. 2013). This paper presented genetic correlations between the five disorders that were the initial foci of the PGC work: attention-deficit hyperactivity disorder (ADHD), BPD, SCZ, MD, and autism spectrum disorder (ASD). Here is the key quote from the abstract:

The genetic correlation calculated using common SNPs was high between schizophrenia and bipolar disorder (0.68 ± 0.04 s.e.), moderate between schizophrenia and major depression (0.43 ± 0.06 s.e.), bipolar disorder and major depression (0.47 ± 0.06 s.e.), and ADHD and major depressive disorder (0.32 ± 0.07 s.e.), low between schizophrenia and ASD (0.16 ± 0.06 s.e.) and non-significant for other pairs of disorders …

The paper dates from the early days of the PGC, when samples sizes were by today's standards relatively small; the number of cases ranged from 3,303 for ASD to 9,087 for SCZ. Five of the ten possible genetic correlations were significant. Nearly everyone was surprised at how high the SCZ–BPD and SCZ–MD correlations were and how relatively low the correlations were between BPD and MD (which were considered by Kraepelin to be subtypes of the same disorder—what he termed manic-depressive insanity).

Considerable debate ensued about possible biases in these figures and the degree to which these results were or were not consistent with prior genetic epidemiological studies. Diagnostic misclassification is one of the major concerns. The phenotypic characterization of many PGC cases was 'shallow' in part because large numbers were needed and resources limited. Early cases of SCZ and BPD can be hard to distinguish. In one published study, these possible errors were modeled using realistic rates of reciprocal misclassification and suggested that the true genetic correlation for BPD and SCZ might be closer to 0.5 than to 0.7 (WRAY et al. 2012).

One statistical point here needs to be noted. If a genetic correlation between two disorders is X, it is often assumed that, as a rough rule of thumb, X% of the risk variants for the two disorders are the same. While the relationship between these two concepts is quite complex, this assumption is almost surely wrong. The true sharing of genetic risk factors between the two disorders is almost certainly closer to X^2 than to X. So if the true genetic correlation between BPD and SCZ is 0.5, then the best guess is that they might share about 25% of their risk variants, which is not such a high number.

In 2018, another paper reported genetic correlations from molecular genetic data between a range of psychiatric disorders and other relevant traits (BRAIN-STORM CONSORTIUM et al. 2018). Genetic correlations were reported for ten instead of six disorders, adding anorexia, post-traumatic stress disorder (PTSD), obsessive-compulsive disorder (OCD) and Tourette's syndrome, and were based on much larger sample sizes (e.g. 75,846 cases of SCZ and 49,367 for BPD). Looking at the highlights, the genetic correlation between BPD and SCZ was again estimated at 0.68, the MD–SCZ correlation declined moderately to 0.34, while the ASD–SCZ correlation increased slightly to 0.21. No big surprises here. Other interesting results were that the genetic correlation between PTSD and MD (0.52) was much higher than that between PTSD and SCZ (0.15), consistent with clinical expectation, while the correlation between MD and OCD was much more modest (0.23).

We conclude this section by noting that both genetic epidemiological and molecular genetic methods indicate that some degree of sharing of genetic risk is common across psychiatric disorders, but the results are by no means uniform. Should we be surprised by these results? Probably not, for two reasons: first, psychiatric disorders are broad syndromes that are defined clinically and not on the basis of well-understood etiological pathways. The idea that each would have a unique set of genetic risk factors is actually not highly plausible. Second, substantial sharing of genetic risk factors is seen in other disease groups. For example, many genetic variants have been found that increase risk for a range of clinically distinct autoimmune disorders and different forms of cancer. This pattern may be more the rule than the exception.

13.5 Mendelian randomization

The genetic models we have been using assume that the correlation between different phenotypes, such as personality and MD, result from the effect of shared genetic (or environmental) factors. But, focusing here only on genetic effects, the correlation between two traits or disorders could result from the causal effect of one phenotype on another. Thinking about psychiatric or substance-use disorders, it is plausible to suggest that in some cases MD could lead to alcoholism, or drug abuse could increase the risk for SCZ.

This idea that genes cause one disorder that in turn causes another has led to an increasingly popular and novel use of genetic variants to clarify the causal relationship between disorders or traits. George Davey Smith, an epidemiologist based in Bristol, has pioneered the method, called Mendelian randomization (DAVEY SMITH AND HEMANI 2014). His background in epidemiology has trained him to consider the possible confounds that would confuse correlation with causation when an association is established. A classic example is in drug trials, where doctors want to be certain that the medication given to one group is the cause of any differences between the treated and non-treated groups. For example, there is a well-known placebo effect in the use of antidepressant drugs, so that if participants know they are being given a drug to treat depression they are more likely to report improvement. Mendelian randomization is a way to avoid such confounding.

Alleles are randomly assigned at birth through the principles of Mendelian segregation so that there is no (or very little) confounding in who, or who does not, receive an allele. This is why the strategy is called Mendelian randomization. Using the drug trial analogy, you can think of people who inherit a risk allele for a trait as receiving a higher-than-average dose of that trait, whereas those who do not inherit the allele are assigned a lower-than-average dose.

Using a genetic variant has two important advantages for epidemiology: first, in genetic association, the direction of causation is from the genetic variant to the trait, and not *vice versa*, providing an anchor on the causal direction. Second, while conventionally measured environmental exposures are often associated with a wide range of behavioral, social, and physiological factors that confound associations with outcomes, genetic variants serve as unconfounded indicators of trait values.

We'll give an example from general medicine of how Mendelian randomization is implemented. Our blood contains a set of proteins that, when needed, interact with each other to cause the blood to clot. This is called the coagulation cascade. When you cut your finger on a kitchen knife, it is a very useful thing.

But sometimes your blood can clot in places where it should not, such as inside your coronary arteries, in which case it can cause a heart attack. As the clotting system can harm as well as protect health, we'd like to know whether altering clotting alters mortality. In particular, if we can show that reducing clotting will reduce mortality, then we'd know that drugs that reduce clotting would be beneficial. How can we find out whether by reducing clotting factors (which we know will increase clotting time) we will cause a decrease in mortality?

Here is where Mendelian randomization comes in. There is a genetic disorder called hemophilia where people are missing one of the key proteins in the clotting cascade. It is a dangerous disorder because even after a slight injury, hemophiliacs are at danger of dying from excessive blood loss because their blood clots so slowly. Most of the genes that produce the key clotting proteins are on the X chromosome, which is why hemophilia is so much more common in males: having only one X chromosome, they only have one copy of the gene, so a mutation will have a more devastating effect than hemophilia in women, whose second copy (women have two X chromosomes) will compensate for the loss of the first.

Referring to Figure 13.2, the genetic variant that does or does not inactivate the clotting protein is variable X. Variable Z is the speed of clotting and variable Y is the overall mortality. Individuals heterozygous for the clotting mutation, who have 50% of the normal level of the clotting protein, clot considerably more slowly than normal individuals. So there is a causal path established from X to Z.

The elegant feature of this study is how they found their heterozygote cases (SRAMEK et al. 2003). The identified all mothers of individuals with hemophilia in the Netherlands, 1,012 in total, thereby avoiding having to genotype very large numbers of individuals (variable X in Figure 13.2). This is smart because, as you recall, males only get one X chromosome from their mother so every mother of a hemophiliac son has to be heterozygous for one of the key clotting proteins. The mothers have increased clotting time (variable Z in Figure 13.2) due to the mutations they carry for hemophilia.

What did the authors find? In the mothers, the overall mortality rate (Y in Figure 13.2) was reduced by 22%, a highly significant difference. Even more impressive, their rate of death from coronary artery disease (i.e. heart attacks) was down by 36%. Rates of

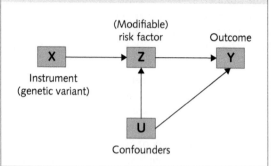

Figure 13.2 A diagram of the basic principles of Mendelian randomization. The genetic variant (or groups of variants) X impacts on one potential trait or disorder (Z), which in turn is a risk factor for disorder or trait Y. For the method to derive proper estimates of the association between X and Y, researchers need to rule out the effect of confounders (U), as well as any direct path from X to Y that does not go through Z.

death from hemorrhage (excess prolonged bleeding) were increased, but this was a very rare event and was more than compensated by the substantial drop in deaths from heart disease.

So, this was a successful implementation of the logic of Mendelian randomization. But there are caveats. One is the problem of confounders (variable U in Figure 13.2). Could there be some other set of factors unknown to the investigators that impacted on clotting time and mortality rate? This doesn't seem likely, as hemophilia mutations are randomly distributed in the Dutch population. But you can never be completely sure; there is always the possibility that some unacknowledged confound, a third or fourth variable, could mediate the association.

Another problem for the interpretation of a Mendelian randomization experiment is pleiotropy, where a single locus influences many phenotypes. Pleiotropy can arise in two ways: a single locus directly influences multiple phenotypes (sometimes called type 1 pleiotropy) or single locus produces a product, a protein for example, that in turn causes changes in multiple phenotypes (type 2 pleiotropy). Type 1 pleiotropy is recognized as problematic for Mendelian randomization: how can we be sure that the genetic variant is having a direct effect on Y, and is not mediated by something else?

Pleiotropy is particularly important in implementations of Mendelian randomization to study behavior. As we note elsewhere in this book, pleiotropy is a distinctive feature of the way genes influence behavior. Furthermore, in psychiatric studies, X is not one genetic variant of large effect (our gene for hemophilia) but many genetic variants of small effect. In the hemophilia example, we could be pretty sure that the only effect of X on our outcome Y was via Z; that is, the only effect of the mutations in the clotting proteins on mortality was through their effect on clotting time. But with complex diseases, it is much more difficult to be sure of this. If any of the genetic variants making up X directly impact on Y (not through their effect on Z), then all bets are off on what is causing what. As an example, if we wanted to know if drug abuse caused SCZ, could we be sure that all the genetic risk variants for drug abuse that we are studying have no direct effect on SCZ and only influence that risk by increasing the risk for drug abuse? Thus, caution is indicated in the interpretation of such studies.

13.6 Genes and sex

In this section, we explore the ways in which sex moderates the impact of genes on behavioral phenotypes. Let's start with a simple example of what geneticists call a sex-limited genetic effect, where genes influence a phenotype in one sex but are silent in the other. A woman might carry one of a variety of risk genes for prostate cancer and pass that risk to her sons, but these genes will have no phenotypic effect on her, as she lacks a prostate gland.

A subtler but related concept is a sex-modified genetic effect. Here, the impact of genetic risk factors is altered but not eliminated by the sex of the body in which they exist. The degree of modification can be estimated by the genetic correlation. In this instance, the genetic correlation means the degree of resemblance of genetic risk factors for a given phenotype in men and women. A value of 1.0 means that exactly the same genetic variants do exactly the same thing in the two sexes. A value of 0.0 would mean that there was absolutely no relationship between the genes that influenced that particular trait in men and the genes that impacted that same trait in women. We have good evidence for such effects in MD.

As we discuss in Chapter 10, section 10.1.4, twin studies have estimated the genetic correlation for MD in men and women, so that with some confidence we can conclude that, while related, the genetic risk factors for MD are not entirely the same in men and women. One plausible (but unconfirmed) scenario illustrates this phenomenon.

Women are at increased risk for developing MD after giving birth. This vulnerability for what is called **postpartum depression** runs in families and is probably influenced by a set of genetic risk factors that is at least partly independent from typical depression. Males might carry the genes to put them at risk for postpartum depression, but they will never get a chance to manifest them because they never give birth, thereby avoiding the profound psychological and biological changes that occur around that time. The prediction is that when the individual genetic variants are found that predispose to MD, some will differ substantially in their impact in men versus women.

MD is not the only psychiatric disorder showing genetic correlations between the sexes that are less than 1. The same thing has been found for alcohol dependence (and is, in this case, backed up by rodent studies) (PRESCOTT AND KENDLER 1999). From a biological and psychological perspective, males and females are different enough that we cannot always assume that genetic effects on behavior will be the same in the two sexes.

13.7 Genes and development

Nearly all of our prior analyses of the impact of genetic risk factors have assumed that genes are static in their effects. In this final section, we hope to show you that, at least for some traits over some time periods in human development, this is a bad assumption to make. We inherit one set of genes from each of our parents, and aside from the occasional somatic mutation, these genes are with us until we die. But this does not mean that the effects of genes remain the same throughout our lives.

One of the most dramatic events in human development is the onset of menstruation in females. This has been studied extensively in humans and is strongly influenced by genetic factors. The last pair of monozygotic twins that I (K.S.K.) interviewed reported starting menstruating on the same day when they were 12 years old. The timing of many other features of human development—the onset of baldness in men, the graying of our hair, or even the paunch in our bellies—is strongly influenced by our genes. So genes can be developmentally dynamic.

We will illustrate this with an example taken from a longitudinal study of 2,490 Swedish twins under the direction of Dr. Paul Lichenstein (LICHTENSTEIN et al. 2007). In this study, symptoms of anxiety and depression were assessed by questionnaires filled out both by the twins and their parents at ages 8–9, 13–14, 16–17, and 19–20 years. The study covered a period of tremendous physical, intellectual, and emotional change: from mid-childhood through puberty and adolescence to young adulthood. In our analyses, we were able to take into account reports on the symptom levels of the twins both from themselves and their parents, and we could examine the continuity and discontinuity of the genetic and environmental influences (KENDLER et al. 2008). The genetic effects are presented graphically in Figure 13.3.

In reading this figure, let's start with section A, which reflects the genetic risk factors for anxiety and depression that are active at 8–9 years. What happens to these genetic effects over time? They attenuate dramatically after age 9 and then sputter along, having a modest effect through to ages 19–20. What this is telling us is that genetic influences on anxiety and depression that occur in childhood decline rapidly in their impact after puberty.

Section B of the figure reflects genetic effects on anxiety and depression that start at ages 13–14, soon after puberty for most of the sample. Compared with the genes that start prepubertally, these effects attenuate much less over time. Although this is already rather complex, it is not the end of the story. There are small spurts of genetic innovation as new genetic effects emerge during development at 16–17 years (section C) and even at 19–20 years (section D).

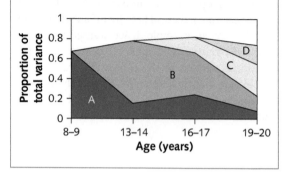

Figure 13.3 Symptoms of anxiety and depression. The four sections reflect the genetic contributions to anxiety and depression from four different age groups (A: 8–9 years; B: 13–14 years; C: 16–17 years; D: 19–20 years).

Adapted with permission from Kendler, K.S., et al. A developmental twin study of symptoms of anxiety and depression: evidence for genetic innovation and attenuation, *Psychological Medicine* 38:1567–75 © Cambridge Journals, published by Cambridge University Press. https://doi.org/10.1017/S003329170800384X.

The bottom line in this short story is that, with respect to a quite common and important human trait—levels of anxiety and depression—genetic effects are strikingly dynamic over the time from childhood to young adulthood. We have evidence for genetic innovation (new genes coming on line) and genetic attenuation (genetic factors that decline in phenotypic impact with development). In thinking about genes and their impact on behavior, time and stage of development matter.

Summary

1. Genes often disrespect diagnostic boundaries between psychiatric disorders. Genetic risk factors for major depression (MD) and generalized anxiety disorder are very closely related. When a range of common psychiatric disorders are examined, genetic risk factors appear to predispose broadly to either internalizing disorders (characterized by anxiety and depression) or externalizing disorders (characterized by acting-out behaviors and/or substance abuse).

2. Genetic factors that influence personality may also alter risk for psychiatric disorders. For example, genes that predispose to high levels of the personality trait of neuroticism impact substantially on the risk for MD.

3. Genotype-based methods of determining genetic correlations between psychiatric diseases indicate the presence of common genetic variants. Genetic correlations are relatively high between different forms of anxiety disorder, between post-traumatic stress disorder and MD, and between

schizophrenia (SCZ) and bipolar disorder (BPD), while more moderate correlations exist between SCZ and MD, BPD and MD, attention-deficit hyperactivity disorder and MD, and MD and obsessive-compulsive disorder.

4. Mendelian randomization is a way to determine causal relationships between diseases and traits, as the alleles segregate randomly at birth (and thus are free of many confounding effects) and the direction of causation is from the genetic variant to the trait, providing an anchor on the causal direction. However, the interpretation of Mendelian randomization in psychiatric genetics can be problematic due to pleiotropy.

5. From a biological and psychological perspective, males and females are different enough that the impact of genetic factors on behavior may not be the same in the two sexes. This is called a sex-modified genetic effect, and occurs in several psychiatric disorders including MD and alcoholism.

6. Genes often act dynamically over time in their effect on psychiatric and psychological traits. For example, the genes that influence levels of anxiety and depression appear to change substantially pre- and postpuberty.

14

Genes and the environment

[On how genetic and environmental factors combine to generate behavioral variation]

In this chapter, we aim to convince you that the impact of genetic risk factors on psychiatric disorders is frequently so closely intertwined with environmental factors that studying genes in isolation is often asking for trouble. We first need to describe two different ways in which genes and environment can interrelate in contributing to the risk of illness. The first mechanism is termed gene–environment (gene by environment) **interaction** (the expression we prefer is: **genetic control of sensitivity to the environment**) and gene–environment **correlation** (or our preferred expression: **genetic control of exposure to the environment**). Then, in the second part of this chapter, we show how adding genetic or familial information to traditional problems in psychiatric or drug-abuse epidemiology can dramatically increase the value of the conclusions we can reach.

14.1 Gene–environment interaction

The common-sense model of how genes and environment combine to contribute to risk for a disease is that they add together; that is, an individual's total risk for a given disorder, such as major depression (MD) or alcoholism, is simply the sum of the risk provided by their genes and the risk created by their various life experiences. This additive model is portrayed in Figure 14.1. The figure depicts the liability to develop a given illness for three hypothetical individuals with low, intermediate, and high genetic liability when exposed to varying levels of environmental risk. Most importantly, the slopes of the lines of total liability are the same for each individual; that is, as individuals move from a low-risk ('protective') to a high-risk ('predisposing') environment, their liabilities increase by the same amount, regardless of their level of genetic risk.

Contrast this additive model with the model depicted in Figure 14.2. Here, as in the additive model, the liability to illness increases from lower to higher genetic risk and with increasing levels of environmental risk. However, the slopes of the lines are now different in individuals with different levels of genetic risk. Individuals with high genetic risk have the steepest slope and those with low genetic risk the flattest slope. In other words, the increase in liability that occurs in moving from a protective to a predisposing environment is itself related to genetic risk.

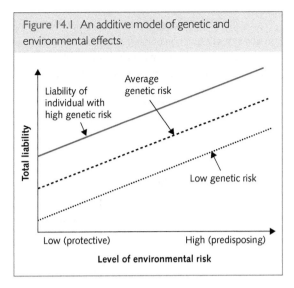

Figure 14.1 An additive model of genetic and environmental effects.

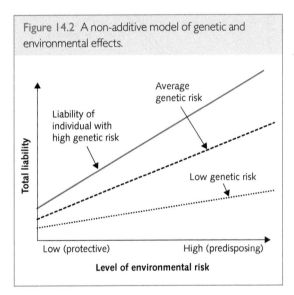

Figure 14.2 A non-additive model of genetic and environmental effects.

This is worth restating. In the model depicted in Figure 14.2, genetic risk contributes to disease liability in two different ways: first, it impacts on overall liability to illness; second, it alters the individual's sensitivity to the pathogenic effects of the environment. The observation that individuals have important differences in their level of stress responsivity is summarized by Burton in his classic text, *Anatomy of Melancholy*, written in 1621:

... according as the humour itself is intended or remitted in men [and] their ... rational soul is better able to make resistance; so are they more or less affected [by adversity]. For that which is but a flea-biting to one, causeth insufferable torment to another; and which one by his singular moderation and well-imposed carriage can happily overcome, a second is no whit able to sustain, but upon every small occasion of misconceived abuse, injury, grief, disgrace [and] loss ... yields so far to passion, that ... his digestion hindered, his sleep gone, his spirits obscured, and his heart heavy ... he himself [is] overcome with melancholy.

BURTON (1621; 2001 edn).

Of the many examples of gene–environment interactions in psychiatry, we will limit ourselves here to examining three, one each involving adoption, twin, and molecular studies. Examining 367 adoptees studied in Iowa, an attempt was made to predict the risk for adolescent antisocial behavior in adolescents from their genetic risk (indexed by a history of antisocial or alcoholic behavior in the biological parents), their environment risk (indexed by an adverse adoptive home environment), and their interaction (CADORET et al. 1983). What was found is shown in Figure 14.3. For those without a genetic risk factor (marked as 'absent' in the figure), adding a high-risk environment led to a small increase in antisocial behavior. However, for those who had a high level of genetic vulnerability (genetic risk factor 'present'), adding a high-risk environment produced a very large increase in antisocial behavior. With respect to the development of antisocial behavior, those at high genetic risk were more sensitive to the pathogenic effects of an adverse adoptive home environment.

Our second example involves MD. In addition to the evidence of genetic risk factors for MD (see Chapter 10, section 10.1.3), adverse stressful life events have been shown to consistently and strongly increase the risk for MD (KESSLER 1997). In the Virginia twin study, we asked whether these two risk factors added together or interacted (KENDLER et al. 1995a). We focused only on the most severe kinds of stressful life events, which in this study included death of a close relative, assault, serious marital problems, and divorce or a romantic break-up.

We were able to assess genetic risk by asking whether the co-twin had a history of MD. We knew, from the twin studies reported in Chapter 10, section 10.1.3, that depression has a heritability of about 40%, which means that the relatives of someone with depression must have an increased genetic risk. The extent of their risk depends on how much genetic material they share with the affected individual:

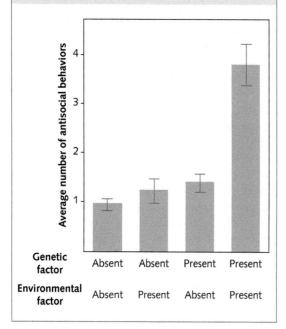

Figure 14.3 Evidence for gene–environment interaction in the etiology of antisocial behavior. Using an adoption sample, this study examined the relationship between the number of rated antisocial behaviors and the presence or absence of genetic and environmental risk factors. Bars show standard error of the mean.

Reproduced with permission from Cadoret, R.J., et al. Evidence for gene-environment interaction in the development of adolescent antisocial behavior. *Behavior Genetics*, 13(3):301–10. Copyright © 1983, Plenum Publishing Corporation. https://doi.org/10.1007/BF01071875.

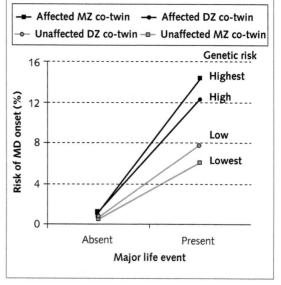

Figure 14.4 Risk of onset of major depression (MD) in a month as a function of inferred genetic risk and the presence or absence of a major stressful life event during that month. MZ, monozygotic; DZ, dizygotic.

Adapted with permission from Kendler, K.S., et al. Stressful life events, genetic liability, and onset of an episode of major depression in women. *The American Journal of Psychiatry*, 152(6):833–842. Copyright © 1995, American Psychiatric Association. https://doi.org/10.1176/ajp.152.6.833.

100% sharing for monozygotic (MZ) twins, 50% sharing for dizygotic (DZ) twins. A person whose MZ co-twin has had an episode of depressive illness will have the highest genetic risk, and those at lowest risk will be twins whose MZ co-twins have never had an episode of depressive illness. As predicted from the interactive model, the impact of a severe stressful life event on the risk for depression was more striking in those at highest genetic risk than in those at lowest genetic risk. In this example, genetic risk factors altered an individual's sensitivity to the depressogenic effects of these very stressful life events.

Focus first on those at lowest genetic risk, indicated by the blue squares in Figure 14.4. In the absence of a stressful life event, they have a 0.5% risk of experiencing a depressive episode each month. If something really bad happens to them, their risk increases,

but even then, only 6.2% of them become clinically depressed. So the increase in risk for depression for these low-risk individuals in moving from a low- to a very high-stress environment is 5.7%.

Now look at the highest genetic risk individuals in Figure 14.4, pictured as the black squares. In low-stress environments, their risk for an onset of MD is 1.1% per month. However, in a month when they experience a severe life event, their risk increases to 14.6%, a gain of 13.5%. When we say that genetic risk can make you more sensitive to the depressogenic effects of stress, what we mean here is the difference in the chance of an onset of depression in going from low- to high-stress environments that is more than twice as great (13.5 versus 5.7%) in those at high versus low genetic risk.

We introduced the CONVERGE study of major depression in Chapter 10, section 10.1.6. This molecular genetic study was organized by two of us (J.F. and K.S.K.) and recruited Han Chinese women with recurrent depression and matched controls. We utilized a deep phenotyping approach that included extensive assessments of environmental risk factors because we wanted to try to understand how genetic and environmental risk factors together contributed to risk for MD.

In our first paper on this topic (CONVERGE CONSORTIUM 2015), we took a simple approach and divided our cases and controls into those who had and had not been exposed to high levels of adversity, either in childhood (child sexual or physical abuse or neglect) and/or adulthood (severe stressful life events). By this definition, cases were over twice as likely to experience environmental adversity than controls. First, we looked for the presence of the type of genetic sensitivity to the environment described above, where the presence of both genetic risk (now indexed by genotypes) and adversity increases the risk of depression more than expected by simply adding the two components. The effect was significant and explained 13% of the total variance.

Next, we asked whether we could see interactions at individual loci (in the same way that has been claimed to occur at the serotonin transporter locus; see Chapter 5, section 5.2.3). We performed two genome-wide association studies (GWAS), one in the exposed and the other in the unexposed subsamples (PETERSON et al. 2018), and suspected three possible outcomes: (i) all loci could show the same sort of interaction; (ii) some loci might show interactions while others do not; or (iii) there could be a more complex pattern of interaction, where the presence and indeed direction of the interaction varies considerably between loci.

Despite the reduction in sample size compared with the entire cohort, three new genome-wide significant loci emerged in the unexposed group (two on chromosome 1 and one on chromosome 8). The interactions were significant at each locus. The results for one locus (indexed by the genetic marker rs7526682 on chromosome 1) are shown in Figure 14.5.

In looking at this figure, you might initially feel confused. The pattern of results is the **opposite** of that seen in our other two examples of gene–environment interaction. In our prior adoption and twin analyses, we saw that genetic effects were stronger in the presence of adversity. Here, we see that in those **without** exposure to adversity, those with the GG genotype have about a 1.4-fold increase in risk for depression compared with those with the CC genotype. By

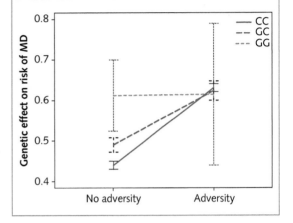

Figure 14.5 Gene–environment interaction detected at a risk locus for major depression (MD). The effects of the three genotypes (GG, GC, and CC) on depression are plotted for those with and without self-reported stressful life events (shown as 'adversity' and 'no adversity'). The error bars at the end of the lines show the 95% confidence intervals of the effect sizes. Data from PETERSON et al. (2018).

contrast, in those exposed to adversity, there is almost no difference between the three genotypes (which means that there is a negligible contribution to depression risk in this group). Similar results were seen at the other two loci.

Finally, we found that the heritability for MD was **higher** in both groups (although more so in those without, than in with exposure to adversity). The genetic correlation between the two groups was 0.64 (although the correlation was not significantly different from 1.0). These two findings are consistent with the hypothesis that **different** genetic effects are present in the two groups.

We can draw three tentative conclusions from these results. First, using molecular tools, we replicated evidence for genetic sensitivity to the environment effects in MD. Second, for depression, genetic sensitivity to the environment probably takes diverse forms; in consequence, the effects obtained from aggregate measures of the entire genome, as assessed by adoption, twin, or single-nucleotide polymorphism heritability designs, need not agree with those from individual markers. Third, genetic effects on MD that operate in one kind of environment are not the same as those that have an impact on risk in another.

14.2 Gene–environment correlation

In this section, we introduce you to an idea that sounds paradoxical: that genetic effects influence environmental experiences. Consider this quote from the summary of an article published in 1949 entitled 'The accident-prone automobile driver' (TILLMANN AND HOBBS 1949).

In this study the existence of accident-prone drivers has been demonstrated in the records of accidents from a bus company extending over a period of 6 years. The frequency of appearance of the same individual in the high accident group in multiple years has been noted. It has been shown that because of their high accident rate their importance in contributing to accidents far exceeds their numbers … A group of high and low accident drivers … have been interviewed and the differences in their personality and background have been noted. It has been demonstrated that the high accident taxi driver most frequently comes from a home marked by parental divorce and instability …. In adulthood, his occupation record is marked by frequent short-term employment and his connections with any firm are frequently terminated by the employer.

One explanation for the existence of accident-prone individuals is that they have an inherited tendency to behave in ways that encourage mishaps. To get you over the surprise of the concept, think of it this way. First, the 'environmental risk factor', traffic accidents, is not randomly distributed throughout the population; second, different personalities experience accidents at different frequencies, some high, some low; and third, personality traits are heritable (see Chapter 12, section 12.2.1). For example personalities that prefer to take risks are more likely to drive fast or to cut corners, and are more likely to get into accidents. So it is not so absurd to think that genes can influence aspects of our environment.

Some of the bad things that happen to us in life are truly just bad luck. You are driving down the highway at the legal speed limit. You are not talking on the phone or trying to text. You are sober and well rested. A chain on a truck in front of you snaps and a log rolls off directly in front of you. There is no way to avoid a serious accident. It's just bad luck.

But for some of these bad things (and more than we would likely want to admit), we have played some role in their occurrence. A number of researchers have examined whether experiencing stressful life events is a heritable trait. The answer has consistently

been 'yes but only modestly', with three representative twin studies finding heritabilities for stressful life event exposure of 20, 25, and 40% (PLOMIN et al. 1990; KENDLER et al. 1993d; BOLINSKEY et al. 2004).

We can divide life events into those whose occurrence we could probably influence (such as interpersonal difficulties or legal problems) and those fateful events such as medical illnesses or the death of relatives (or the log from the truck in front of you) that are probably independent of anything we might do. Estimating heritability for the experience of the two different types of adversity demonstrates what you might expect: much stronger evidence for genetic effects for the events we can influence compared with those that we cannot (KENDLER et al. 1999b). So our genes do influence our chances of exposure to at least some stressful life event experiences.

Another important aspect of our environment is captured by the concept of social support. Social support can be defined as the degree of caring and emotional and physical sustenance an individual obtains from their social environment. Interest in the nature of human social relationships and their impact on health has a long tradition in the field of mental health and the disciplines of sociology, social psychology, and public health. A large body of research has shown that the quality of social relationships predicts general health and mortality, psychiatric symptoms, and emotional adjustment to stress.

You can probably guess what is coming next. A series of twin studies have shown that social support is heritable. In one study where we measured social support twice over a 5-year interval, we found that the heritability of the stable part of social support was over 50% (KENDLER 1997). Our genetically influenced temperament has a lot to do with whether we are exposed to large and caring social networks.

A third example of genetic control over exposure to the environment examines the concept of peer deviance. Peer deviance is important because it is a very strong predictor of future drug use and antisocial behavior. Peer deviance is measured by scales assessing the frequency of deviant behavior of your friend, items like cheating at school, getting into fights, or using drugs. This is the sense in which we are using it here. Thinking back to your teenage years, you can probably remember some well-behaved teenagers who did their homework regularly, didn't smoke or drink, and never got into trouble at school. You also probably met others who rarely did their school assignments, at a young age started smoking and drinking to drunkenness, and enjoyed flouting authority. In assessing peer deviance, we might ask you how much your friends were like the first versus the second group of teenagers?

What influences the deviance of the peers that teenagers like to associate with? Part of it is the family and neighborhood you grew up in, what genetic epidemiologists would call the shared environment. But a longitudinal study of male twins in the Virginia twin study showed that, as twins got older, the genetic influences on peer deviance got stronger and stronger; that is, the similarity in deviance of their peers increased more in MZ than in DZ twins. These results suggest that, as males mature and come to increasingly create their own social worlds, their choice to associate with peers who are low versus high in deviant behavior becomes increasingly influenced by their genetic make-up (KENDLER et al. 2007b).

Our final example of genetic control of exposure to the environment comes from data collected for a GWAS. Figure 14.6 is a Manhattan plot (see Chapter 6, section 6.5.1 for a description if you are not sure how to interpret the plot) presenting the results of a GWAS

for the quantity of cigarettes smoked (TOBACCO AND GENETICS CONSORTIUM 2010). There is a single significant locus on chromosome 15 that contributes to smoking quantity. The P value ($P < 10^{-30}$) is remarkably low. This robust, replicated locus lies in the nicotinic acetylcholine receptor gene cluster.

Figure 14.7 shows a Manhattan plot for a very different kind of trait: lung cancer (TIMOFEEVA et al. 2012). It has a few smaller peaks, but it also has one really tall peak that is in exactly the same place as that in our previous plot on chromosome 15. The large peak in both of these Manhattan plots comes from the same genetic variants in the nicotinic acetylcholine receptor gene cluster.

What is going on here? How could the same variants contribute to both smoking behavior and to lung cancer, two very different kinds of traits?

Nicotine is the addictive substance in cigarette smoke. As found in a range of studies, variants in certain brain nicotine receptors are associated with risk for nicotine dependence and the quantity of cigarettes smoked if you become a regular smoker. These variants work largely by altering the sensitivity of nicotine receptors on key brain cells to the effects of nicotine. If

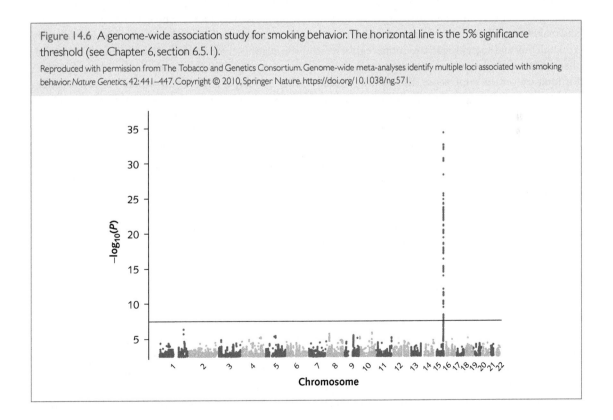

Figure 14.6 A genome-wide association study for smoking behavior. The horizontal line is the 5% significance threshold (see Chapter 6, section 6.5.1).

Reproduced with permission from The Tobacco and Genetics Consortium. Genome-wide meta-analyses identify multiple loci associated with smoking behavior. *Nature Genetics*, 42: 441–447. Copyright © 2010, Springer Nature. https://doi.org/10.1038/ng.571.

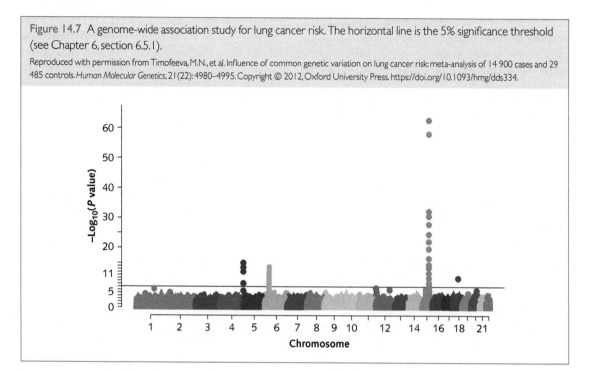

Figure 14.7 A genome-wide association study for lung cancer risk. The horizontal line is the 5% significance threshold (see Chapter 6, section 6.5.1).

Reproduced with permission from Timofeeva, M.N., et al. Influence of common genetic variation on lung cancer risk: meta-analysis of 14 900 cases and 29 485 controls. *Human Molecular Genetics*, 21(22): 4980–4995. Copyright © 2012, Oxford University Press. https://doi.org/10.1093/hmg/dds334.

you have this genetic variant, and become a smoker, you smoke more, only about one cigarette per day more. But over decades, that extra cigarette adds up and makes a big difference to your risk for lung cancer. You will buy and smoke cigarettes more regularly, so that, with repeated exposure, you increase your risk for lung cancer. In some sense, this nicotinic receptor gene is a cancer gene but not in the usual sense of damaging the cell's DNA repair system so that it is more prone to develop mutations. It causes cancer by increasing your probability of exposing yourself to higher levels of a carcinogenic environment: repeated inhalations of cigarette smoke. It directly alters your behavior, and thus indirectly alters your cell's behavior.

By now, we should have convinced you of the plausibility of the claim that genetic effects on exposure to adverse environmental experiences are widespread. One important consequence of these results (as well as findings about genetic sensitivity to the environment) is a blurring of the boundaries between genes and environment (or between nature and nurture). It won't be possible to understand the action of genes on human behavioral disorders and traits without a consideration of the environment.

Before we close this section, we should introduce you to a useful typology of kinds of gene–environmental correlations. The first is passive, where the identified subject does not have to do anything. A common example is that children who get a heavy dose of high-IQ genes typically are also exposed to intellectually stimulating environments. The second is evocative, whereby the world responds to the subject based on their genotype. If the world responds to you in certain ways because you are physically attractive, that would be an example of evocative gene–environment correlation. The third kind is active, where the subject goes out and directly, by their actions (in part under genetic influence), alters their world. This chapter has seen several such examples of active gene–environment correlation: seeking out peers who might resemble you in their level of deviance or, because of the changes in a nicotinic receptor in your brain, seeking out and smoking more cigarettes.

14.3 Using genetics to distinguish correlation from cause

Genetic strategies can be used to sort out the relationship between environmental events that are correlated with disease risk. Suppose you find, as others have reported (HUGHES et al. 1986), that rates of smoking are higher among people with depression than among those without. How do you decide whether depression increases the risk of smoking, whether smoking increases the risk of depression, or whether both arise from some factor that increases the risk of both conditions? One thing that would help is to know the order in which things happen (clearly smoking could not cause depression if it always occurs after the onset of the mood disorder). Analysis of genetic variation is particularly useful here because (except in some very unusual circumstances) genetic variation is established at conception, before birth. Its primacy in the order of events makes genetics a powerful tool for determining what causes what, as we will illustrate with two examples.

In the first example, we look at drug abuse, and examine a situation where some people manage to stop taking drugs, namely during pregnancy. We ask whether pregnancy stops drug abuse, or whether some other factor protects the women who achieve abstinence.

In the second example, we turn to the relationship between psychiatric disorders and intelligence. It's been known for a long time that poor school performance is associated with schizophrenia and depression, but why is that? Could it be that underperforming at school (which typically precedes the onset of disease) causes schizophrenia and depression? Are the psychiatric disorders and academic failure manifestations of the same genetic predisposition? Or is there another factor that explains the relationship?

14.3.1 Genetics reveals that pregnancy protects against drug abuse

Most research on drug abuse is largely focused on the brain mechanisms involved that allow drugs to 'capture' the brain's pleasure systems and warp them so that the only thing important is to be able to use the drug regularly and make sure the supply lasts. But once in a while, someone who decides to stop using drugs manages to become drug free for months or years. Why do some break through the addiction, while others fail?

We asked whether there are experiences that motivate people strongly to avoid drugs, experiences that would provide a natural experiment to investigate how human motivation overcomes the risk for drug abuse. Pregnancy is such an experience. There is evidence for two possible pathways relating drug abuse to pregnancy. In one, drug abuse and/or its risk factors increase the chances of unplanned and early

pregnancy (KIRBY 1999). In the other, pregnancy is related to a reduction in substance abuse, as reported in studies of cohorts of pregnant women (HARRISON AND SIDEBOTTOM 2009). Which pathway is likely to be operating? We set out to answer this question using the registry information available in Sweden (KENDLER et al. 2017). Drug abuse was detected from registration for drug problems in medical and criminal registries or documented abuse of prescription drugs from pharmacy records.

First, we looked at the entire population. As you might expect, during pregnancy the chances that a women was registered for drug abuse decreased moderately. Then, instead of comparing pregnant woman with all other nonpregnant women in the population, we compared them with close female relatives who were not pregnant. Why would we do that?

The answer is simple: relatives share a lot of traits. Think especially about identical twins. About the closest we can come to observing a life lived twice is to compare one identical twin with their co-twin. These two individuals began life as carbon copies of each other. They shared the same womb and the same household, community, school, place of worship, diet, and so on. They are as alike as two human beings can be. But in this case, one of them is pregnant and the other is not. By comparing identical twins, we control not only for those traits we might know about that influence drug abuse (e.g. level of education, school performance, divorced parents) but also for traits that impact on drug abuse that we do not know about.

A small confession before we go further. Even though we had information on the entire country of Sweden, there were not enough MZ twins where only one was pregnant to answer our questions precisely. So we included a large sample of sisters and female cousins along with our rarer MZ twin pairs. We used all these pairs to estimate what the relationship should be among these pairs. Figure 14.8 shows the results we obtained.

In the general population, the risk of onset of drug abuse was about 30% lower when women were pregnant. As Figure 14.8 illustrates, we found a dramatic increase in the protective effects of pregnancy as the degree of genetic and environment sharing between pairs increased. Within pairs of siblings, the reduction in risk for drug use was over 60% and within MZ twin pairs, the risk for drug abuse was over 80% lower in the pregnant twin compared with her nonpregnant twin sister. That is a big difference.

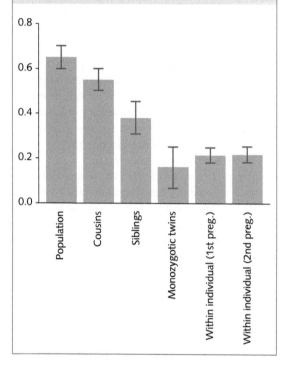

Figure 14.8 Association between pregnancy and risk for drug abuse in the general population, in cousins, siblings, and monozygotic twins discordant for pregnancy, and within individuals (comparing matched periods before and during pregnancy) for a first and second pregnancy. Error bars indicate 95% confidence intervals.
Data from Kendler et al. (2017).

likely reason for the reduction in substance abuse during pregnancy was motivational. Pregnancy preceded the period of abstinence, consistent with its having a causal role. We could go one step further in determining the mechanism for drug abstinence by incorporating genetic analysis. Our genetic modeling rules out (almost) all other explanations, such as the existence of some factor that would explain both pregnancy and drug abstinence. Put simply, studying MZ twins allowed us to analyze the same life, lived twice, and to show that the difference between two women is the motivation of the pregnant one to stop taking drugs.

14.3.2 How genetics explains the relationship between school performance and psychiatric illness

We have known for a long time that low levels of intelligence (or the closely related measure of school performance) predispose to many psychiatric disorders, for example schizophrenia and depression. But we have made less progress in understanding why. By adding a family/genetic perspective to this question, we can gain important new insights into the mechanisms involved.

The Swedes have, for several decades, systematically collected school performance data at age 16 for the entire population. As predicted by prior research, poor school performance predicts an increased risk for nearly every psychiatric disorder. We look first at why this is so for the risk for schizophrenia.

Because we had data on school performance for siblings and cousins, and measures of educational attainment for the parents (how many years of education they completed), we could do a good job of assessing what we called the familial cognitive potential. Intellectual performance runs strongly in families (see Chapter 12, section 12.1.4), so if you know a family's familial cognitive potential, you have a good idea about how well a particular member of that family will perform in school. With this information, we could ask the following critical question: was the risk of developing schizophrenia predicted primarily by poor school performance, or was it predicted by the degree to which an individual's performance met the expectations of their family's cognitive potential? Let me illustrate this critical question with two (made-up) examples.

Imagine Paul who comes from a Swedish family whose members, collectively, performed below

We took one more approach to prove that the impact of pregnancy on risk for drug abuse was causal. We compared pregnant women with themselves; that is, we compared the risk of drug abuse before and during pregnancy (this is called a within-subject design). As you can see from the last two bars on the right of Figure 14.8, the risk for drug abuse was almost the same as our results comparing MZ twins discordant for pregnancy: when pregnant, a woman's risk for drug abuse is about 80% lower than when she is not pregnant.

What is this study doing in a book about genes and behavior? It is to demonstrate that understanding and controlling for the effects of genes can be important in clarifying environmental effects. There are other methods that exploit genetics to assess causal relationships, as we explain in Chapter 13, section 13.5 (on Mendelian randomization) and Chapter 15, section 15.11, but the example of pregnancy is particularly powerful: we were able to demonstrate that the most average at school. His parents ended their education in the eighth grade. His two sisters barely finished high school, struggling to get passing grades. Paul performed at an average level in school, getting what in American schools would be mostly Bs and Cs. For his family, this was exceptional; he did much better than would have been expected given his familial cognitive potential. But if you did not know his background, he would have just been an unremarkable average student.

Now, consider Sara. Sara comes from an intellectually gifted family. Her mother is a lawyer and her father is a physician. All of her three older siblings were at the top of their class in school and got into prestigious universities. Sara, however, was just an average student. On her own, this was not remarkable. However, she fell short of the academic achievements that would have been expected of her, given her family exceptional family background (i.e. her familial cognitive potential).

What would you expect the risk of schizophrenia to be for individuals like Paul versus individuals like Sara? If the risk for schizophrenia is simply a result of a person's academic performance, then their risk should be the same. They both were average students. However, if the risk for schizophrenia is a result of some developmental aberration whereby an individual gets thrown off their cognitive developmental trajectory, then Sara should be at increased risk but not Paul. Indeed, Paul might be at reduced risk. Sara was clearly performing less well at school than most of her relatives. For Paul, it was the opposite.

Figure 14.9 outlines our results for the prediction of schizophrenia from the large Swedish population where we had complete measures of school achievement, good measures of the school achievement of their siblings and cousins, and knew their parents' educational attainment (KENDLER et al. 2018a). Thus, we had a good fix on their familial cognitive potential.

This figure needs some explaining. The vertical axis represents the increased risk for developing schizophrenia, reported as hazard ratios, where a hazard ratio of 1 is average for the population, and a value of 0.5 and 2 represent risks of half and double that of the general population. The horizontal axis reflects the familial cognitive potential divided into five groups: very low (labeled as −2), low (−1), average (0), high (1), and very high (2). Using more generic language, we might say that members of

Figure 14.9 The increased risk (hazard ratio and 95% confidence intervals) for first registration of schizophrenia (SZ) in 1,140,608 Swedish individuals as a function of their school achievement (SA) at age 16 and their familial cognitive aptitude. The horizontal axis reflects the familial cognitive aptitude divided into five groups: very low (2 standard deviations (SD) below the mean), low (1 SD below the mean), average, high (1 SD above the mean), and very high (2 SD above the mean). The vertical axis reflects the school achievement levels (in units of SD) of individuals within each family type. Not all possible combinations are presented because some potential groups (e.g. familial cognitive aptitude of +2 SD and school achievement of −1 or −2 SD) were too rare to calculate with confidence.

Reproduced with permission from Kendler, K.S., et al. The joint impact of cognitive performance in adolescence and familial cognitive aptitude on risk for major psychiatric disorders: a delineation of four potential pathways to illness. Molecular Psychiatry, 23: 1076–83. Copyright © 2017, Springer Nature. https://doi.org/10.1038/mp.2017.78.

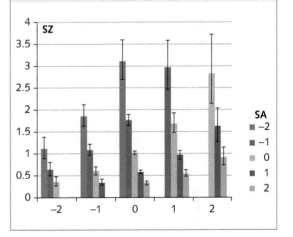

these families on average did very poorly, poorly, average, well, or very well at school. The bars within each group reflect the school achievement scores of the individuals within these families also divided into the same five groups. They go like this (from left to right): very poor school performance, poor school performance, average, good school performance, and very good performance).

With this introduction, what do these results show? Let's start with the average families, those five columns where the average family cognitive potential is 0. In these families, those who performed very poorly at school had an over 300% increased risk for schizophrenia, while a really high-performing individual from an average family had a risk of illness about 65% lower than average.

Now, rather than focusing on a set of bars within a family type, look at all the middle bars in each group of five, which, remember, are for individuals who score about average in their school achievement. What you see is that the risk for schizophrenia for individuals who have the same intellectual ability—average—differs dramatically across family types. If you are like Sara, an average person from a really bright family (the +2 group), you have a risk for schizophrenia increased about 270% over the average. But, at the other extreme (the −2 group), if you have average school achievement but come from a family that performs very poorly at school, like Paul, your risk is reduced more than 60%.

These results answer our question in a rather dramatic way. Indeed, Sara's risk for schizophrenia is over six times higher than Paul's. The risk for schizophrenia is **not** driven primarily by low versus high intelligence. Rather, the risk is driven by the discrepancy between an individual's school performance and that which would be expected of them based on their family background.

Is this true for all psychiatric disorders? From this same study (KENDLER et al. 2018a), we also examined risk for depression. Figure 14.10 is constructed in exactly the same way as the above figure for risk for schizophrenia. The pattern for depression is very different from the one we saw for schizophrenia. On average, risk seems to be a result of an individual's school performance, and the family background does not matter.

We can illustrate this most clearly by comparing Paul and Sara's risk for depression. Paul's risk is captured by those with average school performance who come from families with very poor academic performance. The risk was 1.0, an average risk for depression. Sara's risk is captured by those with an average school performance from very bright families. Her risk was also 1.0.

Think back about what we have learned about the relationship between cognitive ability and risk for psychiatric illness. Before, we knew that doing poorly at school increased the risk for both schizophrenia and depression. It was easy to assume that the underlying mechanism for both schizophrenia and depression might be similar. But our analyses show that this cannot be true. For schizophrenia, risk is **not** related to average cognitive potential but only to the degree that

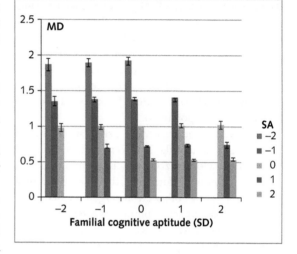

Figure 14.10 The hazard ratio (and 95% confidence intervals) for first registration of major depression (MD) in 1,140,608 Swedish individuals born between 1972 and 1990 as a function of their school achievement at age 16 and their familial cognitive aptitude. Figure components are explained in the legend to Figure 14.9.

Reproduced with permission from Kendler, K.S., et al. The joint impact of cognitive performance in adolescence and familial cognitive aptitude on risk for major psychiatric disorders: a delineation of four potential pathways to illness. *Molecular Psychiatry*, 23: 1076–83. Copyright © 2017, Springer Nature. https://doi.org/10.1038/mp.2017.78.

school achievement differs from that predicted by a person's familial cognitive potential. For depression, the results are the other way around. The only thing that matters for risk for depression is an individual's school performance. The academic performance of the family seems to be immaterial.

14.4 Indirect genetic effects

The relationship between genetic and environmental influences reaches a new level of complexity when we consider indirect genetic effects. Direct genetic effects are those that we have discussed elsewhere in this book, namely effects attributable to genetic variants in our genomes that alter disease risk, behavior, personality, and cognitive ability. Indirect genetic effects are also attributable to genetic variants, but this time not to variants found in our own genomes. Rather, the variants are present in other people's genomes, affecting the behavior of those people who, in their turn, affect us. An example should make things clear.

The writer and activist Darren McGarvey describes how his mother behaved toward him and his siblings (McGarvey 2017):

When I was huddled with my brothers and sisters in a dark room in the middle of the night, while a man screamed death threats through the letter box, it was scary, but not unusual … Whether it was a child being tied to a chair for being cheeky or a baby being booted across the floor for crying by the faceless male drunk she had in occasionally, it all felt bizarrely normal. Not even the sight of her having sex was enough to shock me.

Darren's mother's behavior is a risk factor for Darren's mental health; that is, his mother is an environmental risk, but like him her behavior is in part genetically mediated. The contribution of her genes to her behavior becomes for her son a component of his own environment. Not only does she pass on genetic material to him, but she also exposes her son to the phenotypic manifestation of her own genetic constitution (her propensity for anger, violence, aggression, and drug addiction). This is an indirect genetic effect, the importance of which was first recognized by evolutionary biologists and formulated in a series of papers in the 1980s by Bruce Griffing (e.g. Griffing 1981).

Indirect genetic effects should be distinguished from all the other ways genes influence behavior that we have so far discussed. They are sometimes described as interactions, but they are not the same as the gene–environment interactions we have seen so far. They are a measure of the environment's impact on a person due to the presence of other people. And they affect all animals, not just us, which explains why they have been of great interest to agricultural breeders, as indirect genetic effects influence the reproductive success of farm animals (think of all those chickens, pigs, and cows, contributing, through their genetic constitution, to the farm environment and thus affecting the fecundity of the farmer's stock). As Allen Moore and colleagues point out, the expression of aggression, territoriality, altruism, and courtship depends in part on interactions with other animals (Moore et al. 1997; Schneider et al. 2017; Wolf, 2003).

One powerful way to assess indirect genetic effects is through molecular differentiation of the alleles that parents pass on to their children (which have a direct genetic effect) and those that they do not (which exert indirect genetic effects). Figure 14.11 illustrates how this works. Half of a parent's alleles are transmitted to an offspring and half are not, in a random fashion. The effects of transmitted alleles are those documented throughout

Figure 14.11 Indirect and direct genetic effects on a child. Transmitted alleles influence the phenotype of the offspring, through a direct path. Both transmitted and nontransmitted alleles of the parents can influence the parents' phenotypes, and through them may have an effect on their child. This indirect pathway combines a genetic effect from the mother's and father's phenotype with a nurturing effect on the child.

Reproduced with permission from Kendler, K.S., et al. The joint impact of cognitive performance in adolescence and familial cognitive aptitude on risk for major psychiatric disorders: a delineation of four potential pathways to illness. *Molecular Psychiatry*, 23: 1076–83. Copyright © 2017, Springer Nature. https://doi.org/10.1038/mp.2017.78.

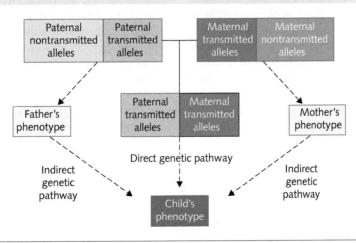

this book: a direct genetic effect on behavior. However, the alleles that the child does **not** receive, the nontransmitted alleles, still have an effect on the child, because they contribute to the parents' phenotype, which in turn has an effect on the children's behavior. This is a 'genetic nurturing' pathway, the indirect way that the alleles can affect behavior. How big is this effect? It turns out, from the limited data we have (just one study so far in humans), that the effect is large.

How can we measure the magnitude of an indirect genetic effect? Augustine Kong, a statistical geneticist working with a large sample of genotyped families from Iceland, solved this problem. He was fortunate to have the genotypes of both parents **and** offspring, rarely collected in large GWAS (KONG et al. 2018). To assess the effect of the nontransmitted alleles, he created polygenic risk scores (PRSs) for Icelandic subjects based on a GWAS of educational attainment (the phenotype is the number of years of education, a measure the genetics of which we described in Chapter 12, section 12.1.6). He constructed PRSs in the usual way (see Chapter 7, section 7.2), that is by taking the genetic effect reported for an allele from one population, and applying it to the same allele in a different population.

Usually the PRS per person is the sum of all the effects attributed to the alleles that the person carries in their genome. Kong did something slightly different. He assigned the scores to each allele, rather than to the genotype (so if the genotype at a locus was AC, he separately assigned an effect to the A allele and to the C allele). From the pedigree data, he knew which alleles had been transmitted to offspring and which had not, so he could assign PRSs to the two classes of allele separately. This allowed him to construct separate PRSs for transmitted and nontransmitted alleles.

Kong compared the effects of the two PRSs on educational attainment. He estimated that the direct effect explains 17.0% of the variance of educational attainment. How much is attributable to nontransmitted effects? Unfortunately, the indirect route is hard to estimate accurately as it includes a number of confounds that make the calculation complex, but here is Kong's summary: '*Assuming that the direct effect alone explains 17.0% of the variance, the variance explained by the transmitted alleles with the nurturing effects included increases to $17.0\% \times 1.74 = 29.6\%$. Additionally including the non-transmitted alleles would increase the variance explained to $17.0\% \times 1.84 = 31.3\%$'* (KONG et al. 2018).

Almost half the total heritability might be attributable to indirect genetic effects! This is surprisingly large, but not inconsistent with studies from other species: an analysis of aggressive behavior in deer mice showed that the assessment of indirect effects on latency to fight was ten times higher than the additive heritability (WILSON et al. 2009). Large effects were also seen on social dominance in red deer (WILSON et al. 2011). Amelie Baud and colleagues estimated that indirect genetic effects explained up to 29% of phenotypic variance, and, consistent with the analysis of aggression in deer mice, found that sometimes the contribution of indirect genetic effects exceeded that of the direct effects (BAUD et al. 2017). An adoption study of alcohol-use disorder in humans showed that the risk for alcohol-use disorder in the adoptee was about equally impacted by the risk for alcohol-use disorder in adoptive parents and the biological parents, suggesting approximately equal effects for genes and the environmental effects from the adoptive parents to whom they were not related (KENDLER et al. 2015a).

The measurement of indirect genetic effects enables the assessment of genetic effects on behavioral measures that previously seemed impervious to analysis outside designs such as adoption studies. The demonstration that indirect genetic effects are real, and substantial, raises new questions about how genes influence behavior. The effects that Kong reported pass from parent to child, but there is no reason why they should not be bidirectional (from child to parent). And with siblings and twins, it's possible that there are reciprocal interactions. Looking more broadly, for predominantly social species, we can ask how do the genotypes of the people we meet influence us? Are the genotypes of colleagues an important influence on our behavior? The flood of data accruing through biobanks and electronic health records is likely to answer these questions.

Summary

1. Genetic and environmental effects often do not simply add together in their impact on risk for psychiatric disorders. Two mechanisms operate: (i) genetic control of sensitivity to the environment, or gene–environment interaction; and (ii) genetic control of exposure to the environment, or gene–environment correlation.

2. Twin and adoption data provide support for the presence of genetic control of sensitivity to the environment. However, results obtained from aggregate measures of the entire genome may not agree with those from individual markers. The presence and direction of interactions may vary considerably between loci.

3. There is also considerable support for the notion that genetic control over exposure to the environment is widespread for humans. This has led to the realization that exposure to some forms of environmental adversity is in part a heritable trait. One important consequence is a blurring of the boundaries between genes and environment. Without taking the environment into consideration, it will not be possible to understand the action of genes on human behavioral disorders and traits.

4. Genetic strategies can be used to examine the relationship between the environment and psychiatric disorders. For example, the association between low intelligence and schizophrenia is driven by the discrepancy between an individual's school performance and that which would be expected of them based on their family background. By contrast, for depression, risk is the result of an individual's school performance, and the family background does not matter.

5. Indirect genetic effects, such as those attributable to parental alleles that a child does not receive (the nontransmitted alleles), are an important contributor to heritability of behavior because they contribute to parents' behavior, which, in turn, impacts on the child's phenotype.

15
Mapping mouse behavior

[On how the genetic architecture of behavior in mice can be detected, and on the characteristics of this architecture]

This chapter, and the ones that follow it, explore what we know about how genes influence the behavior of model organisms. We begin by answering the same questions we addressed in the genetic analysis of human behavior: is animal behavior heritable, and if so by how much? What is its genetic architecture? Polygenic? Mendelian? How big a role does the environment play and how does the environment interact with heritable variation? In answering these questions, we exploit the same raw material, the naturally occurring sequence variation that we examined in human genetics. However, as we will see in CHAPTER 16, the use of model organisms opens up a new possibility: not just observing but also creating genetic variants. We'll start, however, with more familiar fare, and tackle the genetic architecture of behavior in rodents.

Animal geneticists have been asking about the genetic architecture of behavior for a long time. The introduction to a rat genetics paper from the 1920s isn't that different from something that might be written today:

For could we, as geneticists, discover the complete genetic mechanism of a character such as maze-learning ability – i.e., how many genes it involves, how these segregate, what

their linkages are, etc. – we would necessarily, at the same time, be discovering what psychologically, or behavioristically, maze-learning ability may be said to be made up of, what component abilities it contains, whether these vary independently of one another, what their relations are to other measurable abilities, as, say, sensory discrimination, nervousness, etc.

TOLMAN (1924).

In twenty-first-century language, this translates as: an understanding of the genetic architecture of behavior provides an understanding of the behavior itself.

15.1 The heritability of mouse behavior

Recall how we assess heritability in humans, and remember how difficult this can be (finding all those twins or adoptees, dealing with the assumptions about equal environment and nonrandom placement, and so on). Determining whether there is a genetic effect in mice is much simpler, and quicker. The key resource, as in so much else in mouse genetics, is the inbred animal. In an inbred population, there are no genetic differences between individuals. In this, they resemble a pair of twins, but there is an important difference: inbred mice are homozygous at every locus. The DNA they inherit from their mother is the same as that which they inherit from their father. Therefore, no genetic variants segregate in inbred rodent populations (except those that determine sex).

Inbreeding is achieved by encouraging incest. Experimental organisms don't show any restraint about brother–sister (or parent–child) mating, and if you organize this for long enough, you end up with completely inbred animals. If you are a worm geneticist, then your favorite organism actually prefers to have sex with itself, so it's really no problem to get rid of the genetic heterogeneity that so confuses human geneticists. All you do is the leave the worms alone to eat bacteria (a cheap and easily obtainable foodstuff) for a few generations, and they will self-fertilize and end up completely inbred.

Mouse geneticists have to devote more time than worm geneticists to inbreeding their animals; they have to set up brother–sister matings for about 20 generations to create animals that differ genetically only on the sex chromosomes. Once the animals are inbred, they are given some anonymous strain designation (e.g. AKR or DBA, which means nothing to you but means a lot to mouse geneticists). Inbred strains were made as a tool for investigating various aspects of biology: our understanding of immunology has, for example, been critically dependent on having inbred strains that differed in their immune responses, leading to the discovery of regulators of acquired (SNELL 1979) and innate (BEUTLER 2004) immunity.

Access to inbred animals substantially simplifies the genetic analysis of behavior. To show how this works, we'll use the example of the genetics of fearfulness or anxiety in mice. The choice of fearfulness is largely because one of us (J.F.) worked in this area, but it should be stressed that the examples, and conclusions, apply to almost all other traits (including a large literature on drug and alcohol use; CRABBE et al. 2010).

15.2 Locomotor activity

Rodents prefer to live hidden away, in the dark, probably because this is the best way for them to avoid predators. When placed in a brightly lit open arena, a strange, new, and potentially dangerous place, they freeze, or at least move around much less.

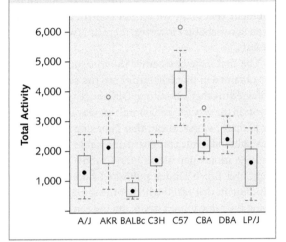

Figure 15.1 A measure of locomotor activity in different inbred strains of mice. The x-axis lists the names of the inbred strains. Total activity was measured as total distance (in cm) travelled over 30 minutes.

The amount of locomotor activity is thus a crude index of the animal's fearfulness.

Figure 15.1 shows a measure of locomotor activity in such a brightly lit arena for eight inbred strains. The strain names are on the x-axis and the y-axis shows how much each strain runs about. There are 12 mice in each group and the figure (called a 'box-and-whisker' plot) represents the distribution of the activity values. The box shows the middle values of the activity score (the black dot is the average), while the whiskers stretch to the upper and lower 95% confidence intervals. You can see there are some outliers (the blue circles), but you can also see that there are big differences between the strains, not all of which are due to the variation in the measure. These differences are due to genetic differences between the strains: as the members of each group are genetically identical, the differences within a group are due to the environment; differences between groups are due to genetic variation. These data tell us that fearfulness (at least as reflected by level of activity in this environment) is heritable.

15.3 Inbred strains: heritability and gene–environment interactions

Inbred-strain comparisons are an effective tool for telling us whether genetic differences exist. Few of the methodological concerns that we dealt with at length in considering human studies are relevant, because we have so much more control over the experiment: there is no selective placement or assortative mating. We don't have to worry (so much) about our genetic effect being an artifact due to some background environment effect that we cannot control and may not even know about. However, there are two important caveats.

The first issue is whether the measures of behavior obtained in one laboratory are the same as those obtained in other laboratory. Differences might reflect errors in the way the behavioral test was administered, or perhaps, worse, mean that behavior is inherently unstable and its measurement unreliable (PFAFF 2001). In 1999, mouse geneticists John Crabbe, Doug Wahlsten, and Bruce Dudek published findings from an experiment that asked whether the same experiment, using the same inbred strain and the same behavioral test, would yield the same results when conducted in three different laboratories (CRABBE et al. 1999). Here's what they reported: 'Despite our efforts to equate laboratory environments, significant and, in some cases, large effects of site were found for nearly all variables.' That's not quite what most people were expecting.

There were direct effects of the environment, such as, for example, 'mice tested in Edmonton were, on average, more active than those tested in Albany or Portland' (something in the laboratory environment in Edmonton made those mice run faster) (CRABBE et al. 1999). But, importantly, some genetic effects were sensitive to the environment in which they were measured: they showed a gene–environment interaction. While genotype was highly significant, accounting for 30–80% of the total variability, a considerable proportion of the variability was accounted for by interaction with the environment. We'll return to the observations about interaction later, when we discuss information obtained from molecular markers.

The second issue is that, although the long tradition of documenting strain differences in behavior provides good evidence that all measurable mouse behavior is heritable (you can access the phenotypes of inbred mice here: https://phenome.jax.org, thanks to the Jackson Laboratory's mouse phenome project), strain differences on their own are not sufficient to provide an estimate of heritability. Here's why: suppose that two genetic loci, on two different chromosomes, contribute to variation in a behavior. Suppose we have just two inbred strains. In strain one, alleles at the first locus increase the behavior, and at the

second locus both alleles decrease the phenotype. In strain two, the configuration is the other way around: alleles at the first locus decrease the behavior, and at the second locus both alleles increase the behavior. Consequently, strains one and two will not show any phenotypic difference, even though there are genetic differences between the two strains that affect the trait. To estimate heritability, we need another design, and the most robust of these, with a venerable history, is artificial selection.

15.4 Selection and realized heritability

Consider what would happen if you compel the most hyperactive people in each generation (let's say the most active 10%) to marry and breed only with each other, and keep doing that for every generation. Would activity in this population eventually reach a plateau, so that everyone ran around at about the same speed? Or would the level of activity continue to climb almost without limit?

Two behavioral geneticists in the USA, John DeFries and John Hegmann, carried out this experiment using mice (DEFRIES AND HEGMANN 1970). Their measure of activity was the total distance a mouse would travel when placed in an open-field arena. They started by mating two inbred lines and then selected the most active male and female from each litter for breeding. They also selected the least active male and female, so that they could see the response to selection in both directions. DeFries and Hegmann included controls, where pairs of animals were selected at random, irrespective of their activity level, to make sure other aspects of the experiment, apart from selection, had no unexpected effects on behavior.

The results were dramatic. After 30 generations of selection, the mice in the high-activity lines were 30 times as active as those in the low-activity lines (DEFRIES et al. 1978). Think about the difference in how far you would travel in a car going 2 versus 60 miles an hour and that will give you a sense of the differences in activity levels of these two groups of mice, all of whom were genetically identical just 30 generations ago.

The rate of the selection response, that is, the slope of the line in the graph, tells us something about the underlying genetic architecture of behavior. If there had been a single genetic variant, or just a few, that substantially increased activity and that accounted for a large proportion of the variation in activity, then applying selection for activity would quickly alter the frequency of these variants. Their frequency would rise rapidly in the group selected for high activity and fall fast in the animals selected for low activity; that is, the selection would quickly reach a floor and a ceiling. However, as Figure 15.2 shows (DeFries et al. 1978), selection resulted in an apparently steady change in both directions (selecting for both high and low

activity). We know from these kinds of curves that single genes with large effects on activity are not present in these two inbred lines. But the low line is reaching a floor by generation 30; the animals are nearly frozen.

Selection provides a measure of heritability that we have not mentioned before: **realized heritability**. Provided there is no nongenetic cause of resemblance between offspring and parents, and no other selective forces operating (e.g. the more active animals should be as equally fertile and viable as the inactive animals),

Figure 15.2 Mean open-field activity in six lines of mice, two selected for high activity (H_1 and H_2), two selected for low activity (L_1 and L_2), and two randomly mated within lines to serve as controls (C_1 and C_2). Mean open-field activity was measured as total distance (in inches) travelled over 5 minutes.

Reproduced with permission from DeFries, J.C., et al. Response to 30 generations of selection for open-field activity in laboratory mice. *Behavior Genetics*, 8(1):3–13. Copyright © 1978, Plenum Publishing Corporation. https://doi.org/10.1007/BF01067700.

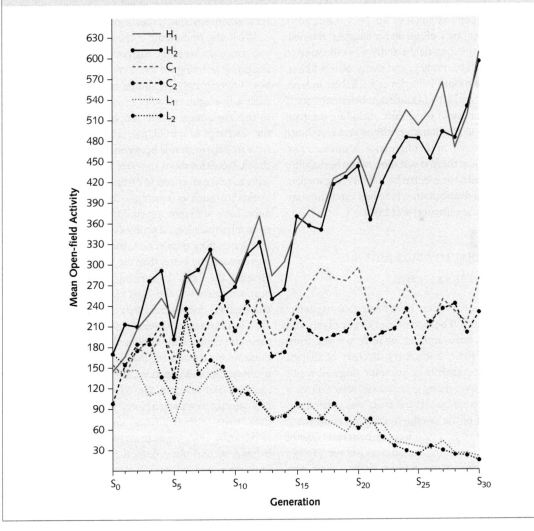

then the ratio of response to selection is equal to the heritability (for technical reasons, realized heritability will nearly always be lower than the kind of heritability we explored in our twin studies).

Realized heritability is calculated as the response to selection divided by the selection differential. It works like this: suppose the mean of the population on the activity score is 100. You select the most active mice to breed the next generation. These parents have a mean of 120, and their offspring have an activity of 110 (higher than the last generation, which means that selection has an effect—the fact that their mean is lower than their parents is an example of regression to the mean, as explained in the Appendix, section A4). The realized heritability is the difference between the activity of the original population (100) and the average of the next generation (110), divided by the difference between the means of the parents (120) and the mean of the original population (100): $10/20 = 0.5$ or 50%. In the experiment by DeFries and colleagues, the realized heritability of open-field activity was estimated to be 0.13 ± 0.02. This finding, and many others like it, confirmed that mouse behavior is heritable. Indeed, heritability estimates for 143 different behavioral traits from a wide variety of organisms, including the fruit fly *Drosophila* and 75 other invertebrate and vertebrate species, show remarkable congruence (MOUSSEAU AND ROFF 1987). While there is variation in the heritability of individual traits, the median heritability (the median is the value in a distribution where half the values are smaller and half are greater) was 25%.

15.5 Genetic models and experimental systems

Animal and plant breeders have been applying selection for millennia. The size of our wheat plants and tomatoes, the large amount of milk we get from our cows, and the extraordinary diversity of shapes, sizes, and temperaments in domestic dogs all result from selection over many generations. Selection for a number of different behavioral traits was carried out in the first half of the twentieth century, confirming the importance of genetic effects on behavior. One of the earliest, and most influential, was Robert Tryon's selection experiments on rats for finding their way through a maze to obtain a food reward, carried out in the 1920s and 1930s (TRYON 1940). Tryon selected rats in a maze and generated two new lines, maze bright

and maze dull rats, which showed heritable, robust, and consistent differences in their performance in the maze-solving task. The creation of selected lines has provided a resource for many subsequent experiments.

Studying rodent genetic models is quite an industry: many inbred strains have been derived from selection experiments, for alcohol consumption (CICCOCIOPPO 2013), anxiety (LANDGRAF AND WIGGER 2002, 2003), depression (EL YACOUBI et al. 2003; WEISS et al. 2008a), aggression (LAGERSPETZ et al. 1968; SANDNABBA 1996), and other phenotypes. There is an enormous literature around these selected lines, attempting to identify physiological correlates for the behavioral changes. The assumption is that selection for anxious behavior or depression should produce animal models of anxiety and depression, and their study should identify biological features (e.g. alterations in neurotransmitter metabolism) that underlie depression or anxiety.

We'll give one example: the Roman high- and low-avoidance strains of rat. Bignami established these two strains by selecting for speed of acquisition in a shuttle box (BIGNAMI 1965). A shuttle box has two chambers, each with a light, and a floor made of electrifiable bars, so that the animal can be taught that the light signals the discharge of a mild, aversive foot shock. Animals have to learn to shuttle between the two boxes to avoid shock. Note that this is an active avoidance (the animals have to make an action to escape) rather than a passive avoidance (such as freezing)—the two forms of avoidance have different psychological, anatomical, and neuropharmacological profiles (GRAY 1982). After only five generations of selection, the high-avoidance strain was consistently better than the low-avoidance strain at escaping the shock (by running into the second chamber) when shown the light (BIGNAMI 1965). Several studies then documented that the behavioral changes are not just due to differences in learning and memory: there are changes in emotional behavior, and indeed these strains are now regarded as models of emotionality, of susceptibility to stress, rather than of learning and memory. For example, the low-avoidance (emotional) rats defecate more when placed in a novel environment than their high-avoidance counterparts (DRISCOLL AND BATTIG 1982; FERRE et al. 1995). A variety of other behavioral and physiological measures support this interpretation (DRISCOLL et al. 1998).

Rodent models of human behavior, such as the Roman high- and low-avoidance rats, face two criticisms. The first is the validity of the phenotype.

The assumption is that selective pressure has created animals that are constitutively depressed, anxious, psychotic, or autistic. Is this claim true? It's a contentious issue (NESTLER AND HYMAN 2010; MONTEGGIA et al. 2018) and pretty much counter to common sense: there are 90 million years of evolutionary history that separate rodents and humans, and we live in significantly different environments. Rodents are active at night, almost blind, have rudimentary social skills, and possess a highly developed sense of smell, while our own species is active (mostly) during daylight hours, is highly social, has a poor sense of smell but great visual skills, and has a highly developed brain that differs significantly in its cells, synapses, and circuits from that of rodents. Nevertheless, it's unarguable that selected models do show behavioral differences. One suggestion is to refer to 'experimental systems' rather than 'animal models' or 'disease models'.

The second issue is genetic. Suppose, after selection, an investigator finds that two traits are correlated. For example, suppose a region of the mouse brain is found to be larger in mice selected for improved performance in a learning task than in control mice. A simple conclusion is that the correlation implies some biological connection, so that an increase in brain size reflects the presence of additional neuronal connections needed for the change in behavior. This is the underlying assumption of the value of studying correlated traits. The problem is that many genetic variants, the substrate for selection, are correlated. The closer they are on a chromosome, the more correlated they are, as measured by linkage disequilibrium (see Chapter 6, section 6.4). So as selection operates to increase the frequency of a variant that influences the selected trait, it will also increase the frequency of variants in linkage disequilibrium, regardless of what those linked variants do. If the linked variants have a phenotypic effect, then the selected animals will show a correlated change in whatever trait is affected. The extent of induced correlation between two traits will depend on a number of features, including the tightness of disequilibrium, the size of the population selected, and the degree of inbreeding (this is described quantitatively in a classic paper by Norman Henderson; HENDERSON 1989). There are ways to alleviate the problem, for example by repeating the selection experiment (as John DeFries did for the selection on open-field activity), using a large population, and extending the selection time, but all these options are expensive and time-consuming, and are usually not undertaken.

15.6 Mapping behavior in rodents

Human geneticists approach genetic mapping using family-based methods (linkage) and association, but mouse geneticists use different strategies, almost all of them involving inbred strains. You might think that this is all just a question of the difference in technique, but this is not so. Understanding the results obtained requires knowledge about how the results were obtained, otherwise you will be misled as to their significance. Like eighteenth-century readers of travelers' tales, amazed at the descriptions of the monstrosities and curiosities that inhabit the corners of the earth (including the USA), many human geneticists are shocked to encounter reports of monstrous effect sizes underlying complex traits in rodents, or the discovery of such strange, unlikely beasts as 'epistasis': multiple interactions between genetic loci. To make sure our readers are not similarly flummoxed, we provide a gentle introduction to the art of genetic mapping in rodents, beginning with a description of how a trait can be mapped to a single chromosome using a special set of inbred strains called chromosomal substitution strains or consomics (NADEAU et al. 2000).

15.6.1 Chromosomal substitution strains

Mice have 20 pairs of chromosomes (19 pairs of autosomes and the two sex chromosomes, X and Y). One way to find out which of these 20 has a gene influencing behavior is to create mice that have one chromosome from one inbred strain, and all the others from a second inbred strain (SINGER et al. 2004). If the strains are called A and B, we generate inbred mice that have chromosome 1 from strain A and the remaining 19 from strain B, mice that have chromosome 2 from strain A (and the remainder from strain B), and so on until we have a complete set of mice that differ from their parents at each of the 20 chromosomes (Figure 15.3). It takes a long time to make a mouse with this chromosomal constitution, but it's possible.

Once the resource, the set of chromosome substitution strains, is available, mapping to a chromosome is robust, quick, and relatively cheap. We compare the behavior (such as activity in the open-field arena) in strain B mice with mice that have chromosome 1 from strain A (but are otherwise all strain B derived). Any difference will be due to genes on chromosome 1. We repeat the process for the other chromosomes. Figure 15.4 shows an example where a locus on

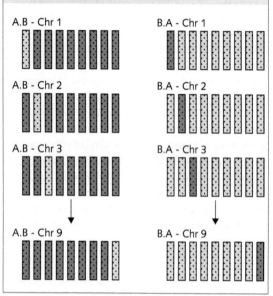

Figure 15.3 Construction of a chromosome substitution strain, achieved by recurrent backcrossing to isolate a single chromosome from one strain with all other chromosomes descended from a second strain. In this example, two such strains are shown. On the left is a chromosome substitution strain (A.B) with single chromosome derived from strain B, while in its reciprocal on the right (B.A), a single chromosome is derived from strain A.

chromosome two contributes to behavioral variation (SINGER et al. 2005).

Because in most cases we measure the trait quantitatively, such as height and weight, and as the genetic effect we are seeking is at a locus on the genome, the process is often called quantitative trait locus (QTL) mapping. QTL mapping with chromosomal substitution strains came onto the market relatively late (in fact, the *Drosophila* geneticists had the idea of making consomics long before the mouse geneticists did; PYLE 1978), but we have described it first as it provides a simple way of seeing what is going on. The drawback is that it does not identify a gene but a chromosome. If we want to find genes, we need better mapping resolution, which we can achieve by generating crosses between inbred strains, a process that will break up chromosomes by recombination, as we explain next (as you will see, gene identification through high-resolution mapping has proved to be just as hard as it has in human genetics).

15.6.2 Intercrosses and backcrosses

Genetic mapping of behavior, or indeed of any quantitative phenotype, such as height or weight, can be carried out through the analysis of offspring of a cross between two inbred strains. The first generation of such a cross (called the F_1, for first filial generation)

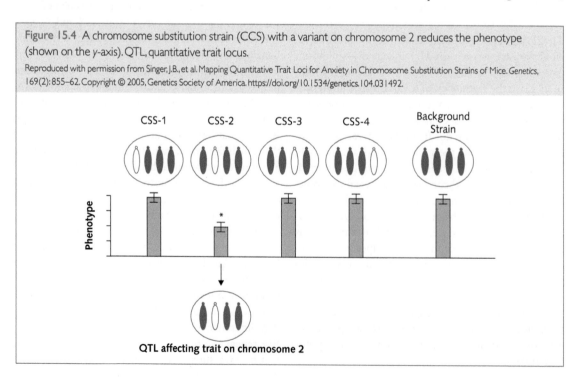

Figure 15.4 A chromosome substitution strain (CCS) with a variant on chromosome 2 reduces the phenotype (shown on the *y*-axis). QTL, quantitative trait locus.

Reproduced with permission from Singer, J.B., et al. Mapping Quantitative Trait Loci for Anxiety in Chromosome Substitution Strains of Mice. *Genetics*, 169(2): 855–62. Copyright © 2005, Genetics Society of America. https://doi.org/10.1534/genetics.104.031492.

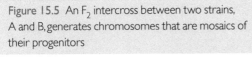

Figure 15.5 An F₂ intercross between two strains, A and B, generates chromosomes that are mosaics of their progenitors

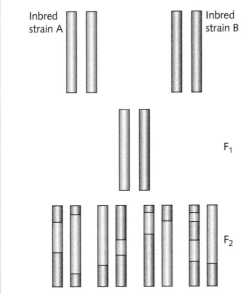

inbred strains, A and B). Genetic mapping requires knowing the origin of each piece of chromosome, in this case which piece of chromosome comes from strain A and which piece comes from strain B. As in human studies, molecular markers give the parental origin. Importantly, the genetic variants used for mapping are not likely to be the genetic variants that contribute to variation in behavior, or to any other phenotype. They need not have any function at all. They are simply markers of whether the chromosome is from strain A or strain B.

With the molecular information in hand, we can ask whether animals that share a chromosomal segment derived from strain A differ in behavior from those with the same segment derived from strain B. Things are a bit more complicated than with the comparison of a single chromosome, as the two groups differ at other regions in the genome, not just at the place we are comparing. But if we test enough mice, these other differences will be randomly distributed between the two groups and their effects will cancel out.

Suppose we have four markers on the chromosome. Each marker has two alleles that we'll call *a* and *b*. Due to recombination, the grandchildren of strain A and B mice can have either two copies of *a*, two copies of *b* or one of each (*ab*). We genotype the mice at the four markers so that for each animal we can say, this animal, at this position on the chromosome, is *aa*, this animal is *ab*, and this animal is *bb*. Figure 15.6 illustrates results for just the *aa* and *bb* animals at the four markers. You can see the differences between the scores reflected in the position of the line in the middle of the box (the mean score).

will have one chromosome from one inbred strain and one from the other another, and so will be genetically identical (no use for mapping), but the second generation (the F₂ intercross) or a backcross (where F₁ animals are bred with the parental strains) will have recombinant chromosomes that are used for mapping (Figure 15.5).

Figure 15.5 shows that the chromosomes of the F₂ generation are mosaics of the two grandparents (the

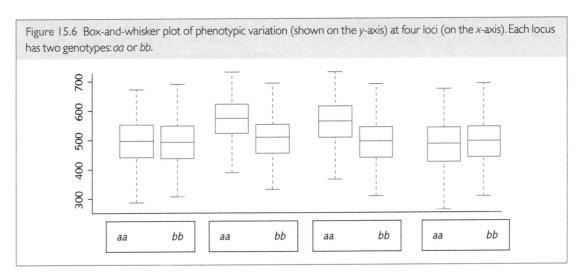

Figure 15.6 Box-and-whisker plot of phenotypic variation (shown on the *y*-axis) at four loci (on the *x*-axis). Each locus has two genotypes: *aa* or *bb*.

Genotypes from 100–300 markers are sufficient to capture variation anywhere in the mouse genome in an F_2. This is a ridiculously small number of markers compared with the millions needed in human studies, but it's because recombinants only occur two or three times per chromosome in a single generation. Figure 15.5 shows this: very large chunks of chromosome are not recombined, so adding additional markers doesn't gain much extra information.

However, a problem with using sparse markers is that it's not straightforward to determine the most likely position of the QTL. For example, in Figure 15.6, there are differences between the two genotypes at the second and third loci (genotype *aa* has higher scores than genotype *bb*), but does this mean the QTL lies exactly between the second and third loci, or closer to one or the other, or possibly outside the interval altogether? The answer is to use linkage information to work out the most likely location. One of the most commonly used methods (maximum likelihood) compares support for a QTL at a particular distance (measured as a recombination fraction) from a marker compared with a model assuming no QTL (LANDER AND BOTSTEIN 1989). The test gives two likelihoods (one supporting a QTL at the specified location, one supporting no QTL), so the result is expressed as a likelihood of odds, or an LOD score, an acronym you will often see in the older literature on genetic mapping. The LOD is the logarithm (base 10) of the likelihood ratio test statistic for a particular value of the recombination fraction. It's worth knowing this, because you will occasionally come across results reported as LOD scores, particularly in mouse genetics. LOD scores are also used to report linkage mapping in human studies (e.g. for results using the sibling-pair method; see Chapter 5, section 5.1.2).

Given that a genome scan uses a few hundred markers (compared with the hundreds of thousands used in human genome-wide association studies (GWAS)), the significance threshold will be lower than the value of $P < 5 \times 10^{-8}$ we introduced for the human analyses (see Chapter 6, section 6.5.2). This is because there is less need to correct for multiple testing (see Chapter 21, section 21.3). We mentioned in Chapter 5, section 5.1.2 that Eric Lander and Leonid Kruglyak derived significance thresholds for human linkage mapping; in the same paper, they reported thresholds for rodent genetic studies (LANDER AND KRUGLYAK 1995). Table 15.1 shows their significance

Table 15.1 Significance thresholds for rodent linkage mapping studies. From LANDER AND KRUGLYAK (1995).

	Suggestive P value	LOD	Significant P value	LOD
Backcross (1 degree of freedom)	3.4×10^{-3}	1.9	1.0×10^{-4}	3.3
Intercross (2 degrees of freedom)	1.6×10^{-3}	2.8	5.2×10^{-5}	4.3

thresholds as both *P* values and LOD scores. Note that for the intercross design, the most commonly used strategy, the base 10 logarithm of the *P* value is the same as the LOD score (a useful fact to remember when you see a result presented as an LOD score, i.e. $-\log_{10}(P) = 5 \times 10^{-5}$ is 4.3, which is the same as the LOD score).

However, don't rely on these standards! They are still useful guides, but there have been developments since Lander and Kruglyak's paper. One of the most important was Gary Churchill and Rebecca Doerge's introduction of a simple-to-use permutation test to determine significance (CHURCHILL AND DOERGE 1994). The permutation test makes fewer assumptions than Lander and Kruglyak's approach so it is more robust, more dependable. Churchill and Doerge proposed repeated shuffling of the trait values at random among the test subjects animals and reanalysis of each shuffled data. For example, to obtain a significance threshold, I take phenotypes I have obtained from my mapping study, randomly assign them to the genotyped animals and reanalyze the data. The randomization means the *P* values I generate must be representative of the *P* values found when there is no true genetic association. I repeat the permutation thousands of times and then ask how often I see, among the thousands of *P* values, results that are equal to, or more significant than, the most significant *P* value found in my data without rearrangement. That result determines what is called an **empirically derived threshold**. The permutation test works well for intercrosses, but it assumes that individuals are equally related, a condition broken in the heterogeneous stocks, and inbred and outbred mice, described below.

15.6.3 Recombinant inbred strains

Recombinant inbreds are derived from crosses between two inbred lines, as described above, but rather than phenotyping and genotyping the F_2 or backcross, the animals are inbred, a slow and expensive process of up to 20 generations. Once created, the lines have the great advantage that they are a permanent repository, providing a potentially limitless supply of animals with the same genetic constitution. As their genetic constitution won't change, they need to be genotyped only once.

Recombinant inbred strains have been a resource for mouse geneticists for decades. Until recently, they were all derived by crossing two parental strains, and the resultant lines are called after their progenitors with an 'X' to denote they are a cross (e.g. the recombinant inbred lines derived from C57BL/6 and DBA are called BXD lines). QTL mapping with these lines is carried out as described above for the intercross, comparing the phenotypes of the two genotypes (as the animals are inbred, there are only homozygotes of the two parental alleles).

An advantage of using inbreds is that many different laboratories can phenotype animals that are genetically identical, and pool their information. For example, traits have been measured over many years for the largest of the classical recombinant inbreds, the BXD panel, providing a dense set of information, including not just behavioral and physiological traits but also metabolic and expression data (e.g. ANDREUX et al. 2012, which can be downloaded from http://www.genenetwork.org). The BXD lines provide an example of the use of multiple forms of 'omic data (as explained in Chapter 4, section 4.7).

15.6.4 Heterogeneous stocks

Intercrosses, backcross strategies, and recombinant inbreds (derived from two lines) are very coarse-grained mapping strategies, typically mapping a locus to not much less than half a chromosome, which, genomically speaking, is a very large distance (DARVASI AND SOLLER 1997). Not only does this make it very hard to identify the specific genes involved, but it can also be misleading about what is present at a locus: for example, two or more closely linked loci whose alleles increase the phenotype will appear as one large effect; conversely, two closely linked loci with alleles that operate in opposite directions might

not be detected at all (this turns out to be surprisingly common).

The mapping methods we explore next provide higher mapping resolution but at the cost of more complex analysis, more markers, and larger sample sizes. The methods have the common feature of exploiting populations with more recombinants than the classical intercrosses and backcrosses. The populations are similar in that their chromosomes are fine-grained mosaics of progenitor stains.

We describe first heterogeneous stock populations, whose constituent animals are descended over many generations (more than 50) from eight inbred strains. The chromosomes of heterogeneous stock mice have a mosaic structure made up of small segments that can be derived back to one of the eight progenitors, and the large number of recombinants that have accrued since the progenitor inbred strains were first crossed increases mapping resolution substantially (Figure 15.7). They provide a mapping resolution of between 2 and 3 Mb (VALDAR et al. 2006a).

15.6.5 The Collaborative Cross

A second major resource for high-resolution mapping, known as the Collaborative Cross, a large set of recombinant inbred strains, is also derived from eight inbred strains (Figure 15.8) (FLINT AND ESKIN 2012). The Collaborative Cross was specifically created for QTL mapping. The idea is similar to the heterogeneous stock populations: eight genetically diverse strains (chosen to maximize the available genetic diversity) were bred over many generations to generate a large number of recombinants. The difference from the heterogeneous stock population is that the animals are inbred, thus providing a permanent resource that any investigator can use for mapping (AYLOR et al. 2011; PHILIP et al. 2011) (there is a stock of Collaborative Cross mice that were not inbred, and this mapping resource, called the diversity outcross, is another example of a heterogeneous stock; CHESLER 2014).

15.6.6 Inbred mice

It's also possible to map QTLs by treating the inbred strains like individuals in a GWAS. Inbred mouse strains are separated from their common ancestors by many generations, and these more distant relationships include many more recombination events, increasing the mapping resolution down to approximately 3 Mb.

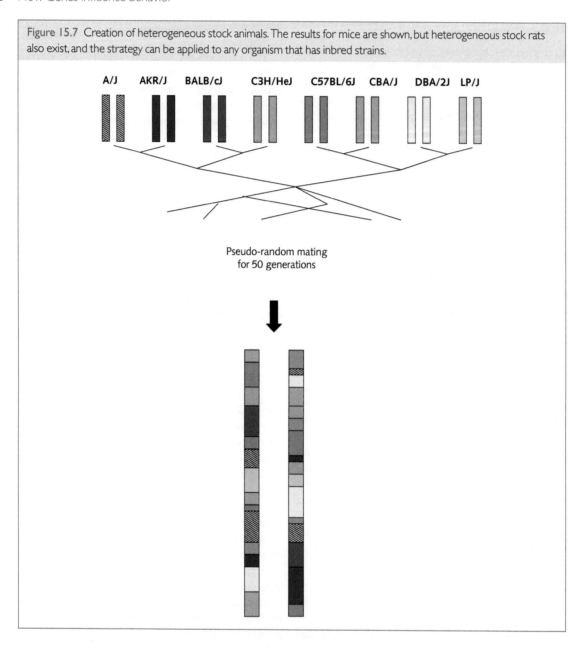

Figure 15.7 Creation of heterogeneous stock animals. The results for mice are shown, but heterogeneous stock rats also exist, and the strategy can be applied to any organism that has inbred strains.

A classical inbred-strain GWAS does not require any breeding steps, as the inbred strains required for phenotyping can be purchased directly from commercial breeders. The inbred-strain GWAS includes many more strains than the two parental strains of the usual F_2 cross, increasing the amount of genetic variation, and the results are completely reproducible as the same strains can be used in a replication study.

However, the first genetic mapping studies using strain comparison were full of false positives. No one had realized that mapping using inbred strains is like mapping using different ethnic groups: strains are not equally related to each other, so the effects of population structure can overwhelm the genetic signal being sought (see Chapter 6, section 6.6.4 for an explanation of why population structure introduces false positives). Furthermore, early studies had limited power because they drew on a small number of strains. More recent larger resources, like the Hybrid Mouse Diversity Panel (BENNETT et al. 2010), have

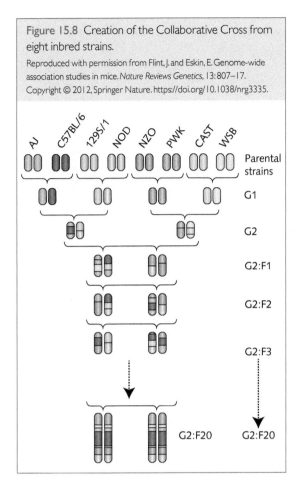

Figure 15.8 Creation of the Collaborative Cross from eight inbred strains.

Reproduced with permission from Flint, J. and Eskin, E. Genome-wide association studies in mice. *Nature Reviews Genetics*, 13:807–17. Copyright © 2012, Springer Nature. https://doi.org/10.1038/nrg3335.

increased power, and the application of methods to deal with population structure has reduced the risk of false positives.

15.6.7 Outbred mice

The last approach is the use of outbred mice, mice not as fully outbred as human populations, or as mice caught in the wild, nor as semi-outbred as the heterogeneous stock, but outbred because they are bred for hundreds of generations and maintained in large population sizes by commercial breeders. The first time I (J.F.) discovered the existence of these commercially outbred mice was on a visit to a local mouse-breeding facility that supplies the standard inbred strains (such as C57BL/6) to university research laboratories. During my brief tour of the facility, we passed an enormous hangar from which came the characteristic smell of mice. 'What's in there?' I asked. 'Oh, that won't be of interest to you. Those are outbred mice. We sell

tens of thousands to the pharmaceutical industry.' Why, I wondered, would pharmaceutical companies need thousands of outbred mice? The answer is that they need to demonstrate drug safety, and do so by giving outbred mice the compounds that they develop. The drug company has no interest in the genetic makeup of the mice; as long as they are 'outbred', nothing else matters. In fact, it turned out that outbred mice have the potential to map at a resolution unobtainable with other methods, down to a few hundred kilobases (YALCIN et al. 2004, 2010).

15.7 Two problems of analysis: population structure and haplotypes

Genetic mapping using the high-resolution mapping strategies presented above faces two problems, one of which, population structure, we have come across in human population genetics (see Chapter 6, section 6.6.4). As mentioned, the effects of population structure vitiated the use of inbred strains, such as the Hybrid Mouse Diversity Panel. Some of the strains are so genetically different that they cannot produce fertile offspring; we are dealing with a species level of difference here. It's a bit like mapping genetic variation in cognitive skills by comparing humans and great apes—we'd see huge differences in both phenotype and genotype, but we can't tell which of the genetic differences are responsible for the cognitive differences and which for all the other ways in which humans and chimps differ.

Figure 15.9 shows the result of mapping in inbred strains without taking into account population structure: you can see there are peaks all over the genome (the horizontal line is the significance threshold) (FLINT AND ESKIN 2012).

The solution in mouse studies has been to estimate the genetic relationship between each individual (you have to create a genetic relationship matrix (GRM)). Put simply, the GRM captures the genetic features that we want to ignore and leaves the features that we want to detect (KANG et al. 2008). Look at Figure 15.10, which shows the same mapping experiment but now analyzed using a GRM (FLINT AND ESKIN 2012). All the peaks are now under the horizontal line with one exception: there is one peak at a locus called *Bwq5* on chromosome 8 (Bwq5 refers to "body weight QTL number 5"). This magic property of the GRM has been exploited in numerous interesting

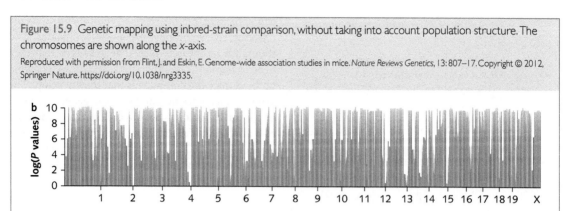

Figure 15.9 Genetic mapping using inbred-strain comparison, without taking into account population structure. The chromosomes are shown along the *x*-axis.

Reproduced with permission from Flint, J. and Eskin, E. Genome-wide association studies in mice. *Nature Reviews Genetics*, 13: 807–17. Copyright © 2012, Springer Nature. https://doi.org/10.1038/nrg3335.

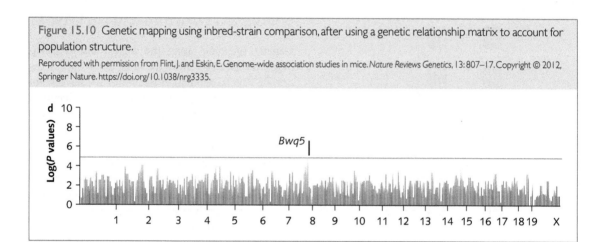

Figure 15.10 Genetic mapping using inbred-strain comparison, after using a genetic relationship matrix to account for population structure.

Reproduced with permission from Flint, J. and Eskin, E. Genome-wide association studies in mice. *Nature Reviews Genetics*, 13: 807–17. Copyright © 2012, Springer Nature. https://doi.org/10.1038/nrg3335.

ways: it captures polygenic variation and can be used to estimate heritability and genetic correlations (see Chapter 7, section 7.5.1).

Population structure affects the heterogeneous stocks, the commercial outbreds, and any strategy that uses animals where the degree of relatedness varies (so that the population includes, for example, first cousins, second cousins twice removed, aunts and uncles, as well as more distantly related individuals). In each case, the application of a GRM can adequately remove population structure and allow the identification of loci.

The second problem affects methods where the ancestral origin of an allele cannot be determined. In an intercross, the origin of every allele is unambiguous: there are only two parental strains, and so an allele has to come from one or the other. In heterogeneous stock or Collaborative Cross mice, there are eight ancestors, and there are almost no loci that have eight alleles. In fact, genotyping arrays interrogate

loci that have two alleles (single-nucleotide polymorphisms that are diallelic). So it's not clear which allele comes from which ancestor. The answer is to use the relationship between markers, exploiting linkage disequilibrium, to estimate which ancestor contributed which allele (Mott et al. 2000). In effect, we estimate the haplotypes at each locus, so this method is often called probabilistic haplotype reconstruction. As just three markers with two alleles each can generate up to eight different haplotypes, you can see how haplotypes can provide sufficient information to tag all of the ancestral chromosomes. The estimated probabilities of descent from each ancestor are expressed as a series of numbers, one for each of the progenitors, and association between trait and the reconstructed haplotypes is tested using the same methods as for standard genotypes (see Appendix, sections A10 and A11).

Why is haplotype analysis important? In some circumstances, reliance on single-marker analysis (the standard method used for human GWAS) will fail to detect a genetic effect. We discuss this later in section 15.11 where we consider how the difference between single-marker and haplotype analyses can be used to identify causative variants; Figure 15.19 gives an example of how haplotype analysis finds a locus that single-marker analysis misses.

15.8 Comparing mapping populations

How do these different mapping resources perform? Figure 15.11 gives an overview, comparing mapping in an F_2 intercross (low resolution), the Hybrid Mouse Diversity Panel (a panel of inbred mice that provide intermediate resolution) and commercially available outbreds (high resolution) (FLINT AND ESKIN 2012). The data are for markers on chromosome 1 and the phenotype is a blood lipid (high-density lipoprotein), that a variant in the *Apoa2* gene is known to affect. You can see that mapping using the commercially available outbreds identifies *Apoa2*. This is currently the highest resolution method we have.

Table 15.2 summarizes the strategies we have described for mapping the naturally occurring genetic basis of behavior in rodents. For the crosses between two inbred strains (F_2 intercrosses), mapping resolution is often so poor that it encompasses half a chromosome. The advantage of using the intercross is that relatively few animals are needed, and relatively few genetic markers need to be genotyped to capture information from across the genome. Heterogeneous stock and Collaborative Cross populations provide higher resolution but still map loci to an interval containing about 20 genes. The Collaborative Cross population has the advantage that no genotyping is required, as every animal is identical to those genotyped previously, and multiple individuals of the same genotype can be genotyped (this is indicated by '* N' in Table 15.2): multiples of the 200 used for mapping can be employed to improve resolution (but not to the point where the 2Mb limit is broken). Commercially available mice provide the highest mapping resolution, but require the most genotyping. No population currently available provides the mapping resolution obtainable in fully outbred populations, such as in humans.

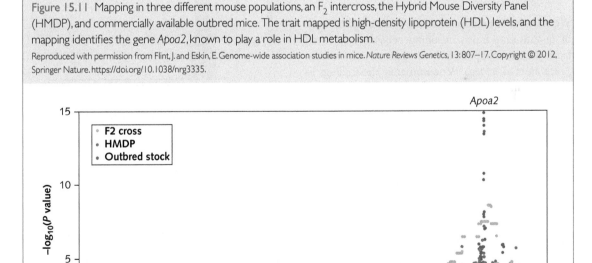

Figure 15.11 Mapping in three different mouse populations, an F_2 intercross, the Hybrid Mouse Diversity Panel (HMDP), and commercially available outbred mice. The trait mapped is high-density lipoprotein (HDL) levels, and the mapping identifies the gene *Apoa2*, known to play a role in HDL metabolism.

Reproduced with permission from Flint, J. and Eskin, E. Genome-wide association studies in mice. *Nature Reviews Genetics*, 13:807–17. Copyright © 2012, Springer Nature. https://doi.org/10.1038/nrg3335.

Table 15.2 Comparison of resources for mapping complex traits in rodents

Mapping population	Number of mice	Number of genetic markers	Resolution	Genes
F_2 intercross	500	100	10–50 Mb	50–250
Heterogeneous Stock	2,000	12,000	3 Mb	20
Commercial outbreds	2,000	100,000	250 Kb	2
Inbred strains	200 * N	0	3 Mb	20
Collaborative Cross	200 * N	0	3 Mb	20

The complexity of the mapping populations correlates with its resolution: the simplest population, the F_2 intercrosses, has two alleles descended from two progenitors with very little variation in allele frequency (each close to 50%), whereas alleles in the commercial outbreds descend from one, or more, of the progenitors and have allele frequencies that vary considerably, most being relatively rare.

15.9 The genetic architecture of behavior in rodents

We start with a description of the genetic architecture of quantitative traits, including behavior, obtained from genetic mapping in heterogeneous stock mice (Valdar et al. 2006a). We summarize the findings in the following sections.

15.9.1 Each QTL accounts for about 2–3% of the variance of the trait

Figure 15.12 plots the effect sizes of about 900 QTLs that influence a range of behavioral, physiological, anatomical, hematological, biochemical, and immunological traits—everything from how much a mouse weighs to how much hemoglobin its red blood cells contain, the size of its adrenal glands, and its behavior in an open-field arena. The distribution of effect sizes for behavior is no different from the distribution of effect sizes of other traits. While there are a few exceptions (the tail of the distribution of effect sizes spreads out on the x-axis) the majority of QTLs are clustered around the median, explaining a bit less than 3% of the variation in a phenotype.

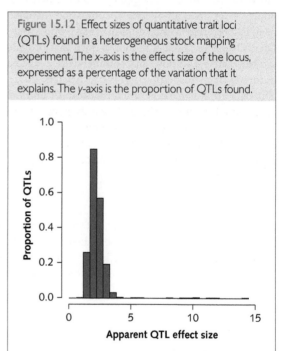

Figure 15.12 Effect sizes of quantitative trait loci (QTLs) found in a heterogeneous stock mapping experiment. The x-axis is the effect size of the locus, expressed as a percentage of the variation that it explains. The y-axis is the proportion of QTLs found.

15.9.2 The number of QTLs detected correlates with the heritability of the trait

Figure 15.13 compares the heritability of each phenotype in heterogeneous stock mice with the number of QTLs detected. There is a clear linear relationship: the larger the heritability, the more QTLs are detected. This finding may not be intuitively obvious: despite what you might think, genetic mapping of a phenotype with a high heritability does not mean that there is an increased chance of discovering a highly significant result. Instead of detecting QTLs of larger effect, you just find more of them. The power to detect

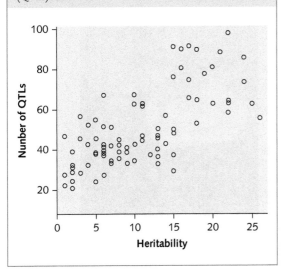

Figure 15.13 Comparison of the heritability of 100 phenotypes with the number of quantitative trait loci (QTLs) identified.

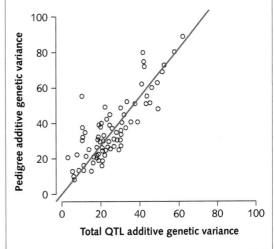

Figure 15.14 The extent to which identified quantitative trait loci (QTLs) account for heritability. Each circle represents the results for a single phenotype. The x-axis is the heritability explained by summing up all QTLs, and the y-axis is an estimate from family data of the heritability of each phenotype.

an effect depends on the locus-specific effect. The effect sizes, as we saw in Figure 15.12, are pretty much the same, so working with a highly heritable phenotype means there will be more effects to find (which does indirectly increase your chances of success).

15.9.3 Detected QTLs account for about three-quarters of the heritability

Figure 15.14 shows that about three-quarters of the heritability (plotted on the y-axis) can be explained by summing the effects of identified QTLs (plotted on the x-axis). In other words, the mapping experiment did a good job of finding most of what was there (to use the language of human genetics, we don't have a much of a 'missing heritability' problem). We are certainly not missing large effects, and while the number of smaller, undetected effects remains unknown, we can be reasonably certain that the conclusions we draw from the loci that we do know about will give us a representative picture of the genetic architecture.

15.9.4 The effect of genetic interactions with the environment is at least as important as that of direct effects

Figure 15.15 shows a mouse balanced, a little precariously, at the end of a short runway, about 1 meter above the ground. Usually, mice will not come out

into such an exposed situation, so the extent to which they do is a measure of lack of anxiety. The apparatus is called an elevated plus maze because the arms are arranged as a cross, with two open and two enclosed arms, and the whole apparatus is elevated. On the left of the figure are the genotypes (AA, AT, and TT) for one marker and the associated effect on the time animals spend in the open arms of the plus maze. There is no significant difference between the three genotypes. On this basis, we would say that there is no QTL near the marker contributing to variation in anxiety.

However, there is another way of analyzing the data, taking into account environmental effects. To carry out the test of anxiety, each mouse has to be removed from a cage, placed in the apparatus, and then returned to the cage. This is not automated; someone has to do the work and testing 2,000 animals needs a team of people (if only to alleviate boredom). Over the course of a 2.5-year project, ten people helped run mice through the elevated plus maze (VALDAR et al. 2006a).

We can ask whether the experimenter has an effect on the mouse's level of anxiety. In Figure 15.16, you can see that there is variation in the anxiety levels measured for each experimenter. This variation is significant (in fact, highly significant), but the effect

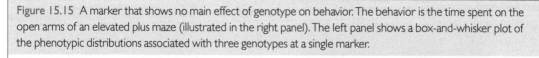

Figure 15.15 A marker that shows no main effect of genotype on behavior. The behavior is the time spent on the open arms of an elevated plus maze (illustrated in the right panel). The left panel shows a box-and-whisker plot of the phenotypic distributions associated with three genotypes at a single marker.

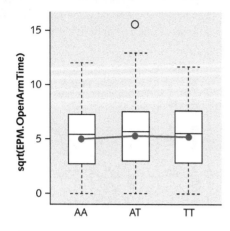

Figure 15.16 Variation in behavior in the same population of mice as measured by ten different experimenters.

is relatively small, accounting for about 5% of the variation.

Now we consider the interaction: does the effect of the experimenter depend on the genotype of the animal? We combine the two figures, plotting the differences in genotype seen for each experimenter (Figure 15.17).

There are quite clearly big differences between experimenters: experimenter 10 elicits a large genetic effect, experimenter 2 none at all. In addition, the direction of effect differs: in the hands of experimenter 5, animals with the TT genotype become more anxious, but they become less anxious in the hands of experimenter 4. Overall, the interaction effect is large, at least twice as large as the effect of experimenter alone.

In fact, on aggregate, the total effect of interactions is at least as important as that of the straight, immediate, unadulterated heritability (VALDAR et al. 2006b).

15.10 Some features of the genetic architecture depend on the population in which the loci are mapped

The picture of the genetic architecture of behavior in heterogeneous stock mice and rats (yes, there is also a rat heterogeneous stock; BAUD et al. 2013) is remarkably similar to that seen in humans: large numbers of

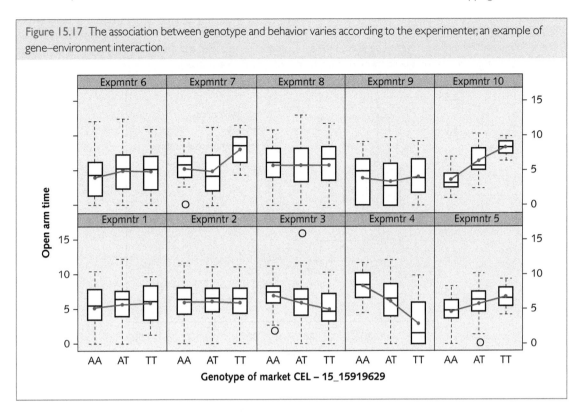

Figure 15.17 The association between genotype and behavior varies according to the experimenter, an example of gene–environment interaction.

loci acting additively and through interactions with the environment to alter traits. But there are differences too, the most egregious of which is the effect size.

A human mapping study of behavioral variation might find loci that each explain 0.1% of the variance, while an equivalent locus in heterogeneous stock mice explains 2%—an enormous difference! Does this mean the genetic architecture of behavior is different in rodents and humans? No, it's just an artifact of the population in which behavior is mapped. We'll explain why this is so.

Compare first the heterogeneous stock results with the F_2 intercrosses and small panels of recombinant inbred strains. In these populations, the effect sizes are even larger. For example, in our study of open-field activity in mice, using John DeFries' selected strains, we found six loci contributing up to 25% of the variance of the behavior (FLINT et al. 1995). That's quite extraordinary, given what we see in human populations. Now compare the heterogeneous stock results with mapping behavior in the commercial outbred population. In the commercial outbreds, the individual effect sizes drop, to below 1% (but still an order of magnitude larger than in a human population)

(NICOD et al. 2016; PARKER et al. 2016). What is going on here?

The differences are due to the varying genetic structures of the populations. In brief, the more the mouse mapping population approximates to human population structure, the more similar the genetic architecture becomes. By contrast, using a simpler mapping population exaggerates the differences: larger effect sizes, smaller numbers of loci, and also locus–locus interactions. You can see this by looking at the allele frequencies. Recall that in the F_2 populations, the simplest of our mapping populations, the allele frequencies are close to 50%. Every allele has this frequency, and there are no rare variants, in the sense we saw in human populations, and the total amount of genetic variability is relatively limited (just two chromosomes are segregating). Figure 15.18 compares the allele frequencies in commercially outbred mice, fully outbred (wild) mice and heterogeneous stocks.

In human populations, the vast majority of genetic variants are rare. Figure 3.10 in Chapter 3 showed that almost half of all variants found in a large sequencing study occurred just once (UK10K Consortium et al. 2015). The distribution of alleles in the wild mouse

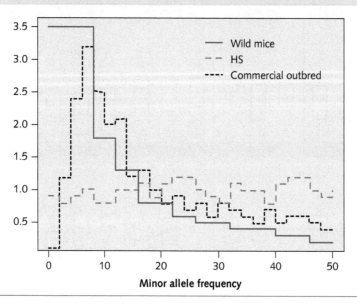

Figure 15.18 Allele frequency distributions of genetic variants in three mouse populations, wild mice, heterogeneous stock (HS), and commercial outbreds. The y-axis shows the proportion of alleles and the x-axis the minor allele frequency (i.e. the frequency in the population of the least common allele at a locus).

population shown in Figure 15.18 is the same as in human populations—the bulk of variation is rare (in fact, there are more alleles segregating in wild mice than in an equivalent human population). By contrast, the distribution of alleles in the heterogeneous stock is almost constant for each allele frequency (the dashed gray line is almost horizontal). The distribution for the commercial outbreds comes closer to that seen for wild mice, with the majority of loci having low-frequency alleles (this approximates to the distribution of alleles in human populations). The consequence is simple: with increasing complexity of the mapping population, larger variation in allele frequency, and many more rare alleles, the average effect of a locus on behavior, or on any trait, decreases.

Many features of genetic architecture are described using relative measures, of which the best known is heritability. Heritability is a property of populations, not of individuals, and as heritability is a reflection of the number of loci and their effect sizes, it's no surprise that these measures depend on where, when, and how they were estimated

In short, as the genetic structure of the mapping population gets closer to that seen in human populations, the genetic architecture of behavior gets closer to that seen in human populations. Perhaps the

remarkable thing is that the picture of genetic architecture of behavior emerging from both research fields is so similar (FLINT AND MACKAY 2009).

15.11 Progressing from locus to gene

Each of the genetic mapping strategies we presented in the previous sections has its own strengths and weaknesses for gene identification; the inbred-strain crosses, due to their simplified genetic architecture, require relatively few animals and provide a robust powerful method to find loci. But the resolution is poor, and in many cases genetic loci identified through coarse-scale mapping are due to multiple, physically linked small effects (TALBOT et al. 1999), making it very hard to disentangle and find the causative genes. Higher-resolution mapping overcomes some of these problems but does not deliver the same resolution as human geneticists enjoy with association mapping (i.e. GWAS), and needs thousands of animals.

One approach to the problem of how to identify genes at a QTL has been to combine genetic data with transcript information (SCHADT et al. 2005). From human GWAS data results, we know that QTLs lie in regulatory regions of the genome and thus are thought to act primarily by altering gene expression. Identifying QTLs that cause changes both in phenotype and in gene expression may thus be a way of finding the genes responsible, a strategy that has paid off in a number of complex diseases (MEHRABIAN et al. 2005). Rather than seeking to identify the sequence variants underlying complex phenotypes, correlations between variations in transcript abundance are used to identify expression networks that, perturbed by the genetic variation at susceptibility loci, contribute to phenotypic variation (CHEN et al. 2008).

The problem with using expression data to identify a causal gene is that the approach doesn't **prove** that the expression changes are due to the same genetic variant that causes the behavioral variation. Furthermore, the expression change does not necessarily lead to the behavioral change: both could be independent outcomes of a genetic variant. The same issues bedevil the transcriptome-wide association analysis approach we described in Chapter 6, section 6.10 and the endophenotype strategy, illustrated in Figure 5.9 in Chapter 5, section 5.3. There have been methodological attempts to permit better inference of causality, but none is totally satisfactory (see PEARL 2000 for a discussion, and also our presentation of Mendelian randomization in Chapter 13, section 13.5). Nevertheless, expression data are still regularly used to support the candidacy of genes. As a general rule, we recommend you treat with suitable caution claims that a gene for behavior has been identified solely on the basis of expression variation.

What if we had sequence data from all the mice in mapping experiments? Surely that should help identify the causative variants and thus help find the causative genes? I (J.F.) used to dream about obtaining complete genome sequences (although I admit I hadn't worked out exactly how to use them). At the time, in 2009, next-generation sequencing was just coming into use, and the Sanger Institute, the large DNA sequencing center outside Cambridge in the UK, had begun using these machines for human sequence analysis. Given that there are thousands of animals in the mapping experiments, the cost of acquiring the genomes seemed prohibitive (then costing about US$10,000 a genome), and a mouse project would certainly take second place to a human one on the new machines. But there is a short cut available to mouse geneticists.

Because the mice used for mapping experiments are descended from known progenitors, it's possible to reconstruct their sequence if you know their ancestors. You can see this from Figures 15.5, 15.7, and 15.8, which show how the mapping populations are a mosaic of the progenitors. Once you know the position of the chromosomal break points, from genotyping, you can reconstruct the entire sequence of each animal if you know the genome sequences of the progenitors. So, by sequencing the genomes of a dozen inbred strains, you get as a byproduct the sequence of literally thousands of mice, including those not yet born (as the sequence can be used by all investigators who use a mapping population derived from the sequenced progenitors). This argument was key to convincing funding agencies to support our attempts to sequence inbred strains.

We had completed the genomes of 17 inbred strains in 2011, by which time we had also worked out how to use the sequence to identify variants that contribute to complex traits, including those influencing behavior (KEANE et al. 2011; YALCIN et al. 2011). The idea we proposed is that the effect of a variant can be seen from two different, complementary angles. The difference between the two viewpoints can be used to determine whether a variant is likely to be functional or not. From one perspective, we look for an association between the phenotype and genotypes in the ways described above: at a causative locus, one allele will increase and the other will decrease the trait value. From the second viewpoint, each allele can be considered as derived from the progenitor strains. In this case, we treat alleles as part of a haplotype descended from one of the sequenced strains. Where there are only two progenitors (as in an F_2), the two viewpoints are identical, but when there are more than two, then the genetic effects can be different. This difference can be very informative.

Suppose, as an extreme case, there are four progenitor strains, of which two (we'll call them A and B) carry a causative allele that increases the trait ('+') and two (C and D) carry the decreasing allele ('−') (Table 15.3). We genotype marker 1, which is very close to the causative variant. At this marker, one allele (A) is linked to '+' and the other allele (G) is linked to the '−' QTL allele. Single-marker analysis (the first point of view) tests for differences between the two alleles. It detects the effect (the three genotypes are associated with different phenotypes). We then genotype marker 2, which is equally close to the QTL, but at this locus, strains A and C carry one allele (C) and strains C and D carry the other (T).

As you can see from Table 15.3, animals with the C allele at marker 2 consist of a mixture of increasing and decreasing QTL alleles. The QTL effects cancel out, as they do for animals carrying the T alleles (the phenotypic scores are all the same, '+−', so no genetic effect can be detected. The failure to detect the QTL effect rules out marker 2 as a causative locus. Haplotype analysis treats animals as descended from the progenitors, so that it detects differences between the haplotypes of the four progenitors in our example: you can see that the phenotypic scores of the haplotype score are correctly associated with the QTL alleles.

Figure 15.19 illustrates the difference between the haplotype and single-marker analysis of phenotype. The black dots, the single-marker analysis in the figure, miss the peak of association found by the haplotype analysis (the solid line in the figure) at about 46 Mb. If we relied solely on the single-marker analysis, we would incorrectly assume that the most likely position for genetic effect on behavior was at about 52 Mb. The difference between the two analyses would evaporate if we had complete sequence information for the heterogeneous stock animals, as we would then be able to genotype the causative sequence variant,

Table 15.3 Single-marker and haplotype analysis of behavior

Strain	Marker 1	QTL allele	Marker 2
A	A	+	C
B	A	+	T
C	G	−	C
D	G	−	T

Genotype at marker 1	Phenotypic score	Genotypes at marker 2	Phenotypic score
AA	++	CC	+−
AG	+−	CT	+−
GG	−−	TT	+−

Haplotype	Phenotypic score
A-C	+
A-T	+
G-C	−
G-T	−

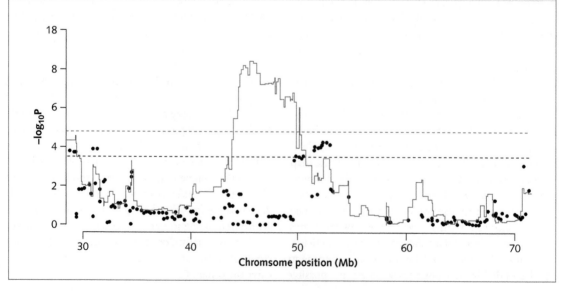

Figure 15.19 Single-marker and haplotype analyses of a trait in a mouse heterogeneous stock. The x-axis is the distance along the chromosome in megabases (Mb) and the y-axis is the strength of the association, measured as the negative logarithm, base 10, of the P value. The black dots are the result of single-marker analysis, and the solid line shows the results from haplotype analysis. Note that the single-marker analysis fails to detect the association at 46 Mb and incorrectly identifies the peak of association at 52 Mb. The two methods have different significance thresholds: the lower dotted line shows the threshold for single-marker analysis and the upper dotted line for haplotype analysis.

and this variant would be highly associated with the trait, agreeing with the haplotype analysis. We exploit the difference between haplotype and single-marker analyses to determine which variants could cause behavioral variation. By comparing haplotype and single-marker analyses, we can determine whether a variant is likely to be causative or not (YALCIN et al. 2005).

Using this method, we can assign to each nucleotide in the genome the probability that it is causative for a trait measured in one of the mapping populations. This analysis confirmed what most of us had suspected—that coding variants had larger effects on behavior (and on other complex traits) than noncoding variants (Figure 15.20).

Our sequence data demonstrated the relative contribution of different classes of variants to traits, but sadly it wasn't able to unequivocally implicate individual genes. As found in human GWAS, the great majority of genetic effects on rodent behavior lie in noncoding regions of the genome whose relationship to genes is hard to disentangle.

In a very small number of cases, coding variants that affect behavior have been identified from mapping, such as the identification of the *Usp46* gene as affecting two measures of depressive like behavior in mice (TOMIDA et al. 2009). One other noteworthy experiment is that from Joe Takahashi's group (his work on identifying genes involved in circadian rhythm is described in Chapter 17, section 17.5). He exploited the relatively small number of sequence differences between two substrains of mice (C67BL6/J and C67BL6/N—you can tell from their names that the strains are closely related) to implicate a non-synonymous mutation of serine to phenylalanine in a gene called cytoplasmic FMRP interacting protein 2 (*Cyfip2*), as a cause of variation in cocaine-response phenotypes (KUMAR et al. 2013) (the mutation has not,

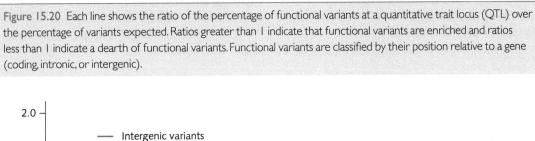

Figure 15.20 Each line shows the ratio of the percentage of functional variants at a quantitative trait locus (QTL) over the percentage of variants expected. Ratios greater than 1 indicate that functional variants are enriched and ratios less than 1 indicate a dearth of functional variants. Functional variants are classified by their position relative to a gene (coding, intronic, or intergenic).

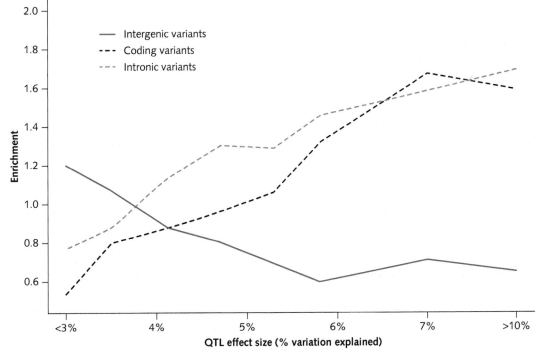

however, been proven to be the causative mutation; this requires genetic engineering described in Chapter 16 to change the allele). But as a general strategy, searching for coding mutations that affect behavior has been unproductive.

If we can't (easily) find a causative variant, what about finding a causative gene? The outbred strategies described above do, sometimes, reduce the likely candidates to one or two genes at a locus. If this happens, which strategies prove that a gene is a candidate? This task is possible when you have access to a mutant, a gene knockout, of the candidate gene. As Trudy MacKay showed in experiments in fruit flies, you can apply a quantitative complementation, or gene–knockout interaction test (LONG et al. 1996). The idea behind the strategy is that a causal variant must have its effect by altering some property of a gene (such as gene expression). If the gene is missing, then the variant can't exert its effect. Consequently, a test for the candidacy of a gene is whether the effect of the variant and the effect of a knockout are independent. By constructing a cross in which both can be examined, separately and together, a statistical test of the interaction between the knockout's effect on the trait, and the genetic variant's effect on the trait, can be used to reveal the gene through which the variant operates.

We performed a gene–knockout interaction test on a locus affecting fear-related behavior and showed that a regulator of G-protein signaling called *Rgs2* was the responsible gene (YALCIN et al. 2004). *Rgs2* belongs to a class of genes that has an important characteristic: they modulate neuronal activity. Genes involved in neuronal modulation, or neuromodulators, emerge from other studies of genetic action on behavior (as we will see in the next chapter), providing a hint that there is a common pathway by which genes influence behavior. However, the experimental design used to identify *Rgs2* requires access to mutations in specified strain backgrounds, which until recently was very hard to obtain. The development of CRISPR (clustered regularly interspaced short palindromic repeats), described in the next chapter, has changed this, although there as yet no applications of CRISPR to finding the genes underlying mouse behavior. We turn next to experiments that directly test the involvement of genes in behavior through the analysis of engineered mutations.

Summary

1. Analysis of inbred strains of mice indicates that behavior is heritable and subject to gene–environment interaction. However, accurate estimates of the amount of heritable variation that is present for a trait are better estimated from selection experiments.

2. Three methods of genetic mapping use progeny of two inbred strains: chromosome substitution lines, back- and intercrosses, and recombinant inbred lines.

3. Higher-resolution mapping is achieved using mice whose chromosomes are descended, usually over many generations, from multiple different strains. The chromosomes are mosaics of the progenitors. Heterogeneous stock and Collaborative Cross mice are examples of mosaic populations. Fully outbred mice provide even higher mapping resolution but not as high as that achieved in human genome-wide association studies.

4. The genetic architecture of behavior in mice is polygenic, with each locus making a small contribution.

5. The genetic architecture of behavior in mice varies according to the mapping population, with few loci of relatively large effect in intercross populations, and many more loci, with relatively small effect, in mosaic and outbred populations.

6. Genetic effects operate in a complex fashion, often in interaction with the environment.

7. While mapping loci that contribute to behavioral variation in rodents is relatively straightforward, it has proved much more difficult to identify the genes. There are very few examples where we unequivocally know the identity of such a behavioral gene. One example is the demonstration that *Rgs2* is involved in fear-related behavior.

16

Reverse genetics

[In which we explain how genetic engineering can be used to create mutations, and what those mutations have told us about how genes influence behavior]

16.1 Forward versus reverse genetics

One of the main justifications for genetic mapping of psychiatric disorders (or indeed of behavioral traits) is the expectation that mapping results will lead to the identification of genes, and from there to an understanding of biological mechanisms, and hence, in the case of disease, to new ideas for treatment. Unfortunately, identifying the molecular variants and the genes that underlie complex traits has proved to be challenging. Mapping was hard, gene identification is harder, and we are a very long way from understanding a mechanism that might allow the development of a novel treatment for schizophrenia or depression.

But what if we turned the problem upside down? Rather than start with a phenotype and try to find its genetic basis, suppose we start with the genes? Our question becomes, which phenotypes are associated with mutations in these genes? Suppose we have genes that might be involved in behavior, and we want to know what these genes do. For example, suppose we

identify all the genes expressed in the brain, or in part of the brain we think is affected by disease, or even in one cell type or one cell constituent. To understand what these genes do, we could observe the consequences of disrupting the gene. But how do we generate the mutations we need? The answer lies in using genetic engineering, a technology that makes it possible to generate at will a mutation in any gene. Genetic engineering opens up a completely new way of doing genetics, going from genotype to phenotype (rather than from phenotype to genotype). This new way of doing things, to distinguish it from the old (which is now called forward genetics), has been called reverse genetics.

This chapter starts by describing the technologies that make reverse genetics possible, and then explores what reverse genetics approaches have told us about how genes influence behavior. As it became clear that genetic effects on behavior depended on the tissue, and on the type of cell in which a mutation occurred, it also became clear that for reverse genetics to illuminate the biology of behavior, genetic engineering was going to have to do more than just generate knock-outs. We include in this chapter a presentation of the technologies that have emerged to provide control over where and when a mutation is turned on and off. These approaches are now standard, and constitute essential tools in teaching us how genes work in the brain (and in any tissue or cell type). We'll introduce you to the revolution in neuroscience that the genetic engineering technologies have enabled and, as an example, show you how it is now possible to create false memories.

16.2 Genetic engineering

Few advances in biology have captured the public imagination more than genetic engineering, the ability to change the genome of an organism, which confers almost divine attributes on the scientists who do this work. To delete or insert genes, to make changes that take away abilities, or improve them, or even endow animals with new faculties, were all impossible before the advances in genetic engineering that occurred in the 1980s. By the turn of the century, the methods had become routine, and introduced a completely new way of using genetics to understand physiology and, more recently, behavior.

Genetic engineering in the mouse had its roots in two lines of investigation: mouse embryo-derived stem cells and homologous recombination. The former made it possible to grow an entire animal from a single cell and the latter made it possible to target a gene. We'll describe both features.

16.2.1 Embryonic stem cells

Stem cells are cells that have the potential to grow into any other cell. Pluripotent stem cells produce all cells of an organism, while multipotent or unipotent stem cells regenerate only specific lineages or tissues. Molecular biologists want pluripotent stem cells for genetic engineering experiments, as these cells have the capacity to grow into an animal. The idea is simple: alter the DNA of the stem cell and then grow it, with the required changes, into a mouse. The idea's execution is less simple.

For a long time, it was thought that pluripotency was a special characteristic of embryonic cells, but proving that this was true was hard. Embryos are busy dividing and developing into an animal, and don't want to sit around in an undifferentiated state while someone tries to grow them in a culture dish. Martin Evans, as part of the work that earned him a Nobel Prize, managed to isolate stem cells from a small mass of dividing cells that had not yet implanted into the wall of the womb. The collection of cells, known as a blastocyst, lies at a stage of development between the fertilized egg cell and the embryo. Evans showed that blastocyst-derived cells could, under the appropriate conditions, continue to grow in an undifferentiated state in the laboratory, and yet retained the capacity to differentiate and, indeed, to grow into mice (EVANS AND KAUFMAN 1981). These cells, embryo-derived stem cells (or embryonic stem cells, often shortened to ES cells), met the criteria for pluripotency.

Two features are important in understanding the use of embryonic stem cells for genetic engineering. The first is their ability to grow in the laboratory. Why this is so important is not immediately obvious. Why is it necessary to culture stem cells? Indeed, why do you need stem cells at all? Why not simply perform the engineering on a fertilized egg? The answer is that the gene-targeting mechanism, homologous recombination (which we describe next), is horrendously inefficient. Only a very small percentage of targeted cells actually contain what you want, so thousands and thousands of cells have to be used to identify the one or two correctly engineered cells. With a growing culture of many hundreds of thousands of cells, you can keep targeting until you are successful.

The second important feature is the animal from which embryonic stem cells are derived. Not all mice can be coaxed into releasing embryonic stem cells. The mice Martin Evans used belong to the 129Sv inbred strain, which frequently develops cancer. To avoid such cancer deaths resulting in the loss of all the genetic engineer's hard work, the 129 mutant is typically mated with another, more physically robust strain. One other strain of mouse (C57BL/6N), a strain that does not have the health problems of the 129 mice, is now routinely used to derive embryonic stem cells.

Attempts to make embryonic stem cells from other strains, and from other species (e.g. rats) used to consume a lot of research hours until a new technology emerged whereby stem cells can be derived not just from blastocysts but from any tissue. A Japanese cell biologist, Shinya Yamanaka, made the remarkable discovery that four key DNA-binding proteins (called Oct3/4, Sox2, Klf4, and c-Myc) can turn a differentiated cell, such as a skin cell (also called a fibroblast), into a stem cell (Takahashi et al. 2007). The ability to create induced pluripotent stem cells, together with further advanced cell-culture techniques to turn them not just into whole animals but into organoids, opened another avenue onto the biology of behavior, making possible experiments carried out on 'brains in a dish', from tissue derived from patients with psychiatric disease (Amin and Pasca 2018). Many of the techniques we describe for genetic engineering are now being carried out in human-derived stem cells and their derived organoids.

16.2.2 Homologous recombination

Homologous recombination is the process by which DNA is exchanged between two similar or identical sequences. Typically, when DNA is injected into cells, a small fraction will integrate, but usually it integrates randomly. Mario Capecchi and Oliver Smithies suggested that many cell types had the machinery to carry out homologous recombination, which would mean that when DNA is injected into a cell, rather than randomly integrating into the genome, it would target a homologous (similar) sequence, recombine, and thereby alter the DNA at that site, and only that site (Capecchi 1989).

As we noted, homologous recombination is not easy to achieve: the ratio of homologous (i.e. targeted) insertions to random insertions at nonhomologous sites is approximately 1 : 1,000. There are now a series of molecular tools to increase the chances that it will happen, favoring homologous over nonhomologous recombination, and this has made gene targeting a relatively straightforward technique. Designing and building mice through homologous recombination became a whole new career option, requiring a mastery of splinkerettes, floxing, recombineering, and more neologisms than Edward Lear ever thought up.

Once targeted, the embryonic stem cells aren't simply grown into a mouse. No, they have to be mixed with other cells from a blastocyst to be incorporated into an embryo, which is thus composed of cells of two different genetic constitutions, normal (wild type) and mutant. The embryo is a chimera, and to generate an animal with a heritable mutation you hope that the cells you spent weeks or months targeting will contribute to the chimera's germ line. Whether that has happened won't be known until the mouse is born and has offspring of its own. Furthermore, once the

Figure 16.1 Generation of mouse germ-line chimeras from embryo-derived stem (ES) cells containing a targeted gene disruption.

Reproduced with permission from Capecchi, M.R. Altering the genome by homologous recombination. *Science*, 244(4910): 1288–92. Copyright © 1989. DOI: 10.1126/science.2660260.

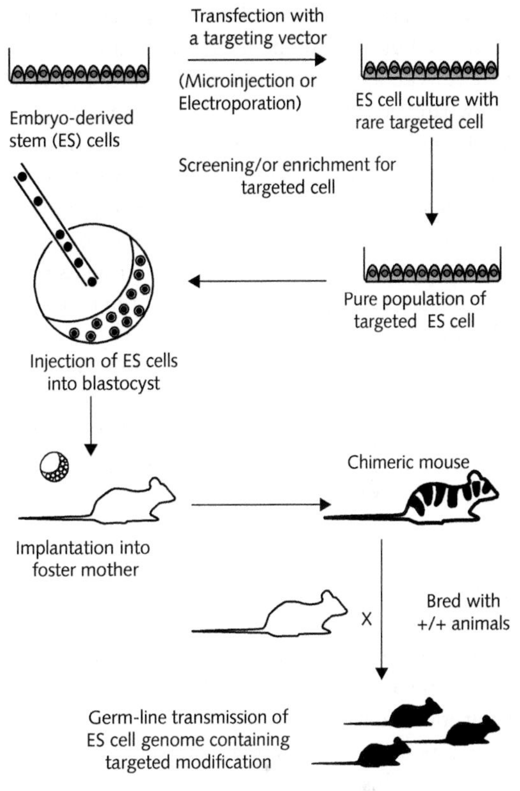

mutation has been passed on to the next generation, it needs to be bred to homozygosity. This is because targeting is so inefficient that you can reliably bet it will never hit both chromosomal copies. Mutants are born as heterozygotes, and it will take at least one round of breeding to generate homozygote knockouts, as shown in Figure 16.1 (Capecchi 1989).

Homologous recombination technology, applied to embryonic stem cells, remains a powerful and effective way to make specific mutations. Mouse knockouts have become standard resources in genetics: more than 25,000 have been generated and you can now order a knockout for almost any gene (Skarnes et al. 2011; Park et al. 2012).

Typically, to make a knockout, the genetic engineer deletes a critical component of a gene, at the same time adding in some extra DNA, such as an antibiotic gene, to identify and select the cells that have succumbed to the engineering. But the technology can do more than just knock out genes; it is possible to replace the DNA—to rewrite it—so that you can change even a single nucleotide. Animals engineered in this way are called knockins, to distinguish them from knockouts. Making a knockin is more challenging than making a knockout, but there are no technical reasons why you can't rewrite any gene, indeed any stretch of DNA, into whatever sequence you want. There is no doubt that homologous recombination is a powerful technology. There are limitations, however, to its use.

Suppose you want to engineer a mutation on a strain other than 129 or C57BL/6N, or to create a mutation in a species other than a mouse. Rodents might be good for studying some features of the genetics of behavior, but with about 75 million years of evolutionary time between mice and humans, they clearly are not good models for many behaviors that psychiatrists and psychologists study (NESTLER AND HYMAN 2010; FERNANDO AND ROBBINS 2011). Homologous recombination works reliably in two mouse strains but not so easily in other strains or in other species. And if you are going to make the mutation yourself, brace yourself for months of painful (and expensive) laboratory work. It is for these reasons that the discovery of CRISPR (clustered regularly interspaced palindromic repeats) is so important. CRISPR has had a democratizing effect, making it possible for (almost) anyone to target mutations, quickly and easily, in any organism. We turn to this revolutionary technology in section 16.2.4, but first let's deal with the distinction between homologous and nonhomologous recombination.

16.2.3 Nonhomologous recombination

Homologous recombination should be compared with its first cousin, nonhomologous recombination. Nonhomologous recombination is also a way to introduce exogenous DNA into a cell, but it's much less fussy. Any organism will do! And it's relatively efficient. It's 1,000-fold more likely to take place than homologous recombination (SMITHIES et al. 1985), which means you can inject the DNA directly into the fertilized embryo, rather than engaging in the fiddly work with embryonic stem cells. Animals with additional material added by this technique are often simply called transgenics (but be careful, as mice with DNA added by homologous recombination may also

be called transgenic). It's possible to inject DNA from any source (jellyfish DNA that contains a fluorescent protein is commonly used), and it will integrate into the host genome.

DNA added by nonhomologous recombination has three important features. First, it enters randomly, integrating anywhere in the genome, sometimes inadvertently creating mutations in genes. Second, the larger the piece of DNA that is injected, the more likely it is to rearrange, breaking up into multiple smaller fragments. Third, multiple copies are often found at the integration site. These features have limited the use of nonhomologous recombination in the creation of genetically engineered animals, but the method is still used for certain applications, such as the generation of mice into which large segments of DNA, about 100 kb or more, have been introduced, sufficiently large to contain gene regulatory elements enabling the study of gene expression and access to specific cell types (GONG et al. 2003).

16.2.4 Genetic engineering: clustered regularly interspaced palindromic repeats (CRISPR)

Many scientific discoveries happen by chance; many discoveries that change entire fields of research emerge from people working on obscure problems. Max Delbrück repeatedly made this point, advising scientists that if they want to make a real breakthrough, they have to work in unfashionable areas. Following his own advice, after his work on phage genetics (leading to a Nobel Prize in 1969), Delbrück devoted the later part of his career to trying to understand the origins of light perception in a fungus (*Phycomyces*). The reliability of Delbrück's advice is, however, open to question. '*Max was marvellous you know. He was always wrong. George Streisinger tells a story in an essay that he once came out of Max's office looking very depressed and someone asked him "What's the matter?" George replied, "Max likes my theory, it must be wrong!"*' (BRENNER AND WOLPERT 2001). As far as I know, not much seems to have been discovered about the phototaxic behavior *Phycomyces*, despite Delbrück's efforts. Nevertheless, the discovery of CRISPR is an example of the value of Delbrück's advice to work in unfashionable areas.

Jennifer Doudna and Emmanuelle Charpentier are the scientists whose 2012 paper in *Science* demonstrated the existence in bacteria of programmable machinery to cut DNA at specific sites, using CRISPR

Figure 16.2 Elements of the clustered regularly interspaced palindromic repeat (CRISPR) system for cutting DNA. (a) A CRISPR array containing repeat regions (black arrows) separated by spacer regions (patterned rectangles). The *cas9* gene encodes a nuclease that cuts DNA matching a spacer; the job of the *cas1*, *cas2*, and *csn2* genes is to acquire new spacers from invading DNA, thus conferring immunity for the cell from DNA containing those sequences. (b) The CRISPR array and the *trans*-activating CRISPR RNA (tracrRNA) is transcribed, processed, and complexed with Cas9. The complex searches for matches to the spacer sequence (the patterned box, shown in the transcribed RNA and in the DNA that it matches). (c) Cas9 binds to the target site and cuts the DNA. The cut is subsequently repaired, resulting in a deletion of the DNA target molecule.

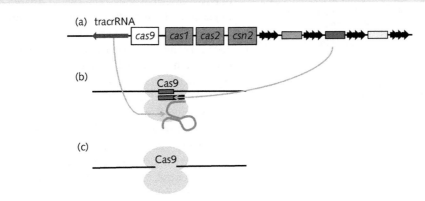

and an enzyme called Cas9 (Figure 16.2) (JINEK et al. 2012). By supplying it with RNA, which is used to target specific sequences, the system could be programmed to recognize, and then cut, any site. CRISPR, multiple copies of a 30 bp repeated sequence separated by spacers of about 36 bp, was first found in a microbe called *Haloferax mediterranei*, which lives in salt marshes on the coast of Spain. Similar repeats occur in other microbes; indeed, about 40% of bacteria and 90% of archaea possess CRISPR sequences (MAKAROVA et al. 2011; KOONIN et al. 2017), in every case associated with homologous (CRISPR-associated or *cas*) genes (JANSEN et al. 2002). The function of the genes and repeats was obscure until it was found that the sequence of the spacers between the repeats is homologous with foreign DNA, such as bacteriophages, which infect and kill bacteria. Philippe Horvath was able to show that spacers in bacteria conferred resistance to phage attack (BARRANGOU et al. 2007). The CRISPR spacer content confers specificity to the protection mechanism, targeting it to a specific phage, while the Cas enzymatic machinery provided resistance to infection by cutting (and hence destroying) the invading DNA (GARNEAU et al. 2010). Doudna and Charpentier showed that the Cas9 endonuclease can be programmed with a single guide RNA, and that the system could thus be used

to cleave any DNA sequence of interest (JINEK et al. 2012). Their paper did not, however, demonstrate that CRISPR worked in mammalian cells, which was shown the following year (CONG et al. 2013; JINEK et al. 2013; MALI et al. 2013).

You can think of CRISPR as consisting of two elements, a cutting system and a memory system. The memory system keeps a record of a small piece of the DNA from an invading organism, while the cutting system uses the recorded DNA as a template to target, cut, and destroy any future invaders. The cutting system is the most notorious, as this is used in genetic engineering applications, but the memory system can also be exploited, for example to record what a cell is transcribing (SHIPMAN et al. 2016). Indeed, it is the recording system that is the most interesting and unique feature of CRISPR.

CRISPR has attracted attention because it dethrones homologous recombination in mouse embryonic stem cells as a way to make mutants. CRISPR methods for genomic engineering are fast and require no extensive molecular biology steps. The system is incredibly efficient, targeting at such a high frequency that quite often both alleles will be cut. This means the targeting can be carried out in fertilized eggs (zygotes), obviating the need for unstable, hard-to-culture embryonic

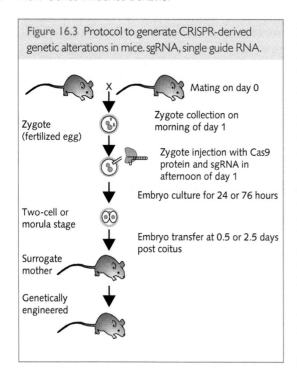

Figure 16.3 Protocol to generate CRISPR-derived genetic alterations in mice. sgRNA, single guide RNA.

X   Mating on day 0

Zygote (fertilized egg)

Zygote collection on morning of day 1

Zygote injection with Cas9 protein and sgRNA in afternoon of day 1

Embryo culture for 24 or 76 hours

Two-cell or morula stage

Embryo transfer at 0.5 or 2.5 days post coitus

Surrogate mother

Genetically engineered

stem cells. All you have to do is inject a mixture of DNA, Cas9 mRNA, and guide RNA into a zygote (Wang et al. 2013a), which can be of any mouse strain, or any species, and you can produce a mutant in a few weeks (Yang et al. 2014), as shown in Figure 16.3.

16.3 Constitutive mutations and the genetic architecture of behavior

The technology we've described above has been used to generate tens of thousands of gene knockouts, that is, animals (mostly mice) that lack the expression of a single gene due to a heritable mutation. These are constitutive mutations, a term used to differentiate them from mutations that are inducible, placed under the control of the experimenter (we'll turn later to how inducible mutations are made and how they are used). What has been learnt about the genetics of behavior from the study of constitutive mutations? We start by looking at the genetic architecture of behavior, as revealed by genetic engineering.

We have repeatedly emphasized that the genetic susceptibility to psychiatric disease and the genetic effects on personality and intelligence consist of

multiple small effects. What does the picture look like for gene knockouts? Is there a small set of genes of large effect that impacts on a behavior? Or does the architecture look like that of naturally occurring variants? To answer these questions, we can use the results emanating from the International Mouse Phenotyping Consortium (IMPC), a collaboration established to systematically examine the consequences of inactivating each one of a large number of genes, and eventually to assign a function to every gene in the genome (Austin et al. 2004; Collins et al. 2007). To find out what mammalian genes do, the consortium employed a standardized set of phenotypes, including some behavioral tests. Three papers summarize the results of the knockout phenotyping experiments (White et al. 2013; de Angelis et al. 2015; Meehan et al. 2017). What have we learnt?

The first lesson is that the two most likely outcomes of a mouse knockout experiment are: (i) a perfectly normal mouse; or (ii) no mouse (White et al. 2013). Either the mutation has no effect (which usually means that the mouse had another copy of the gene, or some other compensatory mechanism), or the mutation is in an essential gene, so that without it the mouse never developed. These lethal mutations occur in about 15% of genes (White et al. 2013). While this is an important figure to bear in mind if you embark on a gene knockout experiment, it doesn't help us understand the genetics of behavior.

In order to take us further, I (J.F.) downloaded the current set of results from the consortium website (http://www.mousephenotype.org/). If you want to do this yourself, I'd recommend you use this link ftp://ftp.ebi.ac.uk/pub/databases/impc/ from which you can obtain the most recent copy of the complete data. When I looked, there were results for 5,937 genes, tested on 1,164 measures (user beware: not all of these are independent tests, and not all knockouts have been tested on everything). As 15% were lethal, then, assuming that there are about 20,000 genes in a mammalian genome, roughly speaking 6,000 is about one-third of all the nonlethal genes, sufficient to give us a reasonably unbiased assessment. Analyzing a consortium's effort carries with it all the provisos that we mentioned in Chapter 6, section 6.6 on genome-wide association studies: effects differ according to who carried out the test and when, as well as to other unacknowledged confounds. So the analysis is not straightforward (the authors use the same methodology that has been successful in dealing with similar problems in human genetics). With these provisos, what was found?

16.3.1 The effects of a genetic knockout on behavior are quantitative

The first important observation is that mutations never simply abolish a behavior, rather they alter it quantitatively, increasing, or more usually decreasing, its occurrence. As the effect is not all or nothing, how big is it? Figure 16.4 gives the answer. The figure shows the effect of knockouts on the behavioral measures that the knockout consortium applied, as well as the effect on other, physiological, measures. Three results stand out. First, the distribution of effect sizes for behavior and physiological traits looks the same. This indicates, as we saw was true for the effect of naturally occurring variants (Chapter 15, section 15.9.1), that there is nothing special about the genetics of behavior. Second, the distribution is highly skewed— the bulk of the effects are small, but there are some outliers. Third, while many mutations do have a large effect the median effect size for a knock out that disrupts behavior is 2.2%. That's not much bigger than

the QTL effect sizes we described in Chapter 15, section 15.9.1!

The relatively small size of the mutation's effect may limit the usefulness of the genetic strategy to determine gene function because, as the effect size gets smaller, it can be lost in background genetic effects; that is, it will be lost among those naturally occurring variants that have been the subject of the previous chapters of this book. This brings us to the second point about how knockouts affect phenotypes.

16.3.2 The phenotype depends on the mouse strain in which it is measured

The consortium went to great lengths to construct mutations in a single strain, C57BL/6N. Why? The 129 strain had been the workhorse for generating mutations, because of the (relative) ease of obtaining stem cells, but, as noted above, the strain tends to develop tumors and may have structural brain abnormalities (not good for behavioral experiments). To avoid these problems, researchers usually cross the mutant animal with a healthier strain (C57BL/6). The offspring of the cross are no longer in a pure strain background.

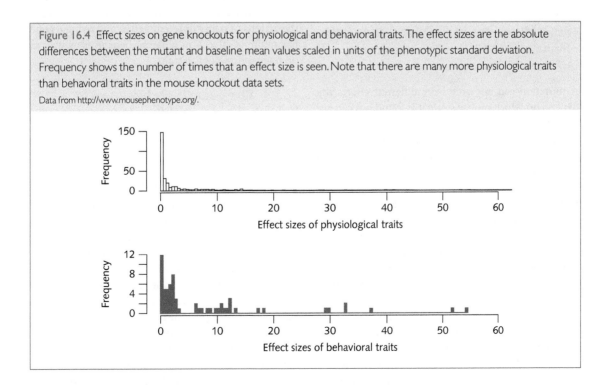

Figure 16.4 Effect sizes on gene knockouts for physiological and behavioral traits. The effect sizes are the absolute differences between the mutant and baseline mean values scaled in units of the phenotypic standard deviation. Frequency shows the number of times that an effect size is seen. Note that there are many more physiological traits than behavioral traits in the mouse knockout data sets.

Data from http://www.mousephenotype.org/.

This matters a lot because, remarkably, the strain background can determine the nature of the phenotypic consequences of the knockout.

This point needs stressing. Although it has been known for decades to mouse geneticists, it still raises eyebrows when people realize what is being said. Mouse geneticists have often found that a mutation lethal in one strain may appear normal in another. Genetic variants that differ between the two strains either compensate for the mutation or aggravate its effect.

Ignoring strain background effects can easily lead to misunderstanding about the effect of a mutation. One common line of thinking goes as follows: I have just identified a gene that, in humans, is involved in a psychiatric disease (let's say bipolar disorder). I'll make a knockout in a mouse and see what it does! Abe Palmer, a mouse geneticist, described what happens if you do this. He tested the effect of a mutation in a gene called *CACNA1C* (believed to be involved with bipolar disorder; Ferreira et al. 2008) on 30 inbred strains (Sittig et al. 2016). As expected, the phenotypes depended on the genetic background, but the extent of variation was surprising. In several cases, the same mutation had completely opposite effects. You could conclude that the genetic mutation either made bipolar disorder worse, or made it better.

16.3.3 Mutations in the same gene have effects on multiple, often apparently unrelated, phenotypes

A mutation can not only have different effects that depend on which mouse strain it is found but can also give rise to different phenotypes. The latter phenomenon is called pleiotropy, a term whose meaning we explain with an example. Mice that are homozygous for deletions of the *Nf1* gene (encoding neurofibromin 1) die before birth, as a result of a cardiac vessel defect. Heterozygous ($Nf1^{+/-}$) mice live, and have learning and memory impairments (Costa et al. 2002). They also develop two forms of cancer (leukemia and pheochromocytoma, a tumor of the adrenal gland), which are different from the neuronal tumors that the mutation also causes. If you look in the IMPC data set (under MGI:97306, the code for the *Nf1* gene), you'll find that the consortium's phenotyping pipeline discovered abnormalities in liver enzymes (aspartate aminotransferase and alanine aminotransferase), indicating that there is something wrong with the liver of these mice. So this single gene affects the development of the cardiac system, the proliferation of nerve cells, the proliferation of blood and adrenal cells, and the liver, as well as impairing learning and memory. And, although we won't continue the list, there are additional phenotypes detected in humans carrying mutations in neurofibromin that have not (so far) been reported in mice. Multiple phenotypes arising out of a single mutation is an example of pleiotropy.

Pleiotropy is the norm for the phenotypic consequences of knockouts. The report of the first 449 knockout genes asserts that '*Sixty-five percent of lines (290/449) had more than one phenotypic hit. Overall, pleiotropy was effectively demonstrated with the pipelines used*' (De Angelis et al. 2015). I counted the number of phenotypes associated with each gene in the IMPC set and found evidence of pleiotropy in almost 80% of genes that altered behavior. This result depends on the power, which with a sample size of 14 for each mutant only detects relatively large effects. With larger sample sizes, it's possible that almost every gene will be found to show pleiotropy. Genetic effects on behavior are no different from those on physiological traits: they are all pleiotropic. In other words, there is no such thing as a gene for learning, a gene for memory, or a gene for fear. In agreement with the human genetic results, gene function does not fit neatly into our classifications of behavior. Behaviors result from the actions of many different genes, and a single gene has multiple effects on different behaviors as well as on many physiological traits.

16.3.4 The genetic architecture of behavior looks no different from the genetic architecture of many other physiological traits

The effects might be a little smaller and the pleiotropy a little greater, but overall there's nothing special about behavior. It's another complex trait, like height and weight, that is attributable to many genes, with the added complications that it's hard to measure, and there are no mutations with all-or-nothing effects.

Overall, the findings from screens of gene function using constitutive knockouts reinforce all the findings we have described for the genetic architecture of behavior: pleiotropy is the norm, a behavior is affected by multiple different genes, genetic effects are quantitative rather than qualitative, and effect sizes are never all or nothing. However, although we've used results from gene knockouts to make points about the genetic architecture of behavior, that's not the reason they were created. Rather, they were generated as a resource to explore brain function, as we explain next.

16.4 Genes involved in learning and memory

Seth Grant, now a neuroscientist in Edinburgh, and Alcino Silva, now at the University of California, Los Angeles, independently introduced gene knockouts to neuroscience. I asked both of them how this happened, and both told me they saw the opportunity, the transformational potential, that molecular genetics was going to have on the learning and memory field. After all, if learning and memory wasn't the key problem in biology, what was?

Silva was working with Ray White, a human geneticist at the University of Utah in Salt Lake City, in charge of an enormous laboratory trying to map the human genome. Mario Capecchi's laboratory was (and still is) in Utah, and so Silva was introduced early to genetic engineering technology, and thought '*This is going to change biology!*' Having no background in neuroscience, no training in making mutants, and certainly none in mouse behavior, Silva decided he was going to use gene knockouts to discover the molecular basis of learning and memory ('*I was naïve!*'). About the same time, Susumu Tonegawa, after receiving the 1987 Nobel Prize for discovering the genetic mechanism underlying the production of antibody diversity, enrolled in an undergraduate course in neuroscience at the Massachusetts Institute of Technology (MIT). Silva heard about this unexpected behavior and decided to write to Tonegawa, telling him he wanted to work in his laboratory and start the knockout work. Tonegawa agreed.

I think he had me in his laboratory for amusement. You know, he called me from the airport about a paper he had just seen in *Science* that claimed genes in the brain were undergoing the same re-arrangements he had described in the immune system. He told me to work on this problem, and was gone for four months. When he got back he asked me what I had found out. I told him 'Susumu, I didn't work on that at all. I've been working on the most important problem in neuroscience'. He looked at me, said 'OK' and never mentioned it again.

A. Silva (personal communication).

Grant approached Eric Kandel with the same suggestion, to use homologous recombination to create knockouts of genes thought to be critical to learning and memory.

Silva and Grant analyzed genetic knockouts in the hope that they could show, one way or another, the role that a gene played in a particular neurobiological process. In some ways, this is just like a candidate-gene study: we think a protein is involved in memory, so we create a gene knockout and test that hypothesis. But there are two differences. First, the prior evidence for the theories is (usually) much stronger than is the case in the human genetic studies, and second the power to detect an effect is much greater, on one hand because the genetic effect is larger and on the other because other confounds can be eliminated (or at least better controlled). For example, one critical advantage of the animal experiments is that they use inbred strains: the comparison is between animals of the same genetic constitution, except for the mutation.

The candidate genes that both Silva and Grant picked encode enzymes called kinases, whose job is to add phosphate groups to protein, and phosphatases, whose job is to remove phosphate groups. The reversibility of the kinase / phosphatase reaction had been a contender as a key component of memory for some time, because of the possibility that it could act in a switch-like manner (Lisman 1985). How, though, could kinases, and other candidate genes, be tied to learning and memory? To understand what was done next, we need to delve into some neurobiology.

Up to now, you haven't needed to know anything about the biology of behavior to appreciate the genetic studies. In fact, the value of the genetic methods is that they make no such assumptions; they are theory free (apart from assuming that the phenotype has a genetic basis). You need to know about segregation, about heritability, about the different types of sequence variation, and the way that genetic association is carried out. But knowledge about what the brain is made of, or how it might work, was not required. However, to appreciate the experiments carried out using knockouts to determine brain function, you do need to know some neuroscience. The next section is a brief primer. Feel free to skip the primer if you are already familiar with long-term potentiation / depression and other cellular correlates of memory.

16.5 A primer in the neurobiology of learning and memory

Our brief history of neuroscience begins with the discovery that a specific type of cell (the neuron) is the functional unit of the brain (the alternative view was that there was a continuous connecting membrane arranged like a set of wires), continues with the discovery that specialized connections exist between neurons (the synapse), that chemicals carry information from one side of the synapse to the other (neurotransmitters), and culminates in possibly the most influential theory in neurobiology, the ionic hypothesis (proposed by Alan Hodgkin, Andrew Huxley, and Bernhard Katz), which explains information flow through neurons as a change in electrical potential, called an action potential (for a wonderful introduction to these ideas, I refer you to ALBRIGHT et al. 2000). Neurons receive input at their dendrites, and then pass on information through their axons, in the form of

action potentials. Neurons have many inputs, the dendrites, but only one output, the axon. Figure 16.5 provides an introduction to the components of a neuron.

Electrical impulses are the language of the nervous system. These tiny signals can be measured directly in neurons by means of microelectrodes, hollow glass needles pulled to a fine tip ($\sim$10 µm) that are inserted into the neuron without killing it. The microelectrode contains a salt solution and a fine wire that connects to an amplifier. With another wire in the fluid surrounding the neuron, the miniscule difference in electrical potential between the inside of the cell and the outside can be accurately measured. The cell's membrane, full of lipids, provides the insulating barrier for this 'membrane potential', which amounts to about -70 millivolts (mV). The negative sign on the membrane potential is due to active pumps in the membrane that extrude positively charged sodium and calcium ions, and to the preponderance of negatively charged proteins kept inside the cell. Most importantly, this membrane potential

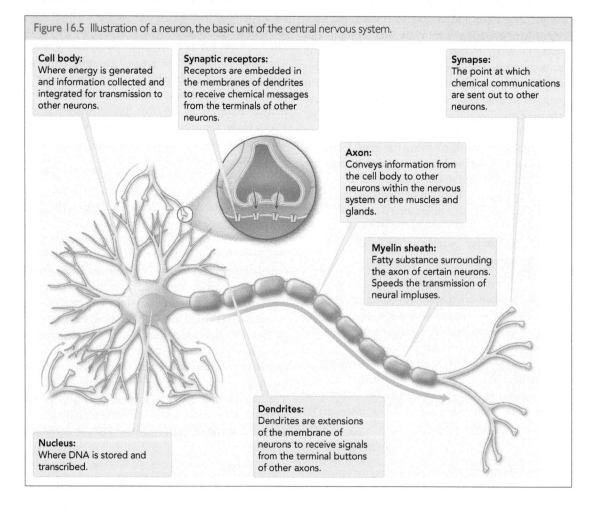

Figure 16.5 Illustration of a neuron, the basic unit of the central nervous system.

Cell body:
Where energy is generated and information collected and integrated for transmission to other neurons.

Synaptic receptors:
Receptors are embedded in the membranes of dendrites to receive chemical messages from the terminals of other neurons.

Synapse:
The point at which chemical communications are sent out to other neurons.

Axon:
Conveys information from the cell body to other neurons within the nervous system or the muscles and glands.

Myelin sheath:
Fatty substance surrounding the axon of certain neurons. Speeds the transmission of neural impulses.

Dendrites:
Dendrites are extensions of the membrane of neurons to receive signals from the terminal buttons of other axons.

Nucleus:
Where DNA is stored and transcribed.

changes when the nerve cell conducts an impulse, an action potential that travels down an axon.

Action potentials move along neurons, but, generally speaking, they don't jump directly from one nerve cell to the next neuron in the chain. The synapse intervenes. Communication between neurons occurs at synapses and uses the same depolarization mechanism. However, this time it is not an electrical signal. Instead, there is a set of small molecules, known as neurotransmitters (e.g. glutamate, serotonin, dopamine, GABA) that are released from the end of the axon and diffuse over to the nearest part of another neuron (often a dendrite) where they bind to receptors. The effect is to open ion channels that initiate new electrical signals, which can then propagate a new action potential. Chemical neurotransmission at synapses is highly regulated and occurs in milliseconds.

Synapses are an obvious place to regulate the flow of information. They are where 'behavioral integration' could occur, a phrase coined by Charles Scott Sherrington, one of the pioneers of neuroscience. Donald Hebb proposed that the information flow between neurons depends on the amount of traffic (HEBB 1949). Hebb's suggestion was that a neuron pays more attention to synapses that are relatively more active. This is a form of cellular memory, and Hebb's rule has become a basic algorithm for learning: '*When an axon of cell A is near enough to excite cell B and repeatedly or persistently takes part in firing it, some growth process or metabolic change takes place in one or both cells such that A's efficiency, as one of the cells firing B, is increased.*' For a long time, there was no evidence of any such change happening in the brain, until in 1973 Tim Bliss and Terje Lømo reported the discovery of long-term potentiation (LTP) (BLISS AND LØMO 1973).

LTP refers to a sustained increase in synaptic transmission following electrical stimulation to a neuronal pathway. High-frequency stimuli produce synaptic strengthening (LTP), while low-frequency stimulation produces synaptic weakening (long-term depression).

LTP and depression have three features that caught the imagination of neuroscientists. First, the timescale over which they occur is long enough to be useful for information storage: LTP lasts days to weeks. Second, inputs that fired at the same time as the stimulus were potentiated, whereas inputs that weren't so coordinated were not potentiated. In other words, it obeyed Hebb's rule that 'cells that fire together wire together'. And finally, Bliss and Lømo identified LTP in the hippocampus, a region of the brain known to be involved in learning and memory (O'KEEFE AND DOSTROVSKY 1971), suggesting that LTP was the cellular correlate of memory.

For LTP to occur, the postsynaptic cell must be depolarized during synaptic activation. In other words, multiple signals arriving at a neuron will make the neuron more likely to fire. To give this behavioral relevance, you can imagine a neuron receiving one input that signals the animal has heard a tone, and immediately afterward a second action potential arrives, triggered by the animal receiving a mild electric foot shock (this is the conditioned fear paradigm that is used a lot to investigate how mice lay down memories). How is LTP induced at synapses that are active during neuronal depolarization? The answer lies in the discovery of a molecular coincidence detector, the NMDA receptor.

Graham Collingridge discovered the importance for LTP of a subtype of receptor for glutamate (glutamate is the major excitatory neurotransmitter in the brain). Glutamate receptors are channels in the cell membrane that open when glutamate binds to them, allowing ions of potassium, sodium, or calcium to travel through a central pore in the receptor. The receptors come in different types, recognized initially by their differing response to drugs. Collingridge discovered that the compound N-methyl-D-aspartate (NMDA) had dramatic effects on hippocampal glutamate receptors (BLISS AND COLLINGRIDGE 1993). Collingridge found that there was another non-NDMA glutamate receptor that also mediated synaptic transmission. This subtype is called the AMPA (α-amino-3-hydroxy-5-methyl-4-isoxazolepropionic acid) receptor. Understanding the relationship between the two receptors, AMPA and NMDA, was to give rise to the discovery of how coincidence detection could occur at a synapse.

It works as follows (Figure 16.6): when an action potential arrives at a synapse, glutamate is released into the synaptic cleft (the gap between the two cells at a synapse) and binds to AMPA receptors that depolarize the postsynaptic membrane. Before depolarization, magnesium blocks the NMDA receptor from opening. Depolarization unblocks the magnesium (MAYER et al. 1984; NOWAK et al. 1984), and now, if another action potential happens to arrive, the glutamate that is released will bind to the NMDA receptor, which allows calcium to enter the cell (MACDERMOTT et al. 1986). In other words, the NMDA receptor opens only when glutamate binds to the receptor and when the neuron is depolarized. This is the critical point: the NMDA receptor acts as coincidence detector (two events are required to make it responsive), exactly the kind of change required for synaptic plasticity, the Hebbian basis of learning and memory.

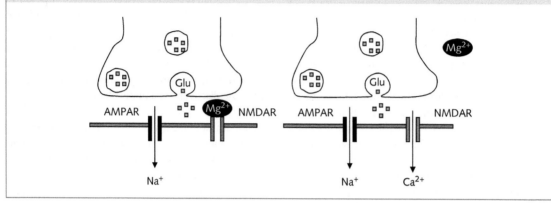

Figure 16.6 A coincidence detector at the synapse: a molecular basis for learning and memory. On the left, the first action potential to arrive at the synapse causes glutamate (Glu) to be released into the synaptic cleft where it depolarizes the membrane through its action on AMPA receptors (AMPAR). Before depolarization, magnesium ions (Mg^{2+}) block the NMDA receptors (NMDAR). After depolarization (right of the figure), the magnesium block is released so that now when glutamate is released it can bind to and activate the NMDA receptor with the consequence that calcium ions (Ca^{2+}) enter the cell.

Calcium influx, through NMDA receptors, is required for the induction of LTP. What calcium does after it enters the cell is complex and not fully understood, but some of the proteins that it binds to are central to our current understanding of the molecular basis of memory. One of these is Ca^{2+}/calmodulin-dependent protein kinase II (CaMKII), which has an extraordinarily high level of expression in neurons, making up about 2% of the total protein (Lisman et al. 2002).

CaMKII is activated by Ca^{2+} entry through the NMDA receptor, and is necessary and sufficient for the induction of LTP. It is enriched at synapses in a region of the neuronal membrane where neurotransmitter receptors sit, together with other proteins involved in converting neurotransmitter binding into signals that alter cellular function. This membrane region, called the postsynaptic density because of its appearance under an electron microscope, is a sort of cellular machine, with hundreds of components, many of which are implicated by genome-wide association signals as involved in autism and schizophrenia.

If one role of the hippocampus is to remember where the animal is, and if the only contender for the cellular basis of memory, LTP, is found in the hippocampus, a critical experiment is to determine whether the two are causally related. To explore the relation between the cellular phenomenon and behavior, Richard Morris developed a spatial memory task, now forever known as the Morris water maze (Morris et al. 1982). The Morris water maze isn't a maze at all (at least, not the sort of maze I think of). Richard Morris's water maze is a large circular pool, about 5 feet in diameter, containing what looks like milk (the water is made opaque so the animal can't see what's hidden under the surface). The sides are vertical and tall enough to stop the animal jumping or climbing out. The only refuge is one small region of the pool where the water is very shallow; here the rat can stand on a hidden platform. Using opaque, milky water to hide the platform means the rat can't see it or smell it.

When a rat is put in the pool, it swims around until by chance it reaches the relative safety of the hidden platform. Replace the rat in the pool and the animal finds the platform a little quicker, as it now has some idea of the hidden location. Keep training the rat in this way and it will eventually learn where the platform is, swimming quickly toward it. By starting the rat randomly at different points in the pool (south, east, north, or west) and providing fixed visual cues (shapes such as stars and squares stuck onto the poolside) as spatial reference points, it's possible to ensure that the only way the rat can solve the maze is by remembering the platform's spatial position. It's as if the animal says to itself, the platform is to the left of the star and to the right of the square, just as we would use a map to triangulate on a position.

The psychologists' way of asking the rat if it has remembered the platform's position in the maze is to remove the platform (which, of course, the rat won't notice because the water is opaque) and then put the rat into the pool. The rat with intact spatial memory will swim to where it thinks the platform is located and persist in swimming over the missing platform (you can imagine what it must be thinking). A rat that doesn't remember, or hasn't learnt the task, may swim over the missing platform but won't waste time swimming in pointless circles around one region.

Richard Morris used AP5, the same drug that Graham Collingridge had used to block NMDA receptors, to turn off LTP in the hippocampus (MORRIS et al. 1986). Using two versions of the maze, one with a hidden platform below the water and one with a visible platform, Morris showed that drugged animals failed to learn the location of the hidden platform only, indicating that NMDA receptor inactivation specifically inhibited spatial learning. This selectivity of effect on learning the task implies a direct action on memory.

Morris's paper was one important piece of evidence for a relationship between LTP and spatial memory. However, as Morris pointed out, these experiments do not prove the assumption that AP5 causes its effects on learning and memory by blocking hippocampal synaptic plasticity. *'The logical difficulty is as follows: blockade of LTP by AP5 is a dependent consequence of the drug treatment. The impairment of learning is also a dependent consequence. However, it is fallacious to presume that that one dependent consequence is necessarily the cause of the other'* (MORRIS et al. 1989). A robust demonstration of the presumed relationship between LTP and memory required another way of interfering with the molecular machinery. This was the attraction of the genetic strategies.

16.6 Memory genes

From the description I've given of learning and memory, you'd be forgiven for thinking that the first candidate for the gene-targeting technology would be one of the glutamate receptors, NMDA receptor or AMPA receptor. But at the time, in the 1980s, very little was known about these receptors, certainly not enough to make it possible to target them for genetic deletion. In any case, as was found later, removing NMDA receptors is lethal as they are used in the midbrain to control breathing. Instead, as we mentioned earlier, the first knockouts were made for enzymes involved in adding and removing phosphate groups (kinases and phosphatases). Mutations were later engineered for NMDA and AMPA receptors, but we don't review that story here.

Seth Grant, believing that kinases had to be important in transforming the signal from the NMDA receptor into LTP, set out to look at mice with deleted kinases.

Working with Eric Kandel and Tom O'Dell, he began to examine the knockouts and found that one mutant, *fyn* kinase, did indeed show a learning abnormality (GRANT et al. 1992). At the same time, Alcino Silva, working with Susumu Tonegawa at MIT, used gene targeting to show that CaMKII was required for the induction of LTP (SILVA et al. 1992a, b). Both Silva and Grant also showed that the knockout animals were deficient in spatial memory, using the Morris water maze.

Among the kinases, CaMKII stood out as a candidate for the role of a memory gene. As its name suggests, the kinase activity of CaMKII depends on the levels of calcium in a cell bound to a protein called calmodulin (recall that the NMDA receptor allows calcium to enter the cell). As NMDA receptors let calcium into a cell, and calcium had to act somewhere to change the cell's responsivity, Silva's demonstration of the effect of the mutation argued that CaMKII played a critical role downstream of the NMDA receptor in turning synaptic activity into a memory.

The experiments that Silva and Grant carried out with the mouse gene knockouts showed that the knockout animals were deficient in LTP and had deficits in memory. Their papers were immensely influential, establishing a revolution in neuroscience so that the ensuing years saw a flood of publications testing the involvement of genes in learning and memory (and, indeed, in almost any behavior that could be measured in mice).

As of 2018, there are some 200–300 gene knockouts that show an effect on LTP and memory, an impressive body of data that confirms that the stable long-lasting changes in synaptic function alter learning and memory. As Alcino Silva stressed: *'That's a BIG deal. How many other theories in neuroscience have such deep support?'* (SILVA, personal communication). But beyond establishing the relationship between LTP and memory, how did these results answer the questions about the mechanism itself? We will take one example, that of a gene called *CREB*, to see what has been learnt, and, at the same time, explore the limitations of the reverse genetics approach.

16.6.1 CREB: an example of a genetic toolkit for behavior

Soon after he had published the results of the first knockouts, Alcino Silva heard about an intriguing result. CaMKII, which he had shown was involved in learning and memory, adds phosphate to a transcription factor called cyclic AMP (cAMP) response element-binding protein (CREB). Transcription factors bind to DNA and usually activate the transcription of many genes, and CREB was no exception: it binds to the cAMP response element (CRE), and CRE sites are found within the regulatory regions of numerous genes. This provided a potential pathway in which events occurring at the neuronal membrane, such as activation of a neurotransmitter receptor, could cause the phosphorylation of CREB and trigger its transcriptional activity, which affects many genes, leading to changes that result in the laying down of a memory.

The involvement of CREB in memory function might explain how activity at the synapse could be transformed into changes in gene expression and illuminate how electrical activity is converted into stable synaptic changes that support memory. Silva found that, unlike the CaMKII mutant, the CREB knockout had no effect on learning and LTP when tested in the first 30 minutes. However, when the knockout animals were tested 1 hour after training, their memory was found to be disrupted (BOURTCHULADZE et al. 1994). This finding was important because it demonstrated that it was possible to disentangle long-term memory processes from both short-term memory and the myriad of sensory processing mechanisms required for the acquisition of information.

The results exemplify the first point we made above about mutants: the genetic effects on behavior are quantitative: the CREB knockouts don't simply fail to learn, they just don't learn so well. Figure 16.7 shows results for controls (wild type) and CREB knockout mice after being trained in a task in which a foot shock is associated with a tone; the y-axis is the amount of freezing as a percentage and the x-axis is the time (in minutes) after the mice are exposed to the tone. Memory for the task is indicated by the amount of time spent freezing when the animal hears the tone after the training session. This is the result of an experiment in which animals were tested 1 hour after training.

Figure 16.7 shows that the control animals freeze more than the knockouts, and the difference is highly significant (the error bars for each data point show the standard error of the mean; see Appendix, section A8 if you are not sure what this means). It's important to realize that the knockouts have not completely lost the freezing response (BOURTCHULADZE et al. 1994). They still show 20% freezing, and the response decays with time, as it often does, if they have learnt the response.

The second lesson about the effect of CREB on memory is that while some investigators agreed

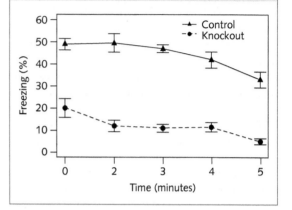

Figure 16.7 Freezing response in CREB knockout mice (circles) compared with controls (triangles). The x-axis is time (in minutes) after the mice are exposed the cue and the y-axis is the percentage of time spent freezing after the cue. Testing was carried out 1 hour after training. Data from BOURTCHULADZE et al. (1994).

that disruption of CREB function impairs LTP and memory (PITTENGER et al. 2002), not everyone did. The title of one such paper tells the story: 'Does cAMP response element-binding protein have a pivotal role in hippocampal synaptic plasticity and hippocampus-dependent memory?' (BALSCHUN et al. 2003). The authors reported no difference between CREB knockout mice and controls. Before you cry out that this looks like another example of the difficulties that plagued the field of candidate-gene studies, there is an important difference between the two studies. The knockouts used were not the same! The first CREB mutation described above was created with classic homologous recombination approaches and resulted in the deletion of two forms of CREB (α and Δ) in all cells of the mouse, including all brain cells. The mice that Balschun and colleagues studied used a strategy we describe in Chapter 16, section 16.7, which results in CREB being deleted in some brain cells, while leaving others unaffected (BALSCHUN et al. 2003). Interestingly, studies that would be published years later showed that CREB not only controls the consolidation of memory but also regulates which cells in a specific brain region go on to store a given memory, a property called memory allocation (SILVA et al. 2009): in a given brain region, cells with higher CREB activity are the easiest to activate (ZHOU et al. 2009), and therefore they are the ones recruited to store memories. So, in the studies by Balschun and colleagues (BALSCHUN et al. 2003), the memories presumably went to the cells with normal CREB function, therefore appearing as if the deletion of CREB had no effect. This example brings us to the third, and most important, lesson.

The third lesson from CREB is that genetic effects of knockouts depend on the tissue, cell type, and developmental stage of the animal. Constitutive deletions are by nature of prolonged duration (essentially throughout the animal's lifetime) and affect all tissues. This not only allows sufficient time for any significant compensatory adaptations to occur but also means that effects in other unrelated systems, such as in other tissues, may confound the effects attributable to deletion in the nervous system. Rob Malenka, a neuroscientist at Stanford, raised this concern, and showed how it could be alleviated by transiently increasing the expression of the *CREB* gene. He did this using a virus containing the *CREB* gene to infect a small region of the brain, and showed that activation had the expected consequences on neurotransmission (MARIE et al. 2005). This led him to conclude that CREB was indeed likely to be involved in synaptic plasticity.

There is one last point to make about the interpretation of the constitutive genetic *CREB* knockouts, and this is that they are part of a larger literature on the role of CREB, considerably strengthening the evidence in favor of the role of the gene in memory. This point touches on the importance of having a strong prior hypothesis for interpreting genetic experiments (which, as we explained in Chapter 5, section 5.2.1, was not true for human candidate-gene studies of psychiatric illness). The molecular function of CREB is conserved across species, in sea slugs and in flies, as we describe next.

16.6.2 *CREB* is a memory gene in sea slugs and flies

When a wave falls on the Californian sea slug, *Aplysia californica*, the animal learns to quickly close its gill opening. The neuronal circuitry is simple—a set of sensory neurons capable of detecting water on the gill connects to a set of motor neurons capable of driving muscles that withdraw the gill. If the gill is repeatedly stimulated, *Aplysia* becomes more blasé and gradually responds less and less. On the other hand, if something alarming happens, such as a bite or a blow to some other part of its body or an electric shock, the snail becomes hypersensitive to any kind of gill stimulation. A single shock gives rise to a memory lasting only a few minutes, but four or five spaced shocks give rise to a memory that lasts several days. Thus, as in mice, and in our own species, there is both short- and long-term memory.

The psychiatrist Eric Kandel showed that a neurotransmitter (serotonin) binding to its receptor activates an enzyme (adenylyl cyclase) to produce a diffusible chemical signal (cAMP), which in turn binds to a protein called protein kinase A (PKA). PKA then phosphorylates yet another enzyme, mitogen-activated protein kinase (MAP kinase), and together the two of them act on CREB. When a phosphate group is added to CREB, the transcription factor enters the cell's nucleus and begins a process of altering which genes are active in that cell. In other words, it changes the proteins that the cell makes, producing lasting changes to the cell's synapses (including growing new ones) and, as a result, induces a long-term form of sensitization. Kandel argued that the formation of these new synapses underlies long-term memory (KANDEL 2001). Kandel won a Nobel Prize for his work on learning and memory in 2000.

CREB is just as important in flies as it is in mice. When an odor (or image) is accompanied by an electric shock, or a heat lamp, flies quickly learn to avoid it; the memory of this association will last for a day at most. If multiple training trials are delivered all at once, the memory still only lasts for a day. But if the training trials are interrupted with 15-minute rest intervals, then the flies will remember it for up to a week (WADDELL AND QUINN 2001). This is a substantial portion of a fruit fly's lifespan. Remarkably, flies with deficiencies of CREB have poor memory (YIN et al. 1994). There is even a human intellectual disability syndrome due to mutations in the CREB-binding protein (*CBP*) gene (PETRIJ et al. 1995).

The story of CREB illustrates an idea about gene function called the genetic toolkit hypothesis. The genetic toolkit for memory consists of a set of genes with specialized functions, coordinated via transcription factors such as CREB, which produce the behavioral phenotype. The idea can be expanded to incorporate a view of behavior as modular. Just as differences in the form and function of limbs arise from the adaptive specialization of limb segments, so too can complex behavior be viewed as being composed of behavioral modules.

One last comment on CREB: there are three forms of CREB (α, β, and Δ—and no, I don't know what happened to γ). When all three forms are deleted, the mutant mice are small and die immediately after birth from respiratory distress (RUDOLPH et al. 1998). How, then, could the creators of a full CREB knockout, with all three forms missing, test its memory? The answer is that the mutation they made is not constitutive, it is inducible, so that the gene was knocked out only in the brain, and it was produced using a technology we describe next.

16.7 Inducible mutations

Inducible mutations are those in which the animal is genetically engineered that a gene is poised, ready to be deleted at the wish of the investigator. Ideally, this can happen in any cell type or tissue, and at any time (although we have not quite realized this ideal yet). We describe two of the most common: Cre-Lox and Tet-inducible mutations.

16.7.1   **The Cre-Lox system**

The Cre-Lox system uses an enzyme that cuts DNA by recognizing a target sequence. By introducing, via homologous recombination, target sequences around

a gene, the enzyme can be directed to cut out that gene (or, indeed, any piece of DNA the investigator wants to remove). The enzyme is called a Cre recombinase, which recognizes 34 bp *loxP* sequences, so you will see this system referred to as the Cre-Lox system; to confuse you further, you will also come across references to 'floxing', which means the same thing, namely an allele is engineered so that it is flanked by two sites that the recombinase recognizes and cuts (Figure 16.8). There is also a system in which a Flp recombinase is used instead of Cre, but the principal is exactly the same.

The trick to making a deletion specific to a tissue is to express the enzyme that does the cutting only in the cell type or tissue of interest. That's hard. In fact, it's still not really been achieved, as we don't know which DNA sequences to use to fully control gene expression. Joe Tsien in Tonegawa's laboratory who published the first incarnation of this approach (there are now lots) used a segment of DNA from in front of the αCaMKII gene, which confines expression to neurons in the forebrain (TSIEN et al. 1996a). Tsien used the approach to target mutations of the NMDA receptor to the forebrain (thus ensuring that the mutation was

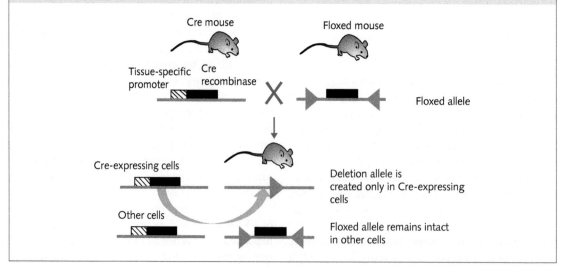

Figure 16.8. Cre-Lox strategy for generating a conditional gene knockout. Two mice with different genetic alterations are required, and the system is thus called a two-component system. The floxed mouse has two 34 bp *loxP* sites, introduced by homologous recombination in embryonic stem cells, which are placed on either side of an essential exon of the gene of interest. The Cre mouse is a transgenic mouse, produced by nonhomologous recombination: injection of a complementary DNA encoding Cre recombinase under the control of a tissue- or cell type-specific promoter. Crossing the floxed mouse with the Cre mouse results in deletion of the floxed exon only in Cre-expressing cells or tissue. Black rectangle, exon of the targeted gene; gray triangles, *loxP* sites; hatched rectangle, promoter driving the time and site of recombination.

Cre mouse

Floxed mouse

Tissue-specific promoter   Cre recombinase

Floxed allele

Cre-expressing cells

Other cells

Deletion allele is created only in Cre-expressing cells

Floxed allele remains intact in other cells

not lethal) and demonstrating that synaptic plasticity in the hippocampus, mediated by NMDA receptors, was essential for the acquisition of spatial memory (TSIEN et al. 1996b).

The idea that a single sequence, such as the promoter of the αCaMKII kinase gene, would provide the required level of control of expression turns out to be naïve (as is clear from our discussion of the regulation of gene expression in Chapter 4, section 4.2). However, while the ability to sequence the RNA of a single cell (TANAY AND REGEV 2017) demonstrated the complexity of gene expression in neurons, single-neuron sequencing has also identified candidate marker genes to classify cells (ZEISEL et al. 2015; SHEKHAR et al. 2016; TASIC et al. 2016). These developments mean that DNA markers sufficient to characterize specific cell types, and possibly to target engineering to them, may become available (ZENG AND SANES 2017).

At the time of writing (beginning of 2019), the ability to identify the DNA sequences in promoters and regulatory elements that confer cell and tissue specificity hasn't yet been achieved, at least not in mammalian systems (gene-expression control is less complex in flies and worms).

An alternative is to put the components of the genetic engineering system, CRISPR/Cas9 or the Cre recombinase, into a virus, which is then injected into the cells of interest in the brain (this is the approach used in Rob Malenka's experiment we described above). Adeno-associated viruses (AAVs) provide long-term expression without genomic integration, are relatively safe and are nonpathogenic, and importantly some forms preferentially infect neurons (such as AAV2) (MURLIDHARAN et al. 2014). However, the amount of foreign DNA that can be incorporated into AAVs is limited, making it difficult, for example, to include Cas9 for AAV-mediated delivery. AAV delivery doesn't provide the precision necessary to target only a specific cell type, and cells in the vicinity of the injection are also going to be infected, so the method is not a panacea. However, cell-specific targeting has in a small number of cases been achieved with viruses carrying short regulatory sequences.

16.7.2 Tet-inducible systems

The 1990s also saw the introduction of the first systems by which mutations could be induced at any time the experimenter wanted. Hermann Bujard and Manfred Gossen developed a system that exploits the ability of an antibiotic (either tetracycline (Tet) or a very similar antibiotic, doxycycline) to complex with a protein (called a transactivator protein, or tTA). A transactivator bound to antibiotic has an altered ability to recognize and bind to a specific DNA sequence (called a tetracycline operator, or response element) (GOSSEN AND BUJARD 1992), which in turn alters the expression of a gene of interest. If that gene of interest is the Cre recombinase, a deletion can be engineered depending on whether or not animals are fed the antibiotic. The two commonly used implementations of the system, called Tet-Off and Tet-On, are illustrated in Figure 16.9.

I should warn you that the system is hard to understand because (at least to me) it always seems back to front: in the Tet-Off system, doxycycline is used to repress transcription of the targeted genes (the presence of the antibiotic ensures the genes are **not** expressed) (Figure 16.9). The antibiotic binds the tTA protein, so that it **cannot** target the tetracycline response element sequences, thereby **preventing** activation of the genes under control of the tetracycline response element. In the Tet-On system (Figure 16.9), a different protein, called reverse tTA, binds the operator only when bound by doxycycline. When animals are fed doxycycline, gene transcription begins. There are alternatives to the Tet system, including tamoxifen to bind to estrogen-response elements, although the system is typically less efficient than Tet and is not so widely used.

16.7.3 Phenotypes of inducible mutants

The ability to generate inducible mutations overcame the limitation that genetic effects on behavior depend on when and where in the brain a gene is expressed. It meant that investigators could demonstrate what a gene did in a specific tissue or cell type at a specific time. For example, if CREB is involved in long-term but not short-term memory, then temporally inducible mutations should be able to distinguish its effects at two time points. In essence, this is what Alcino Silva showed, using a genetically engineered mouse in which CREB could be reversibly inactivated (he used a tamoxifen-responsive element rather than the doxycycline method but the principle is the same) (KIDA et al. 2002). Early inactivation repression, within 2 hours of training an animal to associate a cue with a foot shock, does not interfere with memory, but later repression (up to 12 hours) does.

Figure 16.9 The Tet-inducible system. There are two variations, but the overall design is the same: one animal is engineered to carry a tetracycline (Tet)-controlled protein under the control of a tissue- or a cell type-specific promoter. The second animal is engineered to carry a response element so that the tetracycline-controlled protein will drive the expression of a target gene in the offspring of the two animals. In the Tet-Off system (top half of the figure), the tetracycline-controlled protein is a transactivator (tTA), encoded by a gene called *tetR*. In the absence of the antibiotic (doxycycline (Dox) is used rather than tetracycline), tTA binds to the tetracycline operator sequences (TetO) and activates a promoter to transcribe the target gene. When the animals are fed Dox, tTA undergoes a conformational change and cannot bind to TetO. This turns off expression of the target gene. In the Tet-On system (bottom half of the figure), the tetracycline-controlled protein is a reverse tetracycline-controlled transcriptional activator, which this time binds to TetO only in the presence of Dox. When Dox is not present, rtTA can't bind to TetO so target gene expression is quenched.

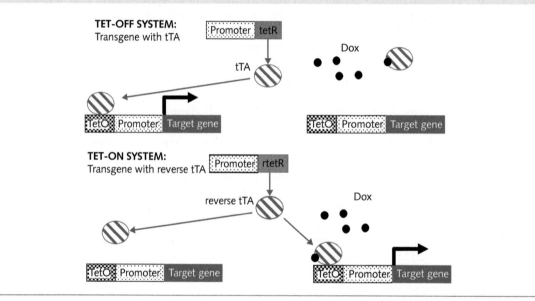

The temporal importance of the effect of mutations was made even clearer in an experiment in which Rene Hen analyzed an inducible mutation of the serotonin$_{1A}$ receptor (GROSS et al. 2002). Drugs that act at the serotonin$_{1A}$ receptor are used to treat anxiety, and mice with constitutive deletions of the receptor (5-HT1AR knockout) have increased anxiety-like behavior (RAMBOZ et al. 1998). Hen asked what would happen if the receptor was deleted after birth. He used the Tet-Off system so that feeding mice doxycycline abolished receptor expression, and omitting doxycycline turned on receptor expression. With an αCaMKII promoter to target the mutation to the forebrain (don't forget the caveats we raised about the efficacy of that strategy in section 16.7.1), he compared the effects of turning the receptor off in the brain during development with turning it off after birth. What he found was striking.

Figure 16.10 shows the results of testing anxiety-like behavior in an elevated plus maze (we described the elevated plus maze in Chapter 15, section 15.9.4 and illustrated it in Figure 15.15) (GROSS et al. 2002). The top part of Figure 16.10 shows what happens when the receptor is turned off during development (to achieve this, the pregnant mothers are fed doxycycline). On the right of the figure, the bars shows that the constitutive knockout (KO) mice are more anxious than the wild-type (WT, normal genotype) mice as they spend less time in the open arms of the elevated plus maze: the KO bar is lower than the WT bar. But the KO bar is the same height as the 'Rescue' bar, the latter representing the behavior of animals in which the mutation is induced after birth. In the lower panel you can see that when the mutation is induced after birth, there is **no** effect on the anxiety-like behavior: the 'Rescue' mice behave the same as the wild-type mice.

Figure 16.10 A mutation whose effects depend on when it occurs. Top panels: to turn off the 5-HT1AR receptor during development, mice were fed doxycycline and their offspring were weaned, and switched to normal food at 3 weeks after birth. The result of testing mice in the elevated plus maze is shown on the right for wild-type (WT), constitutive knockout (KO), and the inducible knockout (Rescue) mice. Bottom panels: to turn off 5-HT1AR in the adult, mice were treated with doxycycline for 2 months and tested in the elevated plus maze (results shown on the right). Asterisks over the bars on the right panel indicate the significance of the comparisons: $*P < 0.05$; $**P < 0.01$.

Reproduced with permission from Gross, C., et al. Serotonin 1A receptor acts during development to establish normal anxiety-like behaviour in the adult. *Nature*, 416: 396–400. Copyright © 2002, Springer Nature. https://doi.org/10.1038/416396a.

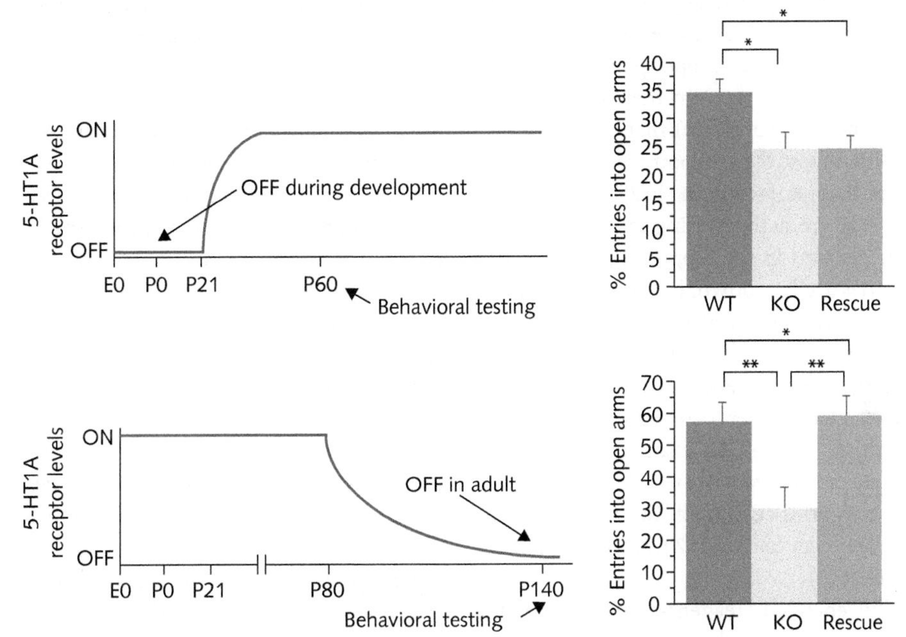

Genetic manipulations of CREB also exemplify the importance of the anatomical location of a mutation. They show how a deletion can have opposite effects, depending on the part of the brain from which the gene is removed.

As a transcription factor, CREB is expected to alter the expression of many genes. Exactly which transcripts are affected depends on the cell type in which CREB is expressed (in addition to the time at which it is expressed). CREB binds to a few thousand gene promoters in any given cell type, and the CREB transcriptome (the complete set of RNA transcripts that CREB influences) differs dramatically from cell type to cell type (CHA-MOLSTAD et al. 2004). When a virus delivers extra copies of CREB to the hippocampus, thereby increasing CREB, the effect is to reduce depressive-like behaviors (CHEN et al. 2001), but elevated CREB activity in the nucleus accumbens (a region of the

brain involved in reward mechanisms) produces depressive-like effects (BARROT et al. 2002). So CREB expression either makes the animal act depressed, or seem cheered up, depending on where in the brain the genetic manipulation is directed.

These examples show how inducible mutations expose the complexity of genetic action on behavior. However, they don't showcase the full power of genetic engineering, which has been immeasurably increased by the introduction of genetic methods that appropriate control over neuronal activity. The remarkable transformation in neuroscience that has grown out of molecular genetics is likely to stand as one of the greatest scientific advances of the twenty-first century. We next explain what the technology can do, and what it is teaching us about how genes influence behavior, in the last section of this chapter.

16.8 Genetically encoded reagents

Genetically encoded tools have revolutionized neuroscience. It's almost no exaggeration to say that genetically encoded reagents made it possible not just to come up with theories about how the brain works but also to test them, with interventions that turn neurons on and off. Using genetically encoded reagents is a bit like replacing a hammer and saw with a scalpel. Before the advent of genetically encoded reagents, neuroscientists were able to image the brain at ever greater resolution and ever deeper penetration, they could invent theories to explain their observations, but they could not carry out experiments with the necessary precision to test these hypotheses. Neuroscientists could see that neurons in the basolateral nucleus of the amygdala are active during fear-related stimuli, and that neurons in the hippocampus are active when memories are laid down, but to prove that the neurons are functioning to process fear or memory requires an interventional experiment. The greatest achievement of genetically encoded reagents is to give neuroscientists that power (LIMA AND MIESENBOCK 2005), an advance that has, for example, made it possible to activate or inactivate neurons at will, and even create false memories (CLARIDGE-CHANG et al. 2009; HAN et al. 2009; GARNER et al. 2012; LIU et al. 2012; RAMIREZ et al. 2013).

Genetically encoded reagents incorporate the methods that we have described so far: any tool that is produced by the transcription and then translation of a piece of DNA is a genetically encoded reagent. Cre recombinase is a genetically encoded reagent, as is the green fluorescent protein that is used to visualize gene expression and protein localization in neurons.

The crown jewel among the genetically encoded reagents are those that have delivered us control over activity of the nervous system, by making neurons fire action potentials (or suppress firing) when exposed to light (optogenetic tools: BANGHART et al. 2004; BOYDEN et al. 2005; LIMA AND MIESENBOCK 2005), heat (thermogenetic tools: KITAMOTO 2001), or chemicals (chemogenetic tools: MAGNUS et al. 2011). Optogenetic technologies have garnered the most interest (judging by the number of publications) due to their capacity to elicit action potentials that are time-locked to light pulses. They involve expressing on the cell surface a light-sensitive protein (derived from microbes). When light strikes the protein, a series of structural changes occur that result in ion transport, channel opening,

or interaction with signaling proteins. By opening a channel, the neuron can be sufficiently depolarized to fire off action potentials (ZHANG et al. 2011). Thus, neuronal firing is dependent on pulses of light (usually given via a laser implanted in the animal's head).

While the patterning of experimentally delivered light pulses may not precisely recreate a physiological neural code, it's closer than anything else currently available and has become a key tool in neuroscience (TYE AND DEISSEROTH 2012). The genetically encoded reagents expressed on neurons to make them light sensitive are called channelrhodopsins. They generate small currents and thus require extremely high expression levels to generate quantities of the channel sufficient to depolarize the neuron. Many copies of the genetically encoded reagent need to be delivered, which can be achieved by injecting the DNA into the cell in a virus (SJULSON et al. 2016). Thermogenetic approaches are useful in insects, where raising the external temperature is sufficient to alter gene expression. Chemogenetic strategies involve expressing novel receptors on the surface of the neuron and then giving drugs to control gene expression. An antibiotic is often used, or sometimes a designer drug. There is a great acronym for this: DREADD, for 'designer receptors exclusively activated by designer drugs' (MAGNUS et al. 2011).

One particularly important application of genetically encoded reagents is the labeling of neurons when they are active. Knowing which neurons are active when you are frightened tells you where in the brain fear is located. Suppose, once you know where those cells are, you label them in such a way that those cells can be turned on, or off. In theory, when the cells are activated, that would make you feel scared, or remove your fear. Or imagine you can label cells when a memory is laid down. Presumably, if those cells are activated then the memory will be resurrected. This is the power that combining cell labeling with optogenetic activation confers.

A central idea underlying the activity labeling approach is that a small set of genes is turned on, often within minutes, after a stimulus arrives at a neuron. A paper from 1986 proposed that following Ca^{2+} entry into the neuron, a calmodulin-sensitive kinase (e.g. CaMKII) would phosphorylate a transcription factor, initially positioned in the cytoplasm, causing it to move into the nucleus to activate gene expression (MORGAN AND CURRAN 1986). In other words, neuronal excitation is directly coupled to transcription, and, indeed, a set of genes has been identified that is expressed very soon after a neuron is activated. The genes are called immediate-early genes (in the literature, they are often referred to simply as IEGs), and they include c-fos and Arc (activity-regulated cytoskeleton-associated protein) (GUZOWSKI et al. 1999). Consequently, if you genetically engineer a mouse so that the promoter of c-fos or Arc is placed in front of a gene of interest, then that gene will be turned on when the cell is active. Neurons under the control of a specific transmitter, such as dopamine, can be captured using a similar approach (LEE et al. 2017). An active neuron can now be marked with fluorescence (if the promoter drives green fluorescent protein) or, more interestingly, you can mark a neuron so that it can be activated later by light (GARNER et al. 2012).

The toolkit of genetic engineering is now so pervasive that there is hardly any area of neuroscience to which it is not applied. Summarizing the findings is tantamount to writing a textbook of molecular neuroscience. For our interest (how genes influence behavior), there are two main threads.

The first is experiments that test the function of genes identified in human studies of psychiatric disease. Gene finding has been hard, so currently there are few such studies, and no general principles or understanding of pathological mechanisms have yet emerged. We give one example below of a mutation found in autism. The second thread includes experiments that use genetic approaches as a tool to discover how the brain works, trying to discover fundamental processes. From the dauntingly large literature, we give examples of what these findings tell us about how genes influence behavior (still relatively little, as you will see), and, perhaps more importantly, we introduce some insights that genetic-based technologies provide about brain function.

16.8.1 A 16p11.2 deletion that causes autism

There is a small deletion, about 600 Kbp in size, toward the end of human chromosome 16p that substantially increases the risk of autism (odds ratio greater than 5) (WEISS et al. 2008b; SANDERS et al. 2011) (see Chapter 19, Figure 19.1, for an illustration of the extent of the rearrangement and the genes it contains). How does it exert its effect? Rob Malenka addressed this question using genetically engineered mice (WALSH et al. 2018). First, he established that the deletion had an effect on mice. He found that homozygous knockout mice were less sociable (sociability is tested by allowing a mouse to explore three chambers, in one of which there is another mouse, constrained within a small wire mesh cup; the proportion of time spent with the other mouse is used to judge how sociable the mouse is). The next question he asked was whether this genetic effect could be localized to a subset of neurons in the brain, whose function would then indicate how the mutation gave rise to the change in sociability. Because in humans the mutation affects all neurons, it was quite possible that the mutation disturbed many different processes, and that no specific region or function would be identified.

In order to answer this question, Malenka drew upon work that pointed to the importance of serotonin in modulating social behavior. He had previously found that activation of oxytocin receptors on serotonin terminals in a part of the brain known as the nucleus accumbens was essential for the processing of social rewards (DOLEN et al. 2013). The neurons that release the serotonin originate in a structure called the dorsal raphé nucleus, of whose 235,000 neurons approximately 165,000 (70%) contain serotonin (BAKER et al. 1990). By injecting channelrhodpsin into the dorsal raphé nucleus of mice that express Cre specifically in serotonin neurons (thus giving Malenka the ability to activity the serotonin neurons in the dorsal raphé nucleus), and in a separate group of mice injecting an inhibitory opsin (which would allow him to silence these neurons), Malenka's group showed that modulation of dorsal raphé serotonin neurons alters sociability: optogenetic activation of dorsal raphé serotonin neurons increased sociability, whereas optogenetic inhibition of dorsal raphé serotonin neurons decreased sociability.

What would happen in the mice with the deletion that increases risk for autism? Malenka found that dorsal raphé serotonin neurons lacking 16p11 had reduced neuronal activity, implying that this impairment contributed to the behavioral deficits. If so, then activating these neurons in the mice with the deletion should rescue the deficit. Indeed, activation of dorsal raphé serotonin neurons rescued the social behavior deficits in mice carrying the deletion in serotonin neurons (WALSH et al. 2018).

Furthermore, enhancing serotonin release specifically in the nucleus accumbens had the same effects. Putting these experimental results together, Malenka demonstrated that the behavioral effect of the deletion happened through its effect on serotonin neurons active in the nucleus accumbens. This set of experiments represents a leap forward in determining where in the brain the deletion has its effect, and which pathways are involved. In Chapter 19, section 19.1, we return to the 16p rearrangement for an evolutionary analysis that casts further light on its function.

16.8.2 Neuromodulation of the stress response

Throughout this book, we avoid suggesting that there are genes for particular behaviors. We report instead a more complex relationship where multiple genes with multiple effects, interacting with the environment, influence behavior. But one class of genes can lay claim, at some level, to be master regulators of behavior. These are neuromodulators, which include a group of protein hormones, and as genes encode proteins, it's possible to argue that some of the cognate genes do function to control behavior.

Most readers will be familiar with the idea that there are molecules circulating in our blood that affect behavior. In fact, this belief is so common that, quite contrary to all evidence, we accept the explanation that someone is excited because of adrenaline, usually described as pumping through the blood stream (injections of adrenaline do no such thing), and we willingly attribute bad mood to changes in the circulating levels of hormones—sex hormones or stress hormones being the likeliest candidates.

The prototypical stress hormone is corticotropin-releasing factor (CRF), a hormone that is present in the brain where it controls the release of other molecules, throughout the body, to alter many features of an animal's physiological state during stress (KORMOS AND GASZNER 2013). When CRF is injected into the brain, it produces some behavioral effects, but it doesn't act like a switch, turning on the behavioral responses that characterize the behavioral manifestations of a state of stress (DUNN AND BERRIDGE 1990). It's not that it has no effect (it makes rodents move around more, and rear up and groom themselves, for instance) but it doesn't turn a normally peaceful relaxed animal into one that (literally) climbs the walls because it is so stressed. How does it work? The short answer is that CRF is a neuromodulator, and while we still don't have a complete answer of its mechanism of action on behavior, we are beginning to make inroads, through the application of genetic technologies.

CRF is one of a large group of neuropeptides whose major role is to modulate the function of neuronal circuitry (MARDER 2012). There are two forms of neuromodulation. An extrinsic neuromodulator (e.g. oxytocin) is released from outside of the neural circuit on which it acts, transported for example in the blood stream, while in intrinsic modulation the neuromodulator is released by one of the circuit components. The important thing about neuromodulators is that they can completely alter the function of a neuronal circuit, making it do totally different things depending, for example, on the environment. Genetic analysis provides some of the evidence for the way that neuromodulators function.

David Anderson, a neuroscientist at CalTech, examined the function of a neuromodulatory neuropeptide called tachykinin2 (Tac2). Anderson found that the amount of Tac2 mRNA is increased in mice made to live alone. When mice are kept in isolation for a few weeks, this induces a set of behaviors that we associate with stress: they appear more anxious and show marked increases in aggressive behavior (WEISS et al. 2004). Anderson wanted to know whether the activity of neurons expressing Tac2 was sufficient for the development of the social isolation-induced behaviors (ZELIKOWSKY et al. 2018). By chemogenetically labeling the Tac2 neurons with a receptor for clozapine (this is the DREADD strategy), he gained control over neurons expressing Tac2, so he could activate them at will (by feeding the animals clozapine).

Increasing Tac2 expression, or activating Tac2 neurons, didn't do much; there was no change in the behavior. But he was able to recapitulate the behavioral effects of social isolation when Tac2 was overexpressed and Tac2 neurons were activated: 80% of mice were aggressive, compared with less than 20% of the control animals (ZELIKOWSKY et al. 2018).

Anderson's work is an example of how neuromodulation combines information from different sources to affect behavior: there has to be an input to increase Tac2 first, before the activation of the neurons will change behavior. Anderson's observation, acquired by exerting genetic control over both the expression of the gene and the activity of the neuron, indicates that neuronal activity, combined with specific gene-expression profiles, is a determinant of behavioral outcomes.

Deployment of genetically encoded reagents, allowing increasingly detailed examination of genetic action, shows the following: the primacy of certain genes, in controlling behavior, reflects an action that depends on when and where in the brain they are expressed. The same gene can have different actions, possibly contradictory, depending on the context.

16.8.3 Genes that determine sex-specific behavior

Innate behaviors are those most likely to be genetically encoded. Rodents display innate sex-specific behaviors, some of which from our perspective are hard to appreciate, such as the behavior of virgin males toward pups: they kill them. After mating, males stop attacking pups, and for a while take a paternal interest, at least from the time of birth of the pups up until weaning. The hardwiring must be genetically determined, but how? Catherine Dulac has been researching the molecular basis of this behavioral switch.

A key insight work came from a genetic study that interfered with the vomeronasal system. Rodents have two different sets of neurons that sniff the outside world, and two brain centers to analyze the results (Figure 16.11). In the back of the nose, there are olfactory receptors to detect chemicals carried in the air, relaying information to the main olfactory bulb and on to the primary olfactory cortex. In humans, this results in the perception of smell. In mice, neurons of the vomeronasal organ (VNO) are also present in the nose and sample water-soluble chemicals collected in nasal mucus. VNO neurons project to the accessory olfactory bulb, which in turn sends fibers to the hypothalamus via the medial amygdala. The VNO is present in most amphibians, reptiles, and nonprimate mammals (KEVERNE 1999).

What does the VNO do? Destroying the VNO eliminates mating behavior in rodents (courtship and copulation) because the animals can't respond to a class of molecule called pheromones. Pheromones are chemicals that animals use to alter the behavior of others; you can think of them as hormones outside the body (STOWERS AND KUO 2015). They can be used to raise an alarm, and to initiate courtship. There are about 600 pheromone receptors in rodents (IBARRA-SORIA et al. 2014), and they interact with an ion channel (a protein in the neuronal membrane that allows ions into the cell) called TPR2. When this gene is deleted, male mice don't fight with other males, and they initiate sexual and courtship behaviors toward both males and females (STOWERS et al. 2002).

This doesn't mean that the ion channel TRP2, which is found only in the VNO, is the behavioral gene. It's just at the start of the pathway. Figure 16.11 shows that VNO neurons connect via the accessory olfactory bulb to a structure called the medial amygdala.

Dulac found that a sex hormone, estrogen, plays an essential role in generating sex-specific patterns of response in the medial amygdala (BERGAN et al. 2014). Estrogen has many jobs. It's most famous as the primary female sex hormone, responsible for the development and regulation of the female reproductive

Figure 16.11 Brain regions in the mouse that process sex-specific social cues and act in the regulation of social behaviors. AOB, accessory olfactory bulb; AVPV, anteroventral periventricular nucleus; BNST, bed nucleus of the stria terminalis; DR, dorsal raphé nucleus; MeA, medial amygdala; MPOA, medial preoptic area; VMH, ventral medial hypothalamus; VNO, vomeronasal organ; VTA, ventral tegmental area.

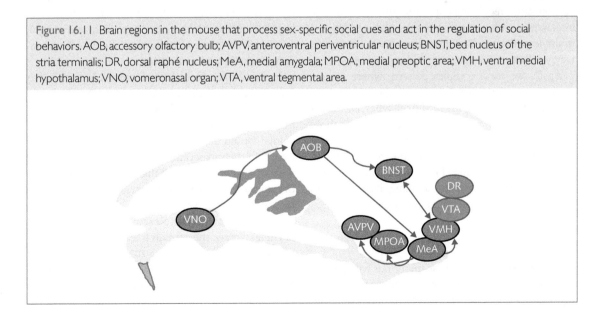

system and secondary sex characteristics. To use the genetic term, it is pleiotropic, and hardly therefore fills the role of a specific behavioral gene. However, with genetically encoded reagents able to determine its temporal and tissue actions, the role of estrogen as a behavioral gene becomes clearer.

A subset of neurons in a structure called the ventral medial hypothalamus, a direct target of the medial amygdala, was put under exogenous control using optogenetic and chemogenetic tools. Activating these neurons makes male mice attack an intruder, while optogenetic inactivation of these neurons suppresses aggression (YANG et al. 2013; LEE et al. 2014). Similarly, optogenetic activation of estrogen receptor-positive neurons in the medial preoptic area elicits a similar amount of male-typical mounting behavior in males and females; inhibiting these neurons disrupts mounting behavior in both sexes.

What might explain the switch that makes male mice stop killing their pups? Dulac is coming close to an answer, one that involves another neuropeptide. Dulac found that neurons in the medial preoptic area are activated during male and female parental behaviors (WU et al. 2014). The critical neurons are those expressing a neuropeptide called galanin. Optogenetic activation of galanin-positive neurons in the medial preoptic area suppresses pup attacks and triggers parental care in virgin males; destroying these neurons (again using genetically encoded reagents to target the galanin-positive neurons) impairs parental behaviors in both fathers and females. Galanin is another classic neuropeptide, regulating numerous processes through interactions with its three receptors and signaling via multiple pathways (including the cAMP/PKA pathway we discussed in the sections about memory).

Again, we see a complex pattern, where no one single gene takes overall control over a behavioral output but instead acts in specific cell types, and under specific contexts. These observations agree with those we reported for the quantitative genetic analysis of behavior in previous chapters. In this chapter, we are, for the first time, getting close to identifying the molecules involved and to understanding the mechanism by which the genes act to alter behavior.

16.8.4 Creating false memories

Yes, it is possible to make false memories with genetic technology, although the sort of memory we are talking about is not what you had for dinner last night,

your boyfriend's birthday, or where you left your car keys. The memory is of the sort we have dealt with throughout this chapter, the one acquired by a mouse when it learns that a tone indicates that it is going to get a foot shock. The task is called fear conditioning, and the way we know that the animal remembers the association is by observing the amount of freezing when it hears, or sees, the cue predictive of shock.

What's going on at the neuron, or more specifically at the synapse, during and after fear conditioning? According to the idea that synaptic plasticity underlies learning and memory (the Hebbian model), synapses at neurons that receive both the tone and the shock inputs will be strengthened. Letting the animal hear the tone (previously associated with foot shock) should result in firing (or increased rates of firing) of fear conditioning-activated neurons. Reactivation of these neurons will recapitulate aspects of the original learning episode; in other words, it will represent the memory of the conditioned fear.

Two groups achieved a *tour de force* demonstration of the power of genetically encoded reagents to create false memory. Mark Mayford's group developed the TetTag mouse to label active neurons. The TetTag system is based on the Tet-Off system. When animals ingest doxycycline, the expression of the tetracycline tTA protein (driven by the *c-fos* promoter) can't do anything: it is bound to the antibiotic. When doxycycline is removed from the animal's diet, the tTA protein will be expressed in active neurons (active neurons express *c-fos*). So, to mark neurons that are laying down a memory, the experimenter stops feeding doxycycline to the mice and then teaches the animals to associate a footshock with a tone (fear conditioning) (GARNER et al. 2012).

In order to create the false memory, Mayford had to identify cells active when the memory was laid down, and then reactivate the labeled neurons. The prediction was that if these cells contained the memory, the animal would behave as if it remembered: it would freeze. In their experiment, Mayford labeled neurons with DREADDs—the cells that had been active while the animal laid down a memory now carried a receptor that, in response to a drug, clozapine, strongly depolarizes the neurons. Reactivation of the neurons did indeed make the animals freeze, just as if they had heard the tone (GARNER et al. 2012).

Susumu Tonegawa used a similar protocol to induce a false memory. Instead of coupling the tTA protein to drive the expression of DREADDs, he genetically engineered the mice so that the labeled neurons (the ones active when the memory was laid down during fear conditioning) contained channel-rhodopsin. This meant that he could now activate the neurons with a laser. When he did so, he had the same result: the animals froze (Liu et al. 2012; Ramirez et al. 2013). He reported that after fear conditioning, during which fear-memory neurons were labeled with channelrhodopsin, freezing levels in response to laser light were higher compared with no light, indicating the induction of a light-induced fear memory (Liu et al. 2012; Ramirez et al. 2013).

Subsequently, Alcino Silva has combined imaging with genetic manipulations of memory to explore how the time at which memories are laid down determines their recall and relationship. The idea is that you expect your memory of who you ate dinner with last night to be linked with whatever else you did after dinner but not to be associated with events and people from weeks before. Using the same genetic manipulations described here, Silva reported that memories acquired close together in time are laid down in an overlapping set of neurons, so that activation of one memory would inevitably lead to the activation of the other linked memory (Cai et al. 2016). Silva also has an explanation for the poorer memory linking he found in older mice, by showing that older mice do not show the overlap, and he even reported that he has a solution: he could rescue the deficit by increasing cellular excitability in a specific subset of neurons

in the hippocampus (Cai et al. 2016). Perhaps, in the not too distant future, we will have a cure for the poor memory that afflicts all three authors of this book.

16.9 Next steps

Many questions in neuroscience that seemed intractable are now open to experimental investigation. Until recently, the spatial and temporal complexity of neurons had made it very hard to grasp the molecular mechanisms that explain how they work. Our knowledge is still limited in comparison with what we know about other cell types, but we now have a suite of genetically encoded reagents that are being used to transform our understanding of how brains work. We can now create any mutation we like, at will, and can express that mutation when we like and almost where we like. The characterization of single cells is giving us unparalleled insights into the components of the nervous system, and is likely soon to grant us genetic access to any tissue, and any cell type. This will mean we can track the activity of cells anywhere in the brain, and bring those cells under our control, so that we can activate or silence them at will. The power of combining genetic and imaging technologies to intervene and to observe the consequences of these interventions is immense. What you have read here is only a foretaste of what we are learning and will soon learn about the molecular basis of behavior.

Summary

1. Reverse genetics means examining the phenotype of a known genotype, for example through testing the effect on behavior of an engineered mutation. Forward genetics maps the location of genetic variants that contribute to phenotypic variation.

2. Gene ablation can be achieved through homologous recombination in embryonic stem cells. Homologous recombination refers to the process by which DNA is exchanged between two similar or identical sequences. By adding DNA that is homologous to part of a gene's sequence, it is possible to target the genomic locus and replace part, or all, of the endogenous gene. Once targeted, the embryonic stem cells

can be incorporated into an embryo, resulting in a mouse with a mutation.

3. DNA can integrate anywhere in the genome through nonhomologous recombination.

4. Genetic engineering using clustered regularly interspaced palindromic repeats (CRISPR) is a highly efficient way of deleting segments of DNA. It does not require the use of embryonic stem cells to introduce a mutation into an animal and can be carried out in any organism.

5. Genetic engineering can be used to create constitutive mutations, heritable mutations present in all cells in the body, and inducible mutations in which the mutation can be

turned on in any tissue and at any time the experimenter wishes.

6. Phenotyping of constitutive mutations in animal shows that their effects on behavior are quantitative, depend on the mouse strain in which they are measured, and are pleiotropic.

7. Analysis of constitutive mutations has identified a number of genes involved in behavior, including *CREB* as a key memory gene.

8. Inducible mutations use two-component systems. Deletions can be obtained with crosses between mice that contain an enzyme that cuts DNA (Cre recombinase) and a mouse that has recognition sites for the enzyme targeted to the gene of interest. The Tet-On and Tet-Off systems use two components to allow experimenters to turn genes on and off at will.

The use of inducible mutations shows that the effects of mutations on behavior depend on the time at which a gene is expressed, the tissue and cell type in which it is expressed, and the environmental context.

9. Genetically encoded reagents have been used to control many aspects of behavior, most dramatically illustrated by the ability optogenetics confers to turn neuronal activity on and off.

10. Genomic engineering technologies are being used to understand how genes influence psychiatric disease and how brains work. For example, they have been used to investigate the role of neuromodulators in regulating behavior and to create false memories in mice.

17

Mutagenesis and the molecular dissection of circadian rhythms

[What we learn about a fly's brain turns out to be useful for learning about our own brain]

17.1 Invertebrate genetics

The questions we ask about the relationship between genes and behavior are to some extent shaped by the experiments we can imagine, which in turn depends on our experience as scientists, what we have done in the past, what we know works, and what we know does not work. Psychiatric geneticists, restricted to working with a model organism whose mating preferences they cannot control or even influence, have struggled with the question of whether in natural populations most behaviors are influenced by many genes or just a few. Mouse geneticists, taking for granted their ability to construct pedigrees of their choosing, also address these questions but answer them in different ways from their colleagues in the psychiatry department.

Invertebrate geneticists live in a different universe. They can do more to their model organism than interview the first-degree relatives about their drinking habits. Flies are routinely exposed to the sort of trauma that no ethical committee would ever allow to be inflicted, even on a rat. In order to discover single genes that impact on behavior, flies are irradiated and given toxic, highly mutagenic chemicals, all with the purpose of disrupting their hereditary material. This approach **creates** genetic variation rather than discovering it in nature.

In the early 1960s, one of the pioneers of modern genetics, Seymour Benzer, inaugurated a new approach to the study of behavior using genetics in the fruit fly *Drosophila melanogaster*. His aim was to dissect behavior into its component genetic parts. In this way, the analysis could go beyond those genes that happen to vary naturally in populations. Benzer's approach represented a distinct departure from traditional behavior genetics, which had focused either on the behavioral influences of natural genetic variation, most of which seemed to result from relatively mild differences in multiple genes, or on spontaneous mutations that occasionally appeared in laboratory strains of flies or mice. He was adapting a paradigm from prokaryotic molecular biology, the aggressive search for single genes, by inducing mutations and testing for altered performance in behavioral tests. The method of inducing mutations was different from that described in Chapter 16. Benzer did not apply targeted mutagenesis but random, genome-wide mutagenesis, a method that in theory can create mutations in any gene. The single-gene, induced-mutation school took no account of whether a gene would exhibit natural variation. Instead, the interest lay in which genes were essential for the behavior, and which neural or biochemical components were selectively altered.

Fruit flies (*D. melanogaster*) are easy to grow in the laboratory, have a short generation time, and have about 100,000 neurons, which, on a logarithmic scale, is almost exactly half way between a single neuron and the number of cells in your brain. More importantly, it is *the* classic invertebrate genetic model system that has been studied intensely for nearly a century.

Of the various behaviors that have been studied with fruit fly mutants, none has been more successfully unraveled than the mechanism of circadian rhythms, the process by which the daily timing of activity and sleep are regulated. While first done in the fruit fly, subsequent identification of genes for circadian rhythms in mice and humans surprised everyone when they turned out to be nearly identical.

17.2 Random mutagenesis

Work in flies had shown that radiation exposure generates mutations, opening up a remarkable opportunity. Hermann Müller, who received the 1946 Nobel Prize in Physiology and Medicine for this work, explains the point:

It was from the first evident that the production of mutations would, as we once stated, provide us with tools of the greatest nicety, wherewith to dissect piece by piece the physiological, embryological, and bio-chemical structure of the organism and to analyze its workings. Already natural mutations … have shown how the intensive tracing of the effects, and interrelations of effects, of just one or a few mutations, can lead to a deeper understanding of the complex processes whereby the genes operate to produce the organism.

https://www.nobelprize.org/prizes/medicine/1946/muller/lecture/.

Radiation-induced mutations turned out to be 'a yardstick of what really random changes should be'. They were found to be of the same essential nature as those arising naturally in the laboratory or field. They

usually occur in one gene without affecting an identical one nearby. The effects may be large or small, and there is a similar ratio of fully lethal to so-called visible gene mutations, that is ones with readily observed phenotypes. There is no evidence that radiation-induced mutations are more deleterious than those that occur spontaneously. In fact, every natural mutation, when searched for long enough, is found to be producible by radiation.

The productive application of mutagenesis had already garnered another Nobel Prize-winning discovery in 1958, for Edward Tatum, Joshua Lederberg, and George Beadle. Their work still dominates biology today, in showing that: (i) all biochemical processes in all organisms are under genetic control; (ii) these overall biochemical processes are resolvable into a series of individual stepwise reactions; (iii) each single reaction is controlled in a primary fashion by a single gene, or in other terms, in every case a 1:1 correspondence of gene and biochemical reaction exists; such that (iv) mutation of a single gene results only in an alteration in the ability of the cell to carry out a single primary chemical reaction.

This model of genetic action, of a single gene controlling a single protein, was worked out using fungi and bacteria bearing variants induced by radiation in the way Müller suggested. There is no doubt that the model has been productive, and that it has resulted in the rigorous demonstration of linear action in many biological processes, but, as has been apparent at a number of points in this book, the model also limits the way we think about genetic action.

We need to clarify one point of disagreement with Müller. While the biological nature of artificially induced mutations may not differ from those that 'arise naturally', in their phenotypic effect such mutations are often quite different from those seen in natural populations. Man-made mutations can produce dramatic effects that are easy to detect in the laboratory but would not survive for many generations in the wild due to the effects of natural selection. We return to consider differences between induced and naturally occurring mutations at the end of this chapter, but for the moment, we need to look in more detail at the origins of an idea that crucially influenced the thinking of many people about the influence of genes on behavior.

The exclusive use of radiation for producing mutations was a short-lived technique. Chemicals that do the same thing were found and are now the tool of choice. A scientist in Edinburgh (J. M. Robson) noted that mustard gas produced effects similar to radiation, leading to a series of experiments that confirmed the mutagenic potential of the compound and similar chemical substances. Two of these, ethyl methanesulfonate and *N*-ethyl-*N*-nitrosourea, were later adopted for precisely the purpose Müller suggested: *'to dissect piece by piece the physiological, embryological, and bio-chemical structure of the organism and to analyze its workings'*.

Assuming that what is true for biochemical pathways is true for all phenotypes, Benzer thought that mutagenesis could be used to understand the biology of behavior. As he wrote in 1973:

The circuit components of behavior, from sensory receptors to the central nervous system to effector muscles, are constructed under direction of the genes. A mutation affecting the structure or function of any of these components may alter behavior. Using classical genetic recombination mapping, the locus of the mutant gene within the one-dimensional chromosomal array can be determined.

BENZER (1973).

It's worth emphasizing the radical nature of this suggestion—to create in the laboratory fly mutants that had abnormal behavior. What on earth would come out of this? Killer flies maybe, the stuff of science fiction horror and summer movie blockbusters. Flies with superior intelligence, able to hold conversations with their creators and even help them design their experiments? Or maybe just a lot of dead flies.

Consider the alternative view, voiced strongly, for example by John Wilcock, a psychologist in Birmingham who had published a long and influential review of gene action on behavior in 1969, 5 years before Benzer's genetics paper quoted above. He dismissed invertebrate behavior as fundamentally uninteresting until *'learning ability is unequivocally demonstrated'* and was scathing about the value of single-gene mutations (WILCOCK 1969).

17.3 Finding a circadian rhythm mutant

Rhythms dominate our bodies, our physiology, and our behavioral patterns. They determine the regularity with which we get up in the morning and go to bed at night. As day and night, neap and spring tides,

seasons and years are enforced upon us all, it seems reasonable to suppose that biological rhythmicity is entrained by environmental cues and to conceive of a system that uses diurnal temperature change to increase or decrease the rate of chemical reactions that in turn lead to waking and sleeping.

In fact, organisms have endogenous rhythms. When placed in laboratory conditions of constant darkness and temperature, rhythmic activities persist in plants, insects, and mammals, and they continue to occur once per day with a periodicity that is close to, but not exactly, 24 hours (Figure 17.1) (HYUN et al. 2005). The period is said to be circadian. Before the days of electric lights, humans were active mainly in the daytime. Mice and rats, on the other hand, are active at night, as anyone who has waited for a subway late at night knows.

Experiments in birds provide a particularly striking example of the importance of circadian rhythms for behavior. Starlings can be trained to fly to a food reward, thereby allowing investigation into the methods they employ to determine which way to fly. One cue they use is the position of the sun in the sky, a cue that can be replaced in the laboratory by an electric light. Unlike the sun, light bulbs are usually stationary, and birds trained to rely on a light bulb as a direction-giver add 15° every hour to the angle relative to the artificial sun. This means that the starlings must have access to a reliable internal clock. Furthermore, the clock can be reset by shifting the daily cycle of light and darkness to which the bird is exposed (equivalent to giving the birds jet lag). After a 6-hour shift, starlings make a 90° error in their estimate of where

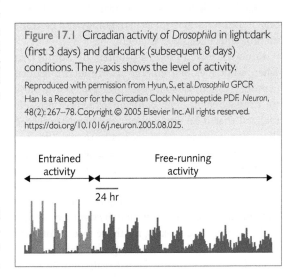

Figure 17.1 Circadian activity of *Drosophila* in light:dark (first 3 days) and dark:dark (subsequent 8 days) conditions. The *y*-axis shows the level of activity.

Entrained activity

Free-running activity

24 hr

the food reward is located; homing pigeons are similarly affected, setting off due east when they should be going north.

Fruit flies are active in the daytime, like us. This is easily observable in their movements. At night, they are mostly immobile. They do not lie down or curl up in a corner, but they actually are asleep when immobile, as shown by the fact that it is harder to rouse them when in this state and if they are forced to be active for prolonged periods during the night, they must recover the lost sleep the next day. The timing of the active/inactive times is not simply due to the lights being on and off. When flies are kept in a regular day/night (light:dark) cycle for several days and then left in constant darkness (Figure 17.1), they will continue to show the same cycle of active/inactive periods as if the lights were coming on in the morning and going off in the evening (HYUN et al. 2005). Moreover, if the timing of the light:dark cycle is shifted by several hours (e.g. from 8 a.m./8 p.m. to 12 p.m./12 a.m.), the flies will gradually be entrained to the new daily cycle over several days, just as it takes us several days (one day per time zone, on average) to adjust to a new time zone when we travel.

The search for circadian rhythm mutants in the fruit fly was one of the first attempts to isolate behavioral mutants in Seymour Benzer's laboratory at Caltech. Ron Konopka had come to the lab as a graduate student already captivated by the idea of studying circadian rhythms. Most of those around him were skeptical. Some thought it would be too difficult to devise a way of screening for alterations in these rhythms. Others thought that such mutations would invariably be lethal; these included Max Delbrück, one of the founders of molecular biology and Benzer's mentor and colleague at Caltech.

Konopka solved the first problem by noting that when adult flies first emerge from metamorphosis (a process called eclosion), they prefer to do so around the time when the lights come on in the morning. He realized that if he dumped out the newly emerged flies every morning from each culture vial after the lights had been on for several hours, few flies would emerge during the rest of the day. This gave him a simple screening test for new mutants. If he started many fly cultures, each consisting of a possible new mutant strain, he could test them by clearing the flies each morning and looking for those rare cultures in which adults continued to emerge around the clock. To induce new mutants, he fed male flies on sugar

water containing a chemical mutagen and then mated them, taking their offspring to start individual cultures that might harbor such mutations. To ensure that all the flies in these cultures would have the same mutant chromosome, he used a genetic trick available in fruit flies that permits the male's X chromosome to be directly transmitted to male offspring and not to females (as usually occurs). The trick involves mating them to females carrying what is known as an 'attached-X' chromosome in which both of the female's X chromosomes are attached to the same centromere and thus go together into the egg. When that egg is fertilized by an X-bearing sperm from the male, the zygote dies. When it is fertilized by a Y-bearing sperm, it forms a normal female, as the Y chromosome does not determine sex in flies.

Konopka solved the second problem by isolating circadian rhythm mutants that were, indeed, alive. After screening through some 2,000 cultures of mutagenized flies, he found three lines in which adult flies had abnormal times of emergence (KONOPKA AND BENZER 1971). Actually, he found the first one after screening only 200 lines. While this may seem like a lot, it is a pittance when stacked up against most others' experience of trying to isolate behavioral mutants, where it is not uncommon to go through more than 1,000 before finding anything.

To confirm that these were true circadian rhythm mutants, he grew the flies in constant darkness to test their internal clocks and found that they continued to show abnormal patterns of eclosion (Figure 17.2). In a completely independent test for a different kind of rhythm, he also tested them for their locomotor activity at different times of day and night. He placed adults in a 12-hour light:12-hour dark cycle to entrain them and then left them in constant darkness while monitoring their movements. Nowadays, there is a commercially available device for doing this, but back then he used the lab's spectrophotometer, a piece of equipment designed to detect light absorption and used for measurements of protein concentration and other molecules. Konopka simply put some flies in the little glass vessel, known as a cuvette, normally used for the liquid to be measured, and placed it into the machine. He then ran the chart recorder slowly over the course of the day and night to record any interruptions of the light beam due to the flies jumping up and down. Normal flies are active during the day and in constant darkness during the time that would be the

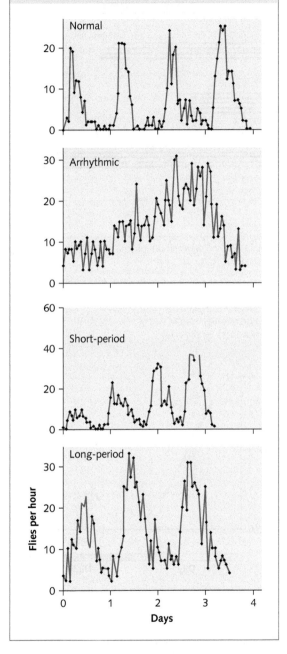

Figure 17.2 Circadian eclosion behavior in normal flies and in three mutants of the *period* gene.

Reproduced from Konopka, R.J., and Bvenzer, S. Clock Mutants of *Drosophila melanogaster*. PNAS, 68(9): 2112–6. https://doi.org/10.1073/pnas.68.9.2112.

The three mutants all differed from the normal 24-hour periodicity. One showed a 'long-day' period of 28 hours, another showed a 'short-day' period of 19 hours, and the third was arrhythmic, showing no discernible periodicity (Figure 17.3) (KONOPKA AND BENZER 1971).

But this was not the only piece of evidence for the fundamental nature of the mutants' effects. Even more convincing was the fact that using the method of linkage analysis, it could be shown that all three mutations mapped to the same place on the fly's X chromosome. Additional genetic tests were performed to see whether they were mutations of the same gene, such as producing female flies in which one X chromosome carried one of the rhythm mutations and the other X carried another of them. In these complementation tests, all three mutations behaved as expected for alleles of the same locus; that is, they reflected different aberrant copies of the same gene. With this, the *period* gene was christened, and the genetic analysis of circadian rhythms was launched.

17.4 Molecular clockworks

The world of biological research changed dramatically with the arrival of gene cloning, and this had an immediate impact on the study of circadian rhythms. The *period* gene of *Drosophila* was among the first genes to be cloned using a strategy that was initially only possible in the fruit fly: positional cloning. Even if nothing was known about what kind of protein was involved or where or when it was produced, one could clone the gene simply by knowing exactly where it is on the chromosome and isolating a series of overlapping pieces of DNA that covered that region of chromosome. This approach is also referred to as 'chromosome walking' because it needs only one starting piece of DNA from somewhere nearby to 'walk' along the chromosome by isolating successive overlapping pieces. The tools of *Drosophila* genetics make it possible to narrow down a gene's location using strains carrying rearranged chromosomes, that is, chromosomes that have been broken by X-rays and reattached incorrectly. The breakpoints of these rearrangements provide a physical marker for a chromosome region that is visible in the microscope and a molecular marker for that region because it changes the DNA sequence (*Drosophila* chromosomes, unlike mammalian chromosomes, can be found in a special

day. If one calculates the periodicity of their activity cycle (represented by the Greek letter τ—pronounced 'tau'), it comes out to very nearly a 24-hour period of activity/inactivity.

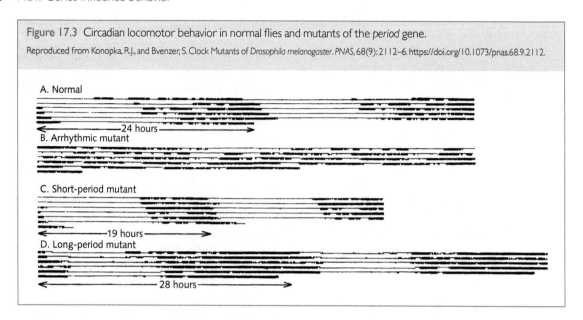

Figure 17.3 Circadian locomotor behavior in normal flies and mutants of the *period* gene.

Reproduced from Konopka, R.J., and Bvenzer, S. Clock Mutants of *Drosophila melanogaster*. PNAS, 68(9): 2112–6. https://doi.org/10.1073/pnas.68.9.2112.

expanded state in the salivary glands. These polytene chromosomes can be as long as half a millimeter, big enough that it is possible to see small rearrangements down a microscope).

Using this strategy, Michael Rosbash and Jeffrey Hall at Brandeis University isolated a piece of DNA containing the *period* gene. Michael Young at Rockefeller University accomplished the same goal in parallel. The definitive proof that this really was the gene came from showing that the isolated piece of DNA could be reintroduced back into a fertilized *Drosophila* egg coming from the arrhythmic mutant strain and rescue its defective circadian rhythm (i.e. restore it back to normal).

Having the gene in hand, however, is not the same thing as understanding what makes the clock tick. Looking at the sequence of amino acids encoded by the gene did not provide any immediate answers, nor did seeing where in the brain the mRNA for *period* was actively transcribed. More interesting was the fact that the period protein is localized in the nucleus. These are all pieces of the puzzle, but not enough. The breakthrough came when they looked at the levels of the *period* gene's product over the course of a 24-hour day. Critically, they found that both the mRNA and the protein cycled (Figures 17.4 and 17.5) (Hardin et al. 1990; Zerr et al. 1990).

Not only that, but the short-day mutant's mRNA cycled earlier in the day and the long-day mutant's mRNA cycled later. This all occurred in a light/dark

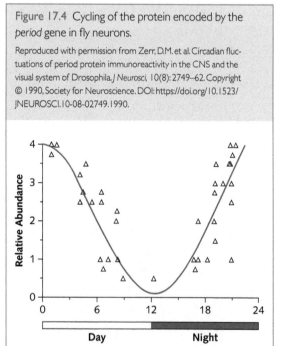

Figure 17.4 Cycling of the protein encoded by the *period* gene in fly neurons.

Reproduced with permission from Zerr, D.M. et al. Circadian fluctuations of period protein immunoreactivity in the CNS and the visual system of Drosophila. *J Neurosci*, 10(8): 2749–62. Copyright © 1990, Society for Neuroscience. DOI: https://doi.org/10.1523/JNEUROSCI.10-08-02749.1990.

cycle. In constant darkness, the wild-type mRNA continued to cycle with a 24-hour cycle , the short-day mRNA with a short-day cycle, and the long-day mRNA with a long-day cycle. The protein levels followed suit. In addition to matching the behavioral period in a light/dark cycle and in constant darkness,

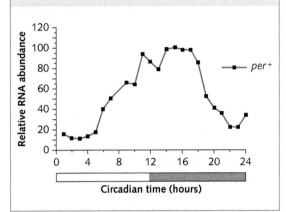

Figure 17.5 Cycling of *period* mRNA in normal (*per*+) brains.

Reproduced with permission from Hardin, P.E., et al. Feedback of the *Drosophila period* gene product on circadian cycling of its messenger RNA levels. *Nature*, 343:536–40. Copyright © 1990, Springer Nature. https://doi.org/10.1038/343536a0.

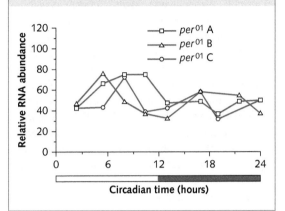

Figure 17.6 Little or no cycling of *period* mRNA in the brains of three different arrhythmic flies due to the *per*° mutation.

Reproduced with permission from Hardin, P.E., et al. Feedback of the *Drosophila period* gene product on circadian cycling of its messenger RNA levels. *Nature*, 343:536–40. Copyright © 1990, Springer Nature. https://doi.org/10.1038/343536a0.

the wild-type protein and mRNA also showed a gradual shift when that cycle was advanced or retreated, as in jet lag. In the mutants, the short-day protein cycled on a shorter time period, the long-day mutant made very little protein and had a very damped oscillation, and the arrhythmic mutant made no detectable protein.

All of this seemed reasonable and consistent, but there was an anomaly. The peak of the period protein comes about 6 hours after the peak of its mRNA. This is much longer than is normally required for the process of producing protein from mRNA (i.e. translation), and was critical later on for clarifying the entire circadian mechanism.

The real breakthrough came when Rosbash, Hall, and their postdoc Paul Hardin looked at the levels of *period* mRNA in the arrhythmic *period* mutant: it remained at a constant level (Figure 17.6) (HARDIN et al. 1990). As this mutant had already shown the genetic characteristics of a null (i.e. knockout) mutation, the result gave the key clue to the rhythm mechanism; that is, in the mutant that makes no protein, the mRNA is transcribed but does not cycle. When this was put together with the fact that there is a 6-hour delay between transcription and translation, and then a further delay between translation and entry back into the nucleus, it suggested a novel interpretation: the period protein feeds back to repress its own transcription over a time course that spans 24 hours. The next

several years' work would bear this out, making these studies on period protein and mRNA cycling the first major breakthrough not only in circadian rhythms but also in the molecular understanding of any behavioral mechanism. The universality of the mechanism also began to surface at this time, when something analogous was found for the circadian clock in the fungus *Neurospora*. But the feedback cycling of the *period* gene product, while compelling, was still incomplete.

17.5 More clock genes

No gene is an island, as amply shown by our many examples in earlier chapters of the interactions among many genes to produce a phenotype. To find more of the key players in the clock mechanism, mutant-hunting strategies similar to Konopka's were set in motion. The first result was an arrhythmic mutant found in flies in Michael Young's laboratory at Rockefeller University, called *timeless*. The *timeless* mRNA cycles over a 24-hour period and its transcriptional profile is flat in both the arrhythmic *period* and the *timeless* mutants. In fact, *timeless* is *period*'s doppelganger. Both are transcribed cyclically, both protein products cycle and are translated with a delay, both are localized in the nucleus, and when either one is mutant, the cycling of both of them grinds to a halt.

In other words, it was a stroke of luck that this gene came up as the next one after *period*. As it turns out, the protein products of the two genes actually bind to each other to repress their own transcription.

At this point, another phylum entered the fray with the isolation by Joe Takahashi of the first clock mutant in the mouse, imaginatively named *Clock* (VITATERNA et al. 1994). When it was cloned, the gene was found to encode a transcription factor (a protein that regulates the transcription of other genes) (ANTOCH et al. 1997; KING et al. 1997). Clock is a member of a family of proteins that are related to each other by having a similar amino acid sequence for a segment of the protein (a 'domain') that binds other proteins with similar domains and also binds to DNA. This is known as a basic-helix–loop–helix (bHLH), a less-than-poetic description of the configuration of amino acids in that segment. The second feature of this family of proteins is that they possess an additional protein-binding domain known as PAS, named with the initials of the first three proteins to have been found with it (Per: period circadian protein; Arnt: Ah receptor nuclear translocator protein; Sim: single-minded protein). We describe more about the mouse clock gene in section 17.6.

The lucky streak continued. The next two mutants to be identified in the fly were the actual targets of the period/timeless complex—the transcriptional complex to which they bind to shut down their own transcription. These were the genes *cycle* and the fly version of *Clock*. Both were found in a Konopka-like screen for mutants with altered daily locomotor rhythms. The *Clock* gene was so named because it was homologous to the recently identified mouse circadian rhythm mutant. Both *Clock* and *cycle* encode transcription factors of the bHLH-PAS family. They associate with each other and also with the protein products of *period* and *timeless* in one big, happy complex that forms the basis for feedback repression of *period* and *timeless* transcription; that is, Clock and cycle drive transcription of *period* and *timeless* when no period and timeless proteins are around but are blocked from doing so in their presence.

The last links in the core mechanism came from further modifications of the original Konopka screen, and all three of the mutants isolated were in genes encoding protein kinases. These enzymes mark or modify other proteins by adding a phosphate group, which in turn affects their susceptibility to being degraded. As the whole cycling mechanism depends on the timely disappearance of the period and timeless proteins, the

tuning of their degradation is important. In contrast to the *period* and *timeless* genes, which can be knocked out without killing the animal, these kinases are more widespread (pleiotropic) in their effects and are lethal when knocked out. To keep flies alive so that they can actually have circadian rhythms, the mutations were screened in heterozygotes or in special strains that selectively produce an excess of the gene product only in the subset of key circadian cells in the brain.

The two kinase genes that were found as heterozygotes were named *Doubletime* and *Andante*. *Doubletime* mutations came in all varieties—long period, short period, and arrhythmic (the original arrhythmic *Doubletime* mutation was lethal in adults!). *Andante* mutants, on the other hand, had narrower effects and only produced a long-period defect. Both kinases are well conserved in virtually all multicellular organisms and are known as casein kinase I (Doubletime) and II (Andante). Both act on the period protein. The third kinase gene, *shaggy*, was originally isolated in a screen for mutants affecting the earliest stages of embryonic development. It is named for the appearance of the mutant embryos, which lack epidermal cells, thus giving the embryo a 'shaggy' look as cells peel off from the outside. It, too, is widely conserved in many phyla, otherwise known as glycogen synthase 3, and is part of a signal transduction pathway. The screen for *shaggy*'s role in rhythmic behavior of adult flies was done by restricting an overabundance of the kinase to the relevant circadian cells by genetic engineering so that only the cells that make timeless protein would make too much shaggy.

By engineering flies so that *shaggy* is expressed only in a small portion of the whole brain, the animals could survive the abnormal enzyme content. At the same time, it revealed the role of the gene by shortening the period. When there is too little enzyme, the period lengthens. The shaggy kinase acts on the timeless protein.

When all of these pieces are put together, a picture emerges of a core clock mechanism, as illustrated in Figure 17.7 (ZERR et al. 1990; HARDIN et al. 1992; BELL-PEDERSEN et al. 2005; HARDIN 2005). One mutant was identified that connected the core clock to the external light stimulus that entrains the fly to the day/night rhythm. This gene, *cryptochrome*, is also highly conserved and found in most higher organisms, where it can serve as a light receptor protein (photoreceptor). When flies make too much cryptochrome, they are supersensitive to light entrainment; that is, their

Figure 17.7 The molecular basis of the circadian clock. Per, period; Tim, timeless; Dbt, Doubletime; Clk, Clock; Cyc, cycle; Sgg, shaggy; PP2a, protein phosphatase 2A; Ck2, casein kinase 2.

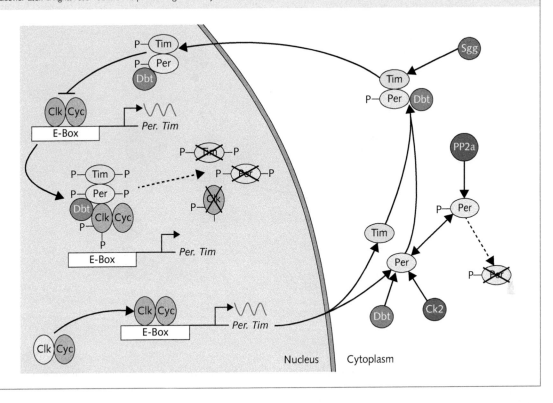

biological clock entrains to low light intensities that are ineffective in normal flies.

When flies are mutant for cryptochrome, they will entrain normally if their visual system is intact, but if they are blind **and** mutant for cryptochrome, then they fail to entrain. On the other hand, blind flies with a normal *cryptochrome* gene entrain normally. This says that flies can use either their visual system or their *cryptochrome* gene to entrain, but they need at least one to be working (Figure 17.8) (HARDIN 2005).

The tie-in to the clock comes from the ability of the cryptochrome protein to bind to the timeless protein. More specifically, when cryptochrome absorbs light, it changes conformation, allowing it to bind to timeless. Complexing with cryptochrome makes timeless ripe for degradation. When timeless levels fall off, the clock is reset. In this way, the fly's clock keeps up with the advance of sunrise following the winter solstice and the retreat of sunrise following the summer solstice. And as if to reinforce the fly's sensitivity to

changes in the time of daybreak, the levels of cryptochrome protein accumulate in the dark and decline in the light so that it is ready when morning comes.

The picture of a mechanism that these findings provide is detailed to an extent that is rare indeed for any aspect of behavior. Imagine everyone's surprise when a nearly identical mechanism was found in our mammalian cousin, the mouse.

17.6 The mouse's clock

The great attraction of mutagenesis is its ability to generate mutants that don't exist naturally. Mammalian geneticists for a long time envied the ability of fly geneticists to use radiation, or more efficiently chemicals, to induce mutations in their model organism, identify large-effect, highly penetrant mutations that affect a phenotype or trait of interest, and then, with that biological handhold, begin to pull apart

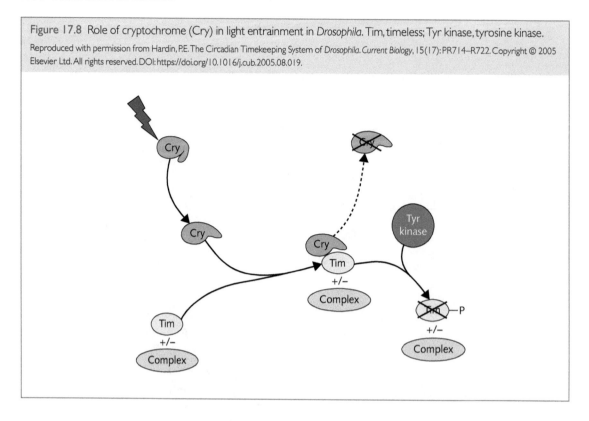

Figure 17.8 Role of cryptochrome (Cry) in light entrainment in *Drosophila*. Tim, timeless; Tyr kinase, tyrosine kinase.

the guts of the phenotype to see what made it work. Mutagenesis had been successfully applied to complex multicellular phenotypes such as embryonic pattern formation (NUSSLEIN-VOLHARD AND WIESCHAUS 1980), and Benzer's success with circadian mutants raised the hope that the same strategy would work in mammals. In the 1990s, mouse geneticists applied mutagenesis to dissect circadian rhythms, using the mutagenic compound N-ethyl-N-nitrosourea (ENU) to create circadian mutations in mice (VITATERNA et al. 1994), leading to the identification of the *Clock* gene (ANTOCH et al. 1997; KING et al. 1997).

Finding mammalian periodicity genes became easier with the cloning of a set of fly rhythm genes. At that point, molecular techniques could be used to identify homologs in the mouse. The proteins are sufficiently similar for the gene of one species to be used as a way of finding the gene in another. In the simplest approach, the DNA from one species can be used to hybridize to the DNA of the other (see Chapter 3, section 3.4 for a description of hybridization). This method avoids the need for a lengthy positional cloning adventure.

As it turns out, mice have four different *period* genes (of which only three are transcribed and translated) and two *cryptochrome* genes, one *timeless* gene, and one *cycle* homolog (*Bmal1*). They also have the same set of kinases present in flies, in some cases in multiple forms. The functional roles of these genes have been tested by engineering knockout mutations (using methods described in Chapter 16, section 16.2) and by biochemical tests for protein interactions.

The remarkable finding is that the mouse's clock is so similar to the fly's (Figure 17.9) (YOUNG 1998; BELL-PEDERSEN et al. 2005). There are some differences: *cryptochrome* substitutes for *timeless* in the transcriptional feedback loop and the mouse has multiple forms for many of the genes. This is a reflection of ancient duplications of the entire vertebrate genome, but the extra copies of the *period* and *cryptochrome* genes are put to good use. These similarities between invertebrate and vertebrate clock mechanisms are remarkable and more than justify Benzer's intuition that the fruit fly could teach us about the genes underlying behavior. Much more has been conserved through evolution than anyone had previously suspected.

Figure 17.9 Homology between molecular clocks in *Drosophila* and mice.
Reproduced with permission from Young, M.W. Marking Time for a Kingdom. *Science*, 288(5465): 451–453. Copyright © 2000, AAAS. DOI: 10.1126/science.288.5465.451.

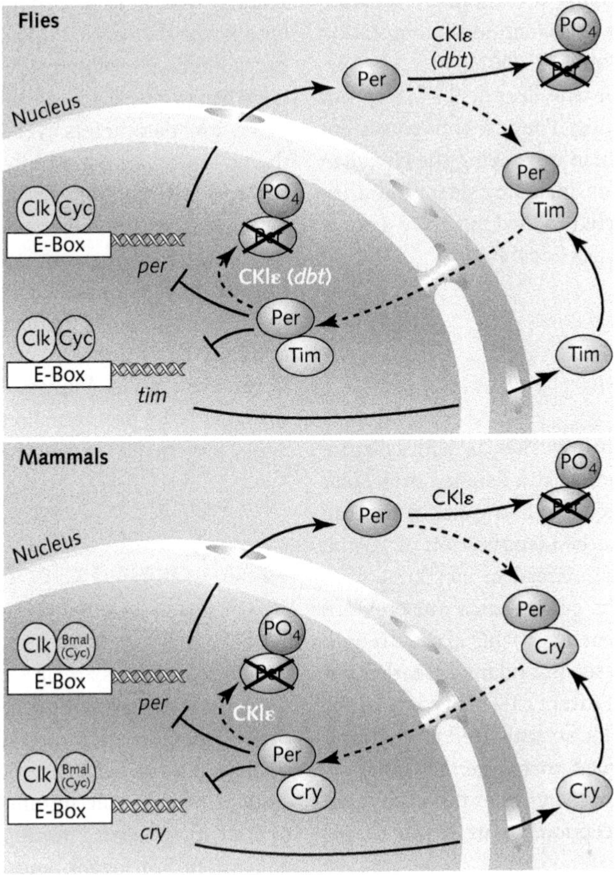

17.7 Human clock genes and unquiet sleep

If we consider the natural variants in the fruit fly, and also the high degree of molecular conservation of the circadian mechanism in mammals, then we should not be surprised to learn that there are genetic polymorphisms that have effects on human circadian behavior.

Certain sleep syndromes have been found that have been associated with variants in human homologs of the *period* gene. One of the multiple human homologs of *Drosophila period*, human *period3* (*hPer3*), has been associated with delayed sleep phase syndrome, in which individuals cannot get to sleep until very late at night. A study of Finnish patients with this syndrome revealed a length polymorphism in the *hPer3* gene, such that the long allele showed a morning preference and the short allele an evening preference; that is, different variants produced different settings of the clock. Another hereditary sleep disorder, familial advanced sleep phase syndrome (FASPS), is a dominant variant in which affected individuals are 'morning larks' with a 4-hour advance of their sleep, temperature, and melatonin rhythms; that is, they go to sleep early and get up early. The *hPER2* gene in affected individuals has a single difference in an amino acid: substitution of glycine for serine in the part of the protein that is phosphorylated by the human homolog of the *Doubletime* gene, casein kinase I epsilon (*CKIε*; Toh et al. 2001). This substitution makes *hPer2* harder to phosphorylate and affects clock timing. Thus, a variant in human sleep behavior is attributable to a missense (i.e. amino acid substitution) mutation in a clock component, hPER2, which alters the circadian period. Consistent with this finding, a completely separate case of FASPS in a family was identified as a mutation in the CKIε gene itself (Toh et al. 2001).

These variants are not true disorders of sleep but rather of its circadian timing. They are thus consistent with the role of the clock in organizing the circadian pattern of many behaviors, including sleep. And they are relatively mild: affected individuals have a functioning clock, but it runs fast or slow.

17.8 Lessons from fly rhythms

The foregoing narrative has provided us with a picture of the core clock mechanism at a detailed molecular level that, in turn, has explained how clocks work in humans and has allowed the identification of several human variants affecting circadian rhythms. The natural variants in rhythm genes match our previous descriptions of natural genetic variants in the strength of their effect. What may seem at odds with earlier discussions, however, is the extent to which some of the circadian genes seem to be so dedicated to rhythms. We had previously argued that a gene's effect on behavior is usually one of many roles played by that gene. Are rhythms an exception? Or are we not seeing the whole picture?

It certainly helps that a fly can survive quite well even if it has no endogenous ability to generate a circadian rhythm. If this were not true, the Benzerian strategy would have been far less successful. On the other hand, many of the genes are widespread in their action. All of the kinases result in lethality when knocked out. Clock and cycle have not been knocked out completely in flies, so we do not know if such mutations are lethal. The quintessential dedicated genes, period and timeless, are not entirely rhythm oriented. This realization comes from studies of cocaine sensitization in fruit flies.

Mutants in period, Clock, cycle, and Doubletime have all lost the sensitization response to repeated exposure to cocaine. This response does not show any circadian regulation in flies either, so it seems that these genes are regulating the induction of an enzyme necessary for sensitization, independently of the clock.

Is the success of this mutant-hunting approach due to some special feature of circadian rhythms? Mutant hunters would take issue with this because of the many other instances in which the strategy has worked well, but there may be a kernel of truth in it. One factor that smiled upon circadian rhythms is the exceedingly low variance of the behavior itself. The circadian period is one of the most tightly controlled behavioral parameters ever found. As a consequence, the ability to see genetically based differences above background variation, the signal:noise ratio, is much better than any other behavior that has been studied.

That said, it does not mean that every single fly shows a beautiful circadian rhythm. Rather, it means that among those flies in the population that do show a good rhythm (most of them), the variance is quite low. There are always clunkers, however, who won't go along with the program. So nature's unstoppable tendency to generate variation shows itself in that way.

17.9 Love on the fly (sex-specific development in the nervous system)

Circadian rhythms are an extreme example of evolutionary conservation of a genetic mechanism. Many other behaviors are more species specific, none more so than mating behavior. Mating behavior has been studied in the fruit fly almost from the beginning of fruit fly genetics, starting with one of the founders, Alfred Sturtevant, and his students early in the twentieth century. It drew the attention of investigators early on for its stereotypical character, exciting interest from generations of fly geneticists.

When a fly makes its way to a fallen rotting peach, or to the rim of a vat of crushed grapes, it is likely to be looking for a mate. When a female heads for that same peach, she is either looking for males or for a suitable place to lay her eggs once she has mated. The next time you see a bruised piece of fruit lying on the ground, be aware that you are looking at an active singles scene.

In the world of fruit flies, females generally have the upper hand when it comes to mate choice. Males must prove themselves worthy before a female will consent. A female may mate only once or twice in her life, because she will fertilize many eggs with sperm from a single mating. Therefore, the choice of a mate matters a lot.

Fallen pieces of fruit are social centers for many different creatures. Various species of fruit fly may end up together on the same peach, but they will not make the mistake of mating with a fly of the wrong species. Males are not always so successful at making such discriminations at first, and although a male *D. melanogaster* may initially try to court a different *Drosophila* species, they generally will not get very far. Recognition is crucial, and much of it is based on chemical sensing.

If the male is in the right mood (i.e. if he is old enough but not too old, the humidity is right, and he has not been cooped up with other males—which will dull his sense of smell from excessive exposure), he will sidle up to a female and tap her abdomen with his foreleg. The stimulation of his various chemoreceptors by the female's pheromones will then trigger his performance of the fruit fly courtship ritual. In males whose sensory cells have been genetically engineered to be unresponsive, courtship will not happen. Conversely, if these cells are genetically engineered to be hyperexcitable, the male will start courting at the drop of a hat.

Much of the circuitry for male courtship is the result of differences in neuronal wiring in the male brain compared with the female brain. Many of the sex-specific brain regions were identified by genetically engineering flies whose brains had been altered to develop as part male and part female.

Surprisingly little of the brain needs to develop as male for the triggering of the first steps in courtship: orienting, tapping, and wing display (in which the male extends one wing laterally and then vibrates it to produce a love song). Only one side of the bilaterally symmetrical brain needs to develop as male for this behavior, and only the dorsal part at that. This part of the brain is where the signal that triggers the courtship sequence originates. Sexually mixed flies do not perform fully normal courtship, but they will display these initial steps.

Later steps in courtship require male development in other parts of the nervous system: production of the song requires male neurons in the thoracic ganglion, licking of the female's genitals requires male brain tissue in the midbrain, attempted copulation requires male neurons in the antennal lobes, midbrain, and abdominal ganglion, and copulation requires additional male neurons in the abdominal ganglion. Much of the circuitry for courtship is not sex specific, however, but simply makes use of circuits common to many behaviors. These common circuits are present in both sexes and have been recruited into the courtship routine in males, which involves a substantial portion of the fly's nervous system.

A gene called *fruitless* is essential for much of this male-specific development. The gene encodes, among other products, multiple forms of a transcription factor, one of which is male specific. In males missing the male-specific product, courtship does not occur. Conversely, when this male-specific product of the *fruitless* gene is genetically engineered to be expressed in a developing female brain, that adult female will perform male-like courtship behavior that is never otherwise seen in females. The male-specific product of the *fruitless* gene is expressed in the brain regions known to undergo sex-specific development for courtship.

However, when flies are genetically engineered to be mutant for *fruitless* in a central part of the brain called the median bundle, they violate the sequencing rule and skip directly to attempting copulation. The result of this experiment means that neurons in the median bundle of males normally inhibit the performance of the late steps in courtship, and illustrate one mechanism for programming timing in a behavior: sequential activation and inactivation of excitatory and inhibitory circuits.

These sequential steps of activation/inactivation are now becoming visible with new techniques for recording activity in brains of flies as they run around freely doing their behavior in a small chamber. The technique works by having a microscope objective move around directly over the head of a fly in which the brain has been engineered to express a transgene that fluoresces when the neurons expressing it are active (Grover et al. 2016). Imaging the active neurons confirms the inference that courtship proceeds by a sequential turning on and off of brain regions known to be required for normal courtship behavior (e.g. regions subserving olfactory sensing, high-level behavioral coordination, and motor output).

17.10 Mutagenesis, and forward and reverse genetics

When Benzer began his explorations of genes and behavior in the fruit fly, traditional neurobiologists told him he was crazy to think that genetics would have anything to contribute to understanding the brain. The circadian rhythms story is the most dramatic, but by no means the only, refutation of the naysayers. The added, completely unexpected, bonus is that not only do we understand more about the fruit fly's brain but also about our own. The extent to which all animals share so many of their genes has been one of the major revelations of genome sequencing, and the circadian clock is one of the many molecular mechanisms highly conserved throughout evolution.

Benzer's success came from the application of mutagenesis to a complex phenotype. Mutagenesis worked so beautifully for mutations that influence circadian rhythm, first in flies and later in mice (VITATERNA et al. 1994), that its success prompted the funding of a number of screens for mutations in other behavioral phenotypes in mice. If the biology of circadian rhythms could fall to mutagenesis, surely other mammalian behaviors would also succumb? Surely mutagenesis would identify the molecular basis of learning and memory, of emotion, and of animal models of psychiatric disease? Based on the success of circadian rhythms, ENU mutagenesis for a time became big business, with centers in Europe and the USA generating thousands of mice.

Unfortunately, while mutagenesis was successful in generating mutations that illuminated the fields of immunology, biochemistry, and development, it has not been nearly so successful when applied to psychology. In screens for fearfulness, anxiety, and memory, no large-effect mutations were found. Why was this?

Effect size here is crucial, because a large effect size means genetic mapping and gene identification is easy: in the ENU strategy, the location of the mutation that affects the phenotype has to be mapped (the chemical introduces mutations randomly), and this is done in the same way as quantitative trait locus (QTL) mapping, with a cross between two strains. A small effect size poses just the same problems as those encountered with QTL mapping. The effect size of a behavioral knockout segregating in a cross between two inbred strains turns out to be, on average, about 10%, only twice the effect size that would be expected of a naturally occurring sequence variant (FLINT AND MOTT 2008). This explains why it has been difficult to identify ENU mutants with an effect on behavior (KEAYS et al. 2007) and reduces the question 'Why did mutagenesis work so well for circadian rhythms but so poorly for anxiety and memory?' to 'Why do some phenotypes arise from large-effect mutations while others only arise from small effects?'

This book is full of descriptions of the genetic architecture of personality, intelligence, anxiety, schizophrenia, depression, and autism. A common thread is that large-effect mutations are (almost) absent as a cause of behavioral variation. Yet here, for circadian rhythm, we find just the opposite: large-effect mutations in different pathways, providing much needed insight into the biological mechanisms, and resulting in Nobel Prizes for the discoverers (as we describe below). Why is circadian rhythm different?

The short answer is we don't know; we don't have a way to predict genetic architecture, but like most things in biology, the answer probably lies in the evolutionary origins of the trait. Like every other phenotype, the genetic architecture of traits, the number of loci, and their effect sizes are subject to evolutionary pressure. For example, the same genes in different species harbor mutations whose origins cannot be attributed to common descent. They have arisen independently, suggesting that in some cases evolution favors a particular subset of genetic variants. Taking this one step further, it's also been observed that once an allele has been selected, it may accumulate additional mutations that further enhance its phenotypic effect, so leading to the appearance of 'superalleles' (McGREGOR et al. 2007; LOEHLIN AND WERREN 2012). This raises the interesting hypothesis that the origin, or the consequences, of naturally occurring genetic variants might be predictable (STERN 2013). We return to the importance of evolution in shaping the relationship between genes and behavior, and give examples of 'supergenes' in Chapter 19.

17.11 A Nobel for the little fly

The 2017 Nobel Prize in Medicine or Physiology went to three scientists who had spent their professional lives studying the mechanism of circadian rhythms in the fruit fly, *D. melanogaster*. Circadian rhythms are a universal feature of living organisms, present in nearly all branches of the evolutionary tree—animal, plant, protozoan, and even some bacteria. The mechanism they discovered is shared widely among animals (including humans), and is present in a somewhat modified form in plants, and even more modified in fungi. One of us (R.J.G.) did his PhD work under the tutelage of one of the winners, Jeffrey C. Hall, who shared the prize with his good friend and long-time collaborator, Michael Rosbash (both Professors of Biology at Brandeis University). As a friend of theirs, I was their guest at the Nobel Prize ceremony in Stockholm in December 2017.

The Nobel Prize ceremony is one of the few remaining formal rituals in the world that still commands attention and respect. It is a very formal affair: men are required to wear white tie and tails, and women formal gowns. There was much speculation as to whether Jeffrey Hall, one of the Medicine or Physiology laureates who has a decidedly rebellious and antiformal streak, would acquiesce to the strictures of the week. The hosts provided him with a 'minder' who made sure that he arrived and departed according to schedule. They were surprisingly successful in this task—their major challenge was convincing him to remove his cap for the ceremony. After an hour of discussion, he finally agreed, but ended up carrying it with him onto the stage and sitting on it during the ceremony.

Held in the Stockholm Concert Hall, the Nobel Prize ceremony is a major event of the social calendar in the Swedish capital, as well as being the major event of the scientific calendar each year. There is much bowing to the King, Queen, and Princess on the part of the recipients. There are regal trumpet fanfares through the long ceremony that includes citations of the basis for the 11 prizes—three in Physics, three in Chemistry, and three in Medicine or Physiology, as well as one each in Economics and Literature. The laureates each have their own time to relate the stories of their discoveries at separate ceremonies held earlier in the week. These events are merely a prelude to the round of lectures, school visits, and other ceremonies (including one at the Royal Palace as guests of the King and Queen) that they will make during the week following the ceremony.

After the prize ceremony, the invitees (some 1,500 all told) move over to the Stockholm City Hall for a banquet, which includes entertainment by various dancers and musicians (traditional and modern), and brief light-hearted speeches by one each of the Physics, Chemistry, and Medicine laureates, and the Economics laureate (the literature laureate gave his lecture in a smaller, more private ceremony at the Swedish Academy.) The Stockholm City Hall is a grand place, built early in the twentieth century but decidedly rococo in style: a grand hall with an equally grand stairway descending into it from an upstairs balcony, down which the entertainers descend and perform, the walls decorated with classical-style murals. The serving staff also enter by way of the grand stairway to take up positions—several at each table—until all are assembled, at which time they step forward on cue to the tables to serve. The strategy for serving this large group involves staging areas in the hallways adjacent to the grand hall, which are filled with long tables of warming dishes. The dinner goes on for several courses and several hours.

A high point in the dinner came in the remarks from Michael Rosbash, the designated speaker for the Medicine or Physiology laureates, who related a story about the 1965 laureates and the congratulatory note written to each of them by their mutual good friend, the mischievous and brilliant Seymour Benzer. Benzer had been nominated himself for the prize more than once in prior years (although he never received it). The note, the same one that Seymour addressed and sent to all three, read, *'Dear Jacques/François/André, Congratulations on your prize, too bad you had to share it with those other two jokers.'* Two of the names (a different pair in each case) had been crossed out on each note.

At the end, the assembled guests repair up the stairway to the grand ballroom on the second floor, where a band plays music for the guests to dance to. The actual Nobel Prize certificates—beautiful calligraphy on paper resembling parchment—were on display in glass cases at one end of the ballroom. The dancing went on past 2 a.m., at which time many of the guests left to attend the after-party thrown by the medical students of the Karolinska Institute. They had decorated the foyer of the institute and many rooms upstairs in various themes relating to clocks (as the prize had been for biological clocks). One room had been decorated as a tavern from the eighteenth century, with students in appropriate period garb, serving from behind the bar. Everyone was still decked out in formal wear through the whole after-party, which did not break up until around 6:30 a.m.

Summary

1. The induction of new genetic variants by mutagenesis, originally with radiation and subsequently with chemicals, opened up a new avenue to analyze biological mechanisms, vastly expanding the range and severity of variants that could be obtained.

2. In behavioral analysis, one of the first uses of this approach was in the study of circadian rhythms in the fruit fly *Drosophila*. The first circadian rhythm mutants were all alleles of the *period* gene, producing short-day, long-day, and arrhythmic phenotypes.

3. Molecular cloning of *period* revealed that its mRNA cycles over a 24-hour period in normal flies, and its protein is translated with a delay and is then transported back into the nucleus. In arrhythmic mutants, it is transcribed continually, suggesting that it represses its own transcription.

4. Additional clock mutants, *timeless, cycle, Clock, Doubletime*, and *shaggy*, fill out a picture of a transcriptional feedback loop in which the protein products of the period and timeless genes repress their own transcription by complexing with the Clock and cycle proteins.

5. Entrainment to the light / dark cycle is mediated by the *cryptochrome* gene, whose product absorbs light, and complexes with the timeless protein and facilitates its degradation.

6. An almost identical mechanism is found in mammals, except that there are multiple versions of each gene and *cryptochrome* takes the place of *timeless*. Similar mechanisms also exist in fungi and bacteria, suggesting that the mechanism is exceedingly ancient.

7. Courtship behavior in *Drosophila* consists of a sequence of hard-wired movements. Much of the circuitry for male courtship is the result of differences in neuronal wiring in the male brain, compared with the female brain. The gene *fruitless* is essential for much of male-specific development. In males missing the male-specific product, courtship does not occur, and when a developing female brain expresses the male-specific product, the adult female performs male-like courtship behavior.

8. Mutagenesis applied to mice successfully identified clock components, but mutagenesis applied to other behaviors in mice (including fearfulness and memory tasks) failed to identify large-effect mutants.

9. Circadian rhythm is unusual—possibly unique—as a behavioral phenotype in that large-effect mutations can be found that disrupt core components of the trait. It is unclear why the genetic architecture of circadian rhythms should be so different from other phenotypes described in this book.

18

Many versus one: genetic variation in flies and worms

[On what we can learn about the genetic basis of natural variation in invertebrate behavior]

How much similarity is there between the genetic influences on behavior in humans as compared with 'simpler' organisms? The so-called simpler organisms that have provided us with the most information on this question are the fruit fly *Drosophila melanogaster*, and the nematode (worm) *Caenorhabditis elegans*. In contrast to the last chapter where we examined induced mutations, in this chapter we ask what natural variation in behavior looks like in these animals, allowing us to compare the results with those reported for humans.

18.1 Evolving behavior in the laboratory: *Drosophila*

A direct test of the heritability of behavior calls for genetic experiments, and for this purpose nothing beats the fruit fly *D. melanogaster*. At first glance, fruit flies might not appear to be the most suitable animals to work with. They are too tiny (around 3 mm long) to pick up easily, they don't live very long (a few weeks), and it's not obvious what their normal behaviors

comprise. Fortunately, enterprising biologists have been studying their behavior since 1905 (GREENSPAN 2008).

It may seem that working with fruit flies is no picnic, but in fact in some ways it resembles a picnic. Fly labs always have anywhere from a few to several hundred stray flies buzzing around at any given time. This was no doubt worse in the early days, when they also kept large bunches of bananas sitting around waiting to be peeled and crushed into the bottom of pint-sized milk bottles and stoppered with cotton. Fly cultures are still transferred by hand, but for all of our modern technology, we still bang the bottle against a rubber mat to force the flies down from the stopper, where they normally congregate, and then dump them into a fresh, clean bottle, and jam a clean cotton stopper into the top before they recover enough to fly out. Leaks are nearly unavoidable, and neighboring labs that require sterile conditions for cell culture are chronically grumpy. Fly traps are another ever-present fixture in fly labs—strategically placed bottles with an inch or so of vinegar in the bottom with a funnel placed in the bottle's opening in the desperate hope that a fly will be attracted by the odor of fermented fruit (their first love) and then be trapped inside so that they eventually drown. These help but only a little. Somewhat more effective is to use red wine instead of vinegar, but then there is the problem of the lab's graduate students drinking the trap's contents, not to mention the need to document that the lab is a 'drug-free workplace.'

These early studies of fruit fly behavior were motivated by the idea that so-called 'lower' animals were primarily governed by innate drives or 'tropisms'. This point of view originated with Jacques Loeb, a German-born zoologist who emigrated to the University of Chicago in 1892 to become one of the founders of modern biological research in the USA (PAULY 1987), and represented the beginning of mechanistic studies of behavior. But it was not until the mid-nineteenth century that researchers took up the challenge of trying to understand how genes might influence these little creatures.

In the late 1950s, Jerry Hirsch undertook such a study, taking fruit flies that he collected from a farm stand on Long Island (HIRSCH AND BOUDREAU 1958; HIRSCH 1959; HIRSCH AND ERLENMEYER-KIMLING 1962; HIRSCH AND KSANDER 1969). He and his students bred them selectively for a behavioral difference: when confronted with the choice of walking up or down in a maze, do they have a preference? This gravitational response ('geotaxis', meaning to 'go toward gravity') is weak in most flies; they will generally choose randomly. Hirsch constructed a multichoice maze so that flies would have 15 chances to decide to go up versus down, and those consistently choosing to go up would be separated from those consistently choosing to go down.

The original maze was hand assembled out of rubber tubing and glass 'T' tubes, and it stood several feet tall. Flies were constantly getting stuck part way through. Washing the maze was a major ordeal. All of this merely served as the obligatory apprenticeship that anyone who works with fruit fly behavior inevitably must go through. You have to prove your seriousness before the flies will perform for you.

Hirsch and his students persevered, collecting flies from the two extremes. They bred the 'Hi' flies to each other and the 'Lo' flies to each other and, after 50 generations of behavioral selection and breeding, they had one strain of flies that consistently showed positive geotaxis (meaning they wanted to go down in the direction of gravity) and another strain of flies that consistently showed negative (up) geotaxis. It was clear that these flies did not need to learn the behavior; they did it first crack out of the starting gate. More

importantly, the behavioral differences had 'evolved' over many generations through breeding. In other words, he had enriched for the genetic differences that would alter the behavior, and which were naturally occurring and already present in the starting fly population.

Hirsch and his students then went on to show that each chromosome carried genes that contributed to the selected phenotype. Flies have four chromosomes, one of which has very few genes, and it is relatively easy, using the breeding strategies we described for rodents, to separate individual chromosomes from the Hi or Lo lines. From these experiments and from the statistical analysis of various hybrids between Hi and Lo, they concluded that there were many genes contributing to the behavioral difference.

Each year, Hirsch would teach a genetics lab course at the University of Illinois and the students in the course would reselect the Hi and Lo lines (Figure 18.1) (RICKER AND HIRSCH 1988). This went on for 25 more years, with occasional neglect (approximately generation 200) and one year accidentally mixing the strains together but then reselecting them (1978, approximately generation 240). Then in 1985, after 550 generations, the strains spontaneously stabilized. You can see this at the right end of the graph in Figure 18.1 where the lines start to show much less variability in the mean score. The stabilization meant they could maintain the Hi and Lo phenotypes without further selection. The most likely explanation for this change in the genetic architecture of the behavior is that there was no longer any genetic variation at the loci contributing to geotaxis; that is, all the relevant genes had become 'fixed' in the Hi or Lo strains. For example, the flies could have become, like the mice described in Chapter 15, section 15.1, inbred, or at least sufficiently inbred to remove the genetic cause of behavioral variation.

In one of his final studies of these flies in the late 1980s, Hirsch reported that when he took the strains apart again, isolating individual chromosomes onto a neutral genetic background as before, the contributions of particular chromosomes had changed in comparison with the study carried out 20 years earlier (RICKER AND HIRSCH 1988). This showed that further genetic changes had occurred during the long-term maintenance of these strains. Mutations had apparently arisen spontaneously and, because of the continued selection pressure, these mutations had also become 'fixed' in the lines.

Fifteen years later, after Hirsch had retired, we (R.J.G. and postdoc Dan Toma) contacted him to request the fly strains. He had been keeping them in

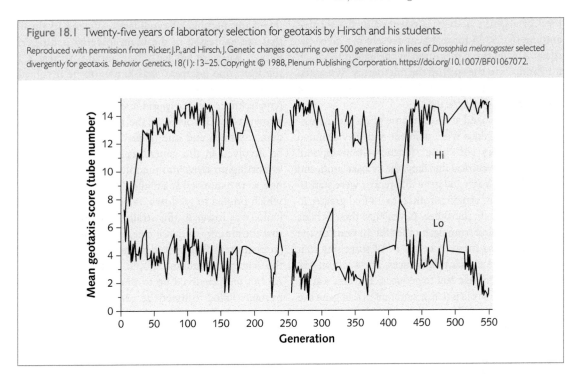

Figure 18.1 Twenty-five years of laboratory selection for geotaxis by Hirsch and his students.

Reproduced with permission from Ricker, J.P., and Hirsch, J. Genetic changes occurring over 500 generations in lines of *Drosophila melanogaster* selected divergently for geotaxis. *Behavior Genetics*, 18(1): 13–25. Copyright © 1988, Plenum Publishing Corporation. https://doi.org/10.1007/BF01067072.

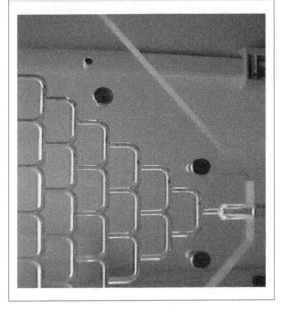

Figure 18.2 Flies from the Hi strain enter the maze and consistently take the 'upward' choice.

Photograph by R. J. Greenspan.

his office all along, and when we tried them out on our spiffy new plexiglass maze (Figure 18.2), they performed like gangbusters.

18.2 Finding the genes

Through the 30-year saga of Hirsch's geotactic flies, the nagging question that always lurked behind the scenes was which genes were responsible for the behavioral divergence. If a naturally occurring difference was primarily due to the strong effect of one gene, then classical techniques could be used to map that gene to a small region of its chromosome. After that, identifying the gene was a matter of luck, depending on whether any of the genes in that region of chromosome were known and whether one of them happened to be the culprit. With the advent of gene cloning in the 1980s, genes in the fruit fly could more easily be identified on the basis of their genetic map position. None of these approaches could be applied in a situation where many genes of small effect were each combining to produce the phenotype. Such was the case for geotaxis. So we (Dan Toma and R.J.G.) adopted another approach, examining the expression of genes in the *Drosophila* genome using microarrays.

A microarray is a slide that has tiny spots of DNA deposited or synthesized onto it, each spot representing one of the animal's genes (see Chapter 3, sections 3.4 and 3.8). In the case of the fruit fly, this comes to approximately 15,000 spots that measure how active each gene is at any given time in any given tissue. The level of a gene's activity is reflected in the relative amount of mRNA for that gene present in a given tissue at a given time, measured by the extent to which it hybridizes to the spot on the microarray. We asked which genes are differentially active in the brains of the Hi and Lo geotactic flies.

When we looked at the outcome of such an experiment, lots of genes were different and they spanned every category of gene function—transcription, chemical and electrical signaling, the cytoskeleton, and metabolism. Nearly 250 gene differences were statistically significant, ranging from 2- to 18-fold greater in one strain versus the other. Given that these strains had been isolated from each other for 50 years by the time we got to them, it is not at all surprising that there should be so many differences. The question was which ones were relevant to geotaxis. The best way to confirm a gene's role is to test a mutant of that gene in a strain that has never been selected for its geotaxis preference. Where do we find such mutants? We inquire at the huge center in Bloomington, Indiana, that maintains over 23,000 genetically defined strains of fruit flies.

Restricting ourselves to those genes for which there were available mutants helped us narrow down the number of candidates. Sticking to those that are expressed in the brain lowered it further; in fact, we were left with four mutants to test. Of the four, one (*prospero*) was known to be involved in development of the sensory nervous system, two (*cryptochrome* and *pigment-dispersing factor*) were known to be involved in the mechanism of circadian rhythms, and one (*importin-α*) was known to be involved in the general cellular mechanism of importing proteins into the nucleus. All but the last one were expressed at a higher level in the Lo strain, which prefers to go down. *Importin-α*, on the other hand, was lower in this strain. The prediction was that the *prospero*, *cryptochrome*, and *pigment-dispersing factor* mutant flies should prefer to go up, whereas the *importin-α* mutant should prefer to do down. All but *prospero* lived up to the prediction: *prospero* mutants failed to meet the prediction because the mutation had no effect whatsoever. In the case of *pigment-dispersing factor*, the degree of negative geotaxis was proportional to the amount of gene product produced, as shown using a series of genetic variants that under- and overproduce it (Figure 18.3) (TOMA et al. 2002).

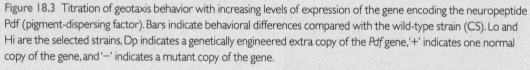

Figure 18.3 Titration of geotaxis behavior with increasing levels of expression of the gene encoding the neuropeptide Pdf (pigment-dispersing factor). Bars indicate behavioral differences compared with the wild-type strain (CS). Lo and Hi are the selected strains, Dp indicates a genetically engineered extra copy of the *Pdf* gene, '+' indicates one normal copy of the gene, and '−' indicates a mutant copy of the gene.

Reproduced with permission from Toma, D.P., et al. Identification of genes involved in *Drosophila melanogaster* geotaxis, a complex behavioral trait. *Nature Genetics*, 31:349–53. Copyright © 2002, Springer Nature. https://doi.org/10.1038/ng893.

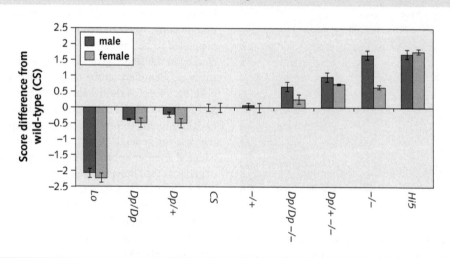

18.3 Genes for geotaxis?

We interpreted these results as saying that three of the genes, *cryptochrome*, *pigment-dispersing factor*, and *importin-α*, were likely to be contributors to the geotaxis phenotype of the selected strains, based on their ability to influence the trait independently. The verdict on *prospero* is equivocal. The result says that it cannot influence geotaxis by itself. But whether that means it is irrelevant to the phenotype or just that it requires other genes to have an effect cannot be resolved with the data at hand.

The most surprising aspect of these results is that the three effective genes are the ones we thought would be the least likely to be relevant. Two are involved in the control of circadian rhythms but apparently are not restricted in their action to that process. Cryptochrome is a light-absorbing protein that is required for entrainment of the fly's circadian rhythm to the rising and setting of the sun (or the incubator lamp, as the case may be). It is expressed in a large set of cells in the brain where it is known to interact with the other components of the circadian clock mechanism, although it has now also been implicated in magnetic field detection. Pigment-dispersing factor is a neuropeptide, a small protein that is secreted by a small set of neurons as a signal that regulates the pattern of daily locomotor activity in the fly. It relays signals from the circadian clock to the rest of the brain to coordinate daily activity. The third effective gene, *importin-α*, is present in nearly all cells of the fly, where it assists in the essential cellular mechanism of transporting proteins into the nucleus. If the *importin-α* gene is eliminated completely, it is lethal for the fly.

None of these genes would have been expected to produce a specific effect on geotaxis. This is not to say that we know much about how geotaxis works. Unlike humans, where gravity is sensed by the movement of little floating particles in our inner ear, insects apparently sense gravity by the postural pressure of their legs on stretch receptors in their joints and also by the weight of their antenna as sensed by stretch receptors at their base.

These results forced us to rethink our ideas about how genes influence behavior. Perhaps there are no such things as 'genes for geotaxis'. Instead, genes may be versatile and able to participate in many different processes. By this way of thinking, *cryptochrome* and *pigment-dispersing factor* are not circadian rhythm genes but genes that can affect many functions, two of which are circadian rhythm and geotaxis. As for *importin-α*, the fact that it can exert such a selective effect while being so widely used for other biological functions provides an important insight into the subtle way gene activities intertwine to affect behavior. It tells us that modest alterations in such widely acting genes, in combination with other similarly altered genes, can produce specific behavioral effects.

18.4 Aggression

Back when Karl von Frisch and Niko Tinbergen were delving into the innateness of the honeybee dance and the herring gull begging response (VON FRISCH 1967; TINBERGEN 1951), their Austrian colleague Konrad Lorenz was speculating on the evolutionary and innate origins of aggressive behavior. He distinguished fighting between species from fighting within a species, and only considered the latter to be the socially 'motivated' behavior of aggression. Interspecies attacks usually involve a one-sided meal at the end. Intraspecies aggression, by contrast, serves various social functions including territoriality and access to mates.

Fruit flies, despite their reputation as the carefree hippies of the insect world, do actually show aggression under the appropriate conditions. This is more in line with the classical neuroethological studies of Konrad Lorenz (LORENZ 1966).

Male flies will defend a food patch from other encroaching males, not because they refuse to share their lunch but rather because that is where they will find females ready to mate (Figure 18.4) (DIERICK AND GREENSPAN 2006). Food patches are attractive to females as the best place to lay their eggs so that the hatching larvae can start eating right away.

Females presumably favor males who are good at defending their turf. So fruit flies combine territoriality with mate choice. Aggression is a more sophisticated behavior than the reflex-like geotaxis, involving perceptual, cognitive, and social elements. Is a complex behavior of this sort innate or, more appropriately, does it have genetic components? Fruit fly biologists have long noted that flies caught in the wild and brought into the laboratory are more aggressive than their cousins who have been living the life of luxury in the lab for many generations. These wild-caught flies

Figure 18.4 Male fruit fly combat.

Reproduced with permission from Dierick, H.A. Molecular analysis of flies selected for aggressive behavior. *Nature Genetics*, 38: 1023–31. Copyright © 2006, Springer Nature. https://doi.org/10.1038/ng1864.

Figure 18.5 Increase in fruit fly aggression after 11 (dark-blue bars) and 21 (light-blue bars) generations of selection. Aggr I and Aggr II were independent lines selected for male aggression. Neutr I and Neutr II were independent lines from which aggressive males were removed.

Reproduced with permission from Dierick, H.A. Molecular analysis of flies selected for aggressive behavior. *Nature Genetics*, 38: 1023–31. Copyright © 2006, Springer Nature. https://doi.org/10.1038/ng1864.

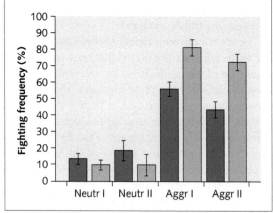

often lose their aggressive tendencies after several generations in the laboratory, but their aggressiveness is easily restored in laboratory flies after some generations of selective breeding. Given our prior success with DNA microarrays to identify genetic differences between selected strains, we (Herman Dierick and R.J.G.) tried the same approach for aggression (DIERICK AND GREENSPAN 2006).

Flies were selected by catching them in the act of fighting in a large (for fruit flies) arena containing many food patches, over 100 males, and roughly half as many females. Males engaged in a fight were snatched up gently but quickly into an aspiration tube by a *deus ex machina* (i.e. Herman) and saved for breeding with sibling females at each generation. Two lines were selected in parallel. As a control, two more lines were selected as neutral or unaggressive by removing all fighting males and picking only the remaining non-fighters for further mating.

After a mere 11 generations of selection, the flies in the 'high-aggression' lines already showed a drastic increase in their tendency to fight and in the intensity of their fighting (Figure 18.5). Even greater increases were seen after 21 generations. Following in our previous tracks, we isolated RNA from the heads of all four strains at generation 21 and compared their levels of gene expression across the whole genome.

18.5 Genes for aggression?

As in the geotaxis flies, the genes that showed up as different in aggressive files included few that might have been expected based on what is known about the biology of aggression. Serotonin is a neurotransmitter that has been associated with aggression in a variety of species, from lobsters and mice to monkeys and humans. We, too, have found that when serotonin levels in flies are pumped up, either by feeding precursors or by genetic engineering of their enzymes so that they produce more than normal, the flies become more aggressive. However, to our surprise, none of the genes relating to serotonin function—its enzymes, receptors, or transporters—are altered in the selected aggressive strains.

The classes of genes that differed between aggressive and nonaggressive strains included the same broad range that we saw for geotaxis—from transcription to metabolism. Some were available as mutants and one of these affected aggression when tested independently on a neutral genetic background. It was a gene known as *cytochrome P450 6a20*, one of a large family of genes involved in detoxification and chemical processing. Among the chemicals known to be processed by cytochrome P450 proteins in insects are pheromones—the chemical signals that insects emit to communicate with each other. One possibility is that the flies have become more aggressive because their perception of other flies has changed ('Just because you're paranoid doesn't mean they aren't out to get you.') This is further reinforced by the finding that an odor-binding protein in the fly's olfactory organ is also one of the genes that differs.

As it turned out, this same *cytochrome P450* gene reared its pugnacious head again in a totally separate study of fly aggression and social experience. When flies are raised in groups, they are less aggressive than when they are raised in solitary confinement. A test of gene expression differences between these isolated and gregarious flies, using the same microarray technology described above, revealed that the selfsame *cytochrome P450 6a20* gene was expressed at a lower level in the more aggressive flies, just as in our selection experiment. Further study showed that this gene is indeed active in the part of the fly's antenna that detects odors.

Although our yield of confirmed genes was smaller this time, perhaps due to the luck of the draw, which left us with fewer appropriate mutants to test, the experiment reinforces the same idea that emerged from the geotaxis experiment: behavior can be significantly and specifically altered by genes that are neither dedicated to that behavior, nor restricted in their action. In this instance, the outcome also showed that major behavioral changes can result from minor differences in gene activities.

18.6 Behavior, the single gene, and the mind of a worm

In 1974, Sydney Brenner was working at the Laboratory of Molecular Biology (LMB) in Cambridge, sharing an office with Francis Crick, co-discoverer of the structure of DNA. Brenner's background was in biochemistry and molecular biology, a discipline in part created by scientists at the LMB. Like Crick, Brenner's interests were turning to the analysis of behavior, but while Crick wanted to tackle the complexities of the mammalian brain including the problem of consciousness, Brenner wanted a far simpler, tractable experimental system.

Brenner thought the fly was too complicated and that Seymour Benzer's search for single-gene mutations would not lead to fundamental insights into the molecular basis of behavior. As *Drosophila* has 100,000 neurons, which at the time was far more than could be enumerated and put into circuit diagrams, it would be too difficult to explain how the mutations had an effect. '*I decided that what was needed was an experimental organism which was suitable for genetical study and in which one could determine the complete structure of the nervous system ... my choice eventually settled on the small nematode,* Caenorhabditis elegans' (BRENNER 1974) (worm geneticists have to learn early to pronounce correctly the name of their favorite organism—as a first approximation try 'see-no-rab-die-tis'. To make life easy, most people refer to it as either *C. elegans* or, with even more simplicity but less accuracy, as 'the worm').

C. elegans lives in soil. It grows to about 1.3 mm and reaches maturity at about 3.5 days old. It eats bacteria and can easily be grown in the laboratory. There are two sexual forms: a self-reproducing hermaphrodite and a much rarer male form, which makes up about 0.1% of a normal worm population. These useful habits make it easy and cheap to look after in the laboratory.

With the worm, the intention was always to do things to completion: to map the position of every neuron; to map all the connections (or, more technically, synapses) between neurons (remember that there are many more connections than there are neurons: a typical mammalian sensory neuron has about 1,200 synapses, but the number is lower in the worm); to know the developmental trajectory of each cell, its pathway from egg to adult; and, eventually, to sequence the entire 100 Mb genome, so that every gene could be identified and every gene assigned a function. As a result, we now know more about the worm than we do about any other multicellular organism on the planet.

18.7 The first complete anatomy and developmental history

The first major achievements were maps of the neuronal circuitry and cell lineage. Brenner's group in Cambridge gradually expanded to take on the challenges. Worms were sliced up into sections small enough to visualize nerve cells at the highest possible resolution (using an electron microscope); every connection, every synapse could be, and was, counted. By 1986, an article of 341 pages was published in the *Philosophical Transactions of the Royal Society* that described the worm's nervous system (WHITE et al. 1986). The title of the paper suggests a purely anatomical intention ('The structure of the nervous system of the nematode *Caenorhabditis elegans*') but open up the article and after a few introductory pages you'll see that each page is headed with something slightly more ambitious: it reads 'The mind of a worm'.

Here's their summary of what they found: '*The nervous system has a total complement of 302 neurons, which are arranged in an essentially invariant structure. Neurons with similar morphologies and connectivities have been grouped into classes; there are 118 such classes. Neurons have simple morphologies with few, if any, branches … Neurons are generally highly locally connected, making synaptic connections with many of their neighbours.*' Hermaphrodite and male worms differ radically in their neuroanatomy: the hermaphrodite has 302 neurons and the male 381, of which 87 are sex specific.

In addition to the circuitry, Brenner and his followers worked on the developmental history of every cell, the origin of every neuron in the worm's body. This was achieved by John Sulston, by systematically observing what happened to each cell as the animal developed from egg through larva to adulthood (SULSTON 1976; SULSTON AND HORVITZ 1977).

18.8 The first complete animal genome

Next came the first complete animal genome. The first step was to make a complete molecular map of the worm's genome by randomly fragmenting the worm DNA into pieces of about 30 kb, circularizing the molecules, and adding sequences that would allow them to be grown in *Escherichia coli* bacteria. Then, selecting an overlapping set of these 30 kb molecules for DNA sequencing, the first complete DNA sequence of a multicellular organism was obtained (C. ELEGANS SEQUENCING CONSORTIUM 1998). Worm geneticists had a wiring diagram, a developmental trajectory, and a complete genome. But if behavior is the goal, just how much can a creature do with 302 neurons?

18.9 Worm behavior genetics

Fly neurobiologists sometimes point out that it makes little sense to try to understand the functioning of a brain that consists of 10 billion neurons when we

don't know how a brain made up of 100,000 works. The same could be said of the worm, which has 302 neurons. The worm could be one animal where we have a good chance of tracking the complete path from gene to behavior. So, what do we know?

One issue, often raised in conversations about the behavior genetics of worms, is the extent to which worms do much of interest. Do worms learn? Can they remember things? And even if they don't do much, is their behavior enough to count as consisting of elements of more complex action, or is their life so different from that of other animals that whatever worms teach us about how genes influence behavior is going to be specific to worms? In other words, will we learn much or indeed anything that applies to other animals?

Let's start with what a worm can do. It can find its own food, drink alcohol (but not as much as flies), and have sex (it prefers to have sex with itself). It likes to live in a dark chaotic world surrounded by decaying objects, specifically a thin film of water between rotting vegetation. It doesn't like being touched, and is prepared to travel quite a distance to find food it likes (and move away from stuff it doesn't). To do all this, it needs to be able to detect traces of chemicals, interpret what they mean, and act appropriately. A lot of worm neurobiology involves tracing the pathway from outside stimuli (e.g. chemical, thermal, pressure) to internal neuronal circuitry through to a final motor output.

Worms **can** learn things. Cathy Rankin showed that tapping the small plastic Petri dishes on which worms live makes them move backward about 1 mm and that with repeated tapping they move backward less (RANKIN et al. 1990). In short, they habituate to the stimulus, just as *Aplysia* (a genus of sea slug) learns to do. It's been shown that this is not just adaptation to the sensory stimulus because the response is reversible (an animal that has habituated can still sense the stimulus but is attending to something else until the stimulus becomes relevant) (BERNHARD AND VAN DER KOOY 2000). And worms even have a rudimentary form of communication, through the secretion and detection of chemicals known as ascarosides. How do genes affect these behaviors?

It has been said that all objects on earth are covered in a thin film of nematode worms. I'm not entirely convinced by that idea, but there certainly are a lot of worms in the world, more than enough for an enormous genome-wide association study, but sadly no one has done that (just imagine collecting all those tiny worms). And no one has carried out a family-based study to assess the inheritance of behavior. But we do have information from genetic-mapping experiments, those with a design similar to inbred-strain crosses carried out in mice.

Cori Bargmann, a worm geneticist at the Rockefeller University in New York, has been using crosses between worm strains to identify the genetic basis of variation in foraging behavior. There are marked, easily recognizable differences in foraging behavior. Solitary foragers move slowly on a bacterial lawn and disperse across it, while social foragers move rapidly and congregate at the edge of a patch of food (this is called 'bordering') and feed together (this is called 'clumping') (Figure 18.6) (DE BONO AND BARGMANN 1998).

Her first discovery was that aggregating strains of *C. elegans* have a single base-pair change in the neuropeptide receptor 1 gene (*npr-1*), introducing a phenylalanine at a position in the protein where solitary feeders have valine. Genetic engineering of the variant form of the *npr-1* gene can turn solitary feeders into social aggregators (DE BONO AND BARGMANN 1998). Worms with the phenylalanine variant in *npr-1* avoid high oxygen environments while feeding, whereas strains with the valine variant do not.

Bargmann found that the behavior could be fully restored in *npr-1* mutants by adding the gene back to a neuron called RMG. Furthermore, specifically killing the RMG neuron (this is possible in *C. elegans* by blasting a laser onto the cell) created the same phenotype as the *npr-1* mutation. RMG is the hub of a network connecting seven classes of neurons, including some oxygen-sensitive neurons (those already implicated in social aggregation) (CHANG et al. 2006). This and other observations led Bargmann to argue that, *'The analysis of RMG suggests a hub-and-spoke model for aggregation behavior, in which distributed sensory inputs are coordinated through gap junctions with the RMG hub to produce distinct, distributed synaptic outputs'* (gap junctions are one way in which neurons connect to each other) (MACOSKO et al. 2009). The discovery of the mutation led, through an understanding of the neuroanatomy, to the delineation of a neuronal circuit mediating behavior. Aggregation requires the integration of different sensory signals, a task that is performed by the RMG neuron that sits at the hub of series of neurons.

Bargmann and her group have also mapped variation in a worm's speed of foraging at different oxygen concentrations. They used a cross, and then recombinant inbred lines, to identify a locus, called *glb-5*, in which they found a duplication in a gene with similarity to globin, a protein that binds oxygen (globins are part of the hemoglobin molecule that transports oxygen in your blood). The duplication reduces gene function by interfering with splicing (PERSSON et al. 2009). The globin acts as a sensor molecule within

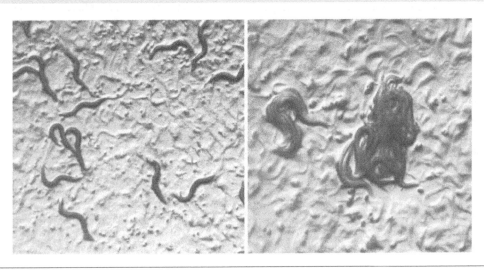

Figure 18.6 Natural variants of the nematode *Caenorhabditis elegans* exhibiting solitary (left) and 'social' or aggregated (right) feeding behavior.

Reproduced with permission from de Bono, M., and Bargmann, C.I. Natural variation in a neuropeptide Y receptor homolog modifies social behavior and food response in *C. elegans*. *Cell*, 94(5): 679–89. Copyright © 1998 Cell Press. All rights reserved. https://doi.org/10.1016/S0092-8674(00)81609-8.

a neuron, so that in response to varying concentrations of oxygen, downstream signaling pathways are modulated within the neuron, eventually leading to a behavioral output. The idea that neurons contain such sensor molecules, rather than relying on signals passed from peripheral sense organs (as mammals do), might seem a bit strange, but for worms this is normal: for example, their head sensory neurons also act as primary carbon dioxide sensors (BRETSCHER et al. 2011).

Glb-5 and *npr-1* are not the only loci that contribute to variation in foraging. Bargmann surveyed chromosome substitution strains and mapped, using recombinant inbred lines, three additional loci affecting aggregation (BENDESKY et al. 2012) (the genetic-mapping tools are described in Chapter 15, section 15.6). Foraging thus emerges as a polygenic trait, but the genetic-mapping experiments identified relatively large-effect variants, altering the coding regions of genes. This, of course, contrasts with the highly polygenic architecture of small effects located within noncoding regions of the genome that we have documented in human studies. Why the difference? Why does foraging not have the same polygenic architecture we have met elsewhere in the book? We posed a similar question at the end of Chapter 17 when we asked why circadian rhythms had such a different genetic architecture to other traits.

The answer, at least for foraging in worms, is almost certainly because of strong selective pressure exerted on the behavior, as became clear when Bargmann tackled the genetic basis of foraging from a different angle, through an analysis of the use of pheromones (a chemical that an animal releases to alter the behavior of others; in worms, these are known as ascarosides). Some ascarosides regulate foraging by suppressing roaming behaviors, and Bargmann used recombinant inbreds to map sensitivity to one of these pheromones. She found a locus (*roam-1*) that accounted for almost half of the genetic variation between the two parental strains, so another large effect (GREENE et al. 2016). Remarkably, the two haplotypes carrying the variant are highly diverged: the two differed at 20% of the positions that they sequenced. Bargmann argued that the difference is due to balancing selection; that is, there are evolutionary pressures to maintain both forms. In some environments, one form confers an advantage, while the second form is advantageous in different environments. Balancing selection may be fairly common throughout the worm genome.

What is worth noting here is that the *roam-1* locus is large, probably contains more than one gene contributing to the foraging behavior, and has some of the features of a 'supergene' (see Chapter 19, section 19.5), a region of DNA in which multiple alleles relevant to the behavior are co-inherited.

There is another observation from worm genetics that parallels findings in other species. Bargmann tracked the behavior of individual worms continuously across development, and explained the variation as due to the action of secreted endocrine or peptide factors, which go by the name of neuromodulators (STERN et al. 2017). These neuropeptides alter the behavior of neuronal circuits, sometimes dramatically so. The classic example of neuromodulation is from Eve Marder's work on the stomatogastric ganglion of the lobster, a small set of interconnected neurons that control feeding rhythms throughout life (MARDER 2012); the sensitivity of the ganglion to neuropeptides depends on developmental stage and environmental context. The importance of neuromodulators appears elsewhere in this book, for example with the role of a regulator of G protein-coupled signaling on fear-related behavior in mice in Chapter 15, section 15.11 (YALCIN et al. 2004). You can find more about neuromodulators in Chapter 16, section 16.8.2.

18.10 Wanderlust in fly larvae

Like *C. elegans*, fly larvae differ in their feeding habits, and again much of this effect can be genetically traced to natural variation at a single locus. Food for fly larvae in the laboratory is a mixture of yeast, sugar, and agar, and they forage for food in one of two ways: as 'rovers' or 'sitters' (SOKOLOWSKI 1980). The 'sitter' strain prefers to remain feeding within a single food patch while 'rovers' move between food patches, and take a straighter path to food than 'sitters'. (A journalist reporting on this finding when the gene was identified likened it to '*a gene for dining out*'.) When there is no food to be found, both variants move at about the same speed, so the difference is not because 'sitters' are intrinsically lazy. 'Rovers' are more common than 'sitters' in nature (70 versus 30%), and selection has been shown to maintain these variants.

Marla Sokolowski in Toronto and her colleagues identified the gene that is responsible for most of this difference: *dg2*, encoding a protein kinase called PKG (for cyclic GMP-dependent protein kinase)

(OSBORNE et al. 1997). As with the worms, it was possible to change larval behavior by genetically engineering the gene into larvae and thus show unambiguously that *dg2* is responsible for the behavioral difference. However, the behavioral change is not entirely due to genetics. After a short period of food deprivation, larval and adult flies with 'rover' alleles can be made to behave as 'sitters' (FITZPATRICK et al. 2007).

The discovery of the 'rover'/'sitter' difference and the identification of its gene preceded that of *npr-1* in *C. elegans*, so when the worm findings came out, both camps were naturally curious to know whether the behavioral similarities in their wanderlust had a common underlying mechanism. This was soon shown to be true when the fly version of *npr-1*, known as *neuropeptide F receptor*, the *C. elegans* version of *dg2*, known as *egl-4*, and the respective neuropeptide-encoding genes in each animal—*npf* in flies and *flp-21* in worms—all turned out to have similar behavioral effects (ROGERS et al. 2003; WU et al. 2003).

Laboratory mutants have confirmed a link between PKG and neuropeptide receptors: the neuropeptide receptors initiate a cascade of reactions in the cell that activate PKG. That's not all. In yet another example of gene action crossing species boundaries, the same PKG gene is responsible for a major developmental transition in foraging behavior in that most social of insects, the honeybee.

Honeybees are probably the most famous social foragers of all time, due in large measure to Karl von Frisch's studies of their dance language (VON FRISCH 1967). When they are growing up in the hive, females start out as 'nurses', tending to the feeding and hygienic needs of each larva as it develops in its little hexagonal cubbyhole. When nurses reach a certain age, they graduate to being foragers, capable of venturing outside the hive, exploring for new sources of nectar, and returning to 'tell' their comrades about where to find the goodies and how much is there. A key step in this transition is the developmental activation of their version of the PKG gene. If their PKG is inhibited, they will not develop into foragers, and those hive members that precociously develop into foragers (as occurs from time to time) have a precociously active PKG. And if that were not coincidental enough, the homolog of this gene in ants is also involved in the shift to foraging behavior (BEN-SHAHAR et al. 2002; LUCAS AND SOKOLOWSKI 2009).

18.11 So which is it? Pluribus or unum?

We began by wanting to compare humans to these apparently simpler animals. How did it come out? Are they genetically like humans? Most natural variation in fly behavior certainly appears to be multigenic, as it seems to be in humans. But rare cases of single-gene natural variation in behavior in flies and worms bear an uncanny similarity to each other in the biochemical pathways that are affected, and in the similarity of the behavioral effects they produce.

In all of these examples, the explanation for what kind of variation can be found must always go back to a consideration of which genes (and which associated biological processes) can be tolerated by natural variation; that is, they do not cause such undue harm that they are eliminated from the population by natural selection. It appears that the pathway consisting of the neuropeptide associated with feeding behavior, its receptor, and PKG activated by it can tolerate a fair degree of variation. This variation is apparently, in at least some environments, sufficiently adaptive to the animal that it maintains itself in natural populations.

Feeding behavior is certainly a very basic behavior. It may be so primordial that the pathways and mechanisms for its performance and regulation are so similar in today's fruit flies and nematodes because their ancient common ancestor, who probably lived 500 million years ago, harbored the same mechanism and passed it on to them.

Summary

1. Laboratory selection experiments in fruit flies can produce major behavioral changes, showing that behavior has a polygenic basis.

2. Assays of gene expression differences in selected lines of fruit flies (using microarrays) followed by analysis of mutants identified genes that influence geotaxis and aggression.

3. The nematode worm *Caenorhabditis elegans* was the first model organism to be comprehensively studied with respect to its anatomy, cell lineage,

and genome sequence, providing a model for interpreting pathways from gene to neuron to behavior.

4. A single-gene mutation in a neuropeptide receptor gene, *npr-1*, found both in laboratory and in wild-caught worms, alters the animal's feeding behavior from solitary to aggregated or 'social'.

5. A locus called *roam-1* with a large effect on worm foraging behavior lies on a haplotype that is under balancing selection (evolutionary pressure maintains the frequency of two haplotypes). Balancing selection may be fairly common throughout the worm genome and might explain the presence of relatively large genetic effects on behavior.

6. A fruit fly gene for cyclic GMP-dependent protein kinase (PKG) also exists as a single-gene behavioral polymorphism in the wild, affecting foraging behavior; it is part of the same signaling system as the worm *npr-1* gene, and also affects analogous behaviors in bees and ants.

7. While naturally occurring variants in single genes can have large impacts on behavior, this is relatively rare. The vast majority of genetic variation in natural populations that impacts on behavior is contained in genetic variants of small effect.

19

Comparative genomics

[On the importance of evolution for explanations of behavioral variation, and on the use of genomes from different species to identify genes and variants that contribute to behavior]

There is a famous essay by the geneticist and evolutionary biologist Theodosius Dobzhansky entitled 'Nothing in biology makes sense except in the light of evolution' (DOBZHANSKY 1973), which is as good an introduction as any I know to the importance of evolution. The aim of this chapter is to introduce you to the insights obtained from examining the genetics of behavior from an evolutionary standpoint. Our starting point is to see how behavioral differences between species can be attributed to their genetic differences. You can think of this as an extension of the approaches we dealt with in earlier chapters, where genetic differences within a species are used to explain behavioral variation **within** a species. In order to examine behavioral differences **between** species, we compare genomes, which is why this chapter is entitled comparative genomics. Creation's hierarchy of species, genera, and taxa, ordered within phylogeny reflecting their genetic relatedness, provides the raw material for comparative genomics.

We start by asking how genetic differences between us and other apes can be used to identify the genetic basis of human behavior (e.g. what makes us able to talk and think). We then look at examples of other interspecies comparisons to determine genetic effects on behavior, using the examples of monogamy, nest building, and social behavior. We continue with the discovery from genetic analyses in ants and birds of supergenes that control behavior.

19.1 Finding out what makes us human: genetic comparisons with great apes

I'd really like to know why I can speak and think in a way that other animals cannot, not even our closest relatives, the chimpanzees. Now, as chimpanzees and humans are separated by about 5 million years of evolutionary history, surely if I had the genomes of humans and chimpanzees, I could look through the sequence, spot the differences, and find out what those differences do? I would then know the genetic basis of language, for example. Of course, there are a lot of other differences between chimpanzees and humans, such as the amount of hair apes have for one, which must also be encoded in our genomes, but surely some careful analysis, and maybe some genetic manipulations in apes, would determine which variants do what? And maybe I could genetically engineer a chimpanzee so that it could talk to me.

Comparative analyses between the genomes of humans and closely related primates have long been expected to identify unique features of human evolution (KING AND WILSON 1975). A chimpanzee genome first became available in 2005 (CHIMPANZEE SEQUENCING AND ANALYSIS CONSORTIUM 2005) from which we learnt that there are about 40 million sequence differences between our genomes and that of our closest relatives. All told, about 1.23% of homologous human and chimpanzee DNA differs at single-nucleotide variants (where, for example, the human genome might have an 'A' and the chimpanzee a 'C' nucleotide). But this is far less than the 90 Mb of unique sequence, that is, sequence that is only found in humans, or only found in chimpanzees. About 3% of the genome is specific in this sense (note that this is higher than the often-quoted figure of a 1% difference between humans and chimpanzees).

The 2005 sequence analysis of human and ape genomes was a first step, but only a first step. Copy-number variation (the type of variation we saw in Chapter 7, section 7.7, Chapter 8, section 8.4, and Chapter 9, section 9.3, which plays a role in the genetic etiology of intellectual disability, schizophrenia, and autism) had evaded detection in the initial great ape genome assemblies. Could structural variants, such as duplicated segments of the genome, include genetic material responsible for the differences between humans and chimpanzees (including behavioral differences)? One geneticist, Evan Eichler, thought they might.

Eichler discovered that there had been a burst of duplication activity during the evolutionary process that gave rise to the human lineage (MARQUES-BONET et al. 2009). These duplication events did not just occur in the human branch. Eichler found gene-containing duplications occurring within the gorilla and chimpanzee lineages that are absent in humans. The relative acceleration of duplication events in the human lineage suggested that genes involved in human evolution might be associated with the appearance of structural variation in the ape genomes. In other words, by focusing on genes affected by the rearrangements, it might be possible to identify those that characterized humankind. But how could the genes be affected? In almost every case, the coding sequences were the same, so the structure of the proteins could not be responsible. What else could it be?

One clue came from an analysis of hundreds of segments of DNA that are deleted in humans but present in chimpanzees, and that are present in other species too. The sequences missing in humans, but highly conserved in other species, occur in noncoding regulatory regions and are enriched near genes involved in neuronal function (MCLEAN et al. 2011). Regulatory regions alter the transcription and splicing of genes, so one hypothesis is that the amount of transcript or isoforms, and presumably the amount and types of protein, is in part responsible for the chimpanzee/human differences.

By 2018, Eichler had generated high-quality assemblies of great ape genomes and had identified almost 6,000 human-specific deletions and 12,000 insertions (KRONENBERG et al. 2018). Ninety structural variants disrupted genes, and an additional 643 affected regulatory regions. Intriguingly, his group also reported the presence of additional, larger structural differences, whose nature and extent they were not able to fully resolve (this is because the sequence identity extended over such large regions that it was impossible to determine how many copies there were and how the sequences were arranged on the genome). The significance of these larger rearrangements is not clear.

In the regions of genomic instability, where copy number and other rearrangements occurred, there are differences in transcript abundance for genes expressed in neurons: 'For example, among the 18 chimpanzee-human inversions, we identified 18 differentially expressed brain genes between chimpanzee and human ... Three of these genes (GLG1, ST3GAL2, and EXOSC6) were significantly up-regulated in human and associated with a 5-Mbp human-specific inversion on chromosome 16q22' (KRONENBERG et al. 2018). In lieu of testing expression in brain tissue, they measured transcript abundance in cerebral organoids, the 'brains in a dish' that can now be grown from stem cells (see Chapter 16, section 16.2.1). They identified several hundred of such differentially expressed genes in neural progenitor cells, all of which became candidates for functional investigation (KRONENBERG et al. 2018).

In an intriguing extension of the use of comparative genomics to explain behavior, Eichler explored the idea of looking at the comparative genomics of psychiatric disease alleles. Recall that some structural variants are risk factors for psychiatric disease. Do any of these rearrangements coincide with the loci that are undergoing accelerated evolution? Eichler looked at one copy-number variant on chromosome 16 that has a relatively high frequency, accounting for approximately 1% of cases of autism (the majority of structural variants are rare) (WEISS et al. 2008b). This is the same rearrangement that we discussed in Chapter 16, section 16.8.1.

Figure 19.1 shows the locus affected by the 16p structural variant. It is 593 kb in size and is flanked by two regions that are almost completely identical (99.5% identity), a feature that is often found in regions of the genome that undergo rearrangement and is believed to be the cause of the structural variant. There are 28 genes in the region, of which 11 are expressed in neurons. Any, or all, could be involved in disease.

Eichler showed that the genetic difference between humans and chimpanzees at the chromosome 16 locus is much larger than predicted by the mean rates of genetic differentiation between chimpanzees and humans (NUTTLE et al. 2016). In other words, this region does indeed appear to be undergoing accelerated evolution. Although the region has undergone extensive rearrangement in all great apes, only within the human lineage are there large (>100 kb) duplications that coincide with the autism critical region. There is a 95 kb segment that was duplicated after human divergence, and this event contributes more sequence that is specific to humans than all combined single-nucleotide variants. Eichler went on to show that the human-specific rearrangement resulted in the fusion of two genes (BOLA2 and SMG1P) to produce a transcript containing 53 residues from *BOLA2* and 164 residues from *SMG1P*. In other words, the rearrangement has created a new gene that is only present in humans. What does this new gene do? Control language? Confer the ability to perform linear algebra? We don't yet know (NUTTLE et al. 2016).

The take-home message is that, while no one has yet found human/chimpanzee genetic differences that can explain, except in the broadest terms, the behavioral differences between species, this is still very much a work in progress. There are indications that genes involved in determining brain development (e.g. the size of the brain) are involved. How these developmental differences translate into behavioral phenotypes remains unknown. As more detailed comparisons occur, and with additional evidence about the function of the genes proposed, it is possible that we will arrive at a genetic explanation for what distinguishes us from our nearest relatives on the evolutionary tree, and this information could explain why we talk and think in the way we do.

19.2 The genetic basis of fidelity

A few years ago the Max Planck Institute for Ornithology identified a 'Casanova gene' in female zebra finches, which show a predisposition for infidelity unrelated to their evolutionary needs.

C. Bailey, New York Review of Books,
September 2018 Issue.

For humans, monogamy is more a life-style choice, or adherence to a social norm, than a biological phenotype. It seems odd to think of it as a characteristic

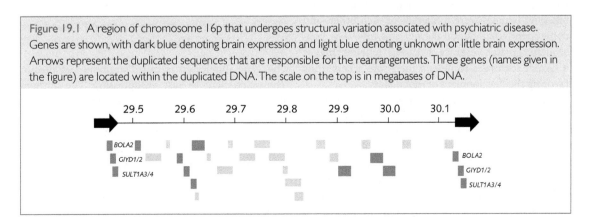

Figure 19.1 A region of chromosome 16p that undergoes structural variation associated with psychiatric disease. Genes are shown, with dark blue denoting brain expression and light blue denoting unknown or little brain expression. Arrows represent the duplicated sequences that are responsible for the rearrangements. Three genes (names given in the figure) are located within the duplicated DNA. The scale on the top is in megabases of DNA.

possessed by only some species, but that's because our view of behavior is too often constrained by our view of ourselves. By looking at the rest of creation, we obtain insights impossible by restricting our observations to one or a few species.

While more than 90% of birds are monogamous (COCKBURN 2006), only between 3 and 9% of mammalian species bond and remain (pretty much) faithful to their partner (KLEIMAN 1977; LUKAS AND CLUTTON-BROCK 2013). Monogamy occurs in 29% of primates, but is absent among whales, dolphins, and porpoises (LUKAS AND CLUTTON-BROCK 2013). Why? This has to have a genetic basis, surely?

From an evolutionary perspective, monogamy in mammals is a puzzle. If you are a male, to maximize the chances of passing on your genes to the next generation it makes no sense to confine your reproductive activities to a single female, particularly as females give birth to and breastfeed the offspring, leaving you with little to do (compared with many fish and arthropods).

There are two explanations for the appearance of monogamy: it evolved because males are unable to defend access to more than one female at a time, or because by forgoing mating opportunities males increase the chances of their offspring surviving (by looking after them better and avoiding attacks by infanticidal competitors). Dieter Lukas and Tim Clutton-Brock looked at the patterns of monogamy in over 2,500 mammalian species and observed an association between monogamy and low population density; monogamy is found in species where females are solitary and the geographical area over which males range (their home range) overlaps those of several females (LUKAS AND CLUTTON-BROCK 2013). They argued that monogamy evolved in mammals where feeding competition between females was intense, breeding females were intolerant of each other, and consequently population density was low. The consequent greater distance between females makes it harder for males to defend their sole access, thus favoring the development of monogamy.

Lukas and Clutton-Brock's argument takes as its premise an evolutionary given, that the major determinant of behavior is its consequences for which genetic variants are passed on to the next generation: monogamy becomes more common when monogamous behavior results in the survival of more offspring of monogamous parents, relative to promiscuous parents. What are these variants? To answer

this question, we turn to yet another nonstandard model organism: the vole.

19.2.1 Prairie voles and meadow voles

On a website at the University of Illinois at Urbana–Champaign you can download 25 years' worth of data (1972–1997) collected by Lowell Getz (http://www.life.illinois.edu/getz/25-year/index.html). Each month, he and his students set traps for rodents: *'Cracked corn was used for prebaiting and as bait in traps. We used vegetation or aluminum shields to protect the traps from the sun during the summer. The wooden traps provided ample insulation in the winter, and making provision of nesting material unnecessary'* (GETZ et al. 2005). The data are for two species, the prairie vole (*Microtus ochrogaster*) and the meadow vole (*Microtus pennsylvanicus*), and they show that adult male and female prairie voles are more often caught as pairs (the same pair is caught together again and again, like a married couple), compared with meadow voles, which are caught as independent individuals. Prairie voles are monogamous, meadow voles less so.

Differences between prairie and meadow voles have been used to argue for the importance of a single gene in determining monogamy, as the title of a 2004 paper makes clear: 'Enhanced partner preference in a promiscuous species by manipulating the expression of a single gene' (LIM et al. 2004). A single gene! And even more impressively, the authors claimed that they had identified the sequence variant itself that is responsible, a repetitive sequence in a region that controls the amount of transcript. You can imagine how this came across in the media, for after all this was a gene that when overexpressed might make your partner less promiscuous.

19.2.2 Vasopressin and oxytocin: modulators of social behavior

The gene implicated is called arginine vasopressin (shortened here to just vasopressin). The authors already had a pretty good idea that vasopressin and its friend, oxytocin, were involved. Vasopressin is a small protein consisting of nine amino acids (a nonapeptide). The gene that encodes vasopressin is always found in a head-to-head arrangement on mammalian chromosomes with the gene that encodes oxytocin, another nonapeptide. This is because both genes are the products of an ancestral duplication that happened about 500 million years ago, when vertebrates first appeared.

In vertebrates, the peptides are arginine vasotocin (vasopressin in mammals) and the oxytocin-like peptides (isotocin in fish, mesotocin in birds, and oxytocin in mammals). They are usually expressed in the brain and gonads, and are influenced by gonadal steroids, seasonality, and sex. Throughout evolution, their two major functions seem to have been regulation of water and salt (osmotic regulation), blood pressure (baroregulation), and reproduction. Oxytocin has been studied for its peripheral role in uterine contractions during parturition and milk ejection from the breasts. However, the molecules are of interest to us because they are also involved in the social behavior of many species, such as pair bonding and maternal care (DULAC et al. 2014).

In the mollusk *Lymnaea stagnalis*, a single ancestral nonapeptide is expressed selectively in neuronal and gonadal cells where it influences male copulatory behavior (VAN KESTEREN et al. 1995). In finches, mesotocin and its receptor influence such social behavior as flock formation (GOODSON et al. 2009). In the midshipman fish (*Porichthys notatus*), reproductive tactics, including the noises the fish make (grunts), are modulated by arginine vasotocin and isotocin (GOODSON AND BASS 2000).

In mammals, the earliest studies linking oxytocin and social behavior demonstrated that oxytocin infused into the brain made rats nurture any pup that ventured into its nest (PEDERSEN AND PRANGE 1979). Oxytocin injected into female sheep makes the ewe accept a foreign lamb, rejecting all others (KENDRICK et al. 1997). Oxytocin plays a central role in inducing parental behavior in mice. When baby mice are pulled out of their nests, they cry for their mothers, although their cries go unheard by humans as the frequency they use is too high for us to hear. While mothers retrieve pups, virgin females don't respond to the pup's cries unless they have injections of oxytocin. We even have some idea of how this happens: Robert Froemke's work indicated that oxytocin changes the properties of neurons in the left auditory cortex (remarkably, the effect is indeed on one side only), so that they respond to the pup's ultrasonic distress calls (MARLIN et al. 2015).

Tom Insel and Larry Young showed that oxytocin and vasopressin have remarkable effects on the monogamous prairie vole (INSEL et al. 1994). Studies showed marked differences in the distribution of the vasopressin receptor 1a (Avpr1a) in the brains of monogamous prairie voles and promiscuous montane voles. What might these differences mean? One suggestion was that the receptor mediated the signal for pair bonding. Insel and colleagues had already shown that vasopressin was involved in forming partner preferences in voles. In his experiments, a male vole was mated and the following day was given a choice. The vole was put in a chamber where yesterday's partner and a new female, one he had not seen before, were tethered. The measure of pair bonding was the amount of time the male spent with his partner. Blocking Avpr1a 24 hours after a vole had mated caused the male to show no preference for yesterday's sexual partner (WINSLOW et al. 1993). So, if you block vasopressin signaling during mating, you block pair-bond formation. And if you give vasopressin to voles, they will form a bond even if they don't mate.

19.2.3 Can a single genetic variant determine fidelity?

Given that vasopressin is involved in pair bonding, it was a reasonable guess that a genetic variant that contributed to the expression of its receptor in the brain might be responsible for the behavioral differences between the promiscuous and monogamous voles. Young and Insel identified genetic variants in a region at the front of the *Avpr1a* gene. There is a repetitive region, a short tandem repeat of GAGAGA, that they found contributes to variation in expression of the gene (YOUNG et al. 1999; HAMMOCK AND YOUNG 2005). The social prairie and pine voles have a long repeat, while the asocial montane and meadow voles have a short one. Young and Insel argued that variation in the length of the repeat was responsible for the species differences in expression patterns of *Avpr1a*, which in turn is responsible for the behavioral differences. This story appeared to provide an example where a behavioral difference was tied to a single gene, and to a single molecular variant. We hope, by this stage in the book, you are skeptical of such simplistic explanations, and if so, then your skepticism is justified.

Here are the things we know: artificially increasing the *Avpr1a* gene in the prairie vole brain confers the ability to form a partner preference in the promiscuous species (LIM et al. 2004). Differences in *Avpr1a* expression within the brain do contribute to both inter- and intraspecies variability in vasopressin-mediated behaviors. But whether individual differences in a single DNA variant (the *Avpr1a* repeat) are sufficient to drive the differences in gene expression, and sufficient therefore to explain monogamy isn't so firmly established.

The evidence comes from an experiment in which Young and his colleagues selectively bred male prairie voles to have different sized *Avpr1a* repeats. They reported that animals with the long alleles are more likely to form partner preferences than males with short alleles (HAMMOCK AND YOUNG 2005). Zoe Donaldson and Larry Young revisited the importance of the repeat and found a more confusing picture than the selection experiment suggested (DONALDSON AND YOUNG 2013). Using the homologous recombination technology described in Chapter 16, section 16.2, they replaced the mouse DNA sequence in front of the *Avpr1a* gene with vole sequence, creating mice lines that had the meadow vole long-repeat or prairie vole short-repeat variant.

This is exactly the right experiment to do—as the rest of the mouse genome was identical between the lines, the experiment determines the effect attributable to the difference in the repeat, exactly what we want to know. When claims are made that a variant contributes to behavior, and when it is possible, as it is in model organisms, to mutate the variant, then there's little excuse for not engineering the DNA to test the hypothesis. Often, however, the requisite experiment is not carried out (e.g. a nonsynonymous mutation of serine to phenylalanine in Cyfip2 is said to contribute to cocaine-response phenotypes, without evidence from an engineering experiment (KUMAR et al. 2013); see Chapter 15, section 15.11).

Donaldson and Young found that *'contrary to our expectation, this region has only a modest influence on differences in expression patterns across rodent species'* (DONALDSON AND YOUNG 2013). I (J.F.) showed my summary of the experiment to Zoe Donaldson. I was writing a textbook, I explained, and I wanted to make sure I had presented her work correctly. She replied, *'Too bad you can't tell them about how this was approximately 7 years of my life, and as I looked at the autoradiograms, I wanted to cry.'*

The variant had only a very modest effect on behavior too. Furthermore, comparison of the repeat sequences of 25 rodent species revealed that the repetitive element thought to cause monogamy is actually widespread in nonmonogamous voles and other rodents (FINK et al. 2006). Oxytocin, and vasopressin, have an effect on mammalian behavior, but the genetic underpinnings of the variation within and between species is not explained by a single regulatory element.

19.2.4. The polygenic basis of fidelity

Most likely, monogamy is the product of multiple genetic effects, and an experiment by Hopi Hoekstra showed this to be so, using a genetic cross between two species of deer mice with different behaviors. *Peromyscus maniculatus* is promiscuous while *Peromyscus polionotus* is monogamous. Furthermore, *P. polionotus* fathers are caring fathers (licking their pups and huddling with them) compared with the far less involved *P. maniculatus* fathers. Unlike the voles, the two species of deer mice can interbreed in the laboratory, which makes it possible to map the loci that contribute to these differences, using the same methods for mapping used in crosses between inbred strains (see Chapter 15, section 15.6).

Bendesky and Hoekstra found 12 loci on 11 chromosomes that contributed to parenting behaviors (they use parenting behaviors as a proxy for monogamy, arguing that the differences in mating behavior mirror the differences in parenting styles) (BENDESKY et al. 2017). Their findings are consistent with a polygenic, pleiotropic genetic architecture. That a relatively small number of loci were found at genome-wide significance thresholds reflects the lower power to detect the many other loci of small effect that are likely to exist. The relatively small number of mice they used (769) draws attention to another issue. If you recall, human studies require thousands of individuals to detect a small number of genome-wide significant loci for behavior. If you are unsure why the deer mouse F_2 cross identified 12 loci with less than 800 individuals, then look at our discussion on inbred strain crosses in Chapter 15, section 15.9. The critical point here is that the experiment demonstrates that monogamy (or more strictly parenting behavior) is a polygenic trait, possessing the genetic architecture that we have seen before in analyses in humans and in standard model organisms.

We started this section by stating the problem that monogamy poses for evolutionary theory. How does the role of vasopressin fit with the argument that monogamy arises when population density is low (where the greater distance between females makes it harder for males to defend sole sexual access)? Intriguingly, there is an answer, from work that Steve Phelps carried out. Phelps found that expression of *Avpr1a* varied considerably **within** a population of montane voles, not just in comparison with the promiscuous prairie voles (PHELPS AND YOUNG 2003). *Avpr1a* expression in areas of the brain implicated in spatial memory correlated with paternity (OPHIR et al. 2008). Spatial memory? That begins to tie the vasopressin story closer to an evolutionary explanation. Phelps went one step further: genetic variation at *Avpr1a* produces variation in memory regions, and this in turn influences space use and sexual fidelity (OKHOVAT et al. 2015). Spatial use and sexual fidelity are thus tightly linked (as the Lukas and Clutton-Brock explanation requires) and are mediated via differences in expression of *Avpr1a*. The story thus comes close to full circle.

19.3 The genetics of real estate

Hopi Hoekstra's research on deer mice contributes to an understanding of the genetic basis of what Richard Dawkins calls the extended phenotype (DAWKINS 1982). An extended phenotype is the product of an animal's behavior, rather than an observation of what an animal is doing. Common examples of extended phenotypes in animals are nests, or burrows, in other words what an animal makes to serve as its home.

The deer mice (*Peromyscus* species) that Hoekstra researches are the most abundant mammal in North America, found from the Canadian arctic to the Colombian border of Panama. Deer mice look like mice with white bellies; they are much the same size as mice, and they have adapted to occupy almost every habitat in the North American continent (*Mus musculus*, which is the model organism mouse, is not native to North America).

One species of deer mouse, *P. polionotus*, digs deep, complex burrows. The burrow consists of a downward-sloping entrance tunnel, a nest chamber, and a second upward-sloping blind-ending tunnel (Figure 19.2) (WEBER et al. 2013). When the mouse is threatened, it uses the upward-sloping tunnel as an escape, breaking through onto the surface. The architecture of the burrow is so stereotyped that once you know where the entrance is you can guess pretty accurately the location of the escape tunnel, making it relatively easy to capture the mice. Hoekstra found you can breed the mice successfully in the laboratory, where they will also make burrows. Her idea was that the burrows reflect something about the animal's behavior, and that this must be in part genetically determined.

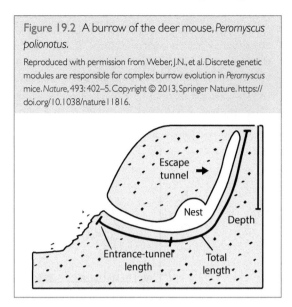

Figure 19.2 A burrow of the deer mouse, *Peromyscus polionotus*.

Reproduced with permission from Weber, J.N., et al. Discrete genetic modules are responsible for complex burrow evolution in *Peromyscus* mice. *Nature*, 493: 402–5. Copyright © 2013, Springer Nature. https://doi.org/10.1038/nature11816.

Not all species of *Peromyscus* make nests, and some prefer much simpler abodes than *P. polionotus*. Hoekstra compared the burrows of *P. polionotus* with those of *P. maniculatus*, which consist of an almost vertical entrance tunnel and a small nest chamber, again highly stereotyped. *P. polionotus* and *P. maniculatus* can mate and produce fertile offspring, making it possible to generate a cross suitable for genetic mapping. Mapping uses the method for crosses between inbred rodent strains, namely finding markers that differentiate the two species and then genotyping sufficient to locate the position of recombinants in an intercross or backcross (see Chapter 15, section 15.6.2).

The offspring dig burrows that vary considerably in shape and size, indicating that the type of nests mice build is a heritable trait. Hoekstra's group poured liquid plastic into the burrows and made polyurethane casts, which they used to measure the total length, as well as the lengths of the components of the burrow (WEBER et al. 2013). They could then map the variation in the nest measurements and in so doing identified three genomic regions that contributed to variation in entrance-tunnel length, and a single region associated with whether the animals constructed escape tunnels or not (WEBER et al. 2013). Figure 19.3 shows the mapping results. As you can see, the shape of the peaks, their broad width, is the same as that seen in the mouse mapping experiments that used crosses between inbred strains (see Figure 15.11 in Chapter 15, section 15.8).

Figure 19.3 Genetic mapping of two measures of deer mice burrows. The vertical axis is the likelihood (expressed as the logarithm of the odds, or LOD; see Chapter 15, section 15.6.2), and the horizontal axis represents each chromosome and linkage group. The vertical ticks are the positions of markers used for the mapping. The dotted line represents the 5% significance threshold and arrows indicate the position of the most significant genetic markers.

Reproduced with permission from Weber, J.N., et al. Discrete genetic modules are responsible for complex burrow evolution in *Peromyscus* mice. *Nature*, 493: 402-5. Copyright © 2013, Springer Nature. https://doi.org/10.1038/nature11816.

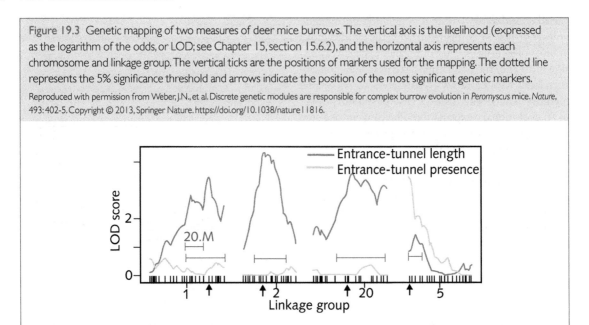

19.4 Social behavior

It seems odd that we should turn to insects for insight into social behavior, but that's what Gene Robinson would have us believe. Insects certainly have some intriguing social behaviors. How about this one?

These insects vomit collectively to defend themselves

The caterpillar-like larva of the common pine sawfly is known for its dramatic self-defense: groups of larvae simultaneously vomit drops of sticky goo at predators. Now, scientists have learnt why some larvae are 'cheats', taking advantage of their comrades' efforts while not joining the fight.

https://www.nature.com/articles/d41586-018-05853-z

Then there's *Nicrophorus vespilloides*, a veritable paragon of social behavior. It's a small beetle, with a pretty orange mark on its back, that finds dead and decaying animals (dead mice), chews up some of the rotting carcass, carries it back to its nest in its mouth, and regurgitates the food into the waiting mouths of its offspring (SMISETH AND MOORE 2004). And finally, most impressively, are the ants, bees, and termites that create entire societies, with workers, nurses, foragers, guards, and queens, communicating with chemical and body languages to act together in a model of social cohesion.

Regarded as one of the major transitions in evolution, right up there with the use of sex as a way to procreate, the highly specialized form of social behavior known as eusociality emerged once in ants, three times in wasps, four times in bees, and once in termites. Eusocial species are marked by a behavioral division of labor; eusocial ants and bees consist of groups that specialize in foraging, brood care, or reproduction. What is the genetic basis for eusociality?

We need to distinguish two separate questions. The first concerns the origins of eusociality, a question that we can frame as, 'What genetic changes took place to enable the appearance of eusocial behavior?' The second concerns the factors that make an insect belong to one of the different castes. These questions overlap, as the factors that determine caste status may also drive the origins of eusociality, as we shall see.

The first question has been addressed by looking at different species of bees, comparing those that do and those that don't have eusocial behavior. Gene Robinson analyzed ten species of bee to identify genomic features that characterized eusocial behavior (KAPHEIM et al. 2015). What did he find? Was there a single gene, a key mechanism? In short, no; in fact, to quote Robinson, '... if it were possible to "replay life's tape", eusociality may arise through different mechanisms each time' (KAPHEIM et al. 2015). His major finding is that changes in gene regulation are key features that enable the transition, rather than the appearance, of novel genes or major shifts in protein structure. He found no signal indicating a common set of genes that defines the emergence of eusociality. His observations complement what has been seen in comparisons between humans and nonhuman primates, where researchers have been looking hard, but with limited success, to find the genes that define human behavior.

What of the second question? How do genes determine the behavioral profiles of the different castes? Insects with highly similar genomes can exhibit striking differences in behavior, so it might seem that genetics does not play an important role, and certainly there are many cases where environmental cues determine the caste of the animal, such as the quality and quantity of food, the social environment during larval development (new queens emerge when the queen is absent), and the temperature during larval development. There is a term for this: 'polyphenism', which means alternative phenotypes are produced by a single genotype in response to environmental cues (SIMPSON et al. 2011).

Polyphenism does not mean genes are not involved. The environmental cues must trigger a biological pathway to exert their effect, acting from the outside to alter the internal state of the insect, and insulin/insulin-like growth factor signaling, juvenile hormone, and vitellogenin pathways are known to be involved. For example, there is an interaction between insulin signaling and juvenile hormone in seed harvester ant queens, which translates environmental cues into changes in vitellogenin production. Vitellogenin is a yolk precursor protein that transports nutrients to the ovaries, affecting egg development. What has this got to do with behavior? It turns out that altering the quantity of vitellogenin injected into eggs influences whether they hatch into queen or worker castes (LIBBRECHT et al. 2013). Furthermore, the parental behavior of the burying beetle (the one that regurgitates rotting rodent meat for its young) has been linked to expression of vitellogenin (ROY-ZOKAN et al. 2015). And in the harvester ant, *Pogonomyrmex barbatus*, vitellogenin has been duplicated and has acquired caste- and behavior-specific expression (NELSON et al. 2007). It's on the basis of these and similar findings that a hypothesis has arisen to explain the evolution of social behavior. The hypothesis is that genes regulating reproduction in a solitary ancestor have been co-opted to function as regulators of eusocial behavior.

However, polyphenism is **not** the rule for insect eusocial behavior. In fact, in some ant species, caste is determined completely by genetics. Indeed, caste determination varies among species, consisting of a complete spectrum of environmental and genetic determinants (SCHWANDER et al. 2010). The picture is one of both genes and environment contributing to behavior, a familiar story that echoes the descriptions we gave in Chapter 14.

That's not quite the end of the story. One major advance has been the ability to apply the genetic tools and resources formally reserved for standard models, like *Drosophila* and mice, to nonstandard organisms. Perhaps the premier example is the work of Daniel Kronauer on ants.

Genetic manipulations, as we explained in Chapter 16, section 16.2, involve introducing alterations into the germ line by targeting genes in embryos, and then crossing the resultant mutants to generate homozygous mutants. This is extremely challenging in species that have complex forms of sexual reproduction that occur only once a year. Kronauer searched for an ant species that wasn't impossible to breed in the laboratory and wasn't going to take years to reproduce. He chose the clonal raider ant, *Ooceraea biroi*, because of its short generation time (about 2 months) and because it reproduces via parthenogenesis. Genetic modifications could be generated using clustered regularly interspaced palindromic repeats (CRISPR), which is so efficient that it targets both alleles, not just one, and stable germ-line mutations were obtained from the progeny of injected individuals, without the need for crosses (TRIBLE et al. 2017) (see Chapter 16, section 16.2.4 for a description of CRISPR).

Kronauer's experiment was to search for genes that were differentially expressed between ants devoted to reproduction and those whose lives were devoted to brood care. In trying to find the genes that distinguished mother ants from nurse ants, he identified one gene, encoding insulin-like peptide 2 (ILP2), that was always significantly upregulated in mothers (this isn't true in bees and wasps, so the finding may only apply to ants) (CHANDRA et al. 2018). Kronauer, working with the raider ant, found that ILP2 is found only in the brain and that it regulates reproduction at least partly by acting on vitellogenin. He was able to demonstrate that social signals mediate insulin signaling independently of internal nutritional state. Using CRISPR's ability to genetically modify ants, he carried out a series of experiments in his nonstandard model organism that led to the following interpretation for the origins of eusociality.

Kronauer's data support the hypothesis that reproductive asymmetry was the start of the path that eventually led to the creation of obligate reproductive queens and the appearance of sterile workers in eusocial species. The reproductive asymmetry arose because of differences in the response of insulin signaling to signals from larvae. At some point, a fraction of larvae emerged as adults with low food reserves, low ILP2 levels, and a greater response to the larval signals, thereby leading them to take on an increasing share of the brood care roles at the expense of reproductive roles (i.e. they acted as nurses). Conversely, adults with good food reserves would have reduced sensitivity to larval signals and were more likely to reproduce. This asymmetry led to the appearance of different castes, which is the basis of eusociality.

19.5 Supergenes

Another phalanx of tiny black ants, three abreast, has just established itself across my (J.F.) kitchen floor, running down the gap between two floor tiles before exiting to the garden under the door. They are indefatigable. I've no idea how many times I've tried to dissuade them. I also have no idea what species they are, but I've no doubt that Laurent Keller would know. Keller lives and works in Lausanne, where he likes to hike and ski in the surrounding mountains, between researching ant genetics. He explained his research strategy to me like this: *'I read the entomology journals until I find some strange observation that has been reported, but never explained, like some phenotype that appears to defy Mendelian segregation, or some species in which sex chromosomes segregate in an apparently impossible manner, and then I go and get some samples and work out what is going on.'* It's a strategy that has paid handsome dividends.

So here is a strange observation about an ant, a fire ant called *Solenopsis invicta*, which originates in South America and is now colonizing North America. There are two forms of colonies: one that has multiple queens (a polygynous colony) and one that has a single reproductive queen (a monogynous colony). As you might expect, the behavior of ants in the two types of colonies differs substantially, including the level of aggression between colonies and how new colonies are set up. A genetic locus is associated with this difference, but the association is unusual.

In the single reproductive colonies, the most common genotype at the locus is a homozygote; in the multiple-queen (polygynous) colonies, queens with this same homozygous genotype never reproduce. Initially, it seemed that the homozygous females were dying prematurely from some illness, some additional but unknown disability, until Keller found out that the queens were being killed by their own workers. In fact, heterozygous workers were primarily responsible, indicating that there existed an allele that made workers kill all queens who did not possess it (KELLER AND ROSS 1998). The gene at this locus is a receptor for a pheromone, a chemical that ants use in communication (KRIEGER AND ROSS 2002), suggesting that the gene has a role in regulating social organization through chemical communication. But something even stranger emerged when Keller discovered that the locus controlling colony type is an inversion of about 10 Mb and occupies more than half of the largest of the ant's 16 chromosomes (WANG et al. 2013b).

The inversion is important because it means a whole set of alleles will be inherited as a single, large unit. When pairs of chromosomes line up at meiosis, at the time when the next generation's chromosomes are created, recombination can only occur when the linear organization of DNA on the two chromosomes is the same (see Figure 5.2 in Chapter 5, section 5.1 for a description of the biology of recombination). When a chromosome with an inversion pairs with a chromosome without that inversion, recombination is suppressed across the inverted region; all the alleles on an inversion are thus inherited together. Keller termed this chromosome in the ant a 'social chromosome'.

Fire ants aren't the only ant species to have a social chromosome. Remarkably, a social chromosome also explains social organization in the Alpine silver ant, *Formica selysi* (PURCELL et al. 2014a). Keller pointed out that social chromosomes have features in common with sex chromosomes: in addition to a large nonrecombining region (most of our X and Y chromosomes don't recombine), the DNA composition is similar, with increased amounts of repetitive elements and deleterious mutations. In mammals, the Y chromosome (which defines male sex) is believed to have arisen after suppression of recombination between a sex locus and a nearby gene that had alleles with beneficial effects for males but harmful effects for females. Over time, the suppression of recombination allowed additional genes with this property to accumulate on the Y chromosome (BACHTROG et al. 2011).

Another way to describe the social chromosome is to call it a 'supergene', a cluster of linked genetic variants that are not separated by recombination. The effect of a supergene on behavior is different from the genetic effects we have considered elsewhere in this book. In human and mouse genetics, behavioral variation typically arises from the joint effects of many loci, each of which makes only a small, independent contribution to trait variation. A supergene usually exists as a single allele, one long haplotype, that marshals the joint effect of many loci to modify behavior (and other traits) in a very plus or minus fashion: if you have the supergene, you behave in one way; if you don't have it, you behave in another.

Supergenes don't only exist in ants; they are found in other species too. White-throated sparrows (*Zonotrichia albicollis*) come in two easy-to-recognize varieties, one with a colored head (tan colored) and one with a white head (Figure 19.4). The white males are aggressive, promiscuous, and devote more time to finding sexual partners than to childcare; tan males are monogamous and stay at home to look after their offspring. Females show similar patterns of behavior (TUTTLE 2003; TAYLOR AND CAMPAGNA 2016). The behavioral difference is inherited, and is associated

with a genetic variant on chromosome 2: white morphs are almost always heterozygous for the chromosome 2 inversion; tan morphs are always homozygous (Figure 19.4). This happens because tan birds only mate with white birds so that white-throated sparrows effectively consist of four sexes (tan male, tan female, white male, and white female). In this species, an individual can mate with only a quarter of the individuals in the population (TUTTLE et al. 2016). The variant is an inversion of about 100 Mb, containing 1,317 genes (TUTTLE et al. 2016).

Ruffs (*Philomachus pugnax*) are a Scandinavian shorebird that displays lekking behavior (Figure 19.5). I (J.F.) wasn't too sure what lekking was. YouTube videos of young men dancing half naked while drinking beer apparently portray lekking in our own species. The lekking behavior that ruffs perform occurs in grassy areas, where they attempt to attract females. There are two behavioral strategies. 'Independent' males are aggressive and defend small territories, while 'satellite' males are nonterritorial and submissive to independent males. Behavioral differences are associated with different plumage patterns: 'satellite' males have white or light-colored plumage, whereas 'independent' males have dark neck ruffs and head tufts (HILL 1991; TAYLOR AND CAMPAGNA 2016). There is also a third form, a 'faeder', which is a male that mimics females, a disguise it uses to get access to females. Just as with the white-throated sparrows, the behavior is inherited (LANK et al. 1995), and again is due to a chromosome inversion, this time a 4.5 Mb region containing about 90 genes (LAMICHHANEY et al. 2016). The genetic changes present on the inversion indicate it has existed for about 4 million years.

The defining feature of a supergene is that it is inherited as one unit, even though it contains many individual genes, a genetic architecture that makes it possible for multiple alleles with advantageous properties to be passed on together. Supergenes don't always involve behavior—they are also known to affect mimicry in butterflies (JORON et al. 2011), and there is a well-known example in mice of an inversion on chromosome 17 that carries a series of mutations that results in an awful crime: the locus breaks Mendelian laws of segregation. The variant (known as the T-haplotype) creates a protein that kills other chromosomes without the inversion, and contains proteins that protect itself from a similar fate (LYON 2003). A mouse carrying a copy of the chromosome 17 inversion will pass

---

Figure 19.4 The two forms of the white-throated sparrow, tan and white morphs, are associated with different behaviors.

Adapted with permission from Taylor, S., and Campagna, L. Avian supergenes. *Science*, 351 (6272): 446–7. Illustration: K. Sutliff/Science. Copyright © 2016, American Association for the Advancement of Science. DOI: 10.1126/science.aae0389.

White -throated sparrow (*Zonotrichia albicollis*)

Tan morph	White morph
♂♀	♂♀

Noninverted
Dull plumage
Submission
High parental care
Infrequent song
  displays
Disassortative mating

Supergene
(~1000 genes)
Bright plumage
High aggression
Low parental care
Frequent song displays
Disassortative mating

Figure 19.5 The three forms of the ruff (*Philomachus pugnax*), which differ in physical appearance and behavior.

Adapted with permission from Taylor, S., and Campagna, L. Avian supergenes. *Science*, 351 (6272): 446–7. Illustration: K. Sutliff/Science. Copyright © 2016, American Association for the Advancement of Science. DOI: 10.1126/science.aae0389.

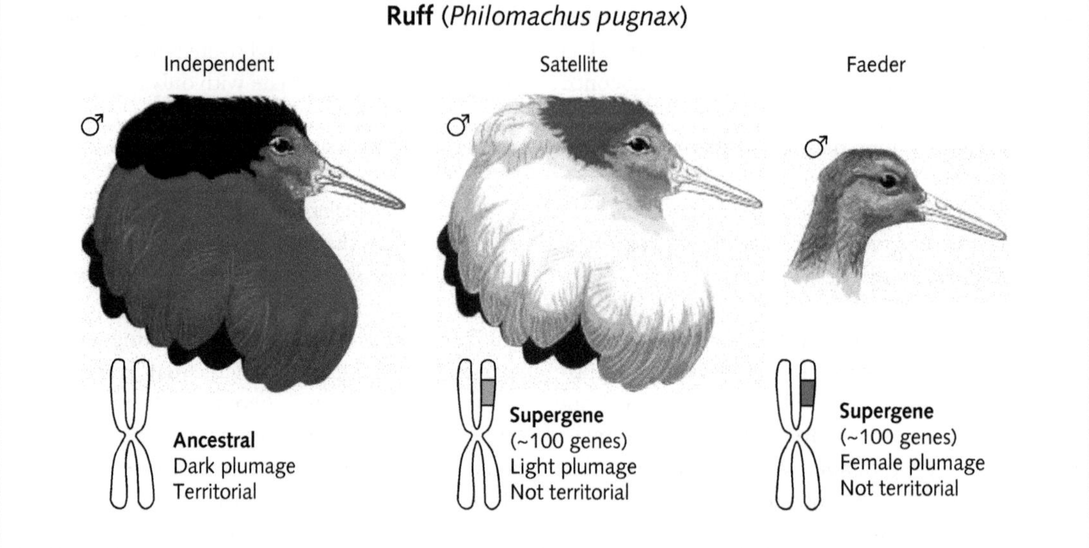

on more than 90% of the T-haplotype, rather than the expected 50%. The inversion encodes both a poison and its cure, neatly exemplifying the advantages that a supergene provides, and also raising in the most acute form the problem of how such a thing could arise in the first place—if a mutation creates the poison first, it would kill itself, and if the antidote were created without a poison, it would be soon lost in the population because it is useless. Explaining the emergence of traits beneficial to an organism that depend on multiple new mutations remains a major problem for biology, not just for behavior genetics (JAY et al. 2018).

Summary

1. The relationship between genes and behavior fits within an evolutionary framework.

2. Differences between the genomes of two or more species that have arisen during the course of evolution can be used to identify the genetic basis of behavioral and other species-specific characteristics.

3. Comparative analyses between the genomes of humans and closely related primates are expected to identify unique features of human evolution. Comparative genomic analyses of human–chimpanzee differences implicate differences in transcript abundance, not protein structure. Several hundred differentially expressed genes in neurons have been identified.

4. A copy-number variant on chromosome 16, a risk factor for schizophrenia and autism, is found at a locus that distinguishes humans from other primates. A human-specific rearrangement produces a new gene that is present only in humans.

5. Monogamy is a behavior that varies considerably among species: 90% of birds but only between 3 and 9% of mammalian species are monogamous. Differences between prairie and meadow voles have been used to argue for the importance of vasopressin in social behavior. Differences in vasopressin receptor expression within the brain contribute to both inter- and intraspecies variability in vasopressin-mediated

behaviors. Spatial use and sexual fidelity are tightly linked and are mediated via differences in expression of the vasopressin receptor.

6. Some genes perform highly conserved roles across different taxa. Oxytocin has an effect on parenting behavior in many different species. Serotonin influences aggressive behavior across a wide range of animals. These are examples of a genetic 'toolkit' hypothesis.

7. Differences in the nesting behavior of different species of deer mice have been mapped to three genomic regions that contribute to variation in entrance-tunnel length, and to a single region associated with whether the animals construct escape tunnels or not.

8. The highly specialized form of social behavior known as eusociality emerged once in ants, three times in wasps, four times in bees, and once in termites. It is thought that changes in gene regulation are key features that enabled the transition, rather than the appearance of novel genes or major shifts in protein structure.

9. Work on ants suggests that reproductive asymmetry was the start of the path that eventually led to the creation of obligate reproductive queens and the appearance of sterile workers in eusocial species.

10. Supergenes, or 'social chromosomes', are clusters of linked genetic variants that are not separated by recombination. Supergenes have been documented in association with chromosomal inversions. They have been shown to account for behavioral differences in ants, sparrows, and ruffs.

20

How genes influence behavior

[In which we try to answer the question of how genes influence behavior, and find ourselves touching on some philosophical and ethical conundrums]

20.1 The work of the behavioral geneticist

When I (J.F.) worked at the University of Oxford, I sometimes received letters written in a sober, supplicatory tone. The writers often used headed notepaper, which included the initials 'H.M.P.', a banner headline for 'Her Majesty's Prison'. Prisoners wrote because they wanted a letter from me, headed 'University of Oxford', stating that, in my view, the behavior that put them in prison was in part (or, better, wholly) attributable to their genes.

This line of reasoning has been tested in court a number of times (BERNET et al. 2007). In 1994, Stephen Mobley was convicted of murder, armed robbery, aggravated assault, and possession of a firearm, and was sentenced to death. His trial occurred soon after the publication of a *Science* paper attributing aggression to a deficiency of monoamine oxidase A (MAOA) (BRUNNER et al. 1993). In this paper, Hans Brunner, a clinical geneticist in the Netherlands, reported an unusual pedigree in which eight affected males had repeated episodes of aggressive, sometimes violent, behavior.

Genetic mapping assigned a mutation that segregated with the criminal behavior to a region of the X chromosome, a region including two genes that are known to metabolize the neurotransmitters serotonin, dopamine, and norepinephrine. The two genes encode two monoamine oxidase (MAO) enzymes: MAOA and MAOB. DNA sequencing of the genes in the violent individuals identified a mutation in MAOA that truncates the protein, making it ineffective. As men have only one copy of the gene (they have one X chromosome), they had almost no functional enzyme. On the face of it, MAOA mutations were making Dr. Brunner's patients aggressive.

Mobley filed a motion seeking funds to hire expert witnesses to assess his potential deficiency in MAOA. The court rejected Mobley's motion and, on 1 March 1 2005, he was executed by lethal injection by the state of Georgia. This has not put others off trying. In Ohio, Dion Wayne Sanders claimed that he was *'incapable of acting with the degree of culpability which the charges of Aggravated Murder involved'* (Court of Appeals of Ohio, Second District, Montgomery County, 21 July 2000). The court found that *'whether an accused acted out of a sense or attitude of rage is not relevant to prove that he acted or did not act "purposefully" and "with prior calculation and design." ... It could not demonstrate that when he shotgunned his grandparents Sanders lacked the specific intent to kill them. It only demonstrates that he was enraged when he did.'* Nevertheless, unlike Mobley, in Sanders' case the jury returned a sentence of life without parole rather than the death penalty. And, in a case heard in Tennessee, the court recognized the possibility that genetic testing of the MAOA locus and the serotonin transporter *'might have influenced the jury's decision in the penalty phase of the trial as to whether to select death as the proper punishment ... The Court finds, as a matter of law, that the expert services sought are necessary to ensure that the constitutional rights of the Defendant are properly protected'* (BERNET et al. 2007).

Genetic testing can save your life and you can have your current DNA configuration analyzed with the help of 23andMe (http://www.23andme.com), *'after producing a saliva sample using an at-home kit'*. You can check out your serotonin transporter status and, indeed, check for the presence of all the genetic characters we have introduced you to in the course of this book. With this information, *'you can use our interactive tools to shed new light on your distant ancestors, your close family and most of all, yourself'*.

All of these claims rest on the idea that genes cause behavior, in a linear pathway from genetic variant, cellular components, and neural circuitry to behavior. In the same way that alcohol disinhibits, cocaine produces euphoria, and LSD induces hallucinations, genes cause our behavior to change. The same assumption lies behind the broadsheet announcements that a 'gene for behavior' has been discovered. It is the hope of the murderer, the promise of the website, the lawyer's check, and the beloved of unquestioning journalists. Academic journals take pains to disavow it but don't deny that it may exist: *'Much research has focused on genes involving major neurotransmitters such as serotonin and dopamine and proteins that affect these neurotransmitters ... research has yet to isolate a specific "crime gene" and probably none ever will. Some of the reasons this is unlikely are incomplete penetrance, genetic heterogeneity, and the complex interaction among environmental factors, development, and gene expression'* (BERNET et al. 2007). In other words, the gene for crime is there, but it's hidden. Is this true?

20.2 Two ways of explaining genetic effects

In this book, we have been taking two different approaches to studying the interrelationship between genes and behavior. On the one hand, we have reviewed many studies that examine individual differences as they exist in natural populations (e.g. humans). These studies work only with what nature provides. On the other hand, we have reported investigations that examine the impact of genes on behavior through genetic engineering in the laboratory, mutating genes, and in other ways perturbing their function, and then seeing if this process changes behavior.

These two approaches toward studying genes and behavior can give very different results. Let us illustrate this. Imagine gene X plays a critical role in producing behavior Y. However, gene X is so important to the organism that even very slight changes in the function of gene X are so poorly tolerated that they are quickly removed from the population by selection. While there are likely to be genetic variants in unimportant parts of this gene, with respect to the work-horse part, the gene is monomorphic—there is no variation. What would happen if we were to study this gene in natural populations by linkage or association studies? We would find that variants in this gene were **not** associated with the behavior. However, if we knocked out the function of this gene in an experimental organism, we could get a dramatic change in the behavior.

So, can we conclude that gene X influences behavior Y or not? It is correct to say that 'variants at gene X in population Z do not influence behavior Y' and to say that 'knockouts of gene X in laboratory Z demonstrate a large effect on behavior Y'. This might be confusing, but it is nonetheless true. One study is not more right than the other; rather, the difference is that the experiments are asking different questions.

These two contrasting approaches, studying naturally occurring variants versus inducing mutations in the lab, began from different premises, and despite the fact that they had the same ultimate goal—an understanding of the genetic underpinnings of behavior—they necessarily had different proximate goals.

The focus of the first or 'quantitative genetic' approach, working with naturally occurring variation, either in the wild or in laboratory selection experiments, asked questions about the effect of all the genetic variants available in a population on a phenotype. Quantitative genetics investigates how phenotypic variation can be explained by its environmental and genetic components, and how evolutionary forces act on these constituents. Much of quantitative genetics is statistical in nature and is based on the work of the statistician Ronald Fisher (FISHER 1918, 1925). Throughout most of the field's history, the individual genetic factors responsible for such naturally occurring, continuously varying traits were unknown, and the analysis consisted largely of estimating the number of contributing loci and the independence (additivity) or nonindependence (epistasis) of their interactions. Often, these results followed from artificial selection experiments in the laboratory.

By contrast, single-gene mutant analysis traces its origins to microbial genetics, which concentrates on cellular mechanisms to the virtual exclusion of all evolutionary or environmental questions. It consists of the induction of mutants to identify relevant genes and the direct demonstration of biochemically definable roles and interactions between them. A particular strength of the approach is the theoretical ability to produce at least one mutation in every relevant gene for a particular trait (saturation mutagenesis) and thereby, presumably, identify all of the genes that are mutable to produce alterations in the selected phenotype. Saturation mutagenesis is a powerful approach in identifying genes involved in some phenotypes (such as how embryos develop; NUSSLEIN-VOLHARD AND WIESCHAUS 1980), although we now know from genome-level analysis that it is actually quite difficult to achieve in practice (GIAEVER et al. 2002). The single-gene, induced-mutation school is not interested in whether a gene would exhibit natural variation. Instead, the interest lies in which genes contribute to the behavior.

A measure of the past separation between single-gene mutagenesis advocates and the quantitative geneticists focusing on natural variation can be seen in their criticisms of each other. The genetic architecture camp criticized their single-gene counterparts (whom they sometimes derisively called 'gene-jocks') for their failure to take into account the subtlety of the various contributions of the genetic background to the phenotype. The monogeneticists disparaged the architecturists for their soft-headedness and their failure to be able to say anything about which genes

might be involved, as well as for the lack of relevance of small-effect genes to underlying mechanisms. One particularly potent put-down is, 'They just care about statistics, not real genes.'

20.3 The single-gene perspective ignores background effects

The severity of a mutant phenotype is a complex function of many factors. It depends, among other things, on the nature and extent of inactivation or alteration of the gene in question, on the gene's role in development and behavior, and on the genetic background in which it is expressed.

Behavioral mutants are notoriously sensitive to variations in genetic background—the natural, genetic heterogeneity in laboratory stocks (also known in the classical genetics literature as 'modifiers'). Whereas quantitative geneticists have long been aware of this problem, single-gene practitioners, in general, have not, sometimes to their embarrassment. A case in point is the erroneous identification of the temperature-sensitive paralytic mutation shibire as the voltage-sensitive sodium channel in flies (KELLY 1974), based on the greater resistance to tetrodotoxin of the mutant relative to a control strain. The control, however, was not of the same genetic background as the mutant. As it turned out, the 'control' strain was exceptionally sensitive to tetrodotoxin (GITSCHIER et al. 1980), whereas the shibire mutant was as resistant to it as most other fly strains (mutant or normal). The genetic basis of resistance in this control strain has never been properly determined.

Mutant phenotypes can fade with time. Their spontaneous disappearance has been reported for mutants affecting learning and brain development in flies (DE BELLE AND HEISENBERG 1996). The ubiquity of the problem has bedeviled mutant studies all along. In a legendary incident at a Drosophila meeting in the early 1970s, one investigator began his talk by saying, 'I would like to announce that Hyperkinetic is now a recessive' (J.C. Hall, personal communication, but not the speaker); that is, the strain no longer showed a dominant mode of inheritance for the mutation, as reported originally.

Such phenomena are presumed to be the result of spontaneous selection for modifying alleles that are present in the population—a distinctly quantitative genetic problem! Moreover, the potency with which a given background can mask or exacerbate the phenotype of a mutation underlines its relevance to the issue of genetic mechanism. A graphic example of the range of these effects was shown in a study of modifiers of the *sevenless* mutation in *Drosophila*, a mutant originally isolated as part of a genetic dissection of phototaxis behavior (the tendency to approach light) (HARRIS et al. 1976) and subsequently studied in great depth for its role in cell fate determination in photoreceptors (BRENNAN AND MOSES 2000). When a moderate allele of *sevenless* (roughly midway between the most severe mutation and the wild type) was placed in a range of different genetic backgrounds, phenotypes were found that ranged from fully wild type to more severe than the most effective enhancer mutations isolated previously (POLACZYK et al. 1998). Clearly, genetic mechanisms cannot be properly understood without paying attention to such background effects; that is, single-gene effects fade into quantitative genetics at the margins.

20.4 Small-effect genes will tell us little about mechanism

Two mouse geneticists, Joe Nadeau and Wayne Frankel, strongly argued against mapping naturally occurring variation as a way to find genes in complex phenotypes (NADEAU AND FRANKEL 2000). They started by pointing out the disappointing record of quantitative trait locus (QTL) methods for identifying genes and argued that alternatives needed to be considered. One problem was that, at least in mouse genetics, most mapping experiments sampled genetic variation segregating in about half a dozen strains so that '*the number of naturally occurring allelic variants that can be evaluated in practice is a relatively small proportion of the total number of genes essential for pathway functions and systems biology*'. By contrast, '*potentially every gene in the genome that affects the trait of interest is a target for mutagenesis*'. QTL methods would, even if they worked, give only a partial insight into the biology of a complex phenotype such as behavior. Suppose 100 genes are necessary for the expression of a behavioral phenotype. Mutations in any of these could potentially affect the behavior, but in any natural population genetic variants will only exist in a subset of those 100 genes, and maybe in only a small subset.

Nadeau and Frankel reserved their strongest argument in favor of mutagenesis for the end: '*... it is the successful identification of the genetic lesions that give rise to mutant phenotypes, compared with the ability to identify the variant that underlies naturally occurring QTLs, which makes mutagenesis the road of choice*'. Mutagenesis produces changes in coding sequence, mutations that truncate genes, mutations that result in null alleles where no functional protein is produced, or proteins of monstrously aberrant function. Such mutations and the genes they influence are easy to identify compared with the naturally occurring noncoding variants that underlie psychiatric disease.

But it is not just clonability that excites mutagenesis aficionados. There is the additional assumption that large-effect mutants are large for a reason: they disrupt key processes. By contrast, mutations with small effects do not, and they are more likely to represent peripheral processes far from the real biological sites of action. However, as we have seen throughout this book, there is little evidence that single-gene mutations directly and specifically influence behavior. Why is this?

20.5 Single genes in genetic architecture terms

Finding genes with a specific role in behavior turns out to be very hard. Three fly geneticists in an article entitled 'Are complex behaviors specified by dedicated regulatory genes? Reasoning from *Drosophila*' could identify only one example in the fly: the *fruitless* gene, which is involved in courtship and related sexual behaviors (BAKER et al. 2001). To be clear, it is not that genes don't influence behavior, it's that they typically don't do it specifically.

Recall the work on geotaxis described in Chapter 18, which made us rethink our ideas about how genes influence behavior—that perhaps there were no such things as 'genes for geotaxis' and that instead genes are versatile and participate in many different processes. We saw that *cryptochrome* and *pigment-dispersing factor* are not circadian rhythm genes but genes that can affect many functions, two of which are circadian rhythms and geotaxis. As for *importin-α*, the fact that it can exert such a selective effect while being so widely used for other biological functions provides an important insight into the subtle ways that gene activities intertwine to affect behavior. It tells us that modest alterations in such widely acting genes, in combination with other similarly altered genes, can produce very specific behavioral effects.

As we noted, a corollary of this nonspecificity is that genetic effects on behavior are typically pleiotropic; that is, a gene has multiple effects on many phenotypes. This is well illustrated by the results of the mouse gene knockout project in which pleiotropy was found to be the norm: '*Sixty-five percent of lines (290/449) had more than one phenotypic hit*' (DE ANGELIS et al. 2015). We described this work in Chapter 16.3.3. In flies, evidence for pervasive pleiotropy comes from correlated responses in transcript expression following selection for copulation latency, aggressive behavior, locomotor startle response, and ethanol resistance, in each case from the same base population (JORDAN et al. 2007).

Pleiotropic action, on the scale we've just described, makes the idea of 'a gene for' behavior impossible to maintain. Single genes do not specify behavior, but in combination with each other and with the environment, they may. Explaining how many, apparently unrelated, molecules are involved in a complex phenomenon is difficult within the causal model adopted from biochemical genetics, in which genes are arranged in linear pathways. Instead, it may be more appropriate to examine genes in networks and assign function to combinations of genes, rather than individual entities.

20.6　Genetic architecture in single-gene terms

In populations, the pleiotropic and network attributes of genes have consequences for how genetic variation can produce behavioral variation. Each allele of a gene can potentially contribute in several ways to a phenotype. These contributions, in turn, depend on the partners with which that gene interacts. Variation can thus occur in a restricted portion of a gene's range of activities if its interacting partners are more sensitive to perturbation in one place than in another. If its interacting partners also come in allelic variants, a further dimension is added.

Phenotypic variation in a population, which is what one measures, is thus not a monotonic function of allelic variation. Instead, it may well represent a more complex fabric than the distribution of alleles alone

might suggest. This may seem to present an even more bewildering picture than the traditional view. Its saving grace, however, is that knowing the network nature of a gene's interactions ultimately makes its contributions to phenotype more comprehensible. Further study of the interacting nature of one gene's variation with that of another, in turn, brings its population genetic architecture within the realm of comprehensibility.

The two perspectives can be distilled into one: many genes for each behavior (*e pluribus unum*), many behaviors from each gene (*ex uno plura*) (GREENSPAN 2004).

20.7　How genes influence behavior

One of us (K.S.K.) has proposed criteria that can be used to judge the validity of a claim that 'X is a gene for Y', where X is a scientist's favorite gene and Y a particular behavior or psychiatric disorder (KENDLER 2005). Four criteria are appropriate in this context: the strength of the association, the specificity of the relationship, the independence of the effect of X on Y, and the causal proximity of X to Y.

We can summarize this as follows: if gene X has a strong, specific association with a behavioral trait or psychiatric disease Y in all known environments, and the physiological pathway from X to Y is short or well understood, then it may be appropriate to speak of X as a gene for Y.

How often might these criteria apply to genes and human behavioral or psychiatric syndromes? The short answer is 'hardly ever'. As we have documented and has been especially clearly shown by the results from genome-wide association studies (GWAS), the strength of the association between the discovered risk genes and behavioral phenotypes are typically weak. They account, at most, for a fraction of a percentage point of the total liability.

Do genes have a specific effect on behavior? Almost certainly not. Variation in gene X hardly ever influences trait Y only, because variants in X affect other phenotypes. Almost never are the genetic influences of trait Y restricted to only gene X; that is, as we just noted, the association between genes and behaviors is not typically one to one, or one to many, or many to one, but rather many to many. Evolution is to blame for this complexity; it is a tinkerer and is promiscuous in the use of genes, taking whatever material is to hand to work with.

Will the relationships we find between genetic variants and psychiatric and behavioral traits be robust, that is, not contingent on other genes or environmental exposures? As we outlined above, this will not always be the case. Many genes that influence behavior are sensitive to the impact of the genetic background and the environment. The causal pathway from gene to behavioral phenotype is rarely direct or imperturbable. Rather, it is commonly indirect and often sensitive to genetic and environmental context.

The last criterion, that of causal proximity, is more complex than those we considered previously and is best illustrated by a brief vignette: a jumbo jet contains about as many parts as there are genes in the human genome. If someone went into the fuselage and removed a 2-foot length of hydraulic cable connecting the cockpit to the wing flaps, the plane could not take off. Is this piece of equipment then a 'cable for flying'?

Most of us would be uneasy answering yes to this question. Why? Because this example violates our conception of **causal proximity**. When we say X is for Y, we expect X to be, to a first approximation, directly and immediately related to Y. This is not the case for the cable and flying. There are many mechanical steps required to get from the function of the cable to a jumbo jet rising off the runway. Saying that the hydraulic cable on a jumbo jet is a cable for flying is probably a lot like saying a particular gene (X) is 'for' behavior. The association is very indirect and will involve many other indirect steps. The 'genes for' behavior hypothesis fails the causal proximity test.

So, on aggregate, this thought experiment indicates that we are almost never justified in using the language or concept of a 'gene for' a behavior or psychiatric trait. The relationship between genes and behavior is too contingent and indirect for such language to be appropriate.

20.8 The relative contributions of genes differ

We've reviewed evidence that throughout the animal kingdom, individual differences in behavior are, almost without exception, influenced by genetic factors. Most commonly, these genetic effects are of modest or moderate rather than overwhelming importance, and sometimes genetic influences are even more modest than that. Across a wide variety of species, including humans, the genetic influences on behavior are typically the result of a large number of individual genes, each of which, on its own, has a small effect on the behavior. In both humans and simpler organisms, the interrelationship between genes and the environment in their impact on behavior is, at least for a number of traits, likely to be complex. Gene–environment interaction, while still much under-researched, may be widespread in its effects. It is equally likely that genes, through 'outside the skin pathways', play critical roles in influencing important aspects of the social environment to which the organism is exposed.

It is possible to explain this complex picture of genetic action from the action of networks of genes. The recognition of the ubiquity of pleiotropy means that each gene has, in effect, its own architecture—a distributed pattern of action through the various stages and tissues of the organism (and even potentially reaching outside the organism in its effects on the environment—think of a beaver's dam). In this sense, the summated action of the genes is not so much a jigsaw puzzle in which each piece fits together with its immediate neighbors in one spot, but rather a flexible, multilayered network—a viewpoint that was implicit in quantitative genetics and that single-gene genetics has slowly been approaching.

Synergism and network flexibility make it easier to conceive how new properties in behavior can emerge: tune an allele up here, tune another one down there, combine them with some other pre-existing variants, allow it all to ripple through the networks, and you have a new behavior. Although no one is yet at the point of being able to demonstrate this in the lab, the threshold effects that are frequently seen in selection experiments—in which the phenotype does not move at all for many generations and then diverges rapidly, or in which the phenotype fluctuates dramatically before diverging—consistently suggests that such effects can occur in the laboratory, where they can be studied in the ways exemplified above. (Note that this discussion does not add the extra layer of complexity that these gene pathways are likely interacting with a range of important environmental influences.)

At the same time, it is also easy to imagine that the number of ways for genes to influence behavior will be manifold. It will depend on the context of other alleles present (i.e. genetic background), as well as on the actual role(s) a given gene plays in this behavior. The impact of one level, the individual gene, on the other, the gene system, is reciprocal: individual genes influence the network, and the network properties, in turn, influence the action of individual genes.

At the beginning of the single-gene era of behavioral studies, Sydney Brenner remarked, '*Understanding the genetic foundations of behavior may well require solving all of the outstanding questions of biology*' (BRENNER 1974). The years that have passed since then suggest that this may not quite be true. But to the extent that we must understand the nature and principles of how gene networks influence complex phenotypes, the synthesis of quantitative and single-gene approaches that is currently under way would seem to be a prerequisite.

20.9 How should we use our new-found knowledge of genetic effects?

As of early 2019, polygenic risk scores (PRSs) are now widely used in research across the social and psychological sciences (see Chapter 7, section 7.2 for an introduction to PRSs). Sample sizes to generate these increasingly PRSs are growing by the month, with some now reaching over 1 million individuals. For moderate costs, multiple companies (not just 23&Me) will perform a GWAS on your DNA, mailing you a prepreprared plastic tube into which you just need to spit a small quantity of saliva. You mail the packet off and get an email in a few weeks with a link to a web site that will allow you to explore the wonders of your genome aided by striking graphics.

You can download your own GWAS data, and web sites are available that will use these data to give you your PRSs for a range of traits including your vulnerability to schizophrenia, bipolar illness, autism, and depression, as well as your scores for propensity to high educational attainment or particular aspects of your personality. The growth in this industry shows no signs of slowing down.

How will we use such information? As many readers may be aware, although illegal, in several countries, widespread selective abortions now occur on the basis of sex. Using the same procedures that are now employed to test for genetic diseases such as Down syndrome or Tay–Sachs disease, it would be very easy to test the PRSs of zygotes in their first trimester of development. If you had a father with schizophrenia, would you be justified in testing your potential offspring early in pregnancy and aborting it if the PRS for schizophrenia was especially high? Given that multiple fertilized human zygotes can now be produced outside a mother's body, would it be acceptable to test 20 of them and pick for implantation the one with the highest PRS for educational attainment? Who would not want a child likely to succeed at school and attain an advanced degree? The first large-scale GWAS for sexual orientation will soon appear. Could such results be used in the future to select against zygotes with PRSs for the tendency to develop a homosexual rather than a heterosexual orientation?

Our task here is to raise these ethical and conceptual concerns, not to provide definitive answers. It is an old saw that the development of human knowledge can outpace the development of human wisdom. This is likely to apply to our rapidly growing knowledge of the genetic contributions to human behavior and psychiatric disorders. But the first step, of course, is to recognize the problems and to begin to struggle seriously with them.

20.10 How to understand an explanation

The goal of science is explanation. Scientists want to understand how things work. In this book, our focus has been on behavior. What can we conclude about what controls behavior? In the last sections, we've contrasted two ways of explaining genetic action, a single-gene versus a quantitative-genetics approach, and we've pointed to the need for their reconciliation, the need to take into account network effects. But a major objection will be, what can a network explain? I understand how a mutation can cause behavior, but a network? This spidery figure of interconnecting lines, what has that ever explained? This problem is the difference between a reductionist and systems approaches to biology.

One of the deep debates in philosophy is between reductionism and emergentism. Reductionism argues that the real causes in our world happen at very basic levels and then work their way upward through increasing levels of complexity. Emergentism takes the position that real causes in our world are more distributed. Some happen at basic levels but others emerge from systems that form out of these more basic constituents. Reductionists see true causes as coming from individual parts, the smaller the better. Emergentists see causes as often arising from interactions between these parts. Reductionists like to think in terms of parts. By contract, emergentists are 'systems thinkers'.

Reductionism views the behavior of complex things as predictable from the behavior of the individual components. So if you understand the little bits, you can predict how the whole thing works. Emergentism disagrees and believes that the behavior of complex things typically does not arise from simply understanding what the parts do in isolation. This is because, as parts interact, new causal effects arise that cannot simply be predicted from the bits themselves.

This is pretty abstract. Let's try to illustrate this with a question about an ecosystem. Imagine a pond with fish, algae, a variety of insects, sunshine, some frogs, and a fisherman or two. A reductionist biologist would examine each individual organism, one at a time, and determine its needs, the waste it produces, what it might eat, and what eats it. They would then measure all the physical properties of the system: the water temperature, pH, degree of sunshine, amount of pollution run-off, and so on. The reductionist would then construct a model of the pond ecosystem putting in all this individual information. They would hope that this model accurately describes how the pond ecosystem works. The emergentist biologist, by contrast, argues that this model is inaccurate because it leaves out important features of the ecosystem. In particular, it does not incorporate the interactions between organisms or between organisms and physical aspects of the environment. Importantly, the emergentist would argue that many of these interactions, which could become quite complex, cannot be predicted from the individual features of the creatures or their physical environment.

Molecular genetic studies aren't inherently superior to twin or adoption studies. They ask different questions and provide distinct but complementary views on how genes relate to behavior. Sometimes, you have to study small bits of the system to understand how they work. Other times, you need to look at whole systems and how they interact with the environment or development. Our job, ultimately, is to put all these perspectives together.

The world of behavioral and psychiatric genetics in particular is, in our view, too 'gene-centric'. If you wanted to be less polite, you could call it 'gene

crazy'. The key question is, how do genes influence behavior? One important step in answering this question is identifying the genes that make a difference, but this is only a start. The real action begins when we try to understand how these genes act within cells, and within organisms who interact with an unpredictable environment that is at best indifferent to their existence and often openly hostile. It will not work, we believe, to assume that the little 'bits' of this puzzle all just add together. We know already that they do not.

One of us (R.J.G.) has publicly spoken out about the excess 'gene-centrism' in our field. At a major meeting on genes and behavior, he began the concluding talk of the conference with the memorable phrase, *'Hello. My name is Ralph Greenspan and I am a former gene-aholic.'* He went on to outline his own serious study of how genes fit into networks, demonstrating the important limitations of our world view of taking one gene at a time.

20.11 Should I write back to the prisoners?

Many people associate strongly the idea of 'genes' and 'determinism'. If something is 'in your genes' then 'God help you' because there is nothing you can do about it. As shown by the science we have reviewed in this book, this is not how the world works for most of the behaviors we care about. Yes, there are what we call Mendelian genes: for some genes, if you have a defective copy you will inevitably develop dementia or another neurological disease that alters behavior. But if you look at our ability to learn, our risk for depression, schizophrenia, or alcoholism, or our personality, genetic influences are probabilistic and not deterministic. The idea that genes 'seal your fate' might be popular in the public imagination, but it is not science.

That said, it is also true that individuals really differ, as a result of the genes they inherit from their parents, in their personalities and their propensity for many behaviors and (at least for people) their risk for psychiatric and drug-use disorders. Genes do matter and, even though we might wish it, we are not all born equal with respect to these vulnerabilities.

Now comes the hard part. How do genetic influences relate to moral responsibility? Assume you are a judge and someone is brought before you for drunk driving because they injured another individual in a resulting car crash. The accused has a strong genetic risk for alcoholism. Is such an individual less culpable for this injury than someone who had little genetic risk?

At root, the question is ethical, not scientific. The system of English common law with which the authors are most familiar is generally intolerant of such justifications. After all, no one forced the suspect to consume the alcohol that made him intoxicated. He should have known about his inability to stop drinking once he started. However, it is naïve to think that we are all born with equal risks for alcoholism or criminal behavior. This is a fact that our judicial system may or may not wish to pay attention to.

In Chapter 12, section 12.3.2, we reviewed evidence that antisocial personality disorder, which substantially increases the risk for criminal behavior, is moderately influenced by genetic factors. Imagine an individual with a strong genetic propensity toward this disorder who has committed a crime. On the one hand, you could argue that because of this high genetic liability, the individual has reduced culpability and therefore ought to receive a shorter sentence. But you could also argue that, because of this same liability, he would be at especially high risk for a repeat offense and therefore ought to receive a longer sentence. These and related issues suggest that we should utilize considerable caution in introducing genes into our courtrooms.

We end with the following quotation from Friedel Weinert, a philosopher of science:

Animals are rigidly controlled by their biology (instincts). But human behavior is largely determined by culture, albeit within limits of biological constraints.

WEINERT (2004).

Weinert expresses a common sentiment. The behavior of animals is highly constrained and largely instinctual. By contrast, we are much freer of our biological roots. This widely held idea has a simple empirical prediction. The degree of genetic influence on animal behavior, that is, the heritability, ought to be much higher than what we see in humans. While perhaps commonsensical, this belief is not supported by the facts. As part of a review article written by two of us (K.S.K. and R.J.G.), we considered the relevant literature in this area: studies comparing the heritability of a wide range of behaviors in humans and experimental organisms (KENDLER AND GREENSPAN 2006). We here paraphrase our conclusion.

Genetic risk factors have been found for every psychiatric condition that has been seriously studied, and significant genetic influences have also been found for more 'normative' human traits such as personality. Heritability varies meaningfully among psychiatric disorders, with relatively consistent results across studies: 20–30% for most anxiety disorders, 30–40% for major depression, 50–60% for alcoholism, and 80% or higher for schizophrenia, bipolar illness, and autism. Genetic variation appears to be present for nearly every behavior ever systematically examined in nonhuman animals. In a review of 57 studies of animal behavior, the mean heritability was reported to be 38% (MEFFERT et al. 2002). These figures are not precisely comparable, as most of the animal studies were done on laboratory strains in artificial conditions while the human studies were, actually, done under more 'natural' conditions, that is, in natural populations in real environments. However, the available evidence does not support our presupposition that our brains and culture have freed us, and our behavior, from the effects of genes. The heritability of human behavior is at least as high as typical behaviors in other animals. This might be an uncomfortable revelation for our readers. Unlike all other creatures on earth, we have the wonderful gifts of language and advanced culture, art and science, poetry, and storytelling. But in our personalities, and our risk for psychiatric and substance-use problems, we are still shaped in important ways by the genes carried in that tiny sperm from our father, swimming for all it is worth, and those waiting in our mother's egg.

Summary

1. Studies of genes and behavior have traditionally fallen into one of two categories: studies of natural variants or studies of mutations induced in the laboratory. The dichotomy also generally corresponds to small versus large effects of individual genes on a behavioral phenotype.

2. Studies of induced mutations have tended to ignore the role of genetic background, assuming it to be irrelevant to fundamental mechanisms.

3. Studies of natural variants have broadened our picture of how wide an array of genes is capable of affecting a given behavior and how pleiotropic these genes are.

4. The dichotomy can be resolved by recognizing that there is a wide spectrum of gene effects, from small to large, and that behaviors can be affected significantly by many of them.

5. A further resolution comes from the realization that genes do not act in isolation but as a part of networks with emergent properties; that is, genetic influences on behavior work through the organism's system of gene interactions and neuronal interactions.

6. Saying that genes influence behavior is not equivalent to saying that genes determine behavior. Thus, the study of genes and behavior will not make obsolete the idea of morality.

21

How do we know a finding is true? quantitative approaches

[In which we explain some basic concepts in statistics]

21.1 The reproducibility crisis

At various junctures in this book, we have raised the question that forms the title of this chapter: how do we know a finding is true? Here, at last, we'll get to answer that question. By now, after encountering examples of where things went wrong, as well as where they went right, and with an introduction to some of the weirder claims from the fringes of our academic discipline, we hope you realize that understanding quantitative approaches is central to the whole endeavor. You need to master the material in this chapter to fully understand what's going on in behavior genetics, and not just behavior genetics. The lessons we teach are relevant also to any discipline that uses statistical methods to establish whether a claim is likely to be true or not. In addition, we hope, should you end up reviewing papers for a journal, you'll find our ideas useful (and hence, indirectly, we hope to reduce the publication of false-positive findings).

You might imagine that the most important tools you will need to judge the truth of a claim involve expertise in choosing and applying an appropriate statistical test. This is certainly a view that has many proponents, and many textbooks, working on this assumption, provide descriptions of what to do and how to do it. In fact, information on how to do all things statistical has never been easier to obtain. There is no lack of advice online (e.g. http://mathworld.wolfram.com/topics/ProbabilityandStatistics.html for formulae and their derivation, and https://www.khanacademy.org/math/statistics-probability for a supply of videos) and, if you are lucky enough to be without access to the internet, we provide a very brief introduction to the relevant statistical methodologies in the Appendix.

So, yes, we think you **should** know when to use a contingency test, when to apply a test of the difference between two means, or when to resort to logistic or linear regression, but we don't think that possessing the skills to run these tests is either sufficient or necessary to avoid the problems that have afflicted the field. Indeed, if it were, it is hard to understand why there has been talk of a 'reproducibility crisis' and warnings of the failure of the scientific method.

The basic tool used to determine whether data support, or don't support, a hypothesis is a statistical test. If a statistical test returns a P value of less than 0.05, then received wisdom has it that a result is likely to be true and supports the hypothesis that the scientist proposed before doing the experiment. In fact, interpretation of a *P* value isn't so straightforward (as we discuss later, in section 21.6), but whether we understand the meaning of a *P* value or not, one thing is obvious: a true result should be true not just once—it should replicate. No one had rigorously tested rates of reproducibility until the advent of Open Science's Reproducibility Project (https://osf.io/). The findings were ugly.

Brian Nosek and colleagues attempted to reproduce findings from 100 psychology studies. They summarize their work, published in *Science* in 2015, as follows: '*Ninety-seven percent of original studies had significant results (P < .05). Thirty-six percent of replications had significant results*' (OPEN SCIENCE COLLABORATION 2015). Psychology was already suffering after fraud allegations surrounding nearly a decade's worth of publications by Dutch social psychologist Diederik Stapel (http://www.apa.org/science/about/psa/2011/12/diederik-stapel.aspx) and Darryl Brem's claims for the existence of extrasensory perception (https://slate.com/health-and-science/2017/06/daryl-bem-proved-esp-is-real-showed-science-is-broken.html). Nosek's report that only a third of claims could be reproduced hardly improved psychology's reputation for robust science, and indeed was interpreted in the press as evidence for a 'reproducibility crisis'.

Not all psychologists agreed with Nosek's findings and a fight broke out, some of it in the pages of *Science* (ANDERSON et al. 2016; GILBERT et al. 2016). One psychologist commented, '*There is some antagonism between those two communities, and both sides each have a perspective that may color the way they're seeing things.*' '*You think it's* slightly *antagonistic?*' Nosek replied (https://www.wired.com/2016/03/psychology-crisis-whether-crisis/).

How should we interpret Nosek's observation that many results do not replicate? Is this a failure of the experiment (e.g. some unknown confound that was not taken into account), of the statistical test (did the researchers use the correct test?), or of the interpretation of the statistical test?

It would be nice to write here that we will explain the failures that led to the 'reproducibility crisis' and that we can provide a simple way to assess 'truth', but that's wishful thinking, as is the idea that belief in astrology, UFOs, and various alternative medical therapies can be dispelled by argument, based on evidence. Some things are '*more "vampirical" than "empirical"—unable to be killed by mere evidence*' (Jeremy Freese, http://boydetective.net/docs/Freese-PredictivePromiscuity.pdf). So, as there is no simple prescription we can teach you, we've structured this chapter around five questions to assess the likelihood that a finding is correct:

1. Was the right test used?
2. Was the correct significance threshold applied?
3. Could the result be attributable to some unacknowledged confound?
4. Did the experiment have sufficient power to deliver the result reported?
5. Was the *P* value interpreted correctly?

21.2 Was the right test used?

With the exception of a couple of embarrassing occasions during a PhD oral examination, I can't think of examples where an incorrect result could be attributed to the use of an inappropriate test. Nevertheless, it's not uncommon to find instances where a slight alteration to the test applied does give a better (i.e. more significant) result. Charitably, you could say that some tests are more powerful than others.

Broadly speaking, the tests used in genetic association fall into two groups: those that deal with categorical phenotypes (e.g. a disease category, diagnosed as schizophrenia or not) and those that deal with quantitative phenotypes (e.g. a score from a personality test). The former include Fisher's exact test, χ^2 (chi-squared) tests, and logistic regression tests. The latter include comparison of two means, analysis of variance, and linear regression (as a rule of thumb, we recommend using logistic regression for categorical measures and linear regression tests for quantitative phenotypes). The Appendix provides a brief introduction to these tests with some worked examples. Here, we provide examples of the (in)accurate application of tests to categorical and quantitative traits.

Consider the data in Table 21.1, taken from a summary of association studies looking at the relationship between serotonin receptors, the serotonin transporter, and affective disorders (ANGUELOVA et al. 2003) (we discuss the role of serotonin receptors and

Table 21.1

Author, year	Group	Receptor	Polymorphism	Origin	N	Allele 1	Allele 2	OR (95% CI)	P value
Fehr, 2000	Control	5HT1B	861 G/C	Germany	108	161	55		
	MDD Patient				74	97	51	1.54 (0.95–2.49)	0.064

transporter with depression in Chapter 5, section 5.2.2). In brief, the story is one in which initial high-profile claims are followed by publication of a mixture of claims for both replication and nonreplication.

Table 21.1 reproduces data from just one study (there are 19 entries in the original table). The phenotype is categorical: subjects either have major depressive disorder (MDD in the table) or they do not (they are controls). So the appropriate test is likely to be a χ^2 or logistic regression test. To start with, can we recapitulate what the authors have done?

Our interest here is on the right (Table 21.1), where the authors have summarized the data for the association analysis. They show the number in each group (the column headed 'N') and then report the allele counts. Allele counts are worked out from genotypes as follows. Suppose we have a marker with two alleles, A_1 and A_2 and we find 59 people with A_1A_1, 43 with A_1A_2 and six with A_2A_2 scores. To work out the alleles, double the number of homozygotes and add on the number of heterozygotes. So for the A_1 allele: $2 \times 59 + 43 = 161$, and for the A_2 allele: $2 \times 6 + 43 = 55$. The control group has 97 A_1 and 51 A_2 alleles. Table 21.2 shows the allele data with additional information about the total count.

The standard analysis for categorical data in this form (a contingency table) is a χ^2 test, in which expected and observed distributions are compared. The observed distribution is obviously the count data in the table, but what is an expected distribution? It is the most likely distribution, given the totals in the margins of the table (totals in rows and columns). There are lots of combinations of numbers that could generate the margin figures, but assuming that the selection is random, then the most likely is row total × column total / total count for each of the cells in the table (so the entry for A_1 cases will be 216 × 258 / 364 = 153.10). Table 21.3 reports the expected data.

In a χ^2 test, the assumption is that the expected distribution and the observed distribution are the same. We calculate for each cell (observed − expected)^2/

Table 21.2

	A_1 allele	A_2 allele	Row total
Cases	161	55	216
Controls	97	51	148
Column total	258	106	364

Table 21.3

	A_1 allele	A_2 allele	Row total
Cases	153.10	62.90	216
Controls	104.90	43.10	148
Column total	258.00	106.00	364

expected and sum the results. For example, the first cell is $(161 − 153.1)^2 / 153.1 = 0.408$ and the total for all cells is: $\chi^2 = 3.44$. Obviously if there are no differences between observed and expected distributions, then each (observed − expected) value equals zero and the final sum is zero. So the larger the value of χ^2, the less likely that the expected and observed distributions are the same. The P value of the χ^2 distribution depends on the number of independent observations (or degrees of freedom). In this case, there is one degree of freedom (calculated as: $(2 − 1) \times (2 − 1)$ for the rows and columns of the table). As a general rule, when a χ^2 exceeds the degrees of freedom, the result is likely to be interesting. In this case, $P = 0.064$, in agreement with the table from the paper (Appendix section A14 provides more information on the analysis of contingency tables).

So the authors used an appropriate test. That's fine, and not particularly surprising. But why do they present, and analyze, their data as alleles? They collect data

Table 21.4

	A_1A_1	A_1A_2	A_2A_2	Row total
Cases	59	43	6	108
Controls	32	33	9	74
Column total	91	76	15	182

as genotypes—why not analyze the association with genotypes? Why go to the bother of converting to alleles? Well, let's analyze the genotypes and see what we get (the genotypes are not given in the summary table from ANGUELOVA et al. 2003, but can be found in the original paper that reported the association: FEHR et al. 2000).Table 21.4 reports the genotypes.

We apply a χ^2 test and obtain $\chi^2 = 3.7$, with $P = 0.16$. That's larger than the P value from the allele test ($P = 0.064$). Why? Recall that the more the χ^2 value exceeds the degrees of freedom, the lower the P value. The allele test has one degree of freedom, while the genotype test has **two** (from $(2 - 1) \times (3 - 1)$ for the rows and columns of the table, respectively). The allele test delivers a lower P value! It's a more powerful test, that is, more likely to find a significant result. But is the choice justified? Are we simply able to generate a more significant result by choosing a 'better' test? We can, in fact, argue for an even **better** way of testing the data.

As far at the χ^2 test is concerned, the number of A_1A_2 genotypes could be more or less than A_1A_1 and A_2A_2. But we know that the allele frequencies are likely to be in Hardy–Weinberg equilibrium (see Appendix, section A.2 if you are not familiar with Hardy–Weinberg equilibrium). In fact, the allele test **assumes** that alleles at the locus are in Hardy–Weinberg equilibrium. When Hardy–Weinberg equilibrium does not hold, the allele

test is not valid. Because sometimes the frequencies diverge from equilibrium, we need a test that is robust to Hardy–Weinberg departures.

It might also be the case that the A_1A_2 genotype is dominant to one or other of the homozygous genotypes (see Appendix, section A3 for a discussion of genetic models). One of three genetic models is possible: additive (in which the effect of the A_1A_2 genotype is halfway between the A_1A_1 and A_2A_2 genotypes), recessive, or dominant (see Table 21.5 for an example that assumes a dominant model). How do we specify a genetic model in our analysis?

One solution is to use a version of the χ^2 test called the Cochran–Armitage trend test (also called the Armitage test and sometimes just the trend test), which is robust to Hardy–Weinberg deviation and allows the specification of a genetic model. Any genetic model specifying a trend in risk with increasing numbers of A_2 alleles (an additive, dominant, or recessive model) can be examined. The calculation includes a weight to multiply the effects of the genotypes, depending on the model. For example, for an additive effect, we use the weights 0, 1, and 2 for the three genotypes, and we obtain the value $P = 0.22$. If, by contrast, we use the weights 0, 0, and 1 for a recessive model we obtain the value $P = 0.048$. So our result is now significant at the 5% threshold! But, other than knowing that a recessive model gives us a significant result, we don't know if this is the true model. Testing lots of models to find the lowest P value isn't good statistics. Rather, it's an example of **P-hacking**—sorting through possible modes of analysis to find a result (see section 21.3 for more discussion of P-hacking). We return to this issue later when we discuss the problem of multiple testing.

To avoid the problem of model choice, the usual strategy is to test only an additive model. This is easily implemented, as genotypes that were originally

Table 21.5

Genotype	n	Neuroticism		Extraversion		Openness		Agreeableness		Conscientiousness	
		Mean	SD	Mean	SD	Mean	SD	Mean	SD	Mean	SD
l/l (group L)	72	53.4	12.0	52.5	10.5	57.2	12.9	45.4	11.3	43.5	11.6
l/s	106	57.8	13.2	53.2	12.2	56.1	12.3	42.6	11.8	40.5	13.5
s/s	43	56.6	11.2	52.3	10.4	55.9	14.1	40.9	11.7	42.1	12.4
l/s + s/s (group S)	149	57.4	12.6	52.9	11.7	56.1	12.8	42.1	11.7	41.0	13.2

obtained as some combination of A, C, G, and T (AA, AC, and CC for example) are just recoded to form a linear series, such as 0, 1, and 2 (or 0, 0.5, and 1), where the heterozygote lies midway between the two homozygous classes.

What about quantitative rather than categorical phenotypes? Here, the choice is comparison of means, analysis of variance, or linear regression. Table 21.5 is from a paper in which the authors claim that a polymorphism in the serotonin transporter gene contributes to variation in the personality factor neuroticism (LESCH et al. 1996). Personality, as discussed in Chapter 12, section 12.2.1, is frequently measured using five scales (O for openness, C for conscientiousness, E for extraversion, A for agreeableness, and N for neuroticism). The scales are all quantitative.

The investigators gave a personality questionnaire to 221 people and worked out the mean score for each of the five personality factors. They reported the mean and the standard deviation of the scores (Table 21.5). Their aim was to establish whether the mean personality scores of group L (homozygous l/l genotypes) are significantly different from the mean scores of genotypes that contain the s allele (group S, comprising l/s and s/s). The null hypothesis tested is that the means are drawn from the same, rather than from two different, distributions. The result of applying a test comparing the two means (a t-test) is a P value of 0.022 (see Appendix, section A8 for a description of the test and for an example). The authors reported a value of $P = 0.024$ in the paper, so they too used the correct test (the difference between our and their results is probably due to rounding errors in the values used).

Although the correct test was applied, there is a problem. Why do the authors use the group L versus group S comparison, rather than the three-genotype group comparison? They do so because they assume a specific genetic model: that the s allele is dominant to the l allele. Consequently, their test is more 'powerful' (i.e. more likely to find a significant P value). The conservative method is to analyze all three genotypes, in which case the appropriate test is an analysis of variance (ANOVA), which tests whether the amount of variation in the three groups is consistent with the variation in the total sample. The larger the ratio of between-group variance to within-group variance, the more the between-group variance contributes to the total variance and hence the less likely it is that the groups belong to the same population. The

ratio is called an F statistic (after Fisher; see Appendix, section A9). I obtained an F statistic of 2.18 with 2 degrees of freedom, and an associated P value of 0.12 (in other words, nonsignificant). This illustrates again how critical the choice of test is for the outcome.

21.3 Was the correct significance threshold applied?

The habit of choosing a statistical test that generates the lowest P value belongs to a larger set of problems that concern the choice of an appropriate significance threshold. To demonstrate, we'll again use the serotonin transporter genetic association studies. They illustrate something that occurs all too frequently in the genetic association literature (and more broadly in statistical analysis): multiple testing.

Typically, the null hypothesis (namely the hypothesis that the genetic association is not true) is rejected when the statistical test returns a P value of less than 0.05. This 5% significance threshold was introduced by Ronald Fisher (FISHER 1925) in the following words: 'If P is between .1 and .9 there is certainly no reason to suspect the hypothesis tested. If it is below .02 it is strongly indicated that the hypothesis fails to account for the whole of the facts. We shall not often be astray if we draw a conventional line at 0.05 ...'

For running a single statistical test, it's usually said that a P value of < 0.05 means that the probability of obtaining the result by chance is less than 1 in 20 (see section 21.6 for a thorough consideration of what the P value means, but for now let's go with this interpretation). In this case, 0.05 is the same as a 5% significance threshold. But when you run multiple tests, it's **not** true that $P = 0.05$ equals the 5% significance threshold. The situation is like this: suppose someone asks you to bet whether they can throw two dice and get a pair of sixes on the first throw. You agree that that seems unlikely and you might agree to the bet, but you'd cry cheat if they kept on throwing the dice until they got what they wanted.

In Table 21.5, there are five personality factors, and the investigators wouldn't have included the data if they hadn't considered testing for the association with all of them. The more tests they carry out, the more likely they are to find, by chance, a few cases where the two means are different, and the difference may meet standard levels of significance ($P < 0.05$). On average, one in every 20 comparisons will differ with $P < 0.05$. Picking five personality factors is really like having five chances at winning on a gambling machine, so it is no longer true that the probability of finding a difference by chance alone is 0.05.

This leads to an important rule: when you are judging whether a claim is true or not, make sure the appropriate significance threshold is applied (significance thresholds are often referred to as 'α' to distinguish them from P values). A threshold (α) of 1 in 20 ($P < 0.05$) is only appropriate for a single test. If many things are tested, the P value has to be lower to keep the threshold at 5%. How much lower?

The simplest way to work out the appropriate P value is to divide by the number of tests you run (this is called a Bonferroni correction, after the Italian mathematician Carlo Emilio Bonferroni (1892–1960) who worked on probability theory). In the case of the association between the serotonin transporter and personality measures from Table 21.5 (LESCH et al. 1996), a 5% significance threshold should take into account the fact that five personality factors were tested. Using the Bonferroni correction method, divide the 0.05 threshold by 5 to get a threshold of 0.01. This corrected threshold means that the result we had above ($P = 0.022$) isn't significant (not at 5% anyway). Thus, just to re-emphasize unless you missed this (many people seem to, I'm not sure why), a P value of less than 0.05 does not necessarily mean a result is significant at a 5% significance threshold.

Failure to take into account multiple testing, as well as the use of an inappropriate ('more powerful') test, are examples of 'P-hacking', a concept introduced in a 2011 article (SIMMONS et al. 2011) where the authors investigated a phenomenon they called 'researcher degrees of freedom', or the freedom of researchers '*to explore various analytic alternatives, to search for a combination that yields "statistical significance", and to then report only what "worked"*'. The most extreme form of this behavior is to keep running experiments, and analyses, until one returns a significant (at $P < 0.05$) result. Just like tossing a coin, if you persist, eventually you will throw heads. Repeated testing under different models increases the chances of finding a false-positive, not a true-positive, result.

The Bonferroni correction assumes that the tests applied are independent. When we test multiple genetic markers, as in a genome-wide association test, many of the loci are in linkage disequilibrium, so they are not entirely independent (linkage disequilibrium means that the genotypes at one locus correlate with the genotypes at another, although not perfectly). This means a Bonferroni correction is overly conservative when applied to genome-wide association results. Its advantage is that it's safe. A result that withstands a Bonferroni correction is likely to be robust.

21.4 Could the result be attributable to some unacknowledged confound?

Geneticists are, usually, less worried than epidemiologists about whether an association is causal, because while genetic variation can cause phenotypic variation, biology is so designed that phenotypes cannot (generally) influence genotypes. Our genomic DNA is 'protected', and while the environment might alter methylation of DNA or histone configuration, it does not in any systematic way ever change the DNA base pair sequence. However, that doesn't excuse geneticists from ruling out the possibility of something else that might explain a genetic association. By far the most important alternative explanation for association analyses are subtle differences in the origins of cases and controls, or unacknowledged genetic relationships in the sample of supposedly genetically unrelated individuals.

For example, among the many association studies of genetic variants in the *DRD2* gene seeking to establish its involvement with schizophrenia, some include subjects of African and European descent. Suppose (although I couldn't find a case this extreme) that the group of individuals affected by schizophrenia is entirely African and the controls are all European. In this case, genetic variants that distinguish cases from controls will be totally confounded with variants that distinguish Europeans from Africans. There will be no way to determine whether a significant difference is due to genetic ancestry or to the risk of schizophrenia. This is the problem of population structure (see Chapter 6, section 6.6.4 for an illustration in Figure 6.4 of this concept and for further discussion). Even a small imbalance in the genetic contribution from one population to the cases can lead to a false positive, especially as risk loci themselves have such small effects.

Population structure can be dealt with using a method called principal component analysis (PRITCHARD et al. 2000). The idea is that the correlations between genotypes reflect the shared descent of the population, and thus can be used to control for the population structure. Alternatively, a genetic relationship matrix, which records pairwise relationships between individuals in the study, can be used to model (and thus correct for) the degree to which each individual is related to every other individual (KANG et al. 2008). Shared relationships reflect descent from common ancestors; risk loci are expected to be present in all groups, regardless of their descent. Both of these approaches require that the information is included in the association analysis, but how can this be done?

The answer is to use regression (linear regression for quantitative phenotypes and logistic regression for categorical traits). Regression models the relationship between phenotype and its predictors as a linear combination of factors. The relationship is typically written as:

$$y = bx + e$$

where y is the phenotype, x is the genotype, and e represents error.

The solution to the incorporation of additional factors in linear regression is simple: each factor is added into the equation as an extra term. In this way any number of factors can be included:

$$y = b_1 x_1 + b_2 x_2 + b_3 x_3 + b_4 x_4 \ldots + e$$

The index after each variable denotes the additional variables and their associated effect. Each of the terms can be any covariate that might influence the phenotype, such as sex or age, as well as the factors that take into account population structure and genetic relatedness. We give examples of regression and its interpretation in the Appendix, sections A10 and A11.

Population structure is not the only confound. Batch effects in genotyping are frequently to blame. For example, experimenters might separately genotype (or sequence) cases and controls, in which case any artifact that alters the genotyping (or sequencing) in one group or the other will result in a difference between the two, which would be incorrectly attributed to genetic differences, when it could simply reflect the batch of reagents used on one particular day. Careful experimental design can limit batch effects but does not guarantee they are eliminated.

21.5 Did the experiment have sufficient power to deliver the result reported?

It might seem surprising that anyone would carry out a study that was not powerful enough to detect what it is supposed to detect, but, well, stuff happens …

One critical factor that determines power in genetic association studies is how big an effect a locus has, that is, by how much it increases the risk of disease or how much it contributes to variation in the phenotype. An effect size in power analyses is a standardized measure that quantifies the size of the difference between two groups, or the strength of an association between two variables (in our case, phenotype and genotype).

A second important factor for power is the sample size we have available: the bigger the sample size, the greater the chance of detecting an effect. However, the relationship between power, effect size, and sample size is not linear. We'll illustrate this as follows.

Assume that the risk locus has an odds ratio of 1.2 (which is a very large effect for genome-wide association studies (GWAS) in behavior genetics), that the risk allele is common (frequency of 0.5), and that we use 1,000 cases and 1,000 controls in our study. Table 21.6 shows the result of running a power analysis.

You can interpret this table as follows: we have 36% power to detect a locus at a P value of < 0.05, and would need 1,538 cases and 1,538 controls to have 80% power (this gives the total in the last column of 3,076). The last threshold in the table ($P < 5 \times 10^{-8}$) is the significance threshold for a GWAS. Power with a sample size of 1,000 cases and 1,000 controls is miniscule to screen the genome (it is just 0.006%). Compare this with trying to find a locus with an odds ratio of 1.1, as shown in Table 21.7. As the effect size reduces, the power drops precipitously: the required sample

Table 21.6

P value threshold	Power	N subjects for 80% power
0.05	0.358	3,076
0.01	0.164	4,576
0.001	0.045	6,691
5×10^{-8}	5.9×10^{-5}	15,518

Table 21.7

P value threshold	Power	N subjects for 80% power
0.05	0.129	11,753
0.01	0.039	17,489
0.001	0.007	25,569
5×10^{-8}	1.86×10^{-6}	59,303

sizes for 80% power are almost fourfold higher than for the odds ratio of 1.2.

Power in genetic studies also depends on the frequency of the allele and the prevalence of the disorder: as you increase the frequency of the allele, the power increases, and as the prevalence of the disease increases, so does the power, although not as much (FLINT AND MUNAFO 2007).

The most important point to remember is that power is related to the square root of the sample size ($\frac{1}{\sqrt{n}}$ where n is the sample size). To put some flesh on this statement, I'll give a numerical example. Suppose we want to detect a genetic effect that increases a trait by 5% in one of two groups (e.g. cases versus controls) and we want to know the sample size to see that effect with a power of 80% (a conventional value for power). How large must the sample be so that we have an 80% chance of achieving statistical significance (at $P < 0.05$)? This means (because of the distribution of standard errors, shown in Figure A6 in the Appendix, section A8) that the difference between the two groups must be at least 2.8 standard errors from zero.

Working with two equally sized groups (cases and controls), the sample size (n) that we need is $\left(\frac{2.8}{0.05}\right)^2$ = 3,136 (i.e. 1,568 in each of the two groups). But of course we know that 5% is a very large genetic effect for a behavioral trait, such as personality. Let's try a smaller effect size, say 1% (which is still enormous for genetic effects on personality). Then the sample size we need for 80% power is 74,800. And for power to detect an effect of 0.5%, closer to what we know is a true genetic effect size for personality, the number we need to genotype is the truly frightening figure of 313,600.

The increase in sample size when we reduce the effect we want to detect from 5% to 0.5% is not ten times but

a hundred times, consistent with the rule that effect size is proportional to $\frac{1}{\sqrt{n}}$. This is counterintuitive for many people, who see it as common sense that the power of a sample should increase linearly with sample size. Alas, it does not scale this way, as we have just shown. We give further details on how power is estimated in the Appendix, section A12.

But so what? If I've just carried out a study and found a result that exceeds a 5% significance threshold (and yes, I did check I had used the correct test, and I did apply the correct significance threshold), why should the power of my study matter? After all, I've found something! I can publish a paper! Why worry about a power analysis?

I'll illustrate why in the way that Jonathan Sterne and George Davey Smith did (STERNE AND DAVEY SMITH 2001). They argued as follows. Suppose that most of the claims in the genetic association literature are incorrect (we have reviewed evidence that this is so in Chapter 5, section 5.2). Let's say 90% of the findings are wrong (but we don't know, from the literature, which 10% are true). This means that out of every thousand experiments, there are only 100 true genetic associations. Let's also say that on average the power of a study is 50% (this is less than most people would be happy to admit: standard practice is to power your study at 80% or more). This means that we will incorrectly say half of the 'true' findings are wrong. Out of 1,000 studies there are 900 where there is no genetic effect, but 5% will show a P value of less than 0.05 by chance (read the multiple testing section again if you are not sure why this is so), so we incorrectly claim that there is a true finding in 45 cases, as shown in Table 21.8.

The second line in the table is critical: it means that only about half of the findings of $P < 0.05$ are true

Table 21.8

Result	No association	True association	Total
Nothing found ($P > 0.05$)	855	50	905
Something found ($P < 0.05$)	45	50	95
Total	900	100	1000

associations. That's not what most people will assume when they find a result with $P < 0.05$. Once we assess power, we find that the 'positive predictive value' (shorthand for 'the finding is a true positive') has dropped to about 50%. Of course, all of this is predicated on the assumption that most of the claims in the genetic association literature are incorrect, because studies are not as well powered as they should be (the assumption is an example of a **prior probability**, a concept we discuss in more detail in section 21.9 on Bayesian methods). As power goes up, so does the positive predictive value. The formula linking positive predictive value to power is:

$$\frac{([1-\beta]R)}{([1-\beta]R + \alpha)}$$

where $(1 - \beta)$ is the power, α is the significance threshold, and R is the odds that effects expected to be true are indeed true (BUTTON et al. 2013).

For example, if we expect five of the genetic loci we test to be true, then $R = 1/(5 - 1) = 0.25$, and with 50% power, 70% of claims will be correct when they have $P < 0.05$.

Low power has three consequences. First, and most obviously, it reduces the chances of discovering genuinely true effects. Second, not at all obviously (as explained above), it lowers the positive predictive value. It debases P values, so that the probability a finding passing a significance threshold is indeed true is less than we expect it to be. Third, an underpowered study will overestimate the magnitude of the effect it detects (IOANNIDIS 2008). This is called the 'winner's curse'. It particularly influences low-powered studies, as can be appreciated from the following argument. Each study produces a different estimate of the genetic effect, due to random sampling. Smaller effects are associated with larger P values so they will be missed by lower-powered studies. Effects larger than average, due to the random sampling, will be over-represented in the lower-powered studies, thus leading to the 'winner's curse'.

There is another intuitive way to understand the association between power, P values and false-positive results. Imagine performing a study that has zero power to detect the effect size you are looking for (because either the sample is very small or the effect you are looking for is nonexistent, or both). You run the study 100 times. You still expect that in 5% of these studies, you will obtain a P value of <0.05. They will all be false positives, because you have no power at all to obtain a true-positive result.

21.6 Was the P value interpreted correctly?

If this is the first time you have read about statistical methods, you likely imagine that the way they work is to determine whether the hypothesis tested (e.g. that genotypes are associated with disease) is true. In fact, counter to most people's intuition, the majority of tests assume the opposite. They test whether the data support a different hypothesis, namely that the genotypes are **not** associated with disease. A low P value means it is unlikely that the genotypes are not associated with the disease, an awkward double-negative way of saying that the genotypes are likely to be associated.

Typically, the null hypothesis (namely, the hypothesis that the genetic association is not true) is rejected when the statistical test returns a P value of less than 0.05 (this is the 5% significance threshold introduced by Ronald Fisher; FISHER 1925). The significance threshold is the probability of rejecting the null hypothesis given that the null hypothesis is true.

This brings us to a most confusing problem, namely the correct interpretation of a P value. If, during an interview for a job at a world famous department of human genetics, you are asked, 'What is a P value?', you can reasonably take this as an indication that things are not going too well. After all, every undergraduate knows that the P value is the probability that the null hypothesis is true. Really? Here is a short test you can give to someone who claims to know the meaning of a P value:

You perform a genetic association test, comparing allele frequencies between cases and controls. Your result is $P = 0.01$. Please decide for each of the statements below whether it is 'true' or 'false'.

1. You have absolutely disproved the hypothesis that there is no difference between the allele frequencies (the null hypothesis).

2. You have found the probability of the null hypothesis being true.

3. You have absolutely proved that there is a difference between the allele frequencies (the experimental hypothesis).

4. You can deduce the probability of the experimental hypothesis being true.

5. You know the probability that you are making a wrong decision if you decide to reject the null hypothesis.

6. You have a reliable experimental finding in the sense that if the experiment were repeated a great number of times, you would obtain a significant result on 99% of occasions.

Adapted from HALLER AND KRAUSS (2002).

This test was given to 30 statistics teachers, 39 professors and lecturers of psychology, and 44 psychology students (HALLER AND KRAUSS 2002). Ninety percent of the professors and lecturers stated that one or more of the six statements were true. In fact, **none** of the statements is true, at least if we apply the definition of a P value from hypothesis testing, and in particular from Neyman–Pearson hypothesis testing. In what follows, we'll explain why.

Putting aside for the moment Bayesian approaches, historically there are two trends of thought on the meaning of the P value, one emanating from Ronald Fisher and the other from Jerzy Neyman and Egon Pearson (Karl Pearson's son). Simplifying, the Fisherian P value indicates 'the strength of the evidence against the hypothesis': the smaller the P value, the better the evidence against the null hypothesis. Fisher felt that the interpretation of the P value was best left up to the researcher (FISHER 1925) and should not provide a definitive guide to the outcome of an experiment. Neyman and Pearson disliked the subjective nature of Fisher's view and wanted to replace it with rules for accepting or rejecting the null hypothesis. They took a different view from Fisher of probability in statistical tests, arguing that *'no test based upon a theory of probability can by itself provide any valuable evidence of the truth or falsehood of a hypothesis'* (NEYMAN AND PEARSON 1933). Instead, *'we may consider some specified hypothesis, as that concerning the group of stars, and look for a method which would hope to tell us, with regard to a particular group of stars, whether they form a system or are grouped "by chance", their mutual distances apart being enormous and their relative movements unrelated. … For however small be the probability that a particular grouping of a number of stars is due to "chance", does this in itself provide any evidence of another "cause" for this grouping but "chance"?'* In this view, probability is a relative frequency, referring to a hypothetical infinite number of events, and does not refer to a single event.

In Neyman–Pearson hypothesis testing, a precise alternative hypothesis has to be decided, and that hypothesis has to be defined **before** the experiment is carried out. It is not enough to say that we expect the genetic variant to increase risk; we have to say by how much we expect it to increase the risk (you can

appreciate why power calculations are essential to hypothesis testing).

Here's an example to explain why the experiment needs to be designed explicitly before the data are collected. The physicist Richard Feynman describes how he was asked to aid a psychologist in assessing the significance of his experimental results.

This man had designed an experiment which would show something which I do not remember, if the rats always went to the right, let's say. I can't remember exactly. He had to do a great number of tests, because, of course, they could go to the right accidentally, so to get it down to one in twenty by odds, he had to do a number of them. And it's hard to do, and he did his number. Then he found that it didn't work. They went to the right, and they went to the left, and so on. And then he noticed, most remarkably, that they alternated, first right, then left, then right, then left. And then he ran to me, and he said, 'Calculate the probability for me that they should alternate, so that I can see if it is less than one in twenty.' I said, 'It probably is less than one in twenty, but it doesn't count.' He said, 'Why?' I said, 'Because it doesn't make any sense to calculate after the event. You see, you found the peculiarity, and so you selected the peculiar case.

FEYNMAN (1998).

In short, *'you cannot judge the probability of something happening after it's already happened'*. In the literature this is called 'HARKing': Hypothesizing After the Result is Known.

Neyman and Pearson recognized two sources of error in statistical testing: false rejection (error I) and false acceptance (error II). The type I error happens when a true null hypothesis is incorrectly rejected (a 'false positive'), while a type II error occurs when a false null hypothesis is incorrectly retained (a 'false negative'). The best test is one that reduces both sources of error to the minimum; however, both *'can rarely be eliminated completely; in some cases it will be more important to avoid the first, in others the second. We are reminded of the old problem considered by Laplace of the number of votes in a court of judges that should be needed to convict a prisoner. Is it more serious to convict an innocent man or acquit a guilty? That will depend upon the consequences of the error; is the punishment death or fine; what is the danger to the community of released criminals; what are the current ethical views on punishment? From the point of view of mathematical theory all that we can do is to show how the risk of error may be controlled and minimized'* (NEYMAN AND PEARSON 1933).

In significance testing: a *P* value is **the probability of the observed data (or of more extreme data points), given that the null hypothesis is true**. The *P* value is **not** the probability of the **hypothesis**, given the data.

Applying this definition, all of the six statements given in the test by HALLER AND KRAUSS (2002) about *P* values are false. Statistics does not prove anything, so (1) and (3) cannot be true. A significance test does not yield the probability of a hypothesis but of observed data, so statements (2), (4), and (5) are incorrect. In order for significant data to reappear in 99% of the replication experiments (as statement (6) requires), the null hypothesis would have to be proven, which is not correct.

As we hope you are beginning to realize, not all *P* < 0.05 are the same ... and be aware that there is a fair degree of emotion surrounding the topic of hypothesis testing (NICKERSON 2000).

I'm currently a PhD student in the social sciences department of a university. I recently got involved with a group of professors working on a project which involved some costly data-collection. None of them have any real statistical prowess, so they came to me to perform their analyses, which I was happy to do. The problem? They want me to p-hack it, and they don't even know it. The project reads like one of your blog posts. The professors want to send this to a high-impact journal (they said *Science*, *Nature*, and *The Lancet* were their first three). There is no research question, and very little underlying theory. They essentially dumped the data on me and told me to email them when 'you find something significant.' The worst part is, there is no malicious intent here and I don't think they even know they're just fishing for p <.05. These are genuinely good, smart people who just want to do a cool study and get some recognition. I don't know if you have any advice to handling this sort of situation.

http://andrewgelman.com/2017/11/11/student-bosses-want-p-hack-dont-even-know/

21.7 Abandon statistical significance!

Critics of Neyman–Pearson hypothesis testing can be surprisingly vehement (LEHMANN 1993; WAGENMAKERS 2007), accusing practitioners of intellectual emptiness in imposing on their results the following 'null ritual':

1. Set up a statistical null hypothesis of 'no mean difference' or 'zero correlation'. Don't specify the predictions of your research hypothesis or of any alternative substantive hypotheses.

2. Use 5% as a convention for rejecting the null. If significant, accept your research hypothesis. Report the result as *P* < 0.05, *P* < 0.01, or *P* < 0.001 (whichever comes next after the obtained *P* value).

3. Always perform this procedure.

Even a brief reading of the statistical methods section of many of the papers we refer to in this book is enough to show that they have a point. It's easy to understand why there have been radical proposals to replace *P* <0.05 with *P*<0.005 (BENJAMIN et al. 2018), or abandon statistical significance (AMRHEIN AND GREENLAND 2018), a goal that 854 scientists signed up to in a commentary in *Nature* (AMRHEIN et al. 2019). Presenting unreliable findings as reliable is a cause of the reproducibility crisis (perhaps **the** cause), so we should avoid giving the false impression that a *P* value above a threshold means there is no association (or no difference), or that because the *P* value is below a given threshold the result is 'true'. How do we do that? What's the best way forward?

The debate hasn't settled (IOANNIDIS 2019), and to be honest it's unlikely that *P* < 0.05 is going to be dethroned any time soon from its role as a gatekeeper to publication, or lose its alchemical property of dividing 'effect' from 'no effect'. However, there are three points worth considering.

The first is that *P* value thresholds are problematic, something that should be evident by now from what we have discussed in this chapter. P-hacking, publication bias, and other unsavory practices mean that *P* values aren't always what they should be. A one-size-fits-all threshold is probably inappropriate, but then it's not always obvious what is appropriate.

One alternative is to **set a threshold so as to limit the number of false positives, or at least state the risk of false positives given the *P* value you find**. *P* < 0.05 is not the same as setting the risk of a false positive to less than 5%; indeed, the false-positive risk is always bigger than a *P* value, although how much bigger depends on the likelihood that there is a real effect (COLQUHOUN 2019).

A key term in this discussion is **prior probability**, meaning the prior probability of a real effect. This is hard to estimate, but it is central to working out whether what you are claiming to be true really is true. We'll return to consider prior probabilities in section 21.9 on Bayesian approaches.

Second, always report effect sizes and consider them as of equal, if not greater, importance than the *P* value. At one level, the effect size can be used to decide whether the effect you are observing is real or not: a nice big fat effect size is hard to miss. Along with sample size, the effect size determines power and thus, like the prior probability, contributes to the probability that a finding is true. However, remember that because of the 'winner's curse' described in section 21.5, the observed effect size is not necessarily the true effect size! It is only the true effect size that determines power.

At another level, however, the effect size should be taken into account because it says something about the importance of the finding. This position sounds reasonable; after all, in genetics we are definitely interested in finding a large-effect mutation that contributes to disease. But the size of the effect may not reflect its utility. A constant complaint about GWAS is that the effect sizes found at individual loci are so small: who cares about an effect that increases the likelihood of disease by less than 1%? GWAS aficionados will immediately take umbrage, and point to the introduction of IL-23 inhibitors for the treatment of inflammatory bowel disease, based in part on GWAS findings (MOSCHEN et al. 2019). Small effects, robustly demonstrated, can be very important indeed.

Third, *P* values and effect sizes should be considered along with existing knowledge, together with the plausibility of the mechanism or finding proposed, and the quality of the research, to come to a view of whether a claim is true or not. In short, applying a statistical test does not absolve a researcher from careful thought.

21.8 Model-fitting approaches

Genetic epidemiology often uses model-fitting approaches, rather than hypothesis testing. The logic of model fitting is different from hypothesis testing. Instead of determining the probability that the results are consistent with the null hypothesis, we instead compare the ability of different models to explain the patterns observed in the data.

This approach requires the calculation of a statistic that evaluates globally how well a particular hypothesis explains the observed data. This is called a 'fit index', and there is a variety to choose from. But this is a detail we do not need to dwell on. The essential

point is that we want to calculate the fit indices for the models we want to evaluate and hope that one of them stands out as providing the best fit to the data.

To illustrate this, we will examine an old twin study of panic disorder (KENDLER et al. 1993c). Lifetime panic disorder had been assessed at personal interview for 2,163 female twins. We wanted to understand the role of genetic and shared environmental effects in explaining the twin resemblance for panic disorder. To do this, we applied three different models to the data. The first assumed that both genetic and shared environmental factors caused twin similarity. The second and third models assumed that twin resemblance could be explained only by shared environment, or only by genes. Our fit index, one commonly used in twin studies, was Akaike's information criterion (AKAIKE 1987) or AIC for short. For the AIC, as for many fit indices, the lower the value, the better the model fits (or explains) the data. The AIC value obtained for models 1, 2, and 3, were, respectively, 0.65, 1.22, and −1.35. The differences were not large, as panic disorder was rare in this sample so our power to discriminate models was modest. However, the pattern of results was clear. The third model, which postulated that twin resemblance was solely a result of genetic factors, explained the data best. According to the third model, heritability of panic disorder was estimated to be 32%.

How do fit indices work? They all try to balance two things that scientists want from their models: explanatory power and simplicity; that is, for the same degree of complexity, everyone would prefer a model that did a better job of explaining the data. But, for the same degree of explanation, everyone would prefer a simple model to a complex one (asserting a preference for the simplest explanation is often called "applying Occam's razor", after William of Occam, a thirteenth-century philosopher). How these two competing goals are balanced differs across specific fit indices, but we won't go into the details here.

21.9 Bayesian approaches

In February 1998, a UK lawyer, Sally Clark, was arrested for the murder of her two infant children, 11-week-old Christopher in 1996 and 8-week-old Harry in 1998. During the ensuing court case, the jury heard from an expert witness (Professor Sir Roy Meadow) about how unlikely it was that the children were the victims of sudden infant death syndrome (SIDS or cot deaths): '[T]he chance of one child dying of cot death is 1 in 8543, so that if you multiply these together, you get the chance of both dying by cot death being 1 in 73 million.' The jury found Sally Clark guilty and she was given a prison sentence.

The statistical evidence was a crucial part of the prosecution's case, and can be summarized as follows: the probability of both children dying from cot deaths is vanishingly small, which, in the words of the prosecutor, made it 'very unlikely that the babies died naturally, and so strongly suggests murder'. However, the probability of two cot deaths tells us nothing about the probability of guilt, a misinterpretation known as the prosecutor's fallacy: a small probability of the evidence given innocence need not mean a small probability of innocence given the evidence. In 2001, the Royal Statistical Society wrote of the evidence presented in court:

Some press reports at the time stated that this (1 in 73 million) was the chance that the deaths of Sally Clark's two children were accidental. This (mis-)interpretation is a serious error of logic known as the Prosecutor's Fallacy. The jury needs to weigh up two competing explanations for the babies' deaths: SIDS or murder. Two deaths by SIDS or two murders are each quite unlikely, but one has apparently happened in this case. What matters is the relative likelihood of the deaths under each explanation, not just how unlikely they are under one explanation (in this case SIDS, according to the evidence as presented).

https://www.rss.org.uk/Images/PDF/influencing-change/2017/SallyClarkRSSstatement2001.pdf

In Sally Clark's case, no one in the court knew how to weigh up the evidence appropriately. By January 2003, three judges had quashed her convictions, but Sally's release came too late: 'Today is not a victory. We are not victorious. There are no winners here. We have all lost out.' Sally died in 2007. John Batt, her solicitor and a member of her defense team, said, 'I spoke to her this week ... she would have good days and bad days, but I don't think you ever can ever recover from something like that. Imagine being in jail where everyone thinks you are the scum of the earth, the lowest human being that walks the earth. The thick end of it is that she lost five to six years of her life in what was state-sponsored torture' https://www.theguardian.com/society/2007/mar/17/childrensservices.uknews.

Sally Clark's case is an extreme example of failure to understand probability theory. We find it hard to assess the probability of one thing given the probability of another. These are problems of conditional probability, which Bayes' theorem addresses.

21.9.1 Bayes' theorem

Bayesian statistics have a bad reputation. It's not just that there always seems to be a lot of hard mathematics, and an incomprehensible language that involves priors and posteriors; it's also that the classical statisticians, the frequentists as they are called, often denigrate their Bayesian colleagues. I hope you can ignore, for the time being, the off-putting characteristics of Bayesian statistics, while I explain why Bayesian approaches are useful. For this, you don't need a mastery of Markov chains, Monte Carlo simulations, and other elements of advanced probability theory. You do, however, need to understand the difference between estimating the probability of independent or dependent events.

Independent events are those where the second event occurs regardless of the outcome of the first event; dependent events involve a choice: either one thing or another. When you throw a die with six faces, you have six possible outcomes: one, two, three, four, five, or six. These are dependent events, and the probability of throwing six or three is the sum of the probability of each: $1/6$ plus $1/6 = 2/6$. If you toss twice, the result of the first throw has no effect on the second throw. The outcome of the two throws is independent, and the probability of obtaining both events is a multiple of the probabilities of each event. The probability of throwing first a 'six' and then a 'three' is $1/6 \times 1/6 = 1/36$.

There's also one piece of terminology worth knowing. To indicate 'given that' in an expression, you'll see the symbol '|' employed. For example, 'probability (data | hypothesis)' means the probability of obtaining the observed data given that the hypothesis is true, a state of affairs we hope applies.

Bayes' theorem provides a way to answer questions of conditional probability (which have the form: the probability of A given the probability of B). In words, Bayes' theorem is:

The probability that the evidence is real (that it supports our hypothesis) is the probability of a true finding from all of the positive findings (both true and false).

To illustrate, we give a classic application of Bayes' theorem: interpreting the results of a diagnostic test.

Imagine you attend a routine medical checkup, and are tested for the presence of schizophrenia. The result comes back positive. How likely is it that you really do have schizophrenia? You need three pieces of information to answer this question. The first is the prevalence of the disease: 1%, in the case of schizophrenia. The second is the true-positive rate of the test (the sensitivity). The third is the true-negative rate (the specificity). These numbers are in the top left and the bottom right corners of Table 21.9. The remaining pieces of information derive from the others: one minus the false positive rate, one minus the false negative rate, and one minus the disease prevalence. Just to be clear about definitions: sensitivity is the proportion of true positives that are correctly identified by the test, or the ability of the test to correctly identify those **with** the disease, while specificity is the proportion of true negatives that are correctly identified by the test, or the ability of the test to correctly identify those *without* the disease (ALTMAN AND BLAND 1994).

Now, using this information, we generate a second table with the probabilities of the four entries in the table, given the disease state. Disease and test result are independent events, so the probability of a true positive, given that you have the disease, is the probability of one times the other. The probability that you have the disease given a positive test result is 0.01 multiplied by 0.8 = 0.008, as shown in the top left of Table 21.10. The other entries in the table are calculated similarly.

Table 21.9

	Disease (1%)	No disease (99%)
Positive test result	0.80	0.05
Negative test result	0.20	0.95

Table 21.10

	If you have the disease	If you don't have the disease
Probability of a positive test	0.008	0.04995
Probability of a negative test	0.002	0.94905

Finally we calculate the probability that the result we obtained is true, that is, the ratio of the probability of a true positive over the probability of getting **any** positive test result. The probability of any positive result means either one or other of the entries in the first row of the table. These are dependent events, so we add them. The ratio we want is thus 0.008/(0.008 + 0.04995) = 0.138. In other words, if you get a positive test result, then the probability that you really do have the disease is 14%.

It's worth considering how the answer is affected by changing the parameters. If the true-positive and true-negative rates are 99%, then the probability that someone with a positive result really does have schizophrenia is still only 50% Why? It's because of the effect of the disease prevalence. Out of 100 people, there will be one person with schizophrenia, and this person will be correctly diagnosed. But 1% of the remaining 99 will be **incorrectly** diagnosed, which means one more person testing positive. So the test will make one correct diagnosis and one incorrect diagnosis. It's wrong half the time.

Exactly the same logic applies in **any** situation where we want to calculate the probability of something being true, given some piece of information. Applying Bayes' theorem to the Sally Clark case is more complex, as we have to compare two conditional probabilities: the probability of a double murder, given that her sons died, and the probability of two cot deaths, given that her children died. But the principle is exactly the same. In this case, the quantity we want is the probability of two cot deaths over the probability of a mother murdering two children. As both are extremely rare events, the ratio lies between 0.1 and 0.5, clearly very different from the 1 in 73 million presented to the jury.

21.9.2 **A Bayesian approach to genetic association tests**

You've carried out a genetic association study of personality—let's say it is a GWAS and you've found a result that exceeds a significance threshold. What is the probability that the finding is true? This is same question as the one for the diagnostic test, but we need different ingredients to answer it. For the diagnostic test, we had the specificity, sensitivity, and the disease prevalence. What are the equivalent ingredients in the case of the genetic association test?

The first two ingredients are easy. The significance threshold provides the estimate of specificity. You can see this by imagining what happens as the significance threshold changes. As the threshold is relaxed, going say from 5 to 20%, more results passing the threshold are incorrect; at a stringent threshold (e.g. going from 5 to 0.1%), it is more likely that a finding passing the threshold is a true positive. This is another way of defining specificity. For sensitivity, we use a power estimate, as sensitivity is the probability that when there is an association we can detect it. Power usually (at least so we hope) is around 80%.

What is the equivalent of prevalence? Here, we have to introduce a new idea: the **prior probability** that the research finding is true (WACHOLDER et al. 2004). In the case of the diagnostic test, when we randomly choose one person out of 100, we have a probability of 1% of selecting someone with schizophrenia. We can estimate the same for the association test in the following way.

Taking what we know about the genetic basis of personality, namely that there are many small-effect loci (the polygenic hypothesis that has recurred throughout this book), it's reasonable to say that there might be 2,000 loci (admittedly, this might be an underestimate), each making a small and equal contribution. It's also reasonable to guess we'll have 2 million loci genotyped for our study of personality. So the prior probability that a genotyped locus is associated with personality is:

$$(2{,}000/2{,}000{,}000)/(2{,}000/2{,}000{,}000 + 1)$$
$$= 0.000999001$$
or, rounding up, $= 0.001$.

We can now make two tables equivalent to those for the diagnostic test analysis. The first looks like Table 21.11.

We then multiply the prior probability of association (0.001) by the entries in the first column and the prior probability of no association (0.999) by the entries in the second column to get our second table (Table 21.12).

Last, we calculate the probability that the result we obtained is true. As before, we calculate the ratio of the

Table 21.12

	If there is an association	If there is no association
Probability of a positive result	0.0008	0.04995
Probability of a negative result	0.0002	0.94905

probability of a true positive (0.0008) over the probability of getting **any** positive test result (the sum of the entries in the first row of the table). The ratio is thus $0.0008/(0.0008 + 0.04995) = 0.0157$. Expressed as a percentage, the probability that the finding is true, given the test result, is 1.6%. The difference with the results from the first table is the low prior probability of association, of 0.1%.

The threshold for a GWAS is, of course, not 0.05, but the standard GWAS threshold of $P < 5 \times 10^{-8}$. Let's see what happens to the probability that our association result is true if we use this estimate of specificity. The change is to the top right entry of the first table. We make this change, and the two tables now look like Tables 21.13 and 21.14. And the ratio is now $(0.0008)/(0.0008 + 4.995 \times 10^{-8}) = 0.999937$. In short, if a finding exceeds the GWAS threshold, the finding is 99.99% likely to be correct. This makes

Table 21.13

	Association (0.1%)	No association (99.9%)
Positive finding	0.80	0.00000005
Negative finding	0.20	0.99999995

Table 21.11

	Association (0.1%)	No association (99.9%)
Positive finding	0.80	0.05
Negative finding	0.20	0.95

Table 21.14

	If there is an association	If there is no association
Probability of a positive result	0.0008	4.995E-08
Probability of a negative result	0.0002	0.99899995

App

A1 Sta

Most gene
line here. \
phenotype
distributio

Normal
of a large n
of its const
heat. Error
thing being
better or w

Variabili
no variabili
between th
and its forn

where Σ m
(the sum of

Squared
standard de
ses). Varian

In geneti
ity in paren
ance; again
square of th

As before, X
and $\bar{Y}$ is the

We are a
Figure A1
Galton coll
parental he
something

The stati
spring heigl
squares. Le

---

sense, and supports the findings reported elsewhere in this book. The Bayesian approach also explains why the candidate-gene tests were not robust: the prior probability of success was low, almost all studies were underpowered, and the significance thresholds were inadequate.

Rather than the rather laborious table construction (which I've introduced to explain the method), it's quicker just to use this formula:

Probability of result being true = (power × prior)/ (power × prior + (1 − prior) × threshold)

This formula is worth remembering, particularly if you have to review a claim that depends on a P value exceeding a given threshold. P values alone aren't the arbiter of whether something is true or not.

We can summarize the Bayesian approach to assessing whether a finding is true or not as follows: a research finding is more likely to be true than false if the power multiplied by the prior probability is greater than the statistical threshold (IOANNIDIS 2005). Suppose our study has a power of 0.8, and our prior is 0.001, the product is 0.0008. Apply the standard genome-wide threshold of $P < 5 \times 10^{-8}$ and then any finding that exceeds the threshold is likely to be true.

But please don't conclude that every GWAS finding that exceeds the 5×10^{-8} threshold is true. The calculation relies not just on the significance threshold, but also on the power. We used a power estimate of 80%, but GWAS are rarely this well powered (refer back to section 21.5 for a consideration of power). Table 21.15 gives the probability that a finding is true for two different thresholds ($P < 0.05$ and $P < 5 \times 10^{-8}$, the GWAS significance threshold) under different assumptions of power. You can think of Table 21.15 as showing how likely your finding is to be correct after

you have carried out the study and found a result that exceeds one of the two given thresholds.

Table 21.5 shows that if we have low power (10%), then at the $P < 0.05$ threshold, the probability that the finding is true (after we have carried out the test) scarcely increases at all over the prior probability. Before we carried out our GWAS of personality with 2 million markers, the prior was 0.001; afterwards, for a P value less than the 0.05 threshold, the probability that the finding is true has increased to just 0.002, hardly any advance on what we knew before! By contrast, at the stringent genome-wide threshold, if we do find something, there is a high probability that the finding is true. Results where the product of power and prior probability exceed the threshold are given in bold, that is, results that are likely to be true.

The approach we have outlined, of assessing whether a finding is true or not was used in a classic paper by John Ioannidis entitled 'Why most published research findings are false' (IOANNIDIS 2005). Ioannidis made his case by pointing out that many studies relied upon achieving a result that was less than $P = 0.05$, without taking into account prior probabilities or power, so that many publications were reporting false positives. You may have noticed that we surreptitiously introduced this approach already, in Table 21.8, where the prior was that most genetic association results for candidate-gene studies were false positives. We hope through these examples that you can appreciate the common sense aspects of Bayesian analysis.

21.9.3 Bayesian hypothesis testing

A Bayesian approach to hypothesis testing agrees with most people's intuitions of what statistics ought to do: it provides support for the belief in a hypothesis, or

---

Table 21.15

Prior probability	Power (threshold)			
	0.1 (0.05)	0.8 (0.05)	0.0001 (5×10^{-8})	0.1 (5×10^{-8})
0.0001	0.0002	0.002	0.167	**0.995**
0.001	0.002	0.016	**0.667**	**1.000**
0.01	0.020	0.139	**0.953**	**1.000**
0.05	0.095	0.457	**0.991**	**1.000**
0.1	0.182	**0.640**	**0.996**	**1.000**

more form
light of the
can recogn
probability
given the d
    Bayesian
probabilitie
late the pro
Bayesian te
It's easy to
we might l
ability (a fl
we have str
to be. Whe
our expecta
able to dem
more signif
be applied?
than 0.01 o
take a Baye
sonable est
ing being tr
the genetic
methodolo
claim. We a
we explain
result, how
    A critical
that, for Ba
esis may va
of the hypc
estimate th
esis. Frequ
that we hav
(e.g. elsewh
approaches
sibility of r
to generate
Bayesian m
of paramete
methods, fc
another hyp
threshold a
positives (a
P-hacking).
tation (Wac
    Such adva
ability of a h
to be true.
not replaced

errors of prediction $\sum (Y - \overline{Y})^2$ is minimal. The slope of the line in the figure is the extent to which X predicts Y and this slope is calculated as:

$$\frac{\sum[(X - \overline{X})(Y - \overline{Y})]}{N\sigma_X^2}$$

where σ_X^2 is the variance of the parental height and N is the sample size.

This measure is referred to as a **regression coefficient**, as it was first derived from analysis of the slope of the line describing the **regression** of offspring on parental values in the height data that Galton collected (GALTON 1886). Regression here means the extent to which the phenotype of offspring reverts, or regresses, to the mean of a population, rather than taking the mean of the parental values. Regression to the mean, a key concept in genetics, is described in more detail in section A4.

The relationship between X and Y is written as a regression equation:

$$Y = bX + e$$

where b is the slope of the regression line and e is an error term.

If there is no relationship between X and Y, then the regression slope (b) is zero and e is the mean value of Y.

Pearson defined another measure of similarity, a **correlation coefficient**, which is scaled from -1 to $+1$. The correlation coefficient (r) is almost, but not quite, the same as the formula for the slope given above. It is defined as the covariance divided by the product of the two standard deviations (rather than $N\sigma_X^2$):

$$r = \frac{\sum[(X - \overline{X})(Y - \overline{Y})]}{s_x s_y}$$

Because the divisor is the product of the two standard deviations, r is related to the variance that is shared by the two measures. Thus, the square of a correlation, r^2, indicates the proportion of variance in one trait that can be predicted by the other.

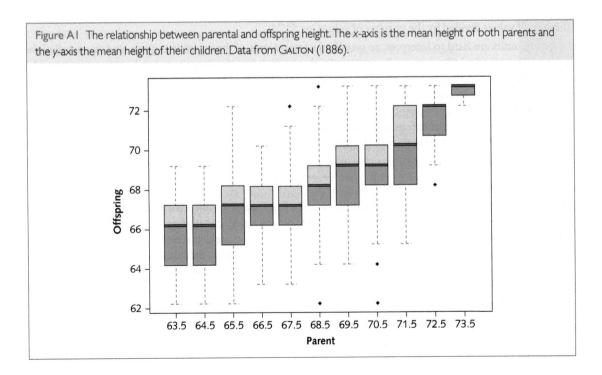

Figure AI The relationship between parental and offspring height. The x-axis is the mean height of both parents and the y-axis the mean height of their children. Data from GALTON (1886).

A2 Genetic models: a single locus and the Hardy–Weinberg principle

In this section, we introduce a central idea in genetics, that of **Hardy–Weinberg equilibrium**. The idea is simple, but its consequences are profound. We start by illustrating Mendelian segregation at a single locus, and then apply the principle to multiple loci and thus explain how complex traits arise from genetic variation (Figure 8.3, in Chapter 8, section 8.10 provides a review of Mendelian inheritance patterns).

The central idea from which most genetic theory can be derived is the random sampling of gametes that generate genotypes in the next generation. This idea is explained by a Punnett square (first proposed by the British geneticist Reginald Punnett (1875–1967)), which shows the contribution of gametes from a locus with two alleles (either A_1 or A_2) from two parents (Table A1). Under Mendelian inheritance, the gametic choice is completely random, so each parent can pass either A_1 or A_2 on to the next generation. This means that the ratio of genotypes in the offspring is given by counting the numbers of genotypes in the boxes in Table A1.

The Punnett square predicts that there will be twice as many genotype A_1A_2 offspring as A_2A_2 or A_1A_1. Table A2 shows a second square where one of the parents is homozygous for the A_1 allele. The genotype ratios in the offspring are two heterozygotes (A_1A_2) to two homozygotes (A_1A_1).

Table A1 A Punnett square, illustrating the contribution of two alleles (A_1 and A_2) from maternal and paternal gametes to create the genotypes of their offspring. In this example, both parents are heterozygotes

		Paternal	
		A_1	A_2
Maternal	A_1	$A_1 A_1$	$A_1 A_2$
	A_2	$A_1 A_2$	$A_2 A_2$

Table A2 A Punnett square, illustrating the contribution of two alleles (A_1 and A_2) when one parent is heterozygote and one is homozygote.

		Paternal	
		A_1	A_1
Maternal	A_1	$A_1 A_1$	$A_1 A_1$
	A_2	$A_1 A_2$	$A_1 A_2$

Table A3 A Punnet square to illustrate how the frequencies of genotypes in an entire population can be predicted from the allele frequencies.

		Paternal	
		pA_1	pA_2
Maternal	pA_1	pA_1^2	$pA_1 \times pA_2$
	pA_2	$pA_1 \times pA_2$	pA_2^2

We can extend this idea to an entire population. Suppose the frequencies of the A_1 and A_2 alleles in a population are pA_1 and pA_2, then laying out the genotypes in the Punnet square format we get the result shown in Table A3. The frequency in the population of the genotype A_1A_1 is pA_1^2, the frequency of A_2A_2 is pA_2^2, and that of A_1A_2 is $2 \times pA_1 \times pA_2$.

To give a numerical example, if $pA_1 = 0.6$ and $pA_2 = 0.4$ (i.e. in the population, 40% of the alleles are A_2 and 60% are A_1), then the genotype frequencies in the population will be 36% A_1A_1, 16% A_2A_2, and 48% A_1A_2. This is an example of the Hardy–Weinberg principle: if the A_1 and A_2 alleles are at frequencies of pA_1 and pA_2, then after one generation of random mating, the frequencies in the population of the three genotypes A_1A_1, A_1A_2, and A_2A_2 are pA_1^2, $2pA_1pA_2$, and pA_2^2, respectively.

Another way to express the Hardy–Weinberg principle is as an expansion of the probabilities for each allele: $(pA_1 + pA_2)^2$. This leads to an easy formulation of any number of alleles. Suppose we have three alleles, a, b, and c. The frequency of genotypes in a randomly mating population is $(pa + pb + pc)^2$. Two general rules follow that govern the frequencies of genotypes in a population:

1. The frequency of any homozygous genotype is the square of the corresponding allele frequency.

2. The frequency of any heterozygous genotype is twice the product of the component allele frequencies.

The remarkable thing is how well this simple prediction holds true. Unlike other equilibria in nature, attainment of Hardy–Weinberg equilibrium is reached in a single generation. We do not require that the parental genotypes are in random-mating proportions, just that the selection of gametes for the next generation is at random. This is the major reason why the Hardy–Weinberg principle underpins analysis of all diploid populations, which includes almost all of the organisms in this book. Deviation from Hardy–Weinberg equilibrium is always a cause of interest (although sadly, in most cases, it reflects error in the measurement of genotypes).

The random sampling of gametes has another important consequence: it causes allele frequencies to gradually drift toward 0 or 1. This means, if nothing else happens to a population over time (no selection or migration), genetic variation will lessen and eventually disappear, with the probability of fixation of an allele being equal to its initial frequency. The fact that this doesn't happen is due to a combination of selection, migration between populations, and the occurrence of new variants due to mutation (the discipline of population genetics is devoted primarily to assessing the relative contribution of these three factors to genetic variation). Subdivision of a population into isolated groups means that over time allele frequencies in different populations will diverge: the greater the degree of isolation, the greater this differentiation will be. This observation has one very important consequence for genetic analysis of behavior (and indeed of any trait): it means that genetic results obtained in one population may not apply in a second because of allele frequency differences.

A3 Genetic models: multiple loci

In Mendelian genetics, we assume that alleles can have dominant, recessive, or additive action (Figure 8.3 in Chapter 8, section 8.10 provides a simple review of Mendelian inheritance patterns). Phenotypes are then predicted to occur based on the segregation of alleles at Mendelian ratios. For phenotypes that are determined by multiple loci, such as almost all those discussed in this book, we also assume that the alleles at each locus behave in dominant, recessive, or additive fashion, but we are now examining their conjoint effect and usually have little idea about their action at an individual locus. Nevertheless, we need to adopt some model of how they work. Below is one such model, with the nomenclature that Falconer proposed (FALCONER AND MACKAY 1996).

We start with a single locus at which we have two alleles: A_1 and A_2. The effect of the homozygote A_1A_1 is to increase the value of the phenotypic score (e.g. personality score, risk of developing schizophrenia, height, or weight) by a quantity 'a' and the effect of the homozygote A_2A_2 is to decrease the score by '$-a$'. Midway between the two homozygotes is the value zero. Under a purely additive model, the heterozygote would occupy this

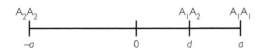

Figure A2 Genotypic effects at a locus. The scaleless horizontal line represents the effect of a genotype on the phenotype. The effect has an additive component (*a*) and a dominance component (*d*).

position. Deviation from the additive model is attributable to dominance of one allele over the other, indicated by calling the heterozygote value '*d*'. This is summarized in Figure A2. Using this nomenclature, we give formulae for the quantities that we use to assess genetic effects.

The mean effect is the average effect of an allele. In a population of individuals, the consequences of genotypic values depend on the frequencies of the genotypes in the population, so we can estimate the effect of the A_1A_1 genotype in the population by counting the number of people with the genotype and multiplying by the effect '*a*'. Summing up the effect of all genotypes and then dividing by the total number gives the mean. This calculation is simplified by knowing that the allele frequencies predict the genotype frequencies (because of Hardy–Weinberg equilibrium). Furthermore, because the allele frequencies must sum to 1, the last step, division by the total number, has already been made in obtaining the frequencies, so the mean effect of the locus is more simply estimated by summing the product of values and frequencies.

Table A4 shows the genotypes, frequencies, values, and the product of the frequencies and values for a single locus with two alleles, using Falconer's notation (Falconer and Mackay 1996).

Adding up the terms in the last column gives the mean (M):

$$M = a(p^2 - q^2) + 2dpq$$

As $p^2 - q^2$ factorizes to $(p + q)(p - q)$ and $p + q$ equals 1 (the allele frequencies must sum to 1), the equation simplifies to:

$$M = a(p - q) + 2dpq$$

Omitting the algebra (see Falconer for a full treatment; Falconer and Mackay 1996), the formula for the variance is:

$$V_a = 2pq[a + d(q - p)]^2$$

We write V_a to make clear that this is the additive variance, where the effects of each locus sum up. Additive heritability explains why relatives are similar, and additive heritability is the usual stuff of most genetic studies. We can obtain a similar formula for the variance attributable to dominance effects:

$$V_d = (2pqd)^2$$

Table A4 Genotypic frequencies and values at a single locus (Falconer and Mackay 1996)

Genotype	Frequency	Value	Freq. × value
A_1A_1	p^2	$+a$	p^2a
A_1A_2	$2pq$	d	$2pqd$
A_2A_2	q^2	$-a$	$-q^2a$

and thus a formula for the entire genetic variance ($V_g = V_d + V_a$):

$$V_g = 2pq[a + d(q-p)]^2 + [2pqd]^2$$

As heritability is the proportion of total phenotypic variance attributable to genetic variation, the above formulae are used to derive measures of heritability. For instance, the phenotypic resemblance between offspring and parent is V_A / V_P, where the subscripts are capitalized to recognize the fact that we are now dealing with the action not of one locus but of many.

Three assumptions justify extending the single-locus model to multiple loci and thus underpin the equations that determine heritability. We review these assumptions briefly here:

1. All genetic effects are additive. Nonadditive effects (dominant or epistatic) are known, but they are assumed to make relatively little contribution to resemblance between individuals. The justification is that, while the presence of any allelic combination has an additive effect, an epistatic effect will only occur in a small proportion of allelic combinations at two (or even fewer at three or more) loci. For example, a two-locus epistatic effect might be found at only one out of the 16 different genotypes, in which the effect not make much of a contribution to heritability; however, if the epistatic effect is large, although rare, its contribution would be nonnegligible.

2. Genetic and environmental effects are independent. This assumption rarely holds. In human studies, because relatives live in similar environments, the genetic correlations are confounded by environmental correlations, which is why human geneticists have relied on twin or adoption studies for estimates of heritability.

3. Loci contributing to the phenotype are in linkage equilibrium. This is a relatively safe assumption (for human genome-wide association studies at least—it is not true for mouse mapping studies that use crosses between inbred strains).

A4 Regression to the mean and the calculation of heritability

We expect tall parents to have tall children and we expect the offspring of short parents to be shorter than average. The same applies for psychological characteristics that can be measured quantitatively, such as personality or intelligence. However, there is another pattern of resemblances within a family that is surprisingly counterintuitive: the children of tall parents are on average shorter than their parents and the children of short parents are on average taller than their parents (the same is again true for psychological traits). The children of parents who score at extremes of a distribution are most affected. When I first met Hans Eysenck, a British psychologist, he handed me a copy of his recently published autobiography (EYSENCK 1990). The dedication page read: 'To my children. May regression to the mean not affect them too severely.'

Regression to the mean was first described by Francis Galton and summarized as follows:

This law tells heavily against the full hereditary transmission of any gift ... The more exceptional the amount of the gift, the more exceptional will be the good fortune of a parent who has a son who equals, and still more if he has a son who overpasses him in that respect. The law is even-handed; it levies the same heavy succession-tax on the transmission of badness as well as of goodness. If it discourages the extravagant expectations of gifted parents that their children will inherit all their powers, it no less discountenances extravagant fears that they will inherit all their weaknesses and diseases.

GALTON (1886).

The idea is demonstrated in Figure A3. The curve shows the distribution of the mean phenotype of pairs of parents, while the arrows below the curve indicate the expected scores of the offspring of parents who score at one extreme of the distribution (on the right of the figure), and the expected scores of offspring of parents whose scores are average, at the middle of the figure. The black arrows indicate the mean scores of offspring and the

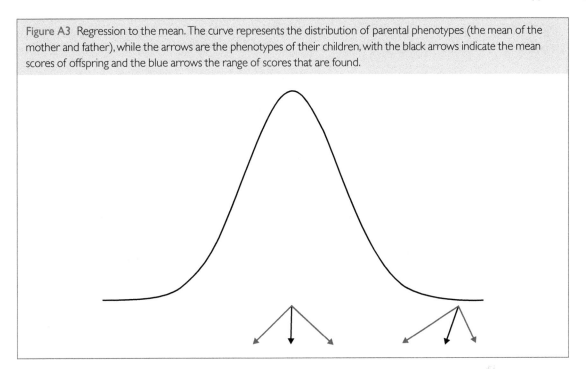

Figure A3 Regression to the mean. The curve represents the distribution of parental phenotypes (the mean of the mother and father), while the arrows are the phenotypes of their children, with the black arrows indicate the mean scores of offspring and the blue arrows the range of scores that are found.

blue arrows the range of scores that are found. The scores of offspring of the extreme-scoring parents are on average lower than the scores of their parents (the black arrow points toward the middle of the distribution). In other words, they regress to the mean.

As the offspring of the extreme-scoring parents regress to the mean, you might expect that over time the total variation in a population will reduce and eventually disappear. But the offspring of the large number of average parents have scores that spread out (as shown in the middle of Figure A3), offsetting the score reduction encountered by the children of the much smaller number of parents at the extremes. Consequently, there is no loss of variability over time.

The degree to which regression to the mean occurs (i.e. the slope of the black arrow in the figure) is a consequence of the heritability of the trait. Suppose there is no heritability and that the extreme scores of the parents are entirely due to some extraneous random factor. In this case, the mean scores of their offspring will be the same as those of the rest of the population. Their mean score will be the population mean. This result is shown by the arrow with the value $H = 0$ in Figure A4. By contrast, if the trait is entirely genetically determined, then the mean will be the same as that of the parents. There will be no regression to the mean, as shown by the vertical arrow ($H = 1$) in Figure A4. Intermediate heritabilities will lie somewhere between these two extremes (arrow marked $H = 0.5$).

Regression to the mean is a key concept not only in genetics but also in statistics, where it has given rise to the term linear regression. It also underlies single-nucleotide polymorphism (SNP)-based heritability, as discussed in Chapter 7, section 7.5.1.

A5 Twin methodology

One commonly used method to work out the contributions of genetic and environmental effects to phenotypic variation in twins is path analysis. The method can be explained with a simple example. Suppose we have two brothers and we have measured their personality with a questionnaire: one brother scores 10 points and the other 14. The two measures are P1 and P2 (here P stands for phenotype).

Figure A4 Regression to the mean for an entirely heritable trait ($H = 1$), a trait with no heritability ($H = 0$), and a moderately heritable trait ($H = 0.5$).

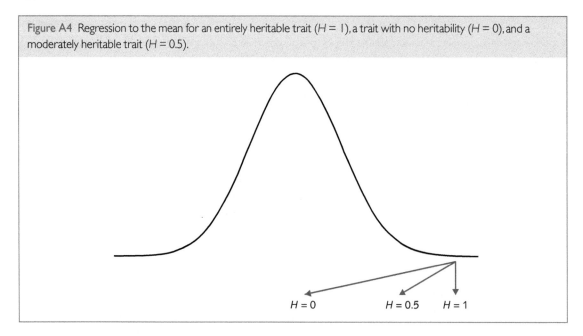

$H = 0$ $H = 0.5$ $H = 1$

We know that two factors contribute to the similarity of the phenotypes: the genetic variants that both siblings share and the environmental factors that make the brothers similar, which we call the common environment factors. In the path diagram illustrated in Figure A5, the common environment is denoted by the letter E and the genotypes of the brothers are denoted A1 and A2 (where A stands for the additive genetic effect). The genotypes are unknown, but if the brothers are dizygotic (DZ) twins, sharing half their genes identical by descent, we know that the correlation between the genotypes must be about 0.5. If they are monozygotic (MZ) twins, sharing all their genes, the correlation is 1.0.

The lower case letters in the path diagram (a and c) alongside the arrows are called path coefficients. By the rules of path analysis, the correlation between P1 and P2 (denoted r) is given by the sum of the two connecting paths. So for DZ twins, the correlation (denoted r_{dz}) is:

$$r_{dz} = 0.5(a \times a) + (c \times c)$$

and for MZ twins the correlation (denoted r_{mz}) is:

$$r_{mz} = 1.0(a \times a) + (c \times c)$$

Figure A5 Path diagrams to decompose environmental and genetic components of correlation in dizygotic and monozygotic twin pairs. Capital letters represent the observed measures of phenotype (P), environment (E), and additive genetic effects (A). Lower case letters are the estimated effects of additive genetic (a) and common environment (c) factors.

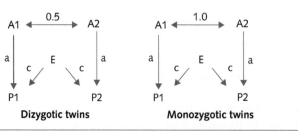

or more simply:

$$r_{dz} = 0.5a^2 + c^2$$

and

$$r_{mz} = 1.0a^2 + c^2$$

To estimate the values of the two terms a^2 and c^2, we use some algebra to give equations for heritability:

$$a^2 = 2(r_{mz} - r_{dz}) \tag{1}$$

and for the environmental effect:

$$c^2 = 2r_{dz} - r_{mz} \tag{2}$$

The individual specific, or nonshared environment is $1 - (a^2 + c^2)$ or:

$$\left(1 - r_{mz}\right) \tag{3}$$

In this example, a^2 reflects the role of additive genetic effects and equals the heritability of the trait (sometimes heritability is termed h^2). This model, which takes into account the genetic, common, and unique environment is thus often referred to as the ACE model, for the additive genetic (A), common environment (E), and nonshared (or unique) environment (E) components that make twins the same or different.

The equations can be explained in words as follows: the correlation in liability for a disorder or trait between MZ twins is the result of all the additive effects of genes (because they share all their genes in common) and their shared environment. The correlation in liability between DZ twins is the result of half of the effect of the additive genes (because they share, on average, half of their genes in common) and their shared environment.

If we have estimates of the values of r_{mz} and r_{dz}, we can calculate a^2 and c^2. A simple estimate of a^2 of a disorder is twice the difference between the MZ and the DZ correlation. An estimate of the effect of shared environment can be obtained by doubling the DZ correlation and subtracting the result from the MZ correlation. These calculations assume that the MZ and DZ correlations are known with equal accuracy, which is never precisely true, so the estimates obtained by the simple method described here will differ somewhat from those obtained using more sophisticated methods, implemented in several statistical software packages (such as Mx; NEALE et al. 2003).

To illustrate this method, let's take data from a twin study of alcohol abuse (as reported by the temperance boards in Sweden; KENDLER et al. 1997). The correlation in liability for alcohol abuse in the male MZ and DZ pairs from that study was estimated at +0.67 and +0.41, respectively. Applying Eqns (1)–(3) above, you should be able to come up with rough estimates for a^2, c^2, and e^2 of 0.52, 0.15, and 0.33, respectively. So, these results suggest that alcohol abuse is strongly affected by genes, but environmental effects of both the shared and individual specific variety are also both important. These rough estimates are quite close to those obtained by estimation procedures implemented in the Mx software package (NEALE et al. 2003): $a^2 = 0.54$, $c^2 = 0.14$, and $e^2 = 0.33$.

The path diagram in Figure A5 is an example of a **structural equation model**, in which unobservable latent constructs (such as the additive genetic (a) and common environmental (c) variables in Figure A5) are assessed. A structural model tests whether variables are interrelated (usually through a set of linear relationships) by examining the variances and covariances of the variables. In brief, the method proceeds as follows. First, a description of how variables are interrelated is drawn up, for example using a path diagram such as Figure A5. The implications of the hypothesized relationships are then worked out for the variances and covariances of the variables. Parameter estimates and standard errors from linear equations are then used to determine whether the variances and covariances fit the model adequately. This typically involves a lot of matrix algebra.

Structural equation modeling has been a mainstay of twin analysis for decades and has evolved into a complex field of analysis. Software to implement and run structural equation models with R can be downloaded from the OpenMx website at https://openmx.ssri.psu.edu. The modeling extends beyond problems in twin analysis; it is applicable to a wide variety of multivariate statistical models, for example in modeling putative causal relationships between variables.

A6 Effect sizes in categorical tests of association

Association tests don't just deliver a *P* value (although you would be forgiven for thinking that this is all geneticists are interested in). The tests also provide an estimate of the effect attributable to the genotype, or allele. For quantitative data, an effect on the trait can be assigned to an allele, so that we can say a particular allele increases height by 0.1 cm, for example. For count data, the estimate is presented as an odds ratio, that is, the probability that the disease is present compared with the probability that it is absent in exposed versus non-exposed individuals.

One way to measure the strength of association is a relative risk, which compares the rates of disease between individuals with different genotypes. Relative risks can only be accurately estimated from prospective cohort studies, in which groups of individuals with different genotypes are followed up to assess which develop disease. In case–control studies, odds ratios—the odds of disease—are calculated from allele frequencies.

The **allelic odds ratio** describes the association between disease and allele by comparing the odds of disease in an individual carrying one allele with the odds of disease in an individual carrying the other allele. The **genotypic odds ratio** describes the association between disease and genotype by comparing the odds of disease in an individual carrying one genotype with the odds of disease in an individual carrying another genotype.

A7 Statistical tests: please use the R statistical programming language!

In the next sections, we provide examples of standard statistical tests used in behavioral genetics. To help you understand them, we provide example datasets and instructions on how to run the tests using the R statistical programming language, which has become a central tool in complex trait genetics (R DEVELOPMENT CORE TEAM 2004). R is open-source software, and you can download and install it on any computer from this website: https://cran.r-project.org.

We provide examples of the use of R use in our book, but we can't teach you how to use R. It's a programming language, and like all languages, you will learn it best by using it. Fortunately, there is a large online community to help you learn the language, and the best way to answer any question is simply to type into a search engine: 'How do I [do this thing] in R?' The level of help you get extends from the basic to the advanced. For online tutorials, we suggest: https://cran.r-project.org/doc/contrib/Lemon-kickstart/ and https://cran.r-project.org/doc/manuals/R-intro.html.

The following link provides a very brief introduction to R, and an even briefer introduction to Bayesian statistics using R: https://media.readthedocs.org/pdf/a-little-book-of-r-for-bayesian-statistics/latest/a-little-book-of-r-for-bayesian-statistics.pdf.

Don't be embarrassed by your ignorance. While writing this section, I checked one of the sources of help, at https://www.r-bloggers.com, and found from the list of most visited articles that most people are asking for basic help. Good luck!

A8 Differences between two samples of quantitative scores (*t*-test)

To assess whether two quantitative measures (such as measures of IQ or personality) are significantly different, we start by assuming that they are normally distributed. Normal distributions arise when a measure arises from many small independent effects. It's worth emphasizing that whenever you see a distribution that looks normal, it's a reasonable guess that whatever causes the variation consists of small independent effects. Height is distributed normally, and given that its heritability is about 80%, it would be reasonable (and correct) to assume that the genetic component consists of many small independent effects.

The second step in working out if the difference in scores is significant is to make the counterintuitive assumption that there really is **no** difference between the mean scores. This step is so counterintuitive but so central to the process of statistical testing that it is needs illustration.

Imagine you measure the height of a group of people and then call them back a few days later to repeat the measurement (assume these are adults, and nothing adverse happens to anyone between the two measures, so their true height does not change). The means of the two measurements would likely differ, but you can discount the possibility that you have inadvertently measured two different things. The statistical test of the difference between two means assumes that the measures you obtained are like this—that they are of the same thing and that they are obtained from the same underlying distribution. The *P* value represents the extent to which that assumption is true: a low *P* value means that it is unlikely that the two results come from the same distribution.

If we obtain lots of different means, by sampling many times from the same population, the distribution of the means has a normal distribution. This is the central limit theorem. Because of the central limit theorem, the distribution of the difference between the two means (drawn from the same population) has a magical property: it's normally distributed with mean = 0 and SD = 1. You can see how the difference between two means must be zero, because of the assumption that the two measures are of the same thing.

Variances, means, and standard deviation are descriptive statistics of the data, and depend on the data, while the standard deviation of the difference between two means is independent of the data. It's not a descriptive statistic, and to make that clear, it's called the standard error of the mean (SEM), or just standard error (SE) (it measures the error in the estimate of the true mean). Regardless of whether we are examining two samples that differ in age, height, weight, personality, variation in mood, or anything that we can measure continuously and that has a distribution that looks (vaguely) normal, the standard error of the difference for the two means will have the same distribution.

The distribution of values of the standard error is shown in Figure A6. You can see that 4.6% of the values $(100 - (34.1 \times 2) + (13.6 \times 2))$ lie further than 2 SE from a mean difference of 0. About 0.2% of values lie further than 2 SE from 0 (this is the basis of the 68–95–99.7% rule: 68% of observations lie within 1 unit, 95% lie within 2 units and 99.7 lie within 3 units). The probability of obtaining an estimate that lies greater than 2 SE from zero is less than 5% (strictly speaking, the 5% threshold occurs at 1.96 SE).

Notice a couple of things about the distribution shown in Figure A6. First, the distance from the mean of zero is given in two directions, positive and negative. As we don't know whether a difference is expected to be greater or less than 0, we consider both tails of the distribution (but there are cases where we are only interested

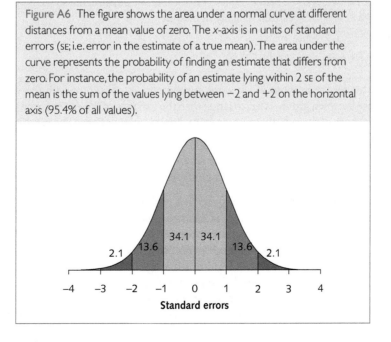

Figure A6 The figure shows the area under a normal curve at different distances from a mean value of zero. The *x*-axis is in units of standard errors (SE; i.e. error in the estimate of a true mean). The area under the curve represents the probability of finding an estimate that differs from zero. For instance, the probability of an estimate lying within 2 SE of the mean is the sum of the values lying between −2 and +2 on the horizontal axis (95.4% of all values).

in differences in one direction, which is referred to as a one-tail test). For example, we might already suspect from a previous experiment that one mean is larger than the other, and so we are only interested in testing that possibility. Usually, however, we include both tails. Second, if the sample sizes are small, then the distribution is **not** the same as that shown in the figure, but not remarkably so. For example, for a sample size of 10, the 5% significance is exceeded at a value of 2.22. So as a rule of thumb, you can pretty much rely on a difference of greater than 1.96 SE as being significant ($P < 0.05$).

The standard error of the difference of two means is calculated by dividing the difference of two means by the variance (due to another property of the normal distribution, the variance of the difference of two independent variates is equal to the sum of their variances). The formula is:

$$\sqrt{\frac{\sigma_1^2}{n1} + \frac{\sigma_2^2}{n2}}$$

which generates a t-statistic.

As an example, we provide a data set containing measures of neuroticism (N) and sex ('ExampleRegression-Data.txt'). The mean of the N scores for men is 7.16 and for women is 9.17. Is this a significant difference?

First, we load our example data into R with the command:

```
data = read.delim ("ExampleRegression Data.txt")
```

Then, we calculate the t-statistic using the t.test command in R:

```
t.test(data$N ~ data$sex)
```

R gives the answer: $t = 14.6$. The t.test command also automatically generates the P value, which in this case is so low that R does not bother to give the exact value ($P < 2.2 \times 10^{-16}$). Suppose we **do** want the exact P value—how in general do we calculate it? The P value is given by working out a portion of the area under the curve in the Figure A6. The pnorm function in R gives the area under the standard normal curve to the **left** of the value you give it: for example, pnorm(1.96) is ~0.975 (i.e. 97.5% of findings lie below 1.96), while pnorm(−1.96) will return the probability of an estimate occurring to the **right** of the value, i.e. above 1.96 (~0.025) (note: for smaller sample sizes (less than 20 or so), use the pt() command rather than pnorm). We want to know the probability of finding the values that lie both above 14.6 and below −14.6. The distribution is symmetric so we simply double the result: $2 \times \text{pnorm}(-14.6) = 2.81 \times 10^{-48}$. Yes, that is highly significant.

A9 Analysis of variance (ANOVA)

An ANOVA tests whether the amount of variation in subgroups is consistent with the variation in the total sample. If the variance between groups is much larger than the variance of the entire group (within-group), then it's likely that one or more of the subgroups come from different populations. The result of an ANOVA test is expressed as a ratio:

$$\frac{\text{between group variance}}{\text{within group variance}}$$

The larger the ratio, the more the between-group variance contributes to the total variance and hence the less likely the groups belong to the same population. The ratio is called an F statistic (after Fisher), and has a distribution that depends on the number of independent observations in the sample (more simply, the degrees of freedom). An F-test gives a P value for the likelihood of observing the F value by chance.

Calculating the components of an ANOVA requires working out the between- and within-group variance. It's not a straightforward calculation. It is best left to a computer, for example the aov function in R. Determining whether there are differences in neuroticism that vary by genotype is a problem appropriate for ANOVA (there are three genotypes for a SNP, so we cannot use the simple comparison of two means, the t-test described above). Referring to our example data set ('ExampleRegressionData.txt'), we can look at the effects of the three genotypes on neuroticism with the command:

```
aov(data$N ~ data$rs35936514)
```

To obtain a P value, we then use the command:

```
summary (aov(data$N ~ data$rs35936514))
```

Usually, ANOVA likes a balanced design, that is, each group has the same number of observations, which won't often apply with genotypes (from Hardy–Weinberg equilibrium, we know that the proportions of genotypes in a population won't ever be the same if, as we hope, the locus is at equilibrium). This doesn't invalidate the method, but it makes it a little more complicated.

A10 Regression analysis

Regression analysis derives from the methods that Galton and then Pearson applied to the problem of regression to the mean. Regression analysis is robust, flexible, and relatively straightforward to use. It incorporates almost all of the ideas presented earlier, so you need to be familiar with variances, covariances, correlations, standard error, and t-statistics. For detailed descriptions, we suggest Andrew Gelman's book (GELMAN AND HILL 2007).

Regression models the relationship between phenotype and its predictors as a linear combination of factors. The relationship is typically written:

$$y = bx + e$$

where y is the phenotype, x the genotype and e represents error.

To see how this works, consider the relationship between genotype and phenotype. Suppose a locus has an effect on neuroticism (N) scores, what do we expect to see when we plot the relationship between the genotypes and N? The simplest model is that each allele contributes a small but equal amount to the N score, so that there is an additive relationship between N score and alleles. Figure A7 gives an example.

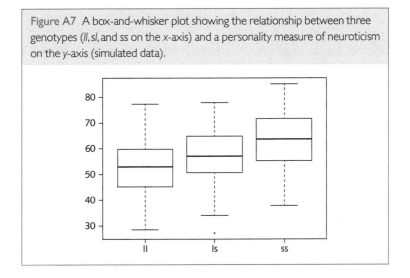

Figure A7 A box-and-whisker plot showing the relationship between three genotypes (*ll*, *sl*, and *ss* on the *x*-axis) and a personality measure of neuroticism on the *y*-axis (simulated data).

Table A5 Genotypes and phenotypes for five individuals with neuroticism (N score)		
Individual	Genotype	Phenotype (N score)
1	0	56.9
2	2	51.1
3	1	58.4
4	2	52.5
5	1	52.6

We ask whether there is a linear correlation between the increase in the number of alleles and the increase in neuroticism score. We score the alleles as 0 for the homozygote *ll*, 1 for the heterozygote *ls*, and 2 for the homozygote *ss*. Table A5 shows the data for five individuals.

The test we apply is to determine whether the correlation between genotype and phenotype is significantly different from zero. To illustrate this, Figure A8 shows two data sets, one on the left where two variables ('a' and 'b') are correlated and one on the right where they are not. A dashed line indicates the slope of the regression line that fits the data best (slope = 1 on the left and 0 on the right).

Estimating the slope is carried out using Pearson's method developed for regression to the mean. The significance is obtained in the same way as significance for a difference of means: we estimate the standard error of the slope and calculate a *t*-statistic:

$$t = b/\mathrm{SE}_b$$

where b is the slope and SE_b is the standard error of b.

As before, the *t*-statistic is a standardized measure so the *P* value is obtained from its distribution (as in Figure A6 and our discussion of using the R pnorm function).

We can also calculate Pearson's correlation coefficient, r, to assess the relationship between the two variables, and again derive a *t*-statistic:

$$t = r\sqrt{\frac{n-2}{1-r^2}}$$

The test of significance for the slope gives the same *P* value as a test of significance for the correlation coefficient. Although the two tests are derived differently, they are algebraically equivalent.

The assumptions under which regression operates are relatively minimal:

1. The relationship between the two variables should be linear. Where it is clearly not so, often the distribution of the variables can be transformed to become linear, by taking the logarithms of the phenotype for example.

2. The error terms (the '*e*' in the equation) should be approximately normally distributed. Importantly, the x or y variables don't have to be normally distributed, which makes the method remarkably robust.

However, regression is sensitive to outliers, so it's safest to remove any outliers and/or to transform the measure so that it becomes normally distributed (e.g. using a quantile normalization procedure). Never believe a regression result unless you've seen the distribution of the data!

As an example, here is the output of a test for a single marker's association with neuroticism (N), in which principal components are included to account for population structure (see Chapter 6, section 6.6.4 for a

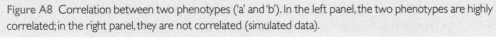

Figure A8 Correlation between two phenotypes ('a' and 'b'). In the left panel, the two phenotypes are highly correlated; in the right panel, they are not correlated (simulated data).

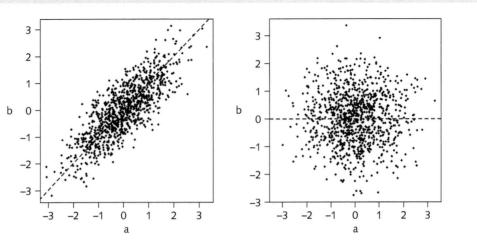

description of the use of principal components in dealing with population structure), as well as age and sex. Load this data set into R:

```
data = read.delim("ExampleRegressionData.txt")
```

and type:

```
summary (lm (N ~ rs35936514 + PC1 + PC2 + PC3 + age + sex, data = data))
```

where 'lm' means linear model.

The genotypes for the marker rs35936514 are coded as 0, 1, and 2, PCs are three principal components, with numerical values that range from −0.05 to 0.04, age is an integer, and sex is either male (M) or female (F). R is tolerant of this mixture of numerical variables (age, PCs) and factors (sex), but be careful: if you assign sex as 2 for females and 1 for males, R will assume that sex is a linear variable where 2 is twice as big as 1. Table A6 gives the result.

The 'estimate' is the slope of the regression, reflecting the size of the effect (often called β). It means that for each additional allele at the rs35936514 locus, the N score will decrease by −0.46 (it would make the result more easily interpretable if the score was standardized (so that the mean = 0 and SD = 1), but we are interested here only in explaining the output of the regression). SE is the standard error of the slope (i.e. of β). Dividing β by the standard error gives a 't-value', which is shown in the next column; for the marker rs35936514, a β value of −0.46 divided by 0.11 gives a t-statistic of −4.2. The associated P value is given in the last column, which in this case is significant at the 5% level.

Note that three additional factors are significant. Increasing age reduces neuroticism (for each year, the score drops by 0.28) and being male reduces the score by 1.6. This has been widely observed in other data sets. On average, we get slightly less emotionally reactive to the tribulations of life as we age. We might wonder whether including these factors alters the genetic association. One simple thing to check is to repeat the analysis dropping the covariates, in which case the result for the marker gives a β value of −0.48, with the same standard error and a t-statistic of −4.3. This small change indicates that including the covariates made little difference. In other cases, the effects are larger.

It's possible to test whether the difference between the two models (with and without the covariates) is significant by using a partial F-test, implemented in R with the anova command. The test is based on the comparison of the residual sum of squares, and thus you need to run two separate regressions and save the results for each one. We select data where we have no missing observations. In R:

```
x = complete.cases(data$sex, data$age, data$N)
```

Table A6 Results of analyzing the relationship between a single genetic marker (rs35936514) and neuroticism, using the R function 'lm'

	Estimate	SE	t-value	Pr(>|t|)
rs35936514	−0.46	0.11	−4.2	2.55×10^{-5}
PC1	−30.73	7.11	−4.3	1.55×10^{-5}
PC2	−9.59	7.29	−1.3	0.188
PC3	−9.18	7.11	−1.3	0.197
Age	−0.28	0.01	−18.2	$<2 \times 10^{-16}$
Sex (M)	−1.64	0.24	−6.9	5.86×10^{-12}

PC1, PC2, and PC3 are the principal components included to control population structure. Age and sex are additional covariates (the effect of sex is reported for males, M). R returns an estimate of the effect (estimate), its standard error (SE), a t-statistic (t-value, the estimate divided by the standard error, or how many standard errors the estimate is distant from zero in Figure A10), and the probability of the t-statistic (Pr(>|t|)).

and then fit two models:

```
fit0 = lm (data$rs35936514[x] ~ data$sex[x] + data$age[x])
```

```
fit1 = lm (data$rs35936514[x] ~ data$sex[d] + data$age[x] + data$N[x])
```

Finally we compare the two fits:

```
anova(fit0, fit1)
```

In this case, the fits are significantly different ($P = 2.874 \times 10^{-5}$).

The flexibility and relative lack of assumptions of linear regression mean that is the method of choice in most applications. In brief, wherever you can, use linear regression. And if you can't use linear regression, try logistic regression.

A11 Logistic regression

Linear regression tests associations for quantitative measures, such as personality scores, which implies that it cannot be used for analyzing categorical measures, such as disease status. The way to overcome this limitation is to use a binary variable, 1 or 0, to indicate who is affected with a disease (e.g. all cases become 1 and all controls 0). We model the probability of the outcome (disease or no disease) as a function of the genotypes and other covariates by assuming that a linear relationship exists between the values on the right of the equation (the x values) and the probability of the outcome (the y values). In other words, any increment in the x values should increase or decrease the probability of the outcome. However, on its own this method won't work, because when the probability is large (or small) a bigger change is needed in the x value compared with when the probability is close to 0.5. Taking the logarithm helps, so that changing the x variable **multiplies** the probability by a fixed amount. However, logarithms are unbounded in only one direction, while linear functions are not. The usual solution to the problem is to use the logistic (or logit) transformation:

$$\log\left(\frac{p}{1-p}\right)$$

This has an unbounded range and can be used as a linear function of the *x* values. This is the basis of logistic regression.

As an example of how to run logistic regression in R, load our 'ExampleRegressionData.txt' data:

```
data = read.delim ('ExampleRegressionData.txt')
```

and test for an association between the marker rs35936514 and major depressive disorder (MDD) as follows:

```
summary (glm(MDD ~rs35936514 + PC1 + PC2 + PC3 + age + sex, data = data,
family = binomial(link = 'logit')))
```

The output is shown in Table A7.

The output has the same format as for linear regression but the coefficients are not so easily interpretable, because of the transformation to the logit scale. For this reason, it is conventional to transform the estimate of the effect size into an odds ratio, by taking the exponent of the estimate (in R, use the exp() command), and to provide 95% confidence limits of the odds ratio by doing the same for the standard error (the upper and lower 95% confidence intervals of the odds ratio are the exponent of the estimate less 1.96 times the standard error and plus 1.96 times the standard error). For the marker, age, and sex, the results are shown in Table A8.

An odds ratio of less than 1 mean the factor reduces risk. So being male reduces the risk of depression. We can take the reciprocal of the odds ratio to obtain the risk for women: being female increases the risk about 1.93-fold.

Table A7 Results of analyzing the relationship between a single genetic marker (rs35936514) and major depressive disorder (MDD), using the R function 'glm'.

	Estimate	SE	*t*-value	Pr(>|t|)
rs35936514	−0.18	0.03	−5.4	4.43×10^{-8}
PC1	−3.43	2.16	−1.6	0.11
PC2	−4.59	2.22	−2.1	0.04
PC3	14.81	2.18	6.8	1.08×10^{-11}
Age	−0.09	0.01	−19.2	$<2 \times 10^{-16}$
Sex (M0	−0.66	0.07	−9.2	$<2 \times 10^{-16}$

PC1, PC2, and PC3 are the principal components included to control population structure. Age and sex are additional covariates (the effect of sex is reported for males, M). R returns an estimate of the effect (estimate), its standard error (SE), a *t*-statistic (*t*-value, the estimate divided by the standard error, or how many standard errors the estimate is distant from zero in Figure A10) and the probability of the *t*-statistic (Pr(>\|t\|).

Table A8 Results of a logistic regression, shown as an odds ratio and 95% confidence intervals (CI) (data from Table A7)

	Odds ratio	95% CI
rs35936514	0.83	0.78–0.89
Age	0.91	0.90–0.92
Sex (M)	0.52	0.45–0.60

A12 Power

Imagine you have just completed a genetic association study of personality in 10,000 people and you have identified a significant effect at one locus (rs35936514, as in Table A6). To your surprise, and annoyance, your colleague in another university then publishes a study in which they fail to replicate your finding ($P > 0.05$). Their study though uses 100 people, not 10,000. What should you make of this? Does their finding throw doubt on yours? Or does the difference in the design, particularly in the sample size, mean that their negative finding doesn't invalidate yours? Can you safely ignore their result or not?

Power analysis answers this question. You need to find out the probability that with a sample size of 100 they could have found the same thing as you did with a sample size of 10,000. If their study was terribly under-powered, the failure to replicate could be attributable to chance.

For the power analysis, in addition to the sample size used, you'll need the effect size you expected them to detect. The bigger the effect, the easier it is to detect and the smaller the sample needed. Your estimate (of β) was −0.46 from a regression analysis (from Table A6). Finally, you'll need to set a significance threshold; let's assume for the time being it is 5% ($P < 0.05$). We'll start by working out your power to detect an estimate of −0.46 with a sample size of 10,000 at $P < 0.05$.

The basic idea upon which power calculations rest is shown in Figure A9. The curve in Figure A9 represents the distribution of estimates we can expect to obtain if we kept on repeating the experiment. Anything more than 1.96 SE above zero (the solid vertical line) is significant ($P < 0.05$). When the distribution of true effect sizes has a mean of at least 2.8 SE from zero (shown by the arrow in the figure), then 80% of the estimates fall to the right of the vertical line, in other words 80% of the estimates have a value $P < 0.05$. This means that we have 80% power to detect a significant result (at 5%) effect. As you might want to change the threshold (we've seen that a 5% threshold is not suitable for all studies), you should know how to estimate power from the standard errors. Subtract 2.8 from 1.96 to give a value of 0.84, and enter this in the R pnorm command to calculate the area under the curve:

```
pnorm(0.84, mean = 0, sd = 1, log = FALSE)
```

This gives 0.8, as expected. If the threshold is higher, you'll need the estimate to be even more standard errors away from zero, in order to maintain the same power.

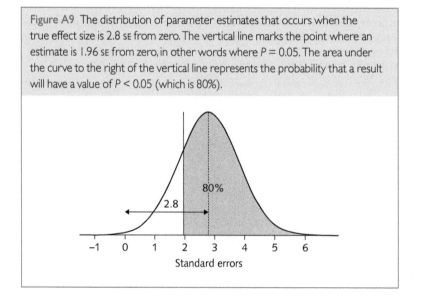

Figure A9 The distribution of parameter estimates that occurs when the true effect size is 2.8 SE from zero. The vertical line marks the point where an estimate is 1.96 SE from zero, in other words where $P = 0.05$. The area under the curve to the right of the vertical line represents the probability that a result will have a value of $P < 0.05$ (which is 80%).

In your case, you used 10,000 subjects to estimate β to be 0.46 with 0.11 SE, and hence observed that β is 4.2 SE away from zero (0.46/0.11). To work out your power, we need to know what proportion of findings would be greater than 1.96 s from zero. In other words, we want the area under the curve whose maximum is now 4.1 (rather than 2.8) and where the cut-off is again 1.96. Using the difference between 4.1 and 1.96 in the pnorm command, the power is 98%, from:

```
pnorm(2.14, mean = 0, sd = 1, log = FALSE)
```

This looks great! However, it does assume that you used a significance threshold of $P < 0.05$, when in fact for a genome-wide association study you'd use 5×10^{-8}, in which case the corrected distance in standard errors from zero is not 1.96 but 5.45, and the difference is now $4.1 - 5.45 = -1.25$, with a power of 9%. Oops, not so good for you, but let's put that to one side, as we really want to test the power of your colleague's finding.

Your colleague had a sample size of 100, so what is their power to detect your finding of a β of 0.46? We can work this out using the fact that standard errors are proportional to 1 over the square root of the sample size n (i.e. proportional to $\frac{1}{\sqrt{n}}$). The expected standard error from the study with $n = 100$ is thus ten times the standard error, or 1.1 (from $\frac{0.11}{x} = \frac{1}{\sqrt{100}}$, which is 0.41 SE from zero. Using the difference between 0.41 and 1.96 in the pnorm command, the power is 6% to detect the effect you found. The chances are that your colleague failed to detect the effect because the study was very underpowered.

Figure A10 shows the power of the two studies, using the same layout as Figure A9. On the left is your study with 10,000 subjects and on the right the study with 100 subjects.

More generally, in a power analysis we want to know, for a given effect size (such as an odds ratio), what sample size will be needed to detect it reliably? How do we work this out? Conventionally, we set power at 80% so we can phrase our question as follows: how large must the sample be so that we have an 80% chance of achieving statistical significance? We apply the same ideas presented above, using the fact that the standard error is proportional to the square root of the sample size.

For a conventional value of 80% power, we need the true estimate to be 2.8 SE from zero. If the estimate we wish to detect is 0.4, then its standard error has to be no greater than $0.4/2.8 = 0.14$. Using our observation that a sample size of 10,000 gives 0.14 SE, the sample size required is $(0.11/0.14) \times 10,000 = 7,857$.

Of course, you might not have a prior estimate of the standard error, and you might not be using linear regression to test the association. In fact, there will a host of assumptions you will have to make in determining

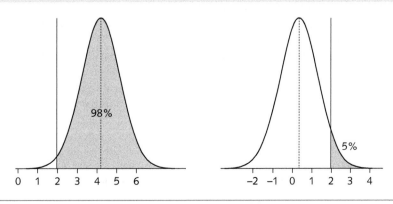

Figure A10 Power in two studies, on the left with 10,000 subjects and on the right with 100 subjects. The curve is the distribution of estimates from an experiment. The solid vertical line is the 5% significance threshold (at 1.96). The area under the curve to the right of the 1.96 thresholds is the power to detect the given effect: 98% on the left and 5% on the right.

power, many of which are likely to be guesses. As experienced statisticians will tell you, you can make a power calculation give you almost any answer you want ('if you torture data long enough, you can get it to admit to anything'). Nevertheless, the basic concepts laid out here will apply.

To work out power we suggest you use one of the many R functions (e.g. power.t.test). Alternatively, for genetic power calculations, use http://zzz.bwh.harvard.edu/gpc/.

A13 Meta-analysis

Suppose that your colleague who found the negative result for a genetic association in a small sample has some friends who are also working on the same genetic variant and who have also collected small samples. Let's say there are 20 such friends, and each has collected sample sizes of between 100 and 600; some of the results are positive ($P < 0.05$), some are negative. In total, the sample size is the same as yours (10,000), but how do you decide if the mixture of results supports or refutes your finding? One thing you cannot do is just count up the positive results and see if they are the majority. This is called the box-score method and it is a mistaken, albeit still employed, approach. Instead, you should combine the data systematically, using a technique called meta-analysis.

While it is possible to combine P values in a meta-analysis, meta-analysis typically combines the effect sizes from each study and then produces from the summed effect sizes a P value reflecting the joint analysis. In general, effect sizes can be reported in three ways:

1. The difference between two means.

2. The amount to which one variable predicts another, for example by a correlation coefficient (r).

3. An odds ratio.

Meta-analysis needs effects to be reported in a standardized way so that they can be combined, usually by standardizing the difference between the means by dividing by the standard deviation. Cohen introduced this measure (Cohen's d), and argued that the standard deviation of either group could be used when the variances of the two groups are homogeneous (COHEN 1988). Cohen defined effect sizes as small ($d = 0.20$, medium ($d = 0.5$), and large ($d = 0.8$), stating that '*there is a certain risk inherent in offering conventional operational definitions for those terms for use in power analysis in as diverse a field of inquiry as behavioral science*'.

Each effect-size estimate is weighted by the sample size of the study that provided it, so that large studies provide better estimates than smaller studies. Where studies report effects using different measures, they have to be converted to one form, using equations given, for example, in BONETT (2007). Not every study reports effect sizes (largely because of the fixation on P values), but their acquisition is necessary for a meta-analysis.

The second critical component for the meta-analysis is a measure of the accuracy with which the effect size is estimated, the standard deviation. Usually, effect sizes are reported with a 95% confidence interval (the interval that contains the true value 95% of the time, assuming you repeated your measurements indefinitely). Because errors are normally distributed (see the discussion above about regression to the mean), the confidence interval can be obtained from our knowledge of the normal distribution in which 95% of the values are contained within 1.96 SD either side of the mean. As long as you know the standard deviation, it is straightforward to calculate the confidence intervals.

The individual effect-size estimates are combined in one of two ways: a fixed-effects model or a random-effects model. Fixed-effects analysis assumes that all study samples are derived from a single population with a common effect size, so that only sampling error is assumed to contribute to the differences between the observed effect-size estimates across individual studies. Random-effects analysis assumes that the study samples included in a meta-analysis can be drawn from a distribution of populations (so that there might be subpopulations for which there is no effect and others for which there is a substantial effect). Effect-size differences between studies are therefore caused both by sampling error and by between-study variance where the effect size for each individual study is assumed to be normally distributed around a mean effect size for the population of studies.

Figure A11   Results of a meta-analysis of four studies. The odds ratio and 95% confidence intervals (CI) are shown as a forest plot on the right of the diagram where the size of the square is proportional to the size of the study and the extent of the horizontal lines represents the confidence interval.

Study name	Statistics for each study					Odds ratio and 95% CI
	Odds ratio	Lower limit	Upper limit	Z Value	p Value	
Caspi	2.181	1.214	3.919	2.608	0.009	
Eley	1.768	0.787	3.971	1.379	0.168	
Surtees	0.708	0.481	1.041	−1.755	0.079	
Kim	1.438	0.672	3.075	0.936	0.349	
	1.119	0.847	1.479	0.791	0.429	

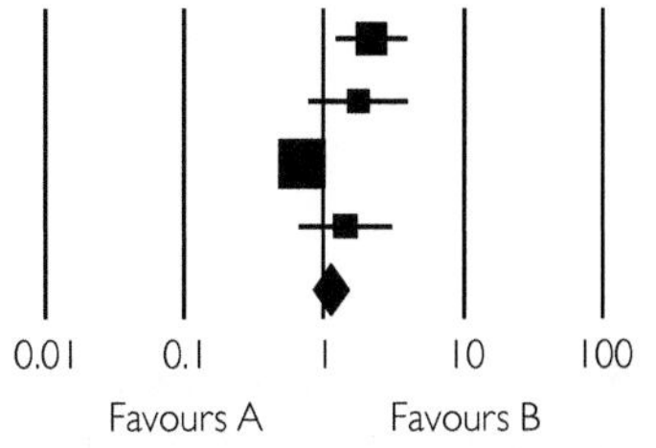

0.01   0.1   1   10   100

Favours A   Favours B

In fixed-effects analysis, increasing the number of studies contributing to the analysis increases power, because additional studies reduce the confidence intervals around the effect-size estimate. However, if there is real between-study heterogeneity, then there is a risk of a false-positive finding. Increasing the number of studies contributing to a random-effects analysis does not necessarily result in an increase in power. The additional studies may also result in larger variance-component estimates if the addition of studies increases the total between-study heterogeneity.

Figure A11 shows an example of data prepared for a meta-analysis. The data are from three studies of the relationship between the serotonin transporter and the risk of depression. In this case, the effect size is an odds ratio (where an odds ratio of 1 means there is no increase in probability, either way). On the right of Figure A11, you can see a forest plot in which the black box represents the size of the study (bigger boxes have bigger sample sizes), and the mean estimate of the odds ratio is in the center of the box. The horizontal lines are the 95% confidence estimates for the mean value, above and below the mean. Look at the results for the Caspi and Surtees studies and you will see that the 95% confidence intervals do not overlap, providing evidence that the results are drawn from a heterogeneous population.

Meta-analysis is never a replacement for adequately powered genetic association studies, and it is only as good as the input data. If the constituent studies are heterogeneous, then it could be argued that meta-analysis is no different from averaging the characteristics of apples and oranges (HUNT 1997). While heterogeneity can be identified by examining the standard deviations of the effect sizes and identifying outliers and clusters, or by using a χ^2 test of heterogeneity, the tests are not powerful. Moreover, heterogeneity is common: analysis of 370 studies addressing 36 genetic associations found significant between-study heterogeneity; the results of a first study often correlated modestly with subsequent reports (IOANNIDIS et al. 2001).

Despite these reservations, meta-analysis is a routine technique in genetic studies because large samples can often only be obtained by combining data from many different research groups. Almost all consortia carrying out research in complex trait genetics report their results as meta-analyses.

One more problem with meta-analysis needs to be considered. Implicitly, a standard meta-analysis assumes that the probability of a study getting published (so it can be accessed for the meta-analysis) is independent of its results. But what happens if negative studies are more likely not to be submitted? This is called the 'file-drawer' problem; i.e. negative studies are more likely to 'stay in the file drawer' than positive studies. While the details need not concern us here, there are statistical methods to estimate this problem and correct for it, if it occurs.

A14 Contingency tables, and Fisher's and χ^2 tests

Genetic data are often presented in the form of contingency tables. We are usually interested in determining whether the frequency of one allele, or genotype, is significantly different in one group (the cases) compared with chance expectations. For example:

	A allele	a allele	Row totals
Cases	161	55	216
Controls	97	51	148
Column totals	258	106	364

Given that I choose 216 alleles at random, what is the probability that 55 or more of them would be among the 106 'a', and 161 or fewer from among the 258 'A' alleles? Importantly, I am not asking about the probability of finding **exactly** 161 and **exactly** 55 but rather enquiring how often we would find 161 **or more** and 55 or **fewer**. To answer this question, we need to know the distribution of probable outcomes. Intuitively, it makes sense that finding a small number of 'a' and a large number of 'A' is unlikely, but our intuition of what is probable and what is not are often wrong.

We generalize the table as follows:

	A allele	a allele	Row totals
Cases	a	b	$a + b$
Controls	c	d	$c + d$
Column totals	$a + c$	$b + d$	$a + b + c + d = n$

Then the probability is:

$$P = (a+b)!(c+d)!(a+c)!(b+d)!/\,!a!b!c!d!n!$$

where '!' is the factorial operator, for example $4! = 4 \times 3 \times 2 \times 1$.

The test is called Fisher's exact test and you can run it using these commands in the R software package:

```
a <- as.table(rbind(c(161,55), c(97,51)))

dimnames(a) <- list (type = c("cases", "controls"), alleles = c("A", "a"))

fisher.test(a)
```

This gives a P value of 0.078, in favor of the hypothesis that the counts arose by randomly sampling from 364 alleles, in the proportions given by the row and column totals.

Computational power limits the use of Fisher's exact test, and with contingency tables of more than two columns and two rows (2×2), especially when one or more of the cells is greater than 100, most computers resort to simulation rather than calculation. So instead of the Fisher test, a χ^2 (chi-squared) test is usually employed, in which expected and observed distributions are compared.

Here's how to analyze it in R:

```
a <- as.table(rbind(c(161,55), c(97,51)))

dimnames(a) <- list (type = c("cases", "controls"), alleles = c("A", "a"))

chisq.test(a, correct = F)
```

If you use R to calculate χ^2 and the P value and you simply type chisq.test(a), you won't get quite the same answer—this is because by default R uses a correction to account for the fact that the distribution of χ^2 is continuous while the data are not. You turn off the correction by using chisq.test(a, correct = F).

Let's assume we have genotype data, again presented as a contingency table:

	AA	Aa	aa	Row totals
Cases	53	35	10	98
Controls	35	21	15	71
Column totals	88	56	25	169

Which we can analyze with the following commands in R:

```
a <- as.table(rbind(c(53,35,10), c(35,21,15)))

dimnames(a) <- list (type = c("cases", "controls"), genotypes = c("AA", "Aa",
"aa"))

chisq.test(a, correct = F)

fisher.test(a)
```

For the two tests, we have similar, but not identical results: $\chi^2 = 3.9695$, $P = 0.1374$, and from the Fisher's test, $P = 0.1499$.

A15 Bayesian terminology

We can't teach you Bayesian statistics, and even an introduction soon hits the problem that Bayesian statistics is full of terminology and equations. Our overview in Chapter 21, section 21.8 is necessarily brief. Here, we provide similarly curt statements about some of the key concepts.

In working with hypotheses (H) and data (D), Bayesians use conditional probabilities. A **conditional probability** is of the form $p(H \mid D)$ where the '|' symbol means 'given that', so $p(H \mid D)$ means the probability of the hypothesis given the data.

The posterior probability that a hypothesis is true, given new data $p(H \mid D)$, is proportional to the product of the likelihood of the new data given the hypothesis $p(H \mid D)$ and the prior probability of the hypothesis $p(H)$. This is Bayes theorem, and it is written:

$$p(H \mid D) = \frac{p(D \mid H)p(H)}{p(D)}$$

When comparing the probability of different hypotheses that use the same data, $p(D)$ is constant and the equation becomes:

$$p(H \mid D) \propto p(D \mid H)\, p(H)$$

This means that $p(H \mid D)$ is proportional to the likelihood of the hypothesis multiplied by the prior probability that the hypothesis is true. We need the likelihood of the hypothesis and a prior probability to obtain the required posterior probability.

Bayesians do not use P values to indicate the support for a hypothesis, making it hard for the uninitiated to evaluate the results reported. However, Bayesians do use 'Bayes factors'. Suppose we have two hypotheses, H_1 and H_0. The likelihood of the hypothesis H_1 divided by the likelihood of the hypothesis H_0 is a Bayes factor.

When H_1 and H_0 are equally probable so that $p(H_1) = p(H_0) = 0.5$, the Bayes factor is equal to the posterior odds of H_1. We can refer to a Bayes factor by B_{10}, where the subscripts indicate which two hypotheses are being compared (in this case H_1 and H_0). So, for example, a Bayes factor of ten for B_{10} means that the prior odds should be updated by a factor of ten in favor of H_1, while a Bayes factor of 0.1 means that prior odds should be updated by a factor of ten in favor of H_0. To interpret a Bayes factor, the following categories are recommended in Table A9 (Kass and Raftery 1995):

Table A9 The interpretation of Bayes factors

Bayes factor	Evidence
1–3	Not worth mentioning
3–20	Positive
20–50	Strong
>150	Very strong

References

1000 Genomes Project Consortium (2010) A map of human genome variation from population-scale sequencing. *Nature* 467: 1061–73.

1000 Genomes Project Consortium (2015) A global reference for human genetic variation. *Nature* 526: 68–74.

Abecasis, G.R., A. Auton, L.D. Brooks, M.A. DePristo, R.M. Durbin, et al. (2012) An integrated map of genetic variation from 1,092 human genomes. *Nature* 491: 56–65.

Agarwal, D.P. and H.W. Goedde (1992) Pharmacogenetics of alcohol metabolism and alcoholism. *Pharmacogenetics* 2: 48–62.

Akaike, H. (1987) Factor-analysis and AIC. *Psychometrika* 52: 317–32.

Akiskal, H.S. and W.T. McKinney, Jr. (1973) Depressive disorders: toward a unified hypothesis. *Science* 182: 20–29.

Albright, T.D., T.M. Jessell, E.R. Kandel, and M.I. Posner (2000) Neural science: a century of progress and the mysteries that remain. *Neuron* 25 Suppl.: S1–55.

Allen, K.M., J.G. Gleeson, S. Bagrodia, M.W. Partington, J.C. MacMillan, et al. (1998) *PAK3* mutation in nonsyndromic X-linked mental retardation. *Nat Genet* 20: 25–30.

Allis, C.D. and T. Jenuwein (2016) The molecular hallmarks of epigenetic control. *Nat Rev Genet* 17: 487–500.

Altman, D.G. and J.M. Bland (1994) Diagnostic tests. 1: Sensitivity and specificity. *Br Med J* 308: 1552.

Altshuler, D., L.D. Brooks, A. Chakravarti, F.S. Collins, M.J. Daly, et al. (2005) A haplotype map of the human genome. *Nature* 437: 1299–320.

Altshuler, D., J.N. Hirschhorn, M. Klannemark, C.M. Lindgren, M.C. Vohl, et al. (2000) The common PPARγ Pro12Ala polymorphism is associated with decreased risk of type 2 diabetes. *Nat Genet* 26: 76–80.

American Psychiatric Association (1987) *Diagnostic and Statistical Manual of Mental Disorders, Third Edition, Revised*. American Psychiatric Association, Washington, DC.

American Psychiatric Association (2013) *Diagnostic and Statistical Manual of Mental Disorders: Fifth Edition*. American Psychiatric Association, Washington, DC.

Amin, N.D. and S.P. Pasca (2018) Building models of brain disorders with three-dimensional organoids. *Neuron* 100: 389–405.

Amir, R.E., I.B. van den Veyver, M. Wan, C.Q. Tran, U. Francke, et al. (1999) Rett syndrome is caused by mutations in X-linked *MECP2*, encoding methyl-CpG-binding protein 2. *Nat Genet* 23: 185–8.

Amrhein, V. and S. Greenland (2018) Remove, rather than redefine, statistical significance. *Nat Hum Behav* 2: 4.

Amrhein, V., S. Greenland, and B. McShane (2019) Scientists rise up against statistical significance. *Nature* 567: 305–7.

Ancelin, M.L. and J. Ryan. (2018) 5-HTTLPR × stress hypothesis: is the debate over? *Mol Psychiatry* 23: 2116–17.

Anderson, C.J., S. Bahnik, M. Barnett-Cowan, F.A. Bosco, J. Chandler, et al. (2016) Response to comment on 'Estimating the reproducibility of psychological science'. *Science* 351: 1037.

Andreux, P.A., E.G. Williams, H. Koutnikova, R.H. Houtkooper, M.F. Champy, et al. (2012) Systems genetics of metabolism: the use of the BXD murine reference panel for multiscalar integration of traits. *Cell* 150: 1287–99.

Anguelova, M., C. Benkelfat, and G. Turecki (2003) A systematic review of association studies investigating genes coding for serotonin receptors and the serotonin transporter: I. Affective disorders. *Mol Psychiatry* 8: 574–91.

Antoch, M.P., E.J. Song, A.M. Chang, M.H. Vitaterna, Y.L. Zhao, et al. (1997) Functional identification of the mouse circadian *Clock* gene by transgenic BAC rescue. *Cell* 89: 655–67.

Austin, C.P., J.F. Battey, A. Bradley, M. Bucan, M. Capecchi, et al. (2004) The knockout mouse project. *Nat Genet* 36: 921–4.

Aylor, D.L., W. Valdar, W. Foulds-Mathes, R.J. Buus, R.A. Verdugo, et al. (2011) Genetic analysis of complex traits in the emerging Collaborative Cross. *Genome Res* 21: 1213–22.

Bachtrog, D., M. Kirkpatrick, J.E. Mank, S.F. McDaniel, J.C. Pires, et al. (2011) Are all sex chromosomes created equal? *Trends Genet* 27: 350–7.

Baio, J., L. Wiggins, D.L. Christensen, M.J. Maenner, J. Daniels, et al. (2018) Prevalence of autism spectrum disorder among children aged 8 years—autism and developmental disabilities monitoring network, 11 sites, United States, 2014. *MMWR Surveill Summ* 67: 1–23.

Baker, B.S., B.J. Taylor, and J.C. Hall (2001) Are complex behaviors specified by dedicated regulatory genes? Reasoning from *Drosophila*. *Cell* 105: 13–24.

Baker, K.G., G.M. Halliday, and I. Tork (1990) Cytoarchitecture of the human dorsal raphe nucleus. *J Comp Neurol* 301: 147–61.

Baker, P., J. Piven, S. Schwartz, and S. Patil (1994) Brief report: duplication of chromosome 15q11–13 in two individuals with autistic disorder. *J Autism Dev Disord* 24: 529–35.

Balschun, D., D.P. Wolfer, P. Gass, T. Mantamadiotis, H. Welzl, et al. (2003) Does cAMP response element-binding protein have a pivotal role in hippocampal synaptic plasticity and hippocampus-dependent memory? *J Neurosci* 23: 6304–14.

Banghart, M., K. Borges, E. Isacoff, D. Trauner, and R.H. Kramer (2004) Light-activated ion channels for remote control of neuronal firing. *Nat Neurosci* 7: 1381–86.

Barak, B. and G. Feng (2016) Neurobiology of social behavior abnormalities in autism and Williams syndrome. *Nat Neurosci* 19: 647–55.

Barcellos, S.H., L.S. Carvalho, and P. Turley (2018) Education can reduce health differences related to genetic risk of obesity. *Proc Natl Acad Sci U S A* 115: E9765–72.

Barker, D.J. (1998) *Mothers, Babies and Health in Later Life*. Churchill Livingstone, Edinburgh, UK.

Barlow, D.P. and M.S. Bartolomei (2014) Genomic imprinting in mammals. *Cold Spring Harb Perspect Biol* 6: a018382.

Baron, M., R. Gruen, J.D. Rainer, J. Kane, L. Asnis, et al. (1985) A family study of schizophrenic and normal control probands—implications for the spectrum concept of schizophrenia. *Am J Psychiatry* 142: 447–55.

Baron, M., N. Risch, R. Hamburger, B. Mandel, S. Kushner, et al. (1987) Genetic linkage between X-chromosome markers and bipolar affective illness. *Nature* 326: 289–92.

Barrangou, R., C. Fremaux, H. Deveau, M. Richards, P. Boyaval, et al. (2007) CRISPR provides acquired resistance against viruses in prokaryotes. *Science* 315: 1709–12.

Barrot, M., J.D. Olivier, L.I. Perrotti, R.J. DiLeone, O. Berton, et al. (2002) CREB activity in the nucleus accumbens shell controls gating of behavioral responses to emotional stimuli. *Proc Natl Acad Sci U S A* 99: 11435–40.

Barry, G. and J.S. Mattick (2012) The role of regulatory RNA in cognitive evolution. *Trends Cogn Sci* 16: 497–503.

Bartley, A.J., D.W. Jones, and D.R. Weinberger (1997) Genetic variability of human brain size and cortical gyral patterns. *Brain* 120: 257–69.

Barton, N., J. Hermisson, and M. Nordborg (2019) Why structure matters. *Elife* 8: e45380.

Baud, A., R. Hermsen, V. Guryev, P. Stridh, D. Graham, et al. (2013) Combined sequence-based and genetic mapping analysis of complex traits in outbred rats. *Nat Genet* 45: 767–75.

Baud, A., M.K. Mulligan, F.P. Casale, J.F. Ingels, C.J. Bohl, et al. (2017) Genetic variation in the social environment contributes to health and disease. *PLoS Genet* 13: e1006498.

Baudouin, S.J., J. Gaudias, S. Gerharz, L. Hatstatt, K. Zhou, et al. (2012) Shared synaptic pathophysiology in syndromic and nonsyndromic rodent models of autism. *Science* 338: 128–32.

Bayes, A., L.N. van de Lagemaat, M.O. Collins, M.D. Croning, I.R. Whittle, et al. (2011) Characterization of the proteome, diseases and evolution of the human postsynaptic density. *Nat Neurosci* 14: 19–21.

Beato, M. (1989) Gene regulation by steroid hormones. *Cell* 56: 335–44.

Bell-Pedersen, D., V.M. Cassone, D.J. Earnest, S.S. Golden, P.E. Hardin, et al. (2005) Circadian rhythms from multiple oscillators: lessons from diverse organisms. *Nat Rev Genet* 26: 544–56.

Belsky, D.W., T.E. Moffitt, D.L. Corcoran, B. Domingue, H. Harrington, et al. (2016) The genetics of success: how single-nucleotide polymorphisms associated with educational attainment relate to life-course development. *Psychol Sci* 27: 957–72.

Ben-Shahar, Y., A. Robichon, M.B. Sokolowski, and G.E. Robinson (2002) Influence of gene action across different time scales on behavior. *Science* 296: 741–4.

Bendesky, A., Y.M. Kwon, J.M. Lassance, C.L. Lewarch, S. Yao, et al. (2017) The genetic basis of parental care evolution in monogamous mice. *Nature* 544: 434–9.

Bendesky, A., J. Pitts, M.V. Rockman, W.C. Chen, M.W. Tan, et al. (2012) Long-range regulatory polymorphisms affecting a GABA receptor constitute a quantitative trait locus (QTL) for social behavior in *Caenorhabditis elegans*. *PLoS Genet* 8: e1003157.

Benito, E., C. Kerimoglu, B. Ramachandran, T. Pena-Centeno, G. Jain, et al. (2018) RNA-dependent intergenerational inheritance of enhanced synaptic plasticity after environmental enrichment. *Cell Rep* 23: 546–54.

Benjamin, D.J., J.O. Berger, M. Johannesson, B.A. Nosek, E.J. Wagenmakers, et al. (2018) Redefine statistical significance. *Nat Hum Behav* 2: 6–10.

Bennett, B.J., C.R. Farber, L. Orozco, H.M. Kang, A. Ghazalpour, et al. (2010) A high-resolution association mapping panel for the dissection of complex traits in mice. *Genome Res* 20: 281–90.

Benzer, S. (1973) Genetic dissection of behavior. *Sci Am* 229: 24–37.

Berg, J.J., A. Harpak, N. Sinnott-Armstrong, A.M. Joergensen, H. Mostafavi, et al. (2019) Reduced signal for polygenic adaptation of height in UK Biobank. *Elife* 8: e39725.

Berg, J.M. and D.H. Geschwind (2012) Autism genetics: searching for specificity and convergence. *Genome Biol* 13: 247.

Bergan, J.F., Y. Ben-Shaul and C. Dulac (2014) Sex-specific processing of social cues in the medial amygdala. *Elife* 3: e02743.

Bergen, S.E., A. Ploner, D. Howrigan, M.C. O'Donovan, J.W. Smoller, P.F. Sullivan, J. Sebat, B. Neale, and K.S. Kendler (2019). Joint contributions of rare copy number variants and common SNPs to risk for schizophrenia. *Am J Psychiatry* 176: 29–35.

Berget, S.M., C. Moore, and P.A. Sharp (1977) Spliced segments at the 5′ terminus of adenovirus 2 late mRNA. *Proc Natl Acad Sci U S A* 74: 3171–5.

Bernet, W., C.L. Vnencak-Jones, N. Farahany and S.A. Montgomery (2007) Bad nature, bad nuture and testimony regarding *MAOA* and *SLC6A4* genotyping at murder trials. *J Forensic Sci* 52: 1362–71.

Bernhard, N. and D. van der Kooy (2000) A behavioral and genetic dissection of two forms of olfactory plasticity in *Caenorhabditis elegans*: adaptation and habituation. *Learn Mem* 7: 199–212.

Bernier, R., C. Golzio, B. Xiong, H.A. Stessman, B.P. Coe, et al. (2014) Disruptive *CHD8* mutations define a subtype of autism early in development. *Cell* 158: 263–76.

Berry-Kravis, E., R. Hagerman, J. Visootsak, D. Budimirovic, W.E. Kaufmann, et al. (2017) Arbaclofen in fragile X syndrome: results of phase 3 trials. *J Neurodev Disord* 9: 3.

Berry-Kravis, E.M., L. Lindemann, A.E. Jonch, G. Apostol, M.F. Bear, et al. (2018) Drug development for neurodevelopmental disorders: lessons learned from fragile X syndrome. *Nat Rev Drug Discov* 17: 280–99.

Bettelheim, B. (1967) *Empty Fortess: Infantile Autism and the Birth of Self.* Free Press, New York.

Beutler, B. (2004) Innate immunity: an overview. *Mol Immunol* 40: 845–59.

Bigdeli, T.B., S.A. Bacanu, B.T. Webb, D. Walsh, F.A. O'Neill, et al. (2014) Molecular validation of the schizophrenia spectrum. *Schizophr Bull* 40: 60–65.

Bignami, G. (1965) Selection for high rates and low rates of avoidance conditioning in the rat. *Anim Behav* 13: 221–7.

Billuart, P., T. Bienvenu, N. Ronce, V. DesPortes, M.C. Vinet, et al. (1998) Oligophrenin-1 encodes a rhoGAP protein involved in X-linked mental retardation. *Nature* 392: 923–6.

Birney, E., J.A. Stamatoyannopoulos, A. Dutta, R. Guigo, T.R. Gingeras, et al. (2007) Identification and analysis of functional elements in 1% of the human genome by the ENCODE pilot project. *Nature* 447: 799–816.

Bis, J.C., C. DeCarli, A.V. Smith, F. van der Lijn, F. Crivello, et al. (2012) Common variants at 12q14 and 12q24 are associated with hippocampal volume. *Nat Genet* 44: 545–51.

Bliss, T.V. and G.L. Collingridge (1993) A synaptic model of memory: long-term potentiation in the hippocampus. *Nature* 361: 31–9.

Bliss, T.V. and T. Lømo (1973) Long-lasting potentiation of synaptic transmission in the dentate area of the anaesthetized rabbit following stimulation of the perforant path. *J Physiol* 232: 331–56.

Blow, M., P.A. Futreal, R. Wooster and M.R. Stratton (2004) A survey of RNA editing in human brain. *Genome Res* 14: 2379–87.

Bolinskey, P.K., M.C. Neale, K.C. Jacobson, C.A. Prescott, and K.S. Kendler (2004) Sources of individual differences in stressful life event exposure in male and female twins. *Twin Res* 7: 33–8.

Bolton, P., H. Macdonald, A. Pickles, P. Rios, S. Goode, et al. (1994) A case–control family history study of autism. *J Child Psychol Psychiatry* 35: 877–900.

Bonaglia, M.C., R. Giorda, R. Borgatti, G. Felisari, C. Gagliardi, et al. (2001) Disruption of the ProSAP2 gene in a t(12;22)(q24.1;q13.3) is associated with the 22q13.3 deletion syndrome. *Am J Hum Genet* 69: 261–68.

Bonett, D.G. (2007) Transforming odds ratios into correlations for meta-analytic research. *Am Psychol* 62: 254–5.

Border, R., E.C. Johnson, L.M. Evans, A. Smolen, N. Berley, et al. (2019) No support for historical candidate gene or candidate gene-by-interaction hypotheses for major depression across multiple large samples. *Am J Psychiatry* 179: 376–87.

Bosker, F.J., C.A. Hartman, I.M. Nolte, B.P. Prins, P. Terpstra, et al. (2011) Poor replication of candidate genes for major depressive disorder using genome-wide association data. *Mol Psychiatry* 16: 516–32.

Botstein, D., R.L. White, M. Skolnick, and R.W. Davis (1980) Construction of a genetic linkage map in man using restriction fragment length polymorphisms. *Am J Hum Genet* 32: 314–31.

Bouchard, T. (2009) Newsmaker interview. Behavioral geneticist celebrates twins, scorns PC science. Interview by Constance Holden. *Science* 325: 27.

Bouchard, T.J., Jr., D.T. Lykken, M. McGue, N.L. Segal, and A. Tellegen (1990) Sources of human psychological differences: the Minnesota Study of Twins Reared Apart. *Science* 250: 223–8.

Bouchard, T.J., Jr. and M. McGue (1981) Familial studies of intelligence: a review. *Science* 212: 1055–9.

Bourtchuladze, R., B. Frenguelli, J. Blendy, D. Cioffi, G. Schutz, et al. (1994) Deficient long-term memory in mice with a targeted mutation of the cAMP-responsive element-binding protein. *Cell* 79: 59–68.

Boyden, E.S., F. Zhang, E. Bamberg, G. Nagel, and K. Deisseroth (2005) Millisecond-timescale, genetically targeted optical control of neural activity. *Nat Neurosci* 8: 1263–8.

Brainstorm Consortium, V. Anttila, B. Bulik-Sullivan, H.K. Finucane, R.K. Walters, et al. (2018) Analysis of shared heritability in common disorders of the brain. *Science* 360: eaap8757.

Brennan, C.A. and K. Moses (2000) Determination of *Drosophila* photoreceptors: timing is everything. *Cell Mol Life Sci* 57: 195–214..

Brenner, S. (1974) The genetics of *Caenorhabditis elegans*. Genetics 77: 71–94.

Brenner, S. and L. Wolpert (2001) *My Life in Science: Sydney Brenner, A Life in Science (Lives in Science)*. BioMed Central, London.

Bretscher, A.J., E. Kodama-Namba, K.E. Busch, R.J. Murphy, Z. Soltesz, et al. (2011) Temperature, oxygen, and salt-sensing neurons in *C. elegans* are carbon dioxide sensors that control avoidance behavior. *Neuron* 69: 1099–113.

Briley, D.A. and E.M. Tucker-Drob (2013) Explaining the increasing heritability of cognitive ability across development: a meta-analysis of longitudinal twin and adoption studies. *Psychol Sci* 24: 1704–13.

Brown, C.J., B.D. Hendrich, J.L. Rupert, R.G. Lafreniere, Y. Xing, et al. (1992) The human *XIST* gene: analysis of a 17 kb inactive X-specific RNA that contains conserved repeats and is highly localized within the nucleus. *Cell* 71: 527–42.

Brownell, J.E., J. Zhou, T. Ranalli, R. Kobayashi, D.G. Edmondson, et al. (1996) Tetrahymena histone acetyltransferase A: a homolog to yeast Gcn5p linking histone acetylation to gene activation. *Cell* 84: 843–51.

Brunner, H.G., M. Nelen, X.O. Breakefield, H.H. Ropers, and B.A. van Oost (1993) Abnormal behavior associated with a point mutation in the structural gene for monoamine oxidase A. *Science* 262: 578–80.

Buiting, K., C. Williams, and B. Horsthemke (2016) Angelman syndrome—insights into a rare neurogenetic disorder. *Nat Rev Neurol* 12: 584–93.

Bulik, C.M., P.F. Sullivan, F. Tozzi, H. Furberg, P. Lichtenstein, et al. (2006) Prevalence, heritability, and prospective risk factors for anorexia nervosa. *Arch Gen Psychiatry* 63: 305–12.

Bulik, C.M., P.F. Sullivan, T.D. Wade, and K.S. Kendler (2000) Twin studies of eating disorders: a review. *Int J Eat Disord* 27: 1–20.

Bulik-Sullivan, B.K., P.R. Loh, H.K. Finucane, S. Ripke, J. Yang, et al. (2015) LD Score regression distinguishes confounding from polygenicity in genome-wide association studies. *Nat Genet* 47: 291–5.

Bundey, S., C. Hardy, S. Vickers, M.W. Kilpatrick, and J.A. Corbett (1994) Duplication of the 15q11–13 region in a patient with autism, epilepsy and ataxia. *Dev Med Child Neurol* 36: 736–42.

Burton, P.R., D.G. Clayton, L.R. Cardon, N. Craddock, P. Deloukas, et al. (2007) Association scan of 14,500 nonsynonymous SNPs in four diseases identifies autoimmunity variants. *Nat Genet* 39: 1329–37.

Burton, R. (1621; this edition 2001) *The Anatomy of Melancholy*. New York Review of Books, New York.

Butterworth, J.S. and E.H. Reppert, Jr. (1960) Auscultatory findings in myocardial infarction. *Circulation* 22: 448–52.

Button, K.S., J.P. Ioannidis, C. Mokrysz, B.A. Nosek, J. Flint, et al. (2013) Power failure: why small sample size undermines the reliability of neuroscience. *Nat Rev Neurosci* 14: 365–76.

Bycroft, C., C. Freeman, D. Petkova, G. Band, L.T. Elliott, et al. (2018) The UK Biobank resource with deep phenotyping and genomic data. *Nature* 562: 203–9.

C. elegans Sequencing Consortium (1998) Genome sequence of the nematode *C. elegans*: a platform for investigating biology. *Science* 282: 2012–18.

Cadoret, R.J., C.A. Cain, and R.R. Crowe (1983) Evidence for gene–environment interaction in the development of adolescent antisocial-behavior. *Behav Genet* 13: 301–10.

Cai, D.J., D. Aharoni, T. Shuman, J. Shobe, J. Biane, et al. (2016) A shared neural ensemble links distinct contextual memories encoded close in time. *Nature* 534: 115–18.

Calvin, C.M., I.J. Deary, D. Webbink, P. Smith, C. Fernandes, et al. (2012) Multivariate genetic analyses of cognition and academic achievement from two population samples of 174,000 and 166,000 school children. *Behav Genet* 42: 699–710.

Cannon, T.D., J. Kaprio, J. Lonnqvist, M. Huttunen, and M. Koskenvuo (1998) The genetic epidemiology of schizophrenia in a Finnish twin cohort. A population-based modeling study. *Arch Gen Psychiatry* 55: 67–74.

Capecchi, M.R. (1989) Altering the genome by homologous recombination. *Science* 244: 1288–92.

Capon, C. and M. Duyme (1989) Assessment of effects of socio-economic status on IQ in a full cross-fostering study. *Nature* 340: 552–4.

Cardno, A.G., E.J. Marshall, B. Coid, A.M. Macdonald, T.R. Ribchester, et al. (1999) Heritability estimates for psychotic disorders: the Maudsley twin psychosis series. *Arch Gen Psychiatry* 56: 162–68.

Carone, B.R., L. Fauquier, N. Habib, J.M. Shea, C.E. Hart, et al. (2010) Paternally induced transgenerational environmental reprogramming of metabolic gene expression in mammals. *Cell* 143: 1084–96.

Caspi, A., K. Sugden, T.E. Moffitt, A. Taylor, I.W. Craig, et al. (2003) Influence of life stress on depression: moderation by a polymorphism in the 5-HTT gene. *Science* 301: 386–9.

Cerda, M., A. Sagdeo, J. Johnson, and S. Galea (2010) Genetic and environmental influences on psychiatric comorbidity: a systematic review. *J Affect Disord* 126: 14–38.

Cha-Molstad, H., D.M. Keller, G.S. Yochum, S. Impey, and R.H. Goodman (2004) Cell-type-specific binding of the transcription factor CREB to the cAMP-response element. *Proc Natl Acad Sci U S A* 101: 13572–7.

Chang, A.J., N. Chronis, D.S. Karow, M.A. Marletta, and C.I. Bargmann (2006) A distributed chemosensory circuit for oxygen preference in *C. elegans*. *PLoS Biol* 4: e274.

Chen, A.C., Y. Shirayama, K.H. Shin, R.L. Neve, and R.S. Duman (2001) Expression of the cAMP response element binding protein (CREB) in hippocampus produces an antidepressant effect. *Biol Psychiatry* 49: 753–62.

Chen, E., M.R. Sharma, X. Shi, R.K. Agrawal, and S. Joseph (2014) Fragile X mental retardation protein regulates translation by binding directly to the ribosome. *Mol Cell* 54: 407–17.

Chen, Q., W. Yan, and E. Duan (2016) Epigenetic inheritance of acquired traits through sperm RNAs and sperm RNA modifications. *Nat Rev Genet* 17: 733–43.

Chen, W.J., E.W. Loh, Y.P.P. Hsu, C.C. Chen, J.M. Yu, et al. (1996) Alcohol-metabolising genes and alcoholism among Taiwanese Han men: independent effect of ADH2, ADH3 and ALDH2. *Br J Psychiatry* 168, 762–7.

Chen, Y., J. Zhu, P.Y. Lum, X. Yang, S. Pinto, et al. (2008) Variations in DNA elucidate molecular networks that cause disease. *Nature* 452: 429–35.

Chesler, E.J. (2014) Out of the bottleneck: the Diversity Outcross and Collaborative Cross mouse populations in behavioral genetics research. *Mamm Genome* 25: 3–11.

Chimpanzee Sequencing and Analysis Consortium (2005) Initial sequence of the chimpanzee genome and comparison with the human genome. *Nature* 437: 69–87.

Church, C., S. Lee, E.A. Bagg, J.S. McTaggart, R. Deacon, et al. (2009) A mouse model for the metabolic effects of the human fat mass and obesity associated *FTO* gene. *PLoS Genet* 5: e1000599.

Churchill, G.A. and R.W. Doerge (1994) Empirical threshold values for quantitative trait mapping. *Genetics* 138: 963–71.

Ciccocioppo, R. (2013) Genetically selected alcohol preferring rats to model human alcoholism. Curr Top Behav Neurosci 13: 251–69.

Claridge-Chang, A., R.D. Roorda, E. Vrontou, L. Sjulson, H. Li, et al. (2009) Writing memories with light-addressable reinforcement circuitry. *Cell* 139: 405–15.

Clarke, T.K., M.K. Lupton, A.M. Fernandez-Pujals, J. Starr, G. Davies, et al. (2016) Common polygenic risk for autism spectrum disorder (ASD) is associated with cognitive ability in the general population. *Mol Psychiatry* 21: 419–25.

Claussnitzer, M., S.N. Dankel, K.H. Kim, G. Quon, W. Meuleman, et al. (2015) *FTO* obesity variant circuitry and adipocyte browning in humans. *N Engl J Med* 373: 895–907.

Cockburn, A. (2006) Prevalence of different modes of parental care in birds. *Proc Biol Sci* 273: 1375–83.

Coffee, B., K. Keith, I. Albizua, T. Malone, J. Mowrey, et al. (2009) Incidence of fragile X syndrome by newborn screening for methylated *FMR1* DNA. *Am J Hum Genet* 85: 503–14.

Coffee Caffeine Genetics Consortium, M.C. Cornelis, E.M. Byrne, T. Esko, M.A. Nalls, et al. (2015) Genome-wide meta-analysis identifies six novel loci associated with habitual coffee consumption. *Mol Psychiatry* 20: 647–56.

Cohen, J. (1988) *Statistical Power Analysis for the Behavioral Sciences*. Lawrence Erlbaum, Hillsdale, NJ.

Cohen, J.C., R.S. Kiss, A. Pertsemlidis, Y.L. Marcel, R. McPherson, et al. (2004) Multiple rare alleles contribute to low plasma levels of HDL cholesterol. *Science* 305: 869–72.

Collins, F.S., R.H. Finnell, J. Rossant, and W. Wurst (2007) A new partner for the International Knockout Mouse Consortium. *Cell* 129: 235.

Collins, F.S., M.S. Guyer, and A. Charkravarti (1997) Variations on a theme: cataloging human DNA sequence variation. *Science* 278: 1580–1.

Collins, P.Y., V. Patel, S.S. Joestl, D. March, T.R. Insel, et al. (2011) Grand challenges in global mental health. *Nature* 475: 27–30.

Colquhoun, D. (2019) The false positive risk: a proposal concerning what to do about *p*-values. *Am Stat* 73: 192–201.

Cong, L., F.A. Ran, D. Cox, S. Lin, R. Barretto, et al. (2013) Multiplex genome engineering using CRISPR/Cas systems. *Science* 339: 819–23.

Converge Consortium (2015) Sparse whole-genome sequencing identifies two loci for major depressive disorder. *Nature* 523: 588–91.

Cooper, G.M., B.P. Coe, S. Girirajan, J.A. Rosenfeld, T.H. Vu, et al. (2011) A copy number variation morbidity map of developmental delay. *Nat Genet* 43: 838–46.

Cooper, J.E., R.E. Kendell, B.J. Gurland, L. Sharpe, J.R.M. Copeland, et al. (1972) *Psychiatric Diagnosis in New York and London (U.S.–U.K. Diagnostic Project)*. Oxford University Press, London.

Corder, E.H., A.M. Saunders, W.J. Strittmatter, D.E. Schmechel, P.C. Gaskell, et al. (1993) Gene dose of apolipoprotein E type 4 allele and the risk of Alzheimer's disease in late onset families. *Science* 261: 921–3.

Costa, R.M., N.B. Federov, J.H. Kogan, G.G. Murphy, J. Stern, et al. (2002) Mechanism for the learning deficits in a mouse model of neurofibromatosis type 1. *Nature* 415: 526–30.

Covington, H.E., 3rd, I. Maze, H. Sun, H.M. Bomze, K.D. DeMaio, et al. (2011) A role for repressive histone methylation in cocaine-induced vulnerability to stress. *Neuron* 71: 656–70.

Crabbe, J.C., T.J. Phillips, and J.K. Belknap (2010) The complexity of alcohol drinking: studies in rodent genetic models. *Behav Genet* 40: 737–50.

Crabbe, J.C., D. Wahlsten, and B.C. Dudek (1999) Genetics of mouse behavior: interactions with laboratory environment. *Science* 284: 1670–2.

Crick, F.H. (1958) On protein synthesis. *Symp Soc Exp Biol* 12: 138–63.

Crick, F.H. (1968) The origin of the genetic code. *J Mol Biol* 38: 367–79.

Cross-Disorder Group of the Psychiatric Genomics, C., S.H. Lee, S. Ripke, B.M. Neale, S.V. Faraone, et al. (2013) Genetic relationship between five psychiatric disorders estimated from genome-wide SNPs. *Nat Genet* 45: 984–94.

Culverhouse, R.C., N.L. Saccone, A.C. Horton, Y. Ma, K.J. Anstey, et al. (2018) Collaborative meta-analysis finds no evidence of a strong interaction between stress and 5-HTTLPR genotype contributing to the development of depression. *Mol Psychiatry* 23: 133–42.

D'Adamo, P., A. Menegon, C. Lo Nigro, M. Grasso, M. Gulisano, et al. (1998) Mutations in *GDI1* are responsible for X-linked non-specific mental retardation. *Nat Genet* 19: 134–9.

Dahoun, S., S. Gagos, M. Gagnebin, C. Gehrig, C. Burgi, et al. (2008) Monozygotic twins discordant for trisomy 21 and maternal 21q inheritance: a complex series of events. *Am J Med Genet A* 146A: 2086–93.

Darmanis, S., S.A. Sloan, Y. Zhang, M. Enge, C. Caneda, et al. (2015) A survey of human brain transcriptome diversity at the single cell level. *Proc Natl Acad Sci U S A* 112: 7285–90.

Darnell, J.C., S.J. van Driesche, C. Zhang, K.Y. Hung, A. Mele, et al. (2011) FMRP stalls ribosomal translocation on mRNAs linked to synaptic function and autism. *Cell* 146: 247–61.

Darvasi, A. and M. Soller (1997) A simple method to calculate resolving power and confidence interval of QTL map location. *Behav Genet* 27: 125–32.

Davenport, C.B. (1915) The feebly inhibited: II. Nomadism or the wandering impulse with special reference to heredity. *Proc Natl Acad Sci U S A* 1: 120–2.

Davey Smith, G. and G. Hemani (2014) Mendelian randomization: genetic anchors for causal inference in epidemiological studies. *Hum Mol Genet* 23: R89–98.

Davies, J.L., Y. Kawaguchi, S.T. Bennett, J.B. Copeman, H.J. Cordell, et al. (1994) A genome-wide search for human type 1 diabetes susceptibility genes. *Nature* 371: 130–6.

Davis, O.S., C.M. Haworth, and R. Plomin (2009) Learning abilities and disabilities: generalist genes in early adolescence. *Cogn Neuropsychiatry* 14: 312–31.

Dawkins, R. (1982) *The Extended Phenotype*. Oxford University Press, New York.

de Angelis, M.H., G. Nicholson, M. Selloum, J.K. White, H. Morgan, et al. (2015) Analysis of mammalian gene function through broad-based phenotypic screens across a consortium of mouse clinics. *Nat Genet* 47: 969–78.

de Belle, J.S. and M. Heisenberg (1996) Expression of *Drosophila* mushroom body mutations in alternative genetic backgrounds: a case study of the mushroom body miniature gene (*mbm*). *Proc Natl Acad Sci U S A* 93: 9875–80.

de Bono, M. and C.I. Bargmann (1998) Natural variation in a neuropeptide Y receptor homolog modifies social behavior and food response in *C. elegans*. *Cell* 94: 679–89.

de Koning, A.P., W. Gu, T.A. Castoe, M.A. Batzer, and D.D. Pollock (2011) Repetitive elements may comprise over two-thirds of the human genome. *PLoS Genet* 7: e1002384.

de la Torre-Ubieta, L., H.J. Won, J.L. Stein, and D.H. Geschwind (2016) Advancing the understanding of autism disease mechanisms through genetics. *Nat Med* 22: 345–61.

de Ligt, J., M.H. Willemsen, B.W. van Bon, T. Kleefstra, H.G. Yntema, et al. (2012) Diagnostic exome sequencing in persons with severe intellectual disability. *N Engl J Med* 367: 1921–9.

de Moor, M.H., P.T. Costa, A. Terracciano, R.F. Krueger, E.J. de Geus, et al. (2012) Meta-analysis of genome-wide association studies for personality. *Mol Psychiatry* 17: 337–49.

de Vladar, H.P. and N. Barton (2014) Stability and response of polygenic traits to stabilizing selection and mutation. *Genetics* 197: 749–67.

de Vladar, H.P. and N.H. Barton (2011) The statistical mechanics of a polygenic character under stabilizing selection, mutation and drift. *J R Soc Interface* 8: 720–39.

de Vries, B.B., R. Pfundt, M. Leisink, D.A. Koolen, L.E. Vissers, et al. (2005) Diagnostic genome profiling in mental retardation. *Am J Hum Genet* 77: 606–16.

Deary, I.J., A. Weiss, and D.G. Batty (2010) Intelligence and personality as predictors of illness and death: how researchers in differential psychology and chronic disease epidemiology are collaborating to understand and address health inequalities. *Psychol Sci Public Interest* 11: 53–79.

Deary, I.J., J. Yang, G. Davies, S.E. Harris, A. Tenesa, et al. (2012) Genetic contributions to stability and change in intelligence from childhood to old age. *Nature* 482: 212–15.

Deaton, A.M. and A. Bird (2011) CpG islands and the regulation of transcription. *Genes Dev* 25: 1010–22.

Deciphering Developmental Disorders Study (2017) Prevalence and architecture of *de novo* mutations in developmental disorders. *Nature* 542: 433–38.

DeFries, J.C., M.C. Gervais, and E.A. Thomas (1978) Response to 30 generations of selection for open field activity in laboratory mice. *Behav Genet* 8: 3–13.

DeFries, J.C. and J.P. Hegmann (1970) Genetic analysis of open-field behavior. In *Contributions to Behavior-Genetic Analysis: The Mouse as a Prototype*, pp. 23–56. Edited by G. Lindzey and D.D. Thiessen. Appleton-Century-Crofts, New York.

Dekker, J., K. Rippe, M. Dekker, and N. Kleckner (2002) Capturing chromosome conformation. *Science* 295: 1306–11.

Demirkan, A., B.W. Penninx, K. Hek, N.R. Wray, N. Amin, et al. (2011) Genetic risk profiles for depression and anxiety in adult and elderly cohorts. *Mol Psychiatry* 16: 773–83.

Dempster, E.L., R. Pidsley, L.C. Schalkwyk, S. Owens, A. Georgiades, et al. (2011) Disease-associated epigenetic changes in monozygotic twins discordant for schizophrenia and bipolar disorder. *Hum Mol Genet* 20: 4786–96.

Devlin, B., M. Daniels, and K. Roeder (1997) The heritability of IQ. *Nature* 388: 468–71.

Dias, B.G. and K.J. Ressler (2014) Parental olfactory experience influences behavior and neural structure in subsequent generations. *Nat Neurosci* 17: 89–96.

Dienes, Z. (2008) *Understanding Psychology as a Science: An Introduction to Scientic and Statistical Inference.* Palgrave MacMillan, New York.

Dierick, H.A. and R.J. Greenspan (2006) Molecular analysis of flies selected for aggressive behavior. *Nat Genet* 38: 1023–31.

Dobzhansky, T. (1973) Nothing in biology makes sense except in the light of evolution. *Am Biol Teach* 35: 125–9.

Dolen, G., A. Darvishzadeh, K.W. Huang, and R.C. Malenka (2013) Social reward requires coordinated activity of nucleus accumbens oxytocin and serotonin. *Nature* 501: 179–84.

Donaldson, Z.R. and L.J. Young (2013) The relative contribution of proximal 5′ flanking sequence and microsatellite variation on brain vasopressin 1a receptor (*Avpr1a*) gene expression and behavior. *PLoS Genet* 9: e1003729.

Down, J.L. (1866) Observations on an ethnic classification of idiots. *Ment Retard* 33: 54–6.

Driscoll, P. and K. Battig (1982) Behavioral, emotional and neurochemical profiles of rats selected for extreme differences in active two way avoidance. In *Genetics of the Brain*, pp. 95–123. Edited by I. Lieblich. Elsevier, Amsterdam.

Driscoll, P., R.M. Escorihuela, A. FernandezTeruel, O. Giorgi, H. Schwegler, et al. (1998) Genetic selection and differential stress responses—the Roman lines/strains of rats. *Ann N Y Acad Sci* 851: 501–10.

Dulac, C., L.A. O'Connell, and Z. Wu (2014) Neural control of maternal and paternal behaviors. *Science* 345: 765–70.

Duman, R.S., G.R. Heninger, and E.J. Nestler (1997) A molecular and cellular theory of depression. *Arch Gen Psychiatry* 54: 597–606.

Duncan, L., Z. Yilmaz, H. Gaspar, R. Walters, J. Goldstein, et al. (2017) Significant locus and metabolic genetic correlations revealed in genome-wide association study of anorexia nervosa. *Am J Psychiatry* 174: 850–8.

Dunn, A.J. and C.W. Berridge (1990) Physiological and behavioral-responses to corticotropin-releasing factor administration—is CRF a mediator of anxiety or stress responses. *Brain Res Rev* 15: 71–100.

Durand, C.M., C. Betancur, T.M. Boeckers, J. Bockmann, P. Chaste, et al. (2007) Mutations in the gene encoding the synaptic scaffolding protein SHANK3 are associated with autism spectrum disorders. *Nat Genet* 39: 25–7.

Eaves, L.J., H.J. Eysenck, and N.G. Martin (1989) *Genes, Culture and Personality: An Empirical Approach.* Academic Press, Harcourt Brace Jovanovich, London.

Eaves, L.J., N.G. Martin and A.C. Heath (1990) Religious affiliation in twins and their parents—testing a model of cultural inheritance. *Behav Genet* 20: 1–22.

Egeland, J.A., D.S. Gerhard, D.L. Pauls, J.N. Sussex, K.K. Kidd, et al. (1987) Bipolar affective disorders linked to DNA markers on chromosome 11. *Nature* 325: 783–7.

El Yacoubi, M., S. Bouali, D. Popa, L. Naudon, I. Leroux-Nicollet, et al. (2003) Behavioral, neurochemical, and electrophysiological characterization of a genetic mouse model of depression. *Proc Natl Acad Sci U S A* 100: 6227–32.

ENCODE (2012) An integrated encyclopedia of DNA elements in the human genome. *Nature* 489: 57–74.

Engreitz, J.M., J.E. Haines, E.M. Perez, G. Munson, J. Chen, et al. (2016a) Local regulation of gene expression by lncRNA promoters, transcription and splicing. *Nature* 539: 452–5.

Engreitz, J.M., N. Ollikainen, and M. Guttman (2016b) Long non-coding RNAs: spatial amplifiers that control nuclear structure and gene expression. *Nat Rev Mol Cell Biol* 17: 756–70.

Ernst, J., P. Kheradpour, T.S. Mikkelsen, N. Shoresh, L.D. Ward, et al. (2011) Mapping and analysis of chromatin state dynamics in nine human cell types. *Nature* 473: 43–9.

Evans, M.J. and M.H. Kaufman (1981) Establishment in culture of pluripotential cells from mouse embryos. *Nature* 292: 154–6.

Eysenck, H. (1990) *Rebel with a Cause*. W.H. Allen & Co., London.

Eysenck, H.J. (1957) *Sense and Nonsense in Psychology*. Penguin, London.

Eysenck, H.J. (1970) The classification of depressive illnesses. *Br J Psychiatry* 117: 241–50.

Falconer, D.S. and T.F.C. Mackay (1996) *Quantitative Genetics*. Longman, Harlow, UK.

Farrell, M.S., T. Werge, P. Sklar, M.J. Owen, R.A. Ophoff, et al. (2015) Evaluating historical candidate genes for schizophrenia. *Mol Psychiatry* 20: 555–62.

Fehr, C., N. Grintschuk, A. Szegedi, I. Anghelescu, C. Klawe, et al. (2000) The HTR1B 861G>C receptor polymorphism among patients suffering from alcoholism, major depression, anxiety disorders and narcolepsy. *Psychiatry Res* 97: 1–10.

Feighner, J.P., E. Robins, S.B. Guze, R.A. Woodruff, Jr., G. Winokur, et al. (1972) Diagnostic criteria for use in psychiatric research. *Arch Gen Psychiatry* 26: 57–63.

Felsenfeld, G. (2014) A brief history of epigenetics. *Cold Spring Harb Perspect Biol* 6: a018200.

Fergusson, D.M., L.J. Horwood, A.L. Miller, and M.A. Kennedy (2011) Life stress, 5-HTTLPR and mental disorder: findings from a 30-year longitudinal study. *Br J Psychiatry* 198: 129–35.

Fernando, A.B. and T.W. Robbins (2011) Animal models of neuropsychiatric disorders. *Annu Rev Clin Psychol* 7: 39–61.

Ferre, P., A. Fernandezteruel, R.M. Escorihuela, P. Driscoll, M.G. Corda, et al. (1995) Behavior of the Roman/Verh high-avoidance and low-avoidance rat lines in anxiety tests: relationship with defecation and self-grooming. *Physiol Behav* 58: 1209–13.

Ferreira, M.A., M.C. O'Donovan, Y.A. Meng, I.R. Jones, D.M. Ruderfer, et al. (2008) Collaborative genome-wide association analysis supports a role for *ANK3* and *CACNA1C* in bipolar disorder. *Nat Genet* 40: 1056–8.

Feynman, R. (1998) *The Meaning of it All: Thoughts of a Citizen Scientist*. Basic Books, New York.

Fink, S., L. Excoffier, and G. Heckel (2006) Mammalian monogamy is not controlled by a single gene. *Proc Natl Acad Sci U S A* 103: 10956–60.

Finucane, H.K., B. Bulik-Sullivan, A. Gusev, G. Trynka, Y. Reshef, et al. (2015) Partitioning heritability by functional annotation using genome-wide association summary statistics. *Nat Genet* 47: 1228–35.

Fischer, E.P. (2007) Max Delbruck. *Genetics* 177: 673–6.

Fischer, J., L. Koch, C. Emmerling, J. Vierkotten, T. Peters, et al. (2009) Inactivation of the *Fto* gene protects from obesity. *Nature* 458: 894–8.

Fisher, R.A. (1918) The correlation between relatives on the supposition of Mendelian inheritance. *Trans R Soc Edinb* 52: 399–433.

Fisher, R.A. (1925) *Statistical Methods for Research Workers*. Oliver & Boyd, London.

Fisher, S.E., F. VarghaKhadem, K.E. Watkins, A.P. Monaco and M.E. Pembrey (1998) Localisation of a gene implicated in a severe speech and language disorder. *Nat Genet* 18: 168–70.

Fitzpatrick, M.J., E. Feder, L. Rowe, and M.B. Sokolowski (2007) Maintaining a behaviour polymorphism by frequency-dependent selection on a single gene. *Nature* 447: 210–12.

Flemming, W., 1882 *Zellsubstanz, Kern und Zelltheilung*. F.C.W. Vogel, Leipzig, Germany.

Flint, J., R. Corley, J.C. DeFries, D.W. Fulker, J.A. Gray, et al. (1995) A simple genetic basis for a complex psychological trait in laboratory mice. *Science* 269: 1432–5.

Flint, J. and E. Eskin (2012) Genome-wide association studies in mice. *Nat Rev Genet* 13: 807–17.

Flint, J. and K.S. Kendler (2014) The genetics of major depression. *Neuron* 81: 484–503.

Flint, J. and T.F. Mackay (2009) Genetic architecture of quantitative traits in mice, flies and humans. *Genome Res* 19: 723–33.

Flint, J. and R. Mott (2008) Applying mouse complex-trait resources to behavioural genetics. *Nature* 456: 724–7.

Flint, J. and M.R. Munafò (2007) The endophenotype concept in psychiatric genetics. *Psychol Med* 37: 163–80.

Flynn, J.R. (1984) The mean IQ of Americans: massive gains 1932 to 1978. *Psychol Bull* 95; 29–51.

Fodor, S.P., J.L. Read, M.C. Pirrung, L. Stryer, A.T. Lu, et al. (1991) Light-directed, spatially addressable parallel chemical synthesis. *Science* 251: 767–73.

Folstein, S. and M. Rutter (1977) Genetic influences and infantile autism. *Nature* 265: 726–8.

Fombonne, E. (2009) Epidemiology of pervasive developmental disorders. *Pediatr Res* 65: 591–8.

Ford, E.B. (1940) Polymorphism and taxonomy. In *The New Systematics*, pp. 493–513. Edited by J. Huxley. Oxford University Press, Oxford, UK.

Foucault, M. (1965) *Madness and Civilization. A History of Insanity in the Age of Reason*. Pantheon Books, New York.

Fraga, M.F., E. Ballestar, M.F. Paz, S. Ropero, F. Setien, et al. (2005) Epigenetic differences arise during the lifetime of monozygotic twins. *Proc Natl Acad Sci U S A* 102: 10604–9.

Francis, D., J. Diorio, D. Liu, and M.J. Meaney (1999) Nongenomic transmission across generations of maternal behavior and stress responses in the rat. *Science* 286: 1155–8.

Francis, D.D., K. Szegda, G. Campbell, W.D. Martin, and T.R. Insel (2003) Epigenetic sources of behavioral differences in mice. *Nat Neurosci* 6: 445–6.

Frayling, T.M., N.J. Timpson, M.N. Weedon, E. Zeggini, R.M. Freathy, et al. (2007) A common variant in the *FTO* gene is associated with body mass index and predisposes to childhood and adult obesity. *Science* 316: 889–94.

Fu, Y.H., D.P. Kuhl, A. Pizzuti, M. Pieretti, J.S. Sutcliffe, et al. (1991) Variation of the CGG repeat at the fragile X site results in genetic instability: resolution of the Sherman paradox. *Cell* 67: 1047–58.

Fulco, C.P., M. Munschauer, R. Anyoha, G. Munson, S.R. Grossman, et al. (2016) Systematic mapping of functional enhancer-promoter connections with CRISPR interference. *Science* 354: 769–73.

Fulco, C.P., J. Nasser, T.J. Jones, G. Munson, D.T. Bergman, et al. (2019) Activity-by-Contact model of enhancer specificity from thousands of CRISPR perturbations. *bioRxiv* https://doi.org/10.1101/529990.

Galton, F. (1869) *Hereditary Genius, an Enquiry into its Laws and Consequences*. Macmillan and Co., London.

Galton, F. (1876) The history of twins, as a criterion of the relative powers of nature and nurture. *J Anthropol Inst GB Irel* 5: 391–406.

Galton, F. (1886) Regression towards mediocrity in hereditary stature. *J Anthropol Inst GB Irel* 15: 246–63.

Gamazon, E.R., H.E. Wheeler, K.P. Shah, S.V. Mozaffari, K. Aquino-Michaels, et al. (2015) A gene-based association method for mapping traits using reference transcriptome data. *Nat Genet* 47: 1091–8.

Gapp, K., A. Jawaid, P. Sarkies, J. Bohacek, P. Pelczar, et al. (2014) Implication of sperm RNAs in transgenerational inheritance of the effects of early trauma in mice. *Nat Neurosci* 17: 667–9.

Garneau, J.E., M.E. Dupuis, M. Villion, D.A. Romero, R. Barrangou, et al. (2010) The CRISPR/Cas bacterial immune system cleaves bacteriophage and plasmid DNA. *Nature* 468: 67–71.

Garner, A.R., D.C. Rowland, S.Y. Hwang, K. Baumgaertel, B.L. Roth, et al. (2012) Generation of a synthetic memory trace. *Science* 335: 1513–16.

Gaugler, T., L. Klei, S.J. Sanders, C.A. Bodea, A.P. Goldberg, et al. (2014) Most genetic risk for autism resides with common variation. *Nat Genet* 46: 881–5.

Gazal, S., H.K. Finucane, N.A. Furlotte, P.R. Loh, P.F. Palamara, et al. (2017) Linkage disequilibrium-dependent architecture of human complex traits shows action of negative selection. *Nat Genet* 49: 1421–7.

Gelernter, J., H.R. Kranzler, R. Sherva, L. Almasy, R. Koesterer, et al. (2014) Genome-wide association study of alcohol dependence: significant findings in African- and European-Americans including novel risk loci. *Mol Psychiatry* 19: 41–9.

Gelman, A. and J. Hill (2007) *Data Analysis Using Regression and multilevel/Hierarchical Models*. Cambridge University Press, Cambridge, UK.

Getz, L.L., K.O. Madan, J.E. Hofmann, B. McGuire, and A. Ozgul (2005) Factors influencing movement distances of two species of sympatric voles. *J Mammal* 86: 647–54.

Giaever, G., A.M. Chu, L. Ni, C. Connelly, L. Riles, et al. (2002) Functional profiling of the *Saccharomyces cerevisiae* genome. *Nature* 418: 387–91.

Gibbons, R.J., D.J. Picketts, L. Villard, and D.R. Higgs (1995) Mutations in a putative global transcriptional regulator cause X-linked mental retardation with α-thalassemia (ATR-X syndrome). *Cell* 80: 837–45.

Gigerenzer, G. (2004) Mindless statistics. *J Socio Econ* 33: 587–606.

Gilbert, D.T., G. King, S. Pettigrew, and T.D. Wilson (2016) Comment on "Estimating the reproducibility of psychological science". *Science* 351: 1037.

Gillespie, R.D. (1929) The clinical differentiation of types of depression. *Guys Hosp Rep* 9: 1109–14.

Gilman, S.R., I. Iossifov, D. Levy, M. Ronemus, M. Wigler, et al. (2011) Rare *de novo* variants associated with autism implicate a large functional network of genes involved in formation and function of synapses. *Neuron* 70: 898–907.

Girard, S.L., J. Gauthier, A. Noreau, L. Xiong, S. Zhou, et al. (2011) Increased exonic *de novo* mutation rate in individuals with schizophrenia. *Nat Genet* 43: 860–3.

Girirajan, S., Z. Brkanac, B.P. Coe, C. Baker, L. Vives, et al. (2011a) Relative burden of large CNVs on a range of neurodevelopmental phenotypes. *PLoS Genet* 7: e1002334.

Girirajan, S., C.D. Campbell, and E.E. Eichler (2011b) Human copy number variation and complex genetic disease. *Annu Rev Genet* 45: 203–26.

Girirajan, S., J.A. Rosenfeld, B.P. Coe, S. Parikh, N. Friedman, et al. (2012) Phenotypic heterogeneity of genomic disorders and rare copy-number variants. *N Engl J Med* 367: 1321–31.

Girirajan, S., J.A. Rosenfeld, G.M. Cooper, F. Antonacci, P. Siswara, et al. (2010) A recurrent 16p12.1 microdeletion supports a two-hit model for severe developmental delay. *Nat Genet* 42: 203–9.

Gitschier, J., G.R. Strichartz, and L.M. Hall (1980) Saxitoxin binding to sodium channels in head extracts from wild-type and tetrodotoxin-sensitive strains of *Drosophila melanogaster*. *Biochim Biophys Acta* 595: 291–303.

Glessner, J.T., K. Wang, G. Cai, O. Korvatska, C.E. Kim, et al. (2009) Autism genome-wide copy number variation reveals ubiquitin and neuronal genes. *Nature* 459: 569–73.

Gluckman, P.D., M.A. Hanson, C. Cooper, and K.L. Thornburg (2008) Effect of in utero and early-life conditions on adult health and disease. *N Engl J Med* 359: 61–73.

Goldsmith, H.H. (1983) Genetic influences on personality from infancy to adulthood. *Child Dev* 54: 331–55.

Gong, S., C. Zheng, M.L. Doughty, K. Losos, N. Didkovsky, et al. (2003) A gene expression atlas of the central nervous system based on bacterial artificial chromosomes. *Nature* 425: 917–25.

Goodfellow, P.N. (1993) Planting alfalfa and cloning the Huntington's disease gene. *Cell* 72: 817–18.

Goodson, J.L. and A.H. Bass (2000) Forebrain peptides modulate sexually polymorphic vocal circuitry. *Nature* 403: 769–72.

Goodson, J.L., S.E. Schrock, J.D. Klatt, D. Kabelik, and M.A. Kingsbury (2009) Mesotocin and nonapeptide receptors promote estrildid flocking behavior. *Science* 325: 862–6.

Gossen, M. and H. Bujard (1992) Tight control of gene expression in mammalian cells by tetracycline-responsive promoters. *Proc Natl Acad Sci U S A* 89: 5547–51.

Gotham, K., S. Risi, A. Pickles, and C. Lord (2007) The Autism Diagnostic Observation Schedule: revised algorithms for improved diagnostic validity. *J Autism Dev Disord* 37: 613–27.

Gottesman, I.I. and T.D. Gould (2003) The endophenotype concept in psychiatry: etymology and strategic intentions. *Am J Psychiatry* 160: 636–45.

Gottesman, I.I. and J. Shields (1967) A polygenic theory of schizophrenia. *Proc Natl Acad Sci U S A* 58: 199–205.

Gould, S. J. (1981) *The Mismeasure of Man*. W. W. Norton & Company, New York.

Grant, B.F., R.B. Goldstein, T.D. Saha, S.P. Chou, J. Jung, et al. (2015) Epidemiology of DSM-5 alcohol use disorder: results from the National Epidemiologic Survey on Alcohol and Related Conditions III. *JAMA Psychiatry* 72: 757–66.

Grant, S.G. N., T.J. O'Dell, K.A. Karl, P.L. Stein, P. Soriano, et al. (1992) Impaired long-term potentiation, spatial learning and hippocampal development in *fyn* mutant mice. *Science* 258: 1903–10.

Gray, J.A. (1982) *The Neuropsychology of Anxiety: an Enquiry into the Function of the Septo-hippocampal System*. Oxford University Press, Oxford, UK.

Greenblatt, E.J. and A.C. Spradling (2018) Fragile X mental retardation 1 gene enhances the translation of large autism-related proteins. *Science* 361: 709–12.

Greene, J.S., M. Brown, M. Dobosiewicz, I.G. Ishida, E.Z. Macosko, et al. (2016) Balancing selection shapes density-dependent foraging behaviour. *Nature* 539: 254–8.

Greenspan, R.J. (2004) E pluribus unum, ex uno plura: quantitative and single-gene perspectives on the study of behavior. *Annu Rev Neurosci* 27: 79–105.

Greenspan, R.J. (2008) The origins of behavioral genetics. *Curr Biol* 18: R192–8.

Greiss, S. and J.W. Chin (2011) Expanding the genetic code of an animal. *J Am Chem Soc* 133: 14196–9.

Griffing, B. (1981) A theory of natural selection incorporating interaction among individuals. I. The modeling process. *J Theor Biol* 89: 635–58.

Griggs, J.L., P. Sinnayah, and M.L. Mathai (2015) Prader-Willi syndrome: from genetics to behaviour, with special focus on appetite treatments. *Neurosci Biobehav Rev* 59: 155–72.

Gross, C., X. Zhuang, K. Stark, S. Ramboz, R. Oosting, et al. (2002) Serotonin1A receptor acts during development to establish normal anxiety-like behaviour in the adult. *Nature* 416: 396–400.

Grove, J., S. Ripke, T.D. Als, M. Mattheisen, R.K. Walters, et al. (2019) Identification of common genetic risk variants for autism spectrum disorder. *Nat Genet* 51: 431–44.

Grover, D., T. Katsuki, and R.J. Greenspan (2016) Flyception: imaging brain activity in freely walking fruit flies. *Nat Methods* 13: 569–72.

GTEx Consortium (2017) Genetic effects on gene expression across human tissues. *Nature* 550: 204–13.

Gupta, R.A., N. Shah, K.C. Wang, J. Kim, H.M. Horlings, et al. (2010) Long non-coding RNA *HOTAIR* reprograms chromatin state to promote cancer metastasis. *Nature* 464: 1071–6.

Gusella, J.F., N.S. Wexler, P.M. Conneally, S.L. Naylor, M.A. Anderson, et al. (1983) A polymorphic DNA marker genetically linked to Huntington's disease. *Nature* 306: 234–8.

Gusev, A., A. Ko, H. Shi, G. Bhatia, W. Chung, et al. (2016) Integrative approaches for large-scale transcriptome-wide association studies. *Nat Genet* 48: 245–52.

Gusev, A., S.H. Lee, G. Trynka, H. Finucane, B.J. Vilhjalmsson, et al. (2014) Partitioning heritability of regulatory and cell-type-specific variants across 11 common diseases. *Am J Hum Genet* 95: 535–52.

Gusev, A., N. Mancuso, H. Won, M. Kousi, H.K. Finucane, et al. (2018) Transcriptome-wide association study of schizophrenia and chromatin activity yields mechanistic disease insights. *Nat Genet* 50: 538–48.

Guttman, M. and J.L. Rinn (2012) Modular regulatory principles of large non-coding RNAs. *Nature* 482: 339–46.

Guy, J., J. Gan, J. Selfridge, S. Cobb, and A. Bird (2007) Reversal of neurological defects in a mouse model of Rett syndrome. *Science* 315: 1143–7.

Guzowski, J.F., B.L. McNaughton, C.A. Barnes, and P.F. Worley (1999) Environment-specific expression of the immediate-early gene *Arc* in hippocampal neuronal ensembles. *Nat Neurosci* 2: 1120–4.

Hagenaars, S.P., S.E. Harris, G. Davies, W.D. Hill, D.C. Liewald, et al. (2016) Shared genetic aetiology between cognitive functions and physical and mental health in UK Biobank (N=112 151) and 24 GWAS consortia. *Mol Psychiatry* 21: 1624–32.

Hall, J.M., M.K. Lee, B. Newman, J.E. Morrow, L.A. Anderson, et al. (1990) Linkage of early-onset familial breast cancer to chromosome 17q21. *Science* 250: 1684–9.

Haller, H. and S. Krauss (2002) Misinterpretations of significance: a problem students share with their teachers? *Methods Psychol Res* 7: 1–20.

Halligan, D.L., F. Oliver, A. Eyre-Walker, B. Harr, and P.D. Keightley (2010) Evidence for pervasive adaptive protein evolution in wild mice. *PLoS Genet* 6: e1000825.

Hammock, E.A. and L.J. Young (2005) Microsatellite instability generates diversity in brain and sociobehavioral traits. *Science* 308: 1630–4.

Han, J.H., S.A. Kushner, A.P. Yiu, H.L. Hsiang, T. Buch, et al. (2009) Selective erasure of a fear memory. *Science* 323: 1492–6.

Han, M. and M. Grunstein (1988) Nucleosome loss activates yeast downstream promoters *in vivo*. *Cell* 55: 1137–45.

Han, S., C. Tai, R.E. Westenbroek, F.H. Yu, C.S. Cheah, et al. (2012) Autistic-like behaviour in Scn1a+/− mice and rescue by enhanced GABA-mediated neurotransmission. *Nature* 489: 385–90.

Harden, K.P., B.W. Domingue, D.W. Belsky, J.D. Boardman, R. Crosnoe, et al. (2019) Genetic associations with mathematics tracking and persistence in secondary school. *bioRxiv* http://dx.doi.org/10.1101/598532.

Hardin, P.E. (2005) The circadian timekeeping system of *Drosophila*. *Curr Biol* 15: R714–22.

Hardin, P.E., J.C. Hall, and M. Rosbash (1990) Feedback of the *Drosophila period* gene product on circadian cycling of its messenger RNA levels. *Nature* 343: 536–40.

Hardin, P.E., J.C. Hall, and M. Rosbash (1992) Circadian oscillations in period gene mRNA levels are transcriptionally regulated. *Proc Natl Acad Sci U S A* 89: 11711–15.

Harris, W.A., W.S. Stark, and J.A. Walker (1976) Genetic dissection of the photoreceptor system in the compound eye of *Drosophila melanogaster*. *J Physiol* 256: 415–39.

Harrison, P.A. and A.C. Sidebottom (2009) Alcohol and drug use before and during pregnancy: an examination of use patterns and predictors of cessation. *Matern Child Health J* 13: 386–94.

Hatzimanolis, A., D. Avramopoulos, D.E. Arking, A. Moes, P. Bhatnagar, et al. (2018) Stress-dependent association between polygenic risk for schizophrenia and schizotypal traits in young army recruits. *Schizophr Bull* 44: 338–47.

Haworth, C.M., M.J. Wright, M. Luciano, N.G. Martin, E.J. de Geus, et al. (2010) The heritability of general cognitive ability increases linearly from childhood to young adulthood. *Mol Psychiatry* 15: 1112–20.

Heard, E. and R.A. Martienssen (2014) Transgenerational epigenetic inheritance: myths and mechanisms. *Cell* 157: 95–109.

Heath, A.C., K.K. Bucholz, P.A.F. Madden, S.H. Dinwiddie, W.S. Slutske, et al. (1997) Genetic and environmental contributions to alcohol dependence risk in a national twin sample: consistency of findings in women and men. *Psychol Med* 27: 1381–96.

Hebb, D.O. (1949) *The Organization of Behavior*. Wiley, New York.

Helbig, I., H.C. Mefford, A.J. Sharp, M. Guipponi, M. Fichera, et al. (2009) 15q13.3 microdeletions increase risk of idiopathic generalized epilepsy. *Nat Genet* 41: 160–2.

Heller, E.A., H.M. Cates, C.J. Pena, H. Sun, N. Shao, et al. (2014) Locus-specific epigenetic remodeling controls addiction- and depression-related behaviors. *Nat Neurosci* 17: 1720–7.

Henderson, C.R., O. Kempthorne, S.R. Searle, and C.M. von Krosigk (1959) The estimation of environmental and genetic trends from records subject to culling. *Biometrics* 15: 192–218.

Henderson, N.D. (1989) Interpreting studies that compare high- and low-selected lines on new characters. *Behav Genet* 19: 473–502.

Herrnstein, R.J. and C. Murray (1994) *The Bell Curve: Intelligence and Class Structure in American Life*. Simon & Schuster, New York.

Heston, L.L. (1966) Psychiatric disorders in foster home reared children of schizophrenic mothers. *Br J Psychiatry* 112: 819–25.

Hettema, J.M., M.C. Neale, and K.S. Kendler (2001) A review and meta-analysis of the genetic epidemiology of anxiety disorders. *Am J Psychiatry* 158: 1568–78.

Hibar, D.P., J.L. Stein, M.E. Renteria, A. Arias-Vasquez, S. Desrivieres, et al. (2015) Common genetic variants influence human subcortical brain structures. *Nature* 520: 224–9.

Hill, W.D., R.C. Arslan, C. Xia, M. Luciano, C. Amador, et al. (2018) Genomic analysis of family data reveals additional genetic effects on intelligence and personality. *Mol Psychiatry* 23: 2347–62.

Hill, W.L. (1991) Correlates of male mating success in the ruff *Philomachus pugnax*, a lekking shorebird. *Behav Ecol Sociobiol* 29: 367–72.

Hindorff, L.A., P. Sethupathy, H.A. Junkins, E.M. Ramos, J.P. Mehta, et al. (2009) Potential etiologic and functional implications of genome-wide association loci for human diseases and traits. *Proc Natl Acad Sci U S A* 106: 9362–7.

Hirsch, J. (1959) Studies in experimental behavior genetics. II. Individual differences in geotaxis as a function of chromosome variations in synthesized *Drosophila* populations. *J Comp Physiol Psychol* 52: 304–8.

Hirsch, J. and J.C. Boudreau (1958) Studies in experimental behavior genetics. I. The heritability of phototaxis in a population of *Drosophila melanogaster*. *J Comp Physiol Psychol* 51: 647–51.

Hirsch, J. and L. Erlenmeyer-Kimling (1962) Studies in experimental behavior genetics: IV. Chromosome analyses for geotaxis. *J Comp Physiol Psychol* 55: 732–9.

Hirsch, J. and G. Ksander (1969) Studies in experimental behavior genetics. V. Negative geotaxis and further chromosome analyses in *Drosophila melanogaster*. *J Comp Physiol Psychol* 67: 118–22.

Holland, A.J., A. Hall, R. Murray, G.F. Russell, and A.H. Crisp (1984) Anorexia nervosa: a study of 34 twin pairs and one set of triplets. *Br J Psychiatry* 145: 414–19.

Holsboer, F. (2000) The corticosteroid receptor hypothesis of depression. *Neuropsychopharmacology* 23: 477–501.

Howie, B.N., P. Donnelly, and J. Marchini (2009) A flexible and accurate genotype imputation method for the next generation of genome-wide association studies. *PLoS Genet* 5: e1000529.

Hughes, J.R., D.K. Hatsukami, J.E. Mitchell, and L.A. Dahlgren (1986) Prevalence of smoking among psychiatric outpatients. *Am J Psychiatry* 143: 993–7.

Hunkapiller, T., R.J. Kaiser, B.F. Koop, and L. Hood (1991) Large-scale and automated DNA sequence determination. *Science* 254: 59–67.

Hunt, M. (1997) *How Science Takes Stock: The Story of Meta-analysis*. Russell Sage Foundation, New York.

Hurst, J.A., M. Baraitser, E. Auger, F. Graham, and S. Norell (1990) An extended family with a dominantly inherited speech disorder. *Dev Med Child Neurol* 32: 347–55.

Huxley, R., A. Neil, and R. Collins (2002) Unravelling the fetal origins hypothesis: is there really an inverse association between birthweight and subsequent blood pressure? *Lancet* 360: 659–65.

Hyde, C.L., M.W. Nagle, C. Tian, X. Chen, S.A. Paciga, et al. (2016) Identification of 15 genetic loci associated with risk of major depression in individuals of European descent. *Nat Genet* 48: 1031–6.

Hyun, S., Y. Lee, S.T. Hong, S. Bang, D. Paik, et al. (2005) *Drosophila* GPCR Han is a receptor for the circadian clock neuropeptide PDF. *Neuron* 48: 267–78.

Ibarra-Soria, X., M.O. Levitin, L.R. Saraiva, and D.W. Logan (2014) The olfactory transcriptomes of mice. *PLoS Genet* 10: e1004593.

Insel, T.R., Z.X. Wang, and C.F. Ferris (1994) Patterns of brain vasopressin receptor distribution associated with social organization in microtine rodents. *J Neurosci* 14: 5381–92.

International Molecular Genetic Study of Autism Consortium (1998) A full genome screen for autism with evidence for linkage to a region on chromosome 7q. *Hum Mol Genet* 7: 571–78.

International Schizophrenia Consortium (2008) Rare chromosomal deletions and duplications increase risk of schizophrenia. *Nature* 455: 237–41.

Ioannidis, J.P. (2005) Why most published research findings are false. *PLoS Med* 2: e124.

Ioannidis, J.P. (2006) Common genetic variants for breast cancer: 32 largely refuted candidates and larger prospects. *J Natl Cancer Inst* 98: 1350–3.

Ioannidis, J.P. (2008) Why most discovered true associations are inflated. *Epidemiology* 19: 640–8.

Ioannidis, J.P. (2019) The importance of predefined rules and prespecified statistical analyses: do not abandon significance. *JAMA* 321: 2067–8.

Ioannidis, J.P., E.E. Ntzani, T.A. Trikalinos, and D.G. Contopoulos-Ioannidis (2001) Replication validity of genetic association studies. *Nat Genet* 29: 306–9.

Iossifov, I., B.J. O'Roak, S.J. Sanders, M. Ronemus, N. Krumm, et al. (2014) The contribution of *de novo* coding mutations to autism spectrum disorder. *Nature* 515: 216–21.

Iossifov, I., M. Ronemus, D. Levy, Z. Wang, I. Hakker, et al. (2012) *De novo* gene disruptions in children on the autistic spectrum. *Neuron* 74: 285–99.

Jacquemont, M.L., D. Sanlaville, R. Redon, O. Raoul, V. Cormier-Daire, et al. (2006) Array-based comparative genomic hybridisation identifies high frequency of cryptic chromosomal rearrangements in patients with syndromic autism spectrum disorders. *J Med Genet* 43: 843–9.

Jamain, S., H. Quach, C. Betancur, M. Rastam, C. Colineaux, et al. (2003) Mutations of the X-linked genes encoding neuroligins NLGN3 and NLGN4 are associated with autism. *Nat Genet* 34: 27–9.

Jandura, A. and H.M. Krause (2017) The new RNA world: growing evidence for long noncoding RNA functionality. *Trends Genet* 33: 665–76.

Jansen, P.R., K. Watanabe, S. Stringer, N. Skene, J. Bryois, et al. (2018) Genome-wide analysis of insomnia (*N*=1,331,010) identifies novel loci and functional pathways. *bioRxiv* http://dx.doi.org/10.1101/214973.

Jansen, R., J.D. Embden, W. Gaastra and L.M. Schouls (2002) Identification of genes that are associated with DNA repeats in prokaryotes. *Mol Microbiol* 43: 1565–75.

Jay, P., A. Whibley, L. Frezal, M.A. Rodriguez de Cara, R.W. Nowell, et al. (2018) Supergene evolution triggered by the introgression of a chromosomal inversion. *Curr Biol* 28: 1839–45.e3.

Ji, W., J.N. Foo, B.J. O'Roak, H. Zhao, M.G. Larson, et al. (2008) Rare independent mutations in renal salt handling genes contribute to blood pressure variation. *Nat Genet* 40: 592–9.

Jinek, M., K. Chylinski, I. Fonfara, M. Hauer, J.A. Doudna, et al. (2012) A programmable dual-RNA-guided DNA endonuclease in adaptive bacterial immunity. *Science* 337: 816–21.

Jinek, M., A. East, A. Cheng, S. Lin, E. Ma, et al. (2013) RNA-programmed genome editing in human cells. *Elife* 2: e00471.

Jing, H., C.R. Vakoc, L. Ying, S. Mandat, H. Wang, et al. (2008) Exchange of GATA factors mediates transitions in looped chromatin organization at a developmentally regulated gene locus. *Mol Cell* 29: 232–42.

John, O.P. and S. Srivastava (1999) The Big-Five trait taxonomy: history, measurement, and theoretical perspectives. In *Handbook of Personality: Theory and Research*, Vol. 2, pp. 102–38. Edited by L.A. Pervin and O.P. John. Guilford Press, New York.

Jordan, K.W., M.A. Carbone, A. Yamamoto, T.J. Morgan, and T.F. Mackay (2007) Quantitative genomics of locomotor behavior in *Drosophila melanogaster*. *Genome Biol* 8: R172.

Jorgenson, E., K.K. Thai, T.J. Hoffmann, L.C. Sakoda, M.N. Kvale, et al. (2017) Genetic contributors to variation in alcohol consumption vary by race/ethnicity in a large multi-ethnic genome-wide association study. *Mol Psychiatry* 22: 1359–67.

Joron, M., L. Frezal, R.T. Jones, N.L. Chamberlain, S.F. Lee, et al. (2011) Chromosomal rearrangements maintain a polymorphic supergene controlling butterfly mimicry. *Nature* 477: 203–6.

Joyce, J. (1914) *Dubliners*. Grant Richards, London.

Kadauke, S. and G.A. Blobel (2009) Chromatin loops in gene regulation. *Biochim Biophys Acta* 1789: 17–25.

Kamin, L.J. (1974) *The Science and Politics of IQ*. John Wiley & Sons, New York.

Kaminsky, Z.A., T. Tang, S.C. Wang, C. Ptak, G.H. Oh, et al. (2009) DNA methylation profiles in monozygotic and dizygotic twins. *Nat Genet* 41: 240–5.

Kandel, E.R. (2001) The molecular biology of memory storage: a dialogue between genes and synapses. *Science* 294: 1030–8.

Kang, H.M., N.A. Zaitlen, C.M. Wade, A. Kirby, D. Heckerman, et al. (2008) Efficient control of population structure in model organism association mapping. *Genetics* 178: 1709–23.

Kanner, L. (1943) Autistic disturbances of affective contact. *Nerv Child* 2: 217–50.

Kanner, L. (1949) Problems of nosology and psychodynamics of early infantile autism. *Am J Orthopsychiatry* 19: 416–26.

Kapheim, K.M., H. Pan, C. Li, S.L. Salzberg, D. Puiu, et al. (2015) Social evolution. Genomic signatures of evolutionary transitions from solitary to group living. *Science* 348: 1139–43.

Karayiorgou, M., M.A. Morris, B. Morrow, R.J. Shprintzen, R. Goldberg, et al. (1995) Schizophrenia susceptibility associated with interstitial deletions of chromosome 22q11. *Proc Natl Acad Sci U S A* 17: 7612–16.

Karg, K., M. Burmeister, K. Shedden, and S. Sen (2011) The serotonin transporter promoter variant (5-HTTLPR), stress, and depression meta-analysis revisited: evidence of genetic moderation. *Arch Gen Psychiatry* 68: 444–54.

Kass, R.E. and A.E. Raftery (1995) Bayes factors. *J Am Stat Assoc* 90: 773–95.

Katayama, S., Y. Tomaru, T. Kasukawa, K. Waki, M. Nakanishi, et al. (2005) Antisense transcription in the mammalian transcriptome. *Science* 309: 1564–6.

Kato, G.J., F.B. Piel, C.D. Reid, M.H. Gaston, K. Ohene-Frempong, et al. (2018) Sickle cell disease. *Nat Rev Dis Primers* 4: 18010.

Kazachenka, A., T.M. Bertozzi, M.K. Sjoberg-Herrera, N. Walker, J. Gardner, et al. (2018) Identification, characterization, and heritability of murine metastable epialleles: implications for non-genetic inheritance. *Cell* 175: 1259–71.e13.

Keane, T.M., L. Goodstadt, P. Danecek, M.A. White, K. Wong, et al. (2011) Mouse genomic variation and its effect on phenotypes and gene regulation. *Nature* 477: 289–94.

Keays, D.A., G. Tian, K. Poirier, G.J. Huang, C. Siebold, et al. (2007) Mutations in α-tubulin cause abnormal neuronal migration in mice and lissencephaly in humans. *Cell* 128: 45–57.

Keller, L. and G. Ross (1998) Selfish genes: a green beard in the red fire ant. *Nature* 395: 573–5.

Kelly, L.E. (1974) Temperature-sensitive mutations affecting the regenerative sodium channel in *Drosophila melanogaster*. *Nature* 248: 166–8.

Kendler, K.S. (1997) Social support: a genetic-epidemiologic analysis. *Am J Psychiatry* 154: 1398–404.

Kendler, K.S. (2005) 'A gene for … ': the nature of gene action in psychiatric disorders. *Am J Psychiatry* 162: 1243–52.

Kendler, K.S. (2016) The phenomenology of major depression and the representativeness and nature of DSM criteria. *Am J Psychiatry* 173: 771–80.

Kendler, K.S., S.H. Aggen, and M.C. Neale (2013) Evidence for multiple genetic factors underlying DSM-IV criteria for major depression. *JAMA Psychiatry* 70: 599–607.

Kendler, K.S., N. Czajkowski, K. Tambs, S. Torgersen, S. Haggen, et al. (2006a) Dimensional representations of DSM-IV Cluster A personality disorders in a population-based sample of Norwegian twins: a multivariate study. *Psychol Med* 36: 1583–91.

Kendler, K.S., C.O. Gardner, M. Gatz and N.L. Pedersen (2007a) The sources of co-morbidity between major depression and generalized anxiety disorder in a Swedish national twin sample. *Psychol Med* 37: 453–62.

Kendler, K.S., C.O. Gardner, and P. Lichtenstein (2008) A developmental twin study of symptoms of anxiety and depression: evidence for genetic innovation and attenuation. *Psychol Med* 38: 1567–75.

Kendler, K.S., C.O. Gardner, M.C. Neale, and C.A. Prescott (2001a) Genetic risk factors for major depression in men and women: similar or different heritabilities and same or partly distinct genes? *Psychol Med* 31: 605–16.

Kendler, K.S., C.O. Gardner, and C.A. Prescott (1999a) Clinical characteristics of major depression that predict risk of depression in relatives. *Arch Gen Psychiatry* 56: 322–7.

Kendler, K.S., M. Gatz, C.O. Gardner, and N.L. Pedersen (2006b) Personality and major depression: a Swedish longitudinal, population-based twin study. *Arch Gen Psychiatry* 63: 1113–20.

Kendler, K.S., M. Gatz, C.O. Gardner, and N.L. Pedersen (2006c) A Swedish national twin study of lifetime major depression. *Am J Psychiatry* 163: 109–14.

Kendler, K.S. and R.J. Greenspan (2006) The nature of genetic influences on behavior: lessons from 'simpler' organisms. *Am J Psychiatry* 163: 1683–94.

Kendler, K.S., A.M. Gruenberg, and D.K. Kinney (1994a) Independent diagnoses of adoptees and relatives as defined by DSM-III in the provincial and national samples of the Danish Adoption Study of Schizophrenia. *Arch Gen Psychiatry* 51: 456–68.

Kendler, K.S., A.C. Heath, M.C. Neale, R.C. Kessler, and L.J. Eaves (1992a) A population-based twin study of alcoholism in women. *JAMA* 268: 1877–82.

Kendler, K.S., K.C. Jacobson, C.O. Gardner, N. Gillespie, S.A. Aggen, et al. (2007b) Creating a social world: a developmental twin study of peer-group deviance. *Arch Gen Psychiatry* 64: 958–65.

Kendler, K.S., J. Ji, A.C. Edwards, H. Ohlsson, J. Sundquist, et al. (2015a) An extended Swedish national adoption study of alcohol use disorder. *JAMA Psychiatry* 72: 211–18.

Kendler, K.S., L.M. Karkowski, and C.A. Prescott (1999b) The assessment of dependence in the study of stressful life events: validation using a twin design. *Psychol Med* 29: 1455–60.

Kendler, K.S. and L. Karkowski-Shuman (1997) Stressful life events and genetic liability to major depression: genetic control of exposure to the environment? *Psychol Med* 27: 539–47.

Kendler, K.S., R.C. Kessler, E.E. Walters, C. MacLean, M.C. Neale, et al. (1995a) Stressful life events, genetic liability, and onset of an episode of major depression in women. *Am J Psychiatry* 152: 833–42.

Kendler, K.S., J. Kuhn, and C.A. Prescott (2004) The interrelationship of neuroticism, sex, and stressful life events in the prediction of episodes of major depression. *Am J Psychiatry* 161: 631–6.

Kendler, K.S., M. McGuire, A.M. Gruenberg, A. O'Hare, M. Spellman, et al. (1993a) The Roscommon Family Study. III. Schizophrenia-related personality disorders in relatives. *Arch Gen Psychiatry* 50: 781–8.

Kendler, K.S., M. McGuire, A.M. Gruenberg, M. Spellman, A. O'Hare, et al. (1993b) The Roscommon Family Study. II. The risk of nonschizophrenic nonaffective psychoses in relatives. *Arch Gen Psychiatry* 50: 645–52.

Kendler, K.S., J. Myers, C.A. Prescott, and M.C. Neale (2001b) The genetic epidemiology of irrational fears and phobias in men. *Arch Gen Psychiatry* 58: 257–65.

Kendler, K.S. and M.C. Neale (2010) Endophenotype: a conceptual analysis. *Mol Psychiatry* 15: 789–97.

Kendler, K.S., M.C. Neale, R.C. Kessler, A.C. Heath and L.J. Eaves (1992b) Major depression and generalized anxiety disorder. Same genes, (partly) different environments? *Arch Gen Psychiatry* 49: 716–22.

Kendler, K.S., M.C. Neale, R.C. Kessler, A.C. Heath and L.J. Eaves (1993c) Panic disorder in women: a population-based twin study. *Psychol Med* 23: 397–406.

Kendler, K.S., M.C. Neale, R.C. Kessler, A.C. Heath, and L.J. Eaves (1993d) A twin study of recent life events and difficulties. *Arch Gen Psychiatry* 50: 589–96.

Kendler, K.S., M.C. Neale, R.C. Kessler, A.C. Heath, and L.J. Eaves (1994b) The clinical characteristics of major depression as indices of the familial risk to illness. *Br J Psychiatry* 165: 66–72.

Kendler, K.S., H. Ohlsson, R.S.E. Keefe, K. Sundquist, and J. Sundquist (2018a) The joint impact of cognitive performance in adolescence and familial cognitive aptitude on risk for major psychiatric disorders: a delineation of four potential pathways to illness. *Mol Psychiatry* 23: 1076–83.

Kendler, K.S., H. Ohlsson, P. Lichtenstein, J. Sundquist, and K. Sundquist (2018b) The genetic epidemiology of treated major depression in Sweden. *Am J Psychiatry* 175: 1137–44.

Kendler, K.S., H. Ohlsson, K. Sundquist, and J. Sundquist (2018c) Sources of parent–offspring resemblance for major depression in a national Swedish extended adoption study. *JAMA Psychiatry* 75: 194–200.

Kendler, K.S., H. Ohlsson, D.S. Svikis, K. Sundquist, and J. Sundquist (2017) The protective effect of pregnancy on risk for drug abuse: a population, co-relative, co-spouse, and within-individual analysis. *Am J Psychiatry* 174: 954–62.

Kendler, K.S., H. Ohtsson, J. Sundquist, and K. Sundquist (2015b) Triparental families: a new genetic-epidemiological design applied to drug abuse, alcohol use disorders, and criminal behavior in a Swedish national sample. *Am J Psychiatry* 172: 553–60.

Kendler, K.S., N. Pedersen, L. Johnson, M.C. Neale, and A.A. Mathe (1993e) A pilot Swedish twin study of affective illness, including hospital- and population-ascertained subsamples. *Arch Gen Psychiatry* 50: 699–700.

Kendler, K.S., N.L. Pedersen, M.C. Neale, and A.A. Mathe (1995b) A pilot Swedish twin study of affective illness including hospital- and population-ascertained subsamples: results of model fitting. *Behav Genet* 25: 217–32.

Kendler, K.S., C.A. Prescott, J. Myers, and M.C. Neale (2003) The structure of genetic and environmental risk factors for common psychiatric and substance use disorders in men and women. *Arch Gen Psychiatry* 60: 929–37.

Kendler, K.S., C.A. Prescott, M.C. Neale, and N.L. Pedersen (1997) Temperance board registration for alcohol abuse in a national sample of Swedish male twins, born 1902 to 1949. *Arch Gen Psychiatry* 54: 178–84.

Kendler, K.S., E. Turkheimer, H. Ohlsson, J. Sundquist, and K. Sundquist (2015c) Family environment and the malleability of cognitive ability: a Swedish national home-reared and adopted-away cosibling control study. *Proc Natl Acad Sci U S A* 112: 4612–17.

Kendrick, K.M., A.P. da Costa, K.D. Broad, S. Ohkura, R. Guevara, et al. (1997) Neural control of maternal behaviour and olfactory recognition of offspring. *Brain Res Bull* 44: 383–95.

Kessler, R.C. (1997) The effects of stressful life events on depression. *Ann Rev Psychol* 48: 191–214.

Kessler, R.C., P. Berglund, O. Demler, R. Jin, D. Koretz, et al. (2003) The epidemiology of major depressive disorder: results from the National Comorbidity Survey Replication (NCS-R). *JAMA* 289: 3095–105.

Kessler, R.C., W.T. Chiu, R. Jin, A.M. Ruscio, K. Shear, et al. (2006) The epidemiology of panic attacks, panic disorder, and agoraphobia in the National Comorbidity Survey Replication. *Arch Gen Psychiatry* 63: 415–24.

Kety, S.S. (1987) The significance of genetic factors in the etiology of schizophrenia: results from the national study of adoptees in Denmark. *J Psychiatr Res* 21: 423–9.

Kety, S.S., D. Rosenthal, P.H. Wender, and F. Schulsinger (1968) The types and prevalence of mental illness in the biological and adoptive families of adopted schizophrenics. *J Psychiatr Res* 6: 345–62.

Kety, S.S., P.H. Wender, B. Jacobsen, L.J. Ingraham, L. Jansson, et al. (1994) Mental illness in the biological and adoptive relatives of schizophrenic adoptees. Replication of the Copenhagen Study in the rest of Denmark. *Arch Gen Psychiatry* 51: 442–55.

Keverne, E.B. (1999) The vomeronasal organ. *Science* 286: 716–20.

Kevles, D. (1985) *In the Name of Eugenics. Genetics and the Uses of Human Heredity.* University of California Press, Los Angeles, CA.

Kida, S., S.A. Josselyn, S. Pena de Ortiz, J.H. Kogan, I. Chevere, et al. (2002) CREB required for the stability of new and reactivated fear memories. *Nat Neurosci* 5: 348–55.

King, D.P., Y. Zhao, A.M. Sangoram, L.D. Wilsbacher, M. Tanaka, et al. (1997) Positional cloning of the mouse circadian *Clock* gene. *Cell* 89: 641–53.

King, M.C. (2014) "The race" to clone *BRCA1*. *Science* 343: 1462–5.

King, M.C. and A.C. Wilson (1975) Evolution at two levels in humans and chimpanzees. *Science* 188: 107–16.

Kirby, D. (1999) *Looking for Reasons Why: The Antecedents of Adolescent Sexual Risk-Taking, Pregnancy, and Child-Bearing.* National Campaign to Prevent Teen Pregnancy, Washington, DC.

Kitamoto, T. (2001) Conditional modification of behavior in *Drosophila* by targeted expression of a temperature-sensitive shibire allele in defined neurons. *J Neurobiol* 47: 81–92.

Kleiman, D.G. (1977) Monogamy in mammals. *Q Rev Biol* 52: 39–69.

Kochinke, K., C. Zweier, B. Nijhof, M. Fenckova, P. Cizek, et al. (2016) Systematic phenomics analysis deconvolutes genes mutated in intellectual disability into biologically coherent modules. *Am J Hum Genet* 98: 149–64.

Kong, A., G. Thorleifsson, M.L. Frigge, B.J. Vilhjalmsson, A.I. Young, et al. (2018) The nature of nurture: effects of parental genotypes. *Science* 359: 424–8.

Konopka, R.J. and S. Benzer (1971) Clock mutants of *Drosophila melanogaster*. *Proc Natl Acad Sci U S A* 68: 2112–16.

Koonin, E.V., K.S. Makarova, and F. Zhang (2017) Diversity, classification and evolution of CRISPR-Cas systems. *Curr Opin Microbiol* 37: 67–78.

Kormos, V. and B. Gaszner (2013) Role of neuropeptides in anxiety, stress, and depression: from animals to humans. *Neuropeptides* 47: 401–19.

Kornberg, R.D. (1974) Chromatin structure: a repeating unit of histones and DNA. *Science* 184: 868–71.

Kornberg, R.D. and J.O. Thomas (1974) Chromatin structure; oligomers of the histones. *Science* 184: 865–8.

Kosmicki, J.A., K.E. Samocha, D.P. Howrigan, S.J. Sanders, K. Slowikowski, et al. (2017) Refining the role of *de novo* protein-truncating variants in neurodevelopmental disorders by using population reference samples. *Nat Genet* 49: 504–10.

Kraepelin, E. (1915) *Clinical Psychiatry. A Text Book for Students and Physicians.* Translated by A. Ross Diefendorf. Macmillan Company, New York.

Krieger, M.J. and K.G. Ross (2002) Identification of a major gene regulating complex social behavior. *Science* 295: 328–32.

Krishnan, V., M.H. Han, D.L. Graham, O. Berton, W. Renthal, et al. (2007) Molecular adaptations underlying susceptibility and resistance to social defeat in brain reward regions. *Cell* 131: 391–404.

Kronenberg, Z.N., I.T. Fiddes, D. Gordon, S. Murali, S. Cantsilieris, et al. (2018) High-resolution comparative analysis of great ape genomes. *Science* 360: eaar6343.

Kruglyak, L. (1999) Prospects for whole-genome linkage disequilibrium mapping of common disease genes. *Nat Genet* 22: 139–44.

Krumm, N., B.J. O'Roak, J. Shendure, and E.E. Eichler (2014) A *de novo* convergence of autism genetics and molecular neuroscience. *Trends Neurosci* 37: 95–105.

Kumar, V., K. Kim, C. Joseph, S. Kourrich, S.H. Yoo, et al. (2013) C57BL/6N mutation in *cytoplasmic FMRP interacting protein 2* regulates cocaine response. *Science* 342: 1508–12.

Lagerspetz, K.Y., R. Tirri, and K.M. Lagerspetz (1968) Neurochemical and endocrinological studies of mice selectively bred for aggressiveness. *Scand J Psychol* 9: 157–60.

Lai, C.S., S.E. Fisher, J.A. Hurst, F. Vargha-Khadem, and A.P. Monaco (2001) A forkhead-domain gene is mutated in a severe speech and language disorder. *Nature* 413: 519–23.

Laing, R.D. and A. Esterson (1964) *Sanity, Madness and the Family.* Tavistock Publications, London.

Lake, R.I., L.J. Eaves, H.H. Maes, A.C. Heath, and N.G. Martin (2000) Further evidence against the environmental transmission of individual differences in neuroticism from a collaborative study of 45,850 twins and relatives on two continents. *Behav Genet* 30: 223–33.

Lam, M., C. Chen, Z. Li, A.R. Martin, J. Bryois, et al. (2019) Comparative genetic architectures of schizophrenia in East Asian and European populations. *Nat Genet* (in press).

Lamichhaney, S., G. Fan, F. Widemo, U. Gunnarsson, D.S. Thalmann, et al. (2016) Structural genomic changes underlie alternative reproductive strategies in the ruff (*Philomachus pugnax*). *Nat Genet* 48: 84–8.

Lander, E. and L. Kruglyak (1995) Genetic dissection of complex traits: guidelines for interpreting and reporting linkage results. *Nat Genet* 11: 241–7.

Lander, E.S. (1996) The new genomics: global views of biology. *Science* 274: 536–9.

Lander, E.S. and D. Botstein (1989) Mapping Mendelian factors underlying quantitative traits using RFLP linkage maps. *Genetics* 121: 185–99.

Landgraf, R. and A. Wigger (2002) High vs low anxiety-related behavior rats: an animal model of extremes in trait anxiety. *Behav Genet* 32: 301–14.

Landgraf, R. and A. Wigger (2003) Born to be anxious: neuroendocrine and genetic correlates of trait anxiety in HAB rats. *Stress* 6: 111–19.

Lane, N. and W. Martin (2010) The energetics of genome complexity. *Nature* 467: 929–34.

Lango Allen, H., K. Estrada, G. Lettre, S.I. Berndt, M.N. Weedon, et al. (2010) Hundreds of variants clustered in genomic loci and biological pathways affect human height. *Nature* 467: 832–8.

Lank, D.B., C.M. Smith, O. Hanotte, T. Burke, and F. Cooke (1995) Genetic polymorphism for alternative mating behaviour in lekking male ruff *Philomachus pugnax*. *Nature* 378: 59–62.

LaPlant, Q., V. Vialou, H.E. Covington, 3rd, D. Dumitriu, J. Feng, et al. (2010) Dnmt3a regulates emotional behavior and spine plasticity in the nucleus accumbens. *Nat Neurosci* 13: 1137–43.

Lee, D., M. Creed, K. Jung, T. Stefanelli, D.J. Wendler, et al. (2017) Temporally precise labeling and control of neuromodulatory circuits in the mammalian brain. *Nat Methods* 14: 495–503.

Lee, H., D.W. Kim, R. Remedios, T.E. Anthony, A. Chang, et al. (2014) Scalable control of mounting and attack by Esr1+ neurons in the ventromedial hypothalamus. *Nature* 509: 627–32.

Lee, J.J., R. Wedow, A. Okbay, E. Kong, O. Maghzian, et al. (2018) Gene discovery and polygenic prediction from a genome-wide association study of educational attainment in 1.1 million individuals. *Nat Genet* 50: 1112–21.

Lee, S.H., S. Ripke, B.M. Neale, S.V. Faraone, S.M. Purcell, et al. (2013) Genetic relationship between five psychiatric disorders estimated from genome-wide SNPs. *Nat Genet* 45: 984–94.

Lee, S.H., J. Yang, M.E. Goddard, P.M. Visscher, and N.R. Wray (2012) Estimation of pleiotropy between complex diseases using single-nucleotide polymorphism-derived genomic relationships and restricted maximum likelihood. *Bioinformatics* 28: 2540–2.

Lehmann, E.L. (1993) The Fisher, Neyman–Pearson theories of testing hypotheses: one theory or two? *J Am Stat Assoc* 88: 1242–9.

Lehrke, R. (1972) Theory of X-linkage of major intellectual traits. *Am J Ment Defic* 76: 611–19.

LeJeune, J., M. Gautier, and R. Turpin (1959) Etudes des chromosomes somatiques de neufs enfants mongolien. *C R Acad Sci Hebd Seances Acad* 248: 1721–2.

Lek, M., K.J. Karczewski, E.V. Minikel, K.E. Samocha, E. Banks, et al. (2016) Analysis of protein-coding genetic variation in 60,706 humans. *Nature* 536: 285–91.

Lepack, A.E., R.C. Bagot, C.J. Pena, Y.E. Loh, L.A. Farrelly, et al. (2016) Aberrant H3.3 dynamics in NAc promote vulnerability to depressive-like behavior. *Proc Natl Acad Sci U S A* 113: 12562–7.

Lesch, K.-P., D. Bengel, A. Heils, S.Z. Sabol, B.D. Greenberg, et al. (1996) Association of anxiety related traits with a polymorphism in the serotonin transporter gene regulatory region. *Science* 274: 1527–30.

Letourneau, A., F.A. Santoni, X. Bonilla, M.R. Sailani, D. Gonzalez, et al. (2014) Domains of genome-wide gene expression dysregulation in Down's syndrome. *Nature* 508: 345–50.

Levy, D., M. Ronemus, B. Yamrom, Y.H. Lee, A. Leotta, et al. (2011) Rare *de novo* and transmitted copy-number variation in autistic spectrum disorders. *Neuron* 70: 886–97.

Lewis, B.P., C.B. Burge, and D.P. Bartel (2005) Conserved seed pairing, often flanked by adenosines, indicates that thousands of human genes are microRNA targets. *Cell* 120: 15–20.

Lewis, E.B. (1996) The bithorax complex: the first fifty years. In *Les Prix Nobel 1995*, pp. 235–60. The Nobel Foundation, Stockholm.

Lewontin, R.C. (1972) The apportionment of human diversity. *Evol Biol* 6: 381–98.

Lewontin, R.C., S. Rose, and L. Kamin (1984) *Not In Our Genes: Biology, Ideology, and Human Nature*. Pantheon Books, New York.

Li, E. and Y. Zhang (2014) DNA methylation in mammals. *Cold Spring Harb Perspect Biol* 6: a019133.

Libbrecht, M.W. and W.S. Noble (2015) Machine learning applications in genetics and genomics. *Nat Rev Genet* 16: 321–32.

Libbrecht, R., M. Corona, F. Wende, D.O. Azevedo, J.E. Serrao, et al. (2013) Interplay between insulin signaling, juvenile hormone, and vitellogenin regulates maternal effects on polyphenism in ants. *Proc Natl Acad Sci U S A* 110: 11050–5.

Lichtenstein, P., C. Tuvblad, H. Larsson, and E. Carlstrom (2007) The Swedish Twin study of CHild and Adolescent Development: the TCHAD-Study. *Twin Res Hum Genet* 10: 67–73.

Lieb, R., B. Isensee, M. Hofler, H. Pfister, and H.U. Wittchen (2002) Parental major depression and the risk of depression and other mental disorders in offspring: a prospective-longitudinal community study. *Arch Gen Psychiatry* 59: 365–74.

Lim, E.T., M. Uddin, S. De Rubeis, Y. Chan, A.S. Kamumbu, et al. (2017) Rates, distribution and implications of postzygotic mosaic mutations in autism spectrum disorder. *Nat Neurosci* 20: 1217–24.

Lim, M.M., Z. Wang, D.E. Olazabal, X. Ren, E.F. Terwilliger, et al. (2004) Enhanced partner preference in a promiscuous species by manipulating the expression of a single gene. *Nature* 429: 754–7.

Lima, S.Q. and G. Miesenbock (2005) Remote control of behavior through genetically targeted photostimulation of neurons. *Cell* 121: 141–52.

Linnér, R.K., P. Biroli, E. Kong, S.F.W. Meddens, R. Wedow, et al. (2018) Genome-wide study identifies 611 loci associated with risk tolerance and risky behaviors. *bioRxiv* http://dx.doi.org/10.1101/261081.

Lisman, J., H. Schulman and H. Cline (2002) The molecular basis of CaMKII function in synaptic and behavioural memory. *Nat Rev Neurosci* 3: 175–90.

Lisman, J.E. (1985) A mechanism for memory storage insensitive to molecular turnover: a bistable autophosphorylating kinase. *Proc Natl Acad Sci U S A* 82: 3055–7.

Liu, J., D.R. Nyholt, P. Magnussen, E. Parano, P. Pavone, et al. (2001) A genomewide screen for autism susceptibility loci. *Am J Hum Genet* 69: 327–40.

Liu, X., S. Ramirez, P.T. Pang, C.B. Puryear, A. Govindarajan, et al. (2012) Optogenetic stimulation of a hippocampal engram activates fear memory recall. *Nature* 484: 381–5.

Lo, M.T., D.A. Hinds, J.Y. Tung, C. Franz, C.C. Fan, et al. (2017) Genome-wide analyses for personality traits identify six genomic loci and show correlations with psychiatric disorders. *Nat Genet* 49: 152–6.

Loehlin, J.C. (1992) *Sage Series on Individual Differences and Development, Vol. 2. Genes and Environment in Personality Development*. Sage Publications, Thousand Oaks, CA.

Loehlin, D.W. and J.H. Werren (2012) Evolution of shape by multiple regulatory changes to a growth gene. *Science* 335: 943–7.

Loehlin, J.C. and R.C. Nichols (1976) *Heredity, Environment and Personality*. University of Texas Press, Austin, TX.

Loesch, D.Z., Q.M. Bui, W. Kelso, R.M. Huggins, H. Slater, et al. (2005) Effect of Turner's syndrome and X-linked imprinting on cognitive status: analysis based on pedigree data. *Brain Dev* 27: 494–503.

Loh, P.R., G. Bhatia, A. Gusev, H.K. Finucane, B.K. Bulik-Sullivan, et al. (2015) Contrasting genetic architectures of schizophrenia and other complex diseases using fast variance-components analysis. *Nat Genet* 47: 1385–92.

Long, A.D., S.L. Mullaney, T.F.C. Mackay, and C.H. Langley (1996) Genetic interactions between naturally occurring alleles at quantitative trait loci and mutant alleles at candidate loci affecting bristle number in *Drosophila melanogaster*. *Genetics* 144: 1497–510.

Loomes, R., L. Hull, and W.P.L. Mandy (2017) What is the male-to-female ratio in autism spectrum disorder? A systematic review and meta-analysis. *J Am Acad Child Adolesc Psychiatry* 56: 466–74.

Lopez-Leon, S., A.C. Janssens, A.M. Gonzalez-Zuloeta Ladd, J. Del-Favero, S.J. Claes, et al. (2008) Meta-analyses of genetic studies on major depressive disorder. *Mol Psychiatry* 13: 772–85.

Lorenz, K. (1966) *On Aggression*. Translated by M.K. Wilson. Harcourt, Brace & World, New York.

Lubke, G.H., J.J. Hottenga, R. Walters, C. Laurin, E.J. de Geus, et al. (2012) Estimating the genetic variance of major depressive disorder due to all single nucleotide polymorphisms. *Biol Psychiatry* 72: 707–9.

Lubs, H.A., R.E. Stevenson, and C.E. Schwartz (2012) Fragile X and X-linked intellectual disability: four decades of discovery. *Am J Hum Genet* 90: 579–90.

Lucas, C. and M.B. Sokolowski (2009) Molecular basis for changes in behavioral state in ant social behaviors. *Proc Natl Acad Sci U S A* 106: 6351–6.

Luciano, M., S.P. Hagenaars, G. Davies, W.D. Hill, T.K. Clarke, et al. (2018) Association analysis in over 329,000 individuals identifies 116 independent variants influencing neuroticism. *Nat Genet* 50: 6–11.

Lukas, D. and T.H. Clutton-Brock (2013) The evolution of social monogamy in mammals. *Science* 341: 526–30.

Luscher, B., Q. Shen, and N. Sahir (2011) The GABAergic deficit hypothesis of major depressive disorder. *Mol Psychiatry* 16: 383–406.

Lynch, M. and J.S. Conery (2003) The origins of genome complexity. *Science* 302: 1401–4.

Lynn, R. (1982) IQ in Japan and the United States shows a growing disparity. *Nature* 297: 222–3

Lyon, M.F. (2003) Transmission ratio distortion in mice. *Annu Rev Genet* 37: 393–408.

Lytton, H., N.G. Martin, and L. Eaves (1977) Environmental and genetical causes of variation in ethological aspects of behavior in two-year-old boys. *Soc Biol* 24: 200–11.

MacArthur, D.G., S. Balasubramanian, A. Frankish, N. Huang, J. Morris, et al. (2012) A systematic survey of loss-of-function variants in human protein-coding genes. *Science* 335: 823–8.

MacArthur, D.G., T.A. Manolio, D.P. Dimmock, H.L. Rehm, J. Shendure, et al. (2014) Guidelines for investigating causality of sequence variants in human disease. *Nature* 508: 469–76.

MacDermott, A.B., M.L. Mayer, G.L. Westbrook, S.J. Smith, and J.L. Barker (1986) NMDA-receptor activation increases cytoplasmic calcium concentration in cultured spinal cord neurones. *Nature* 321: 519–22.

Mackay, T.F., S. Richards, E.A. Stone, A. Barbadilla, J.F. Ayroles, et al. (2012) The *Drosophila melanogaster* Genetic Reference Panel. *Nature* 482: 173–8.

Macosko, E.Z., N. Pokala, E.H. Feinberg, S.H. Chalasani, R.A. Butcher, et al. (2009) A hub-and-spoke circuit drives pheromone attraction and social behaviour in *C. elegans*. *Nature* 458: 1171–5.

Magee, W.J., W.W. Eaton, H.U. Wittchen, K.A. McGonagle, and R.C. Kessler (1996) Agoraphobia, simple phobia, and social phobia in the National Comorbidity Survey. *Arch Gen Psychiatry* 53: 159–68.

Magnus, C.J., P.H. Lee, D. Atasoy, H.H. Su, L.L. Looger, et al. (2011) Chemical and genetic engineering of selective ion channel-ligand interactions. *Science* 333: 1292–6.

Makarova, K.S., D.H. Haft, R. Barrangou, S.J. Brouns, E. Charpentier, et al. (2011) Evolution and classification of the CRISPR-Cas systems. *Nat Rev Microbiol* 9: 467–77.

Mali, P., L. Yang, K.M. Esvelt, J. Aach, M. Guell, et al. (2013) RNA-guided human genome engineering via Cas9. *Science* 339: 823–6.

Manolio, T.A., F.S. Collins, N.J. Cox, D.B. Goldstein, L.A. Hindorff, et al. (2009) Finding the missing heritability of complex diseases. *Nature* 461: 747–53.

Marder, E. (2012) Neuromodulation of neuronal circuits: back to the future. *Neuron* 76: 1–11.

Marie, H., W. Morishita, X. Yu, N. Calakos and R.C. Malenka (2005) Generation of silent synapses by acute *in vivo* expression of CaMKIV and CREB. *Neuron* 45: 741–52.

Marioni, R.E., G. Davies, C. Hayward, D. Liewald, S.M. Kerr, et al. (2014) Molecular genetic contributions to socioeconomic status and intelligence. *Intelligence* 44: 26–32.

Marlin, B.J., M. Mitre, A. D'Amour J, M.V. Chao, and R.C. Froemke (2015) Oxytocin enables maternal behaviour by balancing cortical inhibition. *Nature* 520: 499–504.

Marques-Bonet, T., J.M. Kidd, M. Ventura, T.A. Graves, Z. Cheng, et al. (2009) A burst of segmental duplications in the genome of the African great ape ancestor. *Nature* 457: 877–81.

Marshall, C.R., D.P. Howrigan, D. Merico, B. Thiruvahindrapuram, W. Wu, et al. (2017) Contribution of copy number variants to schizophrenia from a genome-wide study of 41,321 subjects. *Nat Genet* 49: 27–35.

Maurano, M.T., R. Humbert, E. Rynes, R.E. Thurman, E. Haugen, et al. (2012) Systematic localization of common disease-associated variation in regulatory DNA. *Science* 337: 1190–5.

Mayer, M.L., G.L. Westbrook, and P.B. Guthrie (1984) Voltage-dependent block by Mg2+ of NMDA responses in spinal cord neurones. *Nature* 309: 261–3.

Maze, I., H.E. Covington, 3rd, D.M. Dietz, Q. LaPlant, W. Renthal, et al. (2010) Essential role of the histone methyltransferase G9a in cocaine-induced plasticity. *Science* 327: 213–16.

McCarthy, S.E., V. Makarov, G. Kirov, A.M. Addington, J. McClellan, et al. (2009) Microduplications of 16p11.2 are associated with schizophrenia. *Nat Genet* 41: 1223–7.

McClellan, J. and M.C. King (2010) Genetic heterogeneity in human disease. *Cell* 141: 210–17.

McClintock, B. (1951) Mutable loci in maize. *Year B Carnegie Inst Wash* 50: 174–81.

McGarvey, D. (2017) *Poverty Safari: Understanding the Anger of Britain's Underclass.* Luath Press, Edinburgh, UK.

McGowan, P.O., A. Sasaki, A.C. D'Alessio, S. Dymov, B. Labonte, et al. (2009) Epigenetic regulation of the glucocorticoid receptor in human brain associates with childhood abuse. *Nat Neurosci* 12: 342–8.

McGrath, J. and D. Solter (1984) Completion of mouse embryogenesis requires both the maternal and paternal genomes. *Cell* 37: 179–83.

McGregor, A.P., V. Orgogozo, I. Delon, J. Zanet, D.G. Srinivasan, et al. (2007) Morphological evolution through multiple *cis*-regulatory mutations at a single gene. *Nature* 448: 587–90.

McGuffin, P., F. Rijsdijk, M. Andrew, P. Sham, R. Katz, et al. (2003) The heritability of bipolar affective disorder and the genetic relationship to unipolar depression. *Arch Gen Psychiatry* 60: 497–502.

McLaren, J. and S.E. Bryson (1987) Review of recent epidemiological studies of mental retardation: prevalence, associated disorders, and etiology. *Am J Ment Retard* 92: 243–54.

McLean, C.Y., P.L. Reno, A.A. Pollen, A.I. Bassan, T.D. Capellini, et al. (2011) Human-specific loss of regulatory DNA and the evolution of human-specific traits. *Nature* 471: 216–19.

Meehan, T.F., N. Conte, D.B. West, J.O. Jacobsen, J. Mason, et al. (2017) Disease model discovery from 3,328 gene knockouts by The International Mouse Phenotyping Consortium. *Nat Genet* 49: 1231–8.

Meffert, L.M., S.K. Hicks, and J.L. Regan (2002) Nonadditive genetic effects in animal behavior. *Am Nat* 160 (Suppl. 6): S198–213.

Mehrabian, M., H. Allayee, J. Stockton, P.Y. Lum, T.A. Drake, et al. (2005) Integrating genotypic and expression data in a segregating mouse population to identify 5-lipoxygenase as a susceptibility gene for obesity and bone traits. *Nat Genet* 37: 1224–33.

Mendes Soares, L.M. and J. Valcarcel (2006) The expanding transcriptome: the genome as the 'Book of Sand'. *EMBO J* 25: 923–31.

Mendlewicz, J. and J.D. Rainer (1977) Adoption study supporting genetic transmission in manic—depressive illness. *Nature* 268: 327–9.

Meyer-Lindenberg, A. and D.R. Weinberger (2006) Intermediate phenotypes and genetic mechanisms of psychiatric disorders. *Nat Rev Neurosci* 7: 818–27.

Michaelson, J.J., Y. Shi, M. Gujral, H. Zheng, D. Malhotra, et al. (2012) Whole-genome sequencing in autism identifies hot spots for *de novo* germline mutation. *Cell* 151: 1431–42.

Michalon, A., M. Sidorov, T.M. Ballard, L. Ozmen, W. Spooren, et al. (2012) Chronic pharmacological mGlu5 inhibition corrects fragile X in adult mice. *Neuron* 74: 49–56.

Middeldorp, C.M., D.C. Cath, R. van Dyck, and D.I. Boomsma (2005) The co-morbidity of anxiety and depression in the perspective of genetic epidemiology. A review of twin and family studies. *Psychol Med* 35: 611–24.

Miki, Y., J. Swensen, D. Shattuck-Eidens, P.A. Futreal, K. Harshman, et al. (1994) A strong candidate for the breast and ovarian cancer susceptibility gene BRCA1. *Science* 266: 66–71.

Mitchell, J.A. and P. Fraser (2008) Transcription factories are nuclear subcompartments that remain in the absence of transcription. *Genes Dev* 22: 20–5.

Moffitt, T.E. and A. Caspi (2014) Bias in a protocol for a meta-analysis of 5-HTTLPR, stress, and depression. *BMC Psychiatry* 14: 179.

Mohn, F., M. Weber, M. Rebhan, T.C. Roloff, J. Richter, et al. (2008) Lineage-specific polycomb targets and *de novo* DNA methylation define restriction and potential of neuronal progenitors. *Mol Cell* 30: 755–66.

Monaco, A.P., C.J. Bertelson, W. Middlesworth, C.A. Colletti, J. Aldridge, et al. (1985) Detection of deletions spanning the Duchenne muscular dystrophy locus using a tightly linked DNA segment. *Nature* 316: 842–5.

Monteggia, L.M., H. Heimer, and E.J. Nestler (2018) Meeting report: can we make animal models of human mental illness? *Biol Psychiatry* 84: 542–5.

Moore, A.J., E.D. Brodie, 3rd, and J.B. Wolf (1997) Interacting phenotypes and the evolutionary process: I.

Direct and indirect genetic effects of social interactions. *Evolution* 51: 1352–62.

Morgan, H.D., H.G. Sutherland, D.I. Martin, and E. Whitelaw (1999) Epigenetic inheritance at the agouti locus in the mouse. *Nat Genet* 23: 314–18.

Morgan, J.I. and T. Curran (1986) Role of ion flux in the control of c-*fos* expression. *Nature* 322: 552–5.

Morgan, T.H. (1911) The origin of nine wing mutations in *Drosophila*. *Science* 33: 496–9.

Morris, J.K., D.E. Mutton, and E. Alberman (2002) Revised estimates of the maternal age specific live birth prevalence of Down's syndrome. *J Med Screen* 9: 2–6.

Morris, R.G. (1989) Synaptic plasticity and learning: selective impairment of learning rats and blockade of long-term potentiation in vivo by the N-methyl-d-aspartate receptor antagonist AP5. *J Neurosci* 9: 3040–57.

Morris, R.G., E. Anderson, G.S. Lynch, and M. Baudry (1986) Selective impairment of learning and blockade of long-term potentiation by an *N*-methyl-D-aspartate receptor antagonist, AP5. *Nature* 319: 774–6.

Morris, R.G., P. Garrud, J.N. Rawlins, and J. O'Keefe (1982) Place navigation impaired in rats with hippocampal lesions. *Nature* 297: 681–3.

Moschen, A.R., H. Tilg, and T. Raine (2019) IL-12, IL-23 and IL-17 in IBD: immunobiology and therapeutic targeting. *Nat Rev Gastroenterol Hepatol* 16: 185–96.

Mott, R., C.J. Talbot, M.G. Turri, A.C. Collins and J. Flint (2000) A method for fine mapping quantitative trait loci in outbred animal stocks. *Proc Natl Acad Sci U S A* 97: 12649–54.

Mott, R., W. Yuan, P. Kaisaki, X. Gan, J. Cleak, et al. (2014) The architecture of parent-of-origin effects in mice. *Cell* 156: 332–42.

Mousseau, T.A. and D.A. Roff (1987) Natural selection and the heritability of fitness components. *Heredity* 59: 181–97.

Munafò, M.R., C. Durrant, G. Lewis, and J. Flint (2009) Gene × environment interactions at the serotonin transporter locus. *Biol Psychiatry* 65: 211–19.

Munafò, M.R., I.J. Matheson and J. Flint (2007) Association of the *DRD2* gene Taq1A polymorphism and alcoholism: a meta-analysis of case–control studies and evidence of publication bias. *Mol Psychiatry* 12: 454–61.

Muotri, A.R. and F.H. Gage (2006) Generation of neuronal variability and complexity. *Nature* 441: 1087–93.

Murgatroyd, C., A.V. Patchev, Y. Wu, V. Micale, Y. Bockmuhl, et al. (2009) Dynamic DNA methylation programs persistent adverse effects of early-life stress. *Nat Neurosci* 12: 1559–66.

Murlidharan, G., R.J. Samulski, and A. Asokan (2014) Biology of adeno-associated viral vectors in the central nervous system. *Front Mol Neurosci* 7: 76.

Nadal, M., S. Moreno, M. Pritchard, M.A. Preciado, X. Estivill, et al. (1997) Down syndrome: characterisation of a case with partial trisomy of chromosome 21 owing to a paternal balanced translocation (15;21) (q26;q22.1) by FISH. *J Med Genet* 34: 50–4.

Nadeau, J.H. and W.N. Frankel (2000) The roads from phenotypic variation to gene discovery: mutagenesis versus QTLs. *Nat Genet* 25: 381–4.

Nadeau, J.H., J.B. Singer, A. Matin, and E.S. Lander (2000) Analysing complex genetic traits with chromosome substitution strains. *Nat Genet* 24: 221–5.

Najmabadi, H., H. Hu, M. Garshasbi, T. Zemojtel, S.S. Abedini, et al. (2011) Deep sequencing reveals 50 novel genes for recessive cognitive disorders. *Nature* 478: 57–63.

Neale, B.M., Y. Kou, L. Liu, A. Ma'ayan, K.E. Samocha, et al. (2012) Patterns and rates of exonic *de novo* mutations in autism spectrum disorders. *Nature* 485: 242–5.

Neale, M.C., S.M. Boker, G. Xie, and H.H. Maes (2003) *Mx: Statistical Modeling*, 6th edn. Department of Psychiatry, VCU, Richmond, VA.

Nejentsev, S., N. Walker, D. Riches, M. Egholm, and J.A. Todd (2009) Rare variants of *IFIH1*, a gene implicated in antiviral responses, protect against type 1 diabetes. *Science* 324: 387–9.

Nelson, C.M., K.E. Ihle, M.K. Fondrk, R.E. Page, and G.V. Amdam (2007) The gene vitellogenin has multiple coordinating effects on social organization. *PLoS Biol* 5: e62.

Nestler, E.J. (2014a) Epigenetic mechanisms of depression. *JAMA Psychiatry* 71: 454–6.

Nestler, E.J. (2014b) Epigenetic mechanisms of drug addiction. *Neuropharmacology* 76: 259–68.

Nestler, E.J. and S.E. Hyman (2010) Animal models of neuropsychiatric disorders. *Nat Neurosci* 13: 1161–9.

Network and Pathway Analysis Subgroup of Psychiatric Genomics Consortium (2015) Psychiatric genome-wide association study analyses implicate neuronal, immune and histone pathways. *Nat Neurosci* 18: 199–209.

Neyman, J. and E.S. Pearson (1933) On the problem of the most efficient tests of statistical hypotheses. *Philos Trans R Soc Lond A* 231: 289–337.

Nicholls, R.D., J.H. Knoll, M.G. Butler, S. Karam, and M. Lalande (1989) Genetic imprinting suggested by maternal heterodisomy in nondeletion Prader–Willi syndrome. *Nature* 342: 281–5.

Nickerson, R.S. (2000) Null hypothesis significance testing: a review of an old and continuing controversy. *Psychol Methods* 5: 241–301.

Nicod, J., R.W. Davies, N. Cai, C. Hassett, L. Goodstadt, et al. (2016) Genome-wide association of multiple complex traits in outbred mice by ultra-low-coverage sequencing. *Nat Genet* 48: 912–18.

Nielsen, J. and M. Wohlert (1991) Chromosome abnormalities found among 34,910 newborn children: results from a 13-year incidence study in Arhus, Denmark. *Hum Genet* 87: 81–3.

Nirenberg, M.W. and J.H. Matthaei (1961) The dependence of cell-free protein synthesis in *E. coli* upon naturally occurring or synthetic polyribonucleotides. *Proc Natl Acad Sci U S A* 47: 1588–602.

Novarino, G., P. El-Fishawy, H. Kayserili, N.A. Meguid, E.M. Scott, et al. (2012) Mutations in *BCKD-kinase* lead to a potentially treatable form of autism with epilepsy. *Science* 338: 394–7.

Novembre, J., T. Johnson, K. Bryc, Z. Kutalik, A.R. Boyko, et al. (2008) Genes mirror geography within Europe. *Nature* 456: 98–101.

Nowak, L., P. Bregestovski, P. Ascher, A. Herbet, and A. Prochiantz (1984) Magnesium gates glutamate-activated channels in mouse central neurones. *Nature* 307: 462–5.

Nusslein-Volhard, C. and E. Wieschaus (1980) Mutations affecting segment number and polarity in *Drosophila*. *Nature* 287: 795–801.

Nuttle, X., G. Giannuzzi, M.H. Duyzend, J.G. Schraiber, I. Narvaiza, et al. (2016) Emergence of a *Homo sapiens*-specific gene family and chromosome 16p11.2 CNV susceptibility. *Nature* 536: 205–9.

O'Keefe, J. and J. Dostrovsky (1971) The hippocampus as a spatial map. Preliminary evidence from unit activity in the freely-moving rat. *Brain Res* 34: 171–5.

O'Roak, B.J., L. Vives, W. Fu, J.D. Egertson, I.B. Stanaway, et al. (2012a) Multiplex targeted sequencing identifies recurrently mutated genes in autism spectrum disorders. *Science* 338: 1619–22.

O'Roak, B.J., L. Vives, S. Girirajan, E. Karakoc, N. Krumm, et al. (2012b) Sporadic autism exomes reveal a highly interconnected protein network of *de novo* mutations. *Nature* 485: 246–50.

Oates, N.A., J. van Vliet, D.L. Duffy, H.Y. Kroes, N.G. Martin, et al. (2006) Increased DNA methylation at the *AXIN1* gene in a monozygotic twin from a pair discordant for a caudal duplication anomaly. *Am J Hum Genet* 79: 155–62.

Oberle, I., F. Rousseau, D. Heitz, C. Kretz, D. Devys, et al. (1991) Instability of a 550 base pair DNA segment and abnormal methylation in fragile X-syndrome. *Science* 252: 1097–102.

Okbay, A., B.M. Baselmans, J.E. de Neve, P. Turley, M.G. Nivard, et al. (2016) Genetic variants associated with subjective well-being, depressive symptoms, and neuroticism identified through genome-wide analyses. *Nat Genet* 48: 624–33.

Okhovat, M., A. Berrio, G. Wallace, A.G. Ophir and S.M. Phelps (2015) Sexual fidelity trade-offs promote regulatory variation in the prairie vole brain. *Science* 350: 1371–4.

Open Science Collaboration (2015). Estimating the reproducibility of psychological science. *Science* 349: aac4716.

Ophir, A.G., J.O. Wolff, and S.M. Phelps (2008) Variation in neural V1aR predicts sexual fidelity and space use among male prairie voles in semi-natural settings. *Proc Natl Acad Sci U S A* 105: 1249–54.

Oppenheim, J.S., J.E. Skerry, M.J. Tramo, and M.S. Gazzaniga (1989) Magnetic resonance imaging morphology of the corpus callosum in monozygotic twins. *Ann Neurol* 26: 100–4.

Orgel, L.E. and F.H. Crick (1980) Selfish DNA: the ultimate parasite. *Nature* 284: 604–7.

Osborne, K.A., A. Robichon, E. Burgess, S. Butland, R.A. Shaw, et al. (1997) Natural behavior polymorphism due to a cGMP-dependent protein kinase of *Drosophila*. *Science* 277: 834–6.

Otowa, T., K. Hek, M. Lee, E.M. Byrne, S.S. Mirza, et al. (2016) Meta-analysis of genome-wide association studies of anxiety disorders. *Mol Psychiatry* 21: 1391–9.

Pääbo, S. (2014) *Neanderthal Man: In Search of Lost Genomes*. Basic Books, New York.

Pardinas, A.F., P. Holmans, A.J. Pocklington, V. Escott-Price, S. Ripke, et al. (2018) Common schizophrenia alleles are enriched in mutation-intolerant genes and in regions under strong background selection. *Nat Genet* 50: 381–9.

Park, C.Y., L.T. Jeker, K. Carver-Moore, A. Oh, H.J. Liu, et al. (2012) A resource for the conditional ablation of microRNAs in the mouse. *Cell Rep* 1: 385–91.

Parker, C.C., S. Gopalakrishnan, P. Carbonetto, N.M. Gonzales, E. Leung, et al. (2016) Genome-wide association study of behavioral, physiological and gene expression traits in outbred CFW mice. *Nat Genet* 48: 919–26.

Parker, G. (2000) Classifying depression: should paradigms lost be regained? *Am J Psychiatry* 157: 1195–203.

Pasaniuc, B. and A.L. Price (2017) Dissecting the genetics of complex traits using summary association statistics. *Nat Rev Genet* 18: 117–27.

Pauly, P.J. (1987) *Controlling Life. Jacques Loeb and the Engineering Ideal in Biology.* Oxford University Press, New York.

Pe'er, I., R. Yelensky, D. Altshuler, and M.J. Daly (2008) Estimation of the multiple testing burden for genomewide association studies of nearly all common variants. *Genet Epidemiol* 32: 381–5.

Pearl, J. (2000) *Causality: Models, Reasoning and Inference.* Cambridge University Press, Cambridge, UK.

Pedersen, C.A. and A.J. Prange, Jr. (1979) Induction of maternal behavior in virgin rats after intracerebroventricular administration of oxytocin. *Proc Natl Acad Sci U S A* 76: 6661–5.

Pedersen, N.L., R. Plomin, J.R. Nesselroade, and G.E. McClearn (1992) A quantitative genetic analysis of cognitive abilities during the second half of the life span. *Psychol Sci* 3: 346–53.

Pena, C.J., H.G. Kronman, D.M. Walker, H.M. Cates, R.C. Bagot, et al. (2017) Early life stress confers lifelong stress susceptibility in mice via ventral tegmental area OTX2. *Science* 356: 1185–8.

Pena, C.J. and E.J. Nestler (2018) Progress in epigenetics of depression. *Prog Mol Biol Transl Sci* 157: 41–66.

Perris, C. (1966) A study of bipolar (manic-depressive) and unipolar recurrent depressive psychoses. I. Genetic investigation. *Acta Psychiatr Scand Suppl* 194: 15–44.

Persson, A., E. Gross, P. Laurent, K.E. Busch, H. Bretes, et al. (2009) Natural variation in a neural globin tunes oxygen sensing in wild *Caenorhabditis elegans. Nature* 458: 1030–3.

Peterson, R.E., N. Cai, T.B. Bigdeli, Y. Li, M. Reimers, et al. (2016) The genetic architecture of major depressive disorder in han chinese women. *JAMA Psychiatry* 74: 162–8.

Peterson, R.E., N. Cai, A.W. Dahl, T.B. Bigdeli, A.C. Edwards, et al. (2018) Molecular genetic analysis subdivided by adversity exposure suggests etiologic heterogeneity in major depression. *Am J Psychiatry* 175: 545–54.

Petrij, F., H.R. Giles, H.G. Dauwerse, J.J. Saris, R.C.M. Hennekam, et al. (1995) Rubinstein–Taybi syndrome caused by mutations in the transcriptional co-activator CNP. *Nature* 376: 348–51.

Petrovski, S., Q. Wang, E.L. Heinzen, A.S. Allen, and D.B. Goldstein (2013) Genic intolerance to functional variation and the interpretation of personal genomes. *PLoS Genet* 9: e1003709.

Pfaff, D. (2001) Precision in mouse behavior genetics. *Proc Natl Acad Sci U S A* 98: 5957–60.

Phelps, S.M. and L.J. Young (2003) Extraordinary diversity in vasopressin (V1a) receptor distributions among wild prairie voles (*Microtus ochrogaster*): patterns of variation and covariation. *J Comp Neurol* 466: 564–76.

Philip, V.M., G. Sokoloff, C.L. Ackert-Bicknell, M. Striz, L. Branstetter, et al. (2011) Genetic analysis in the Collaborative Cross breeding population. *Genome Res* 21: 1223–38.

Philippe, A., M. Martinez, M. Guilloud-Bataille, C. Gillberg, M. Rastam, et al. (1999) Genome-wide scan for autism susceptibility genes. *Hum Mol Genet* 8: 805–12.

Phillips, M.R., J. Zhang, Q. Shi, Z. Song, Z. Ding, et al. (2009) Prevalence, treatment, and associated disability of mental disorders in four provinces in China during 2001–05: an epidemiological survey. *Lancet* 373: 2041–53.

Pickens, R.W., D.S. Svikis, M. Mcgue, D.T. Lykken, L.L. Heston, et al. (1991) Heterogeneity in the inheritance of alcoholism. A study of male and female twins. *Arch Gen Psychiatry* 48: 19–28.

Pieretti, M., F.P. Zhang, Y.H. Fu, S.T. Warren, B.A. Oostra, et al. (1991) Absence of expression of the *FMR-1* gene in fragile X syndrome. *Cell* 66: 817–22.

Pinto, D., A.T. Pagnamenta, L. Klei, R. Anney, D. Merico, et al. (2010) Functional impact of global rare copy number variation in autism spectrum disorders. *Nature* 466: 368–72.

Pittenger, C., Y.Y. Huang, R.F. Paletzki, R. Bourtchouladze, H. Scanlin, et al. (2002) Reversible inhibition of CREB/ATF transcription factors in region CA1 of the dorsal hippocampus disrupts hippocampus-dependent spatial memory. *Neuron* 34: 447–62.

Plomin, R. and D. Daniels (1987) Why are children in the same family so different from one another? *Behav Brain Sci* 10: 1–16.

Plomin, R. and I.J. Deary (2015) Genetics and intelligence differences: five special findings. *Mol Psychiatry* 20: 98–108.

Plomin, R., N.L. Pedersen, P. Lichtenstein, G.E. Mcclearn and J.R. Nesselroade (1990) Genetic influence on life events during the last half of the lifespan. *Psychol Aging* 5: 25–30.

Polaczyk, P.J., R. Gasperini, and G. Gibson (1998) Naturally occurring genetic variation affects *Drosophila* photoreceptor determination. *Dev Genes Evol* 207: 462–70.

Pollak, R. (1998) *The Creation of Dr. B.: A Biography of Bruno Bettelheim*. Simon & Schuster, New York.

Posthuma, D., E.J. de Geus, W.F. Baare, H.E. Hulshoff Pol, R.S. Kahn, et al. (2002) The association between brain volume and intelligence is of genetic origin. *Nat Neurosci* 5: 83–4.

Prandini, P., S. Deutsch, R. Lyle, M. Gagnebin, C. Delucinge Vivier, et al. (2007) Natural gene-expression variation in Down syndrome modulates the outcome of gene-dosage imbalance. *Am J Hum Genet* 81: 252–63.

Prescott, C.A. and K.S. Kendler (1999) Genetic and environmental contributions to alcohol abuse and dependence in a population-based sample of male twins. *Am J Psychiatry* 156: 34–40.

Pritchard, J.K., M. Stephens, and P. Donnelly (2000) Inference of population structure using multilocus genotype data. *Genetics* 155: 945–59.

Ptashne, M. (2007) On the use of the word 'epigenetic'. *Curr Biol* 17: R233–6.

Pulver, A.E., M. Karayiorgou, P.S. Wolyniec, V.K. Lasseter, L. Kasch, et al. (1994) Sequential strategy to identify a susceptibility gene for schizophrenia: report of potential linkage on chromosome 22q12–q13.1: Part 1. *Am J Med Genet* 54: 36–43.

Purcell, J., A. Brelsford, Y. Wurm, N. Perrin and M. Chapuisat (2014a) Convergent genetic architecture underlies social organization in ants. *Curr Biol* 24: 2728–32.

Purcell, S.M., J.L. Moran, M. Fromer, D. Ruderfer, N. Solovieff, et al. (2014b) A polygenic burden of rare disruptive mutations in schizophrenia. *Nature* 506: 185–90.

Purcell, S.M., N.R. Wray, J.L. Stone, P.M. Visscher, M.C. O'Donovan, et al. (2009) Common polygenic variation contributes to risk of schizophrenia and bipolar disorder. *Nature* 460: 748–52.

Pyle, D. (1978) A chromosome substitution analysis of geotactic maze behavior in *Drosophila melanogaster*. *Behav Genet* 8: 53–64.

R Development Core Team (2004) *R: A Language and Environment for Statistical Computing*. R Foundation for Statistical Computing, Vienna.

Radford, E.J., E. Isganaitis, J. Jimenez-Chillaron, J. Schroeder, M. Molla, et al. (2012) An unbiased assessment of the role of imprinted genes in an intergenerational model of developmental programming. *PLoS Genet* 8: e1002605.

Ramakrishnan, V. (2018) *Gene Machine: The Race to Decipher the Secrets of the Ribosome*. Basic Books, New York.

Ramboz, S., R. Oosting, D.A. Amara, H.F. Kung, P. Blier, et al. (1998) Serotonin receptor 1A knockout: an animal model of anxiety-related disorder. *Proc Natl Acad Sci U S A* 95: 14476–81.

Ramirez, S., X. Liu, P.A. Lin, J. Suh, M. Pignatelli, et al. (2013) Creating a false memory in the hippocampus. *Science* 341: 387–91.

Rando, O.J. (2012) Daddy issues: paternal effects on phenotype. *Cell* 151: 702–8.

Rankin, C.H., C.D. Beck, and C.M. Chiba (1990) *Caenorhabditis elegans*: a new model system for the study of learning and memory. *Behav Brain Res* 37: 89–92.

Rassoulzadegan, M., V. Grandjean, P. Gounon, S. Vincent, I. Gillot, et al. (2006) RNA-mediated non-mendelian inheritance of an epigenetic change in the mouse. *Nature* 441: 469–74.

Rauch, A., D. Wieczorek, E. Graf, T. Wieland, S. Endele, et al. (2012) Range of genetic mutations associated with severe non-syndromic sporadic intellectual disability: an exome sequencing study. *Lancet* 380: 1674–82.

Raznahan, A., D. Greenstein, N.R. Lee, L.S. Clasen, and J.N. Giedd (2012) Prenatal growth in humans and postnatal brain maturation into late adolescence. *Proc Natl Acad Sci U S A* 109: 11366–71.

Rea, S., F. Eisenhaber, D. O'Carroll, B.D. Strahl, Z.W. Sun, et al. (2000) Regulation of chromatin structure by site-specific histone H3 methyltransferases. *Nature* 406: 593–9.

Regier, D.A., W.E. Narrow, D.E. Clarke, H.C. Kraemer, S.J. Kuramoto, et al. (2013) DSM-5 field trials in the United States and Canada, Part II: test–retest reliability of selected categorical diagnoses. *Am J Psychiatry* 170: 59–70.

Renthal, W., I. Maze, V. Krishnan, H.E. Covington, 3rd, G. Xiao, et al. (2007) Histone deacetylase 5 epigenetically controls behavioral adaptations to chronic emotional stimuli. *Neuron* 56: 517–29.

Reynolds, C.A., D. Finkel, J.J. McArdle, M. Gatz, S. Berg, et al. (2005) Quantitative genetic analysis of latent growth curve models of cognitive abilities in adulthood. *Dev Psychol* 41: 3–16.

Rhee, S.H. and I.D. Waldman (2002) Genetic and environmental influences on antisocial behavior: a meta-analysis of twin and adoption studies. *Psychol Bull* 128: 490–529.

Ricker, J.P. and J. Hirsch (1988) Genetic changes occurring over 500 generations in lines of *Drosophila melanogaster* selected divergently for geotaxis. *Behav Genet* 18: 13–25.

Rietveld, C.A., S.E. Medland, J. Derringer, J. Yang, T. Esko, et al. (2013) GWAS of 126,559 individuals identifies genetic variants associated with educational attainment. *Science* 340: 1467–71.

Ripke, S., A.R. Sanders, K.S. Kendler, D.F. Levinson, P. Sklar, et al. (2011) Genome-wide association study identifies five new schizophrenia loci. *Nat Genet* 43: 969–76.

Risch, N., R. Herrell, T. Lehner, K.Y. Liang, L. Eaves, et al. (2009) Interaction between the serotonin transporter gene (*5-HTTLPR*), stressful life events, and risk of depression: a meta-analysis. *JAMA* 301: 2462–71.

Risch, N. and K. Merikangas (1996) The future of genetic studies of complex human diseases. *Science* 273: 1516–17.

Risch, N., D. Spiker, L. Lotspeich, N. Nouri, D. Hinds, et al. (1999) A genomic screen of autism: evidence for a multilocus etiology. *Am J Hum Genet* 65: 493–507.

Roberts, L. (1990) Huntington's gene: so near, yet so far. *Science* 247: 624–7.

Robinson, E.B., K.E. Samocha, J.A. Kosmicki, L. McGrath, B.M. Neale, et al. (2014) Autism spectrum disorder severity reflects the average contribution of *de novo* and familial influences. *Proc Natl Acad Sci U S A* 111: 15161–5.

Roeleveld, N., G.A. Zielhuis and F. Gabreels (1997) The prevalence of mental retardation: a critical review of recent literature. *Dev Med Child Neurol* 39: 125–32.

Rogers, C., V. Reale, K. Kim, H. Chatwin, C. Li, et al. (2003) Inhibition of *Caenorhabditis elegans* social feeding by FMRFamide-related peptide activation of NPR-1. *Nat Neurosci* 6: 1178–85.

Ronald, A. and R.A. Hoekstra (2011) Autism spectrum disorders and autistic traits: a decade of new twin studies. *Am J Med Genet B Neuropsychiatr Genet* 156B: 255–74.

Rosenthal, D., P.H. Wender, S.S. Kety, J. Welner, and F. Schulsinger (1971) The adopted-away offspring of schizophrenics. *Am J Psychiatry* 128: 307–11.

Roy, M.-A., M.C. Neale, N.L. Pedersen, A.A. Mathé and K.S. Kendler (1995) A twin study of generalized anxiety disorder and major depression. *Psychol Med* 25: 1037–49.

Roy-Zokan, E.M., C.B. Cunningham, L.E. Hebb, E.C. McKinney, and A.J. Moore (2015) Vitellogenin and vitellogenin receptor gene expression is associated with male and female parenting in a subsocial insect. *Proc Biol Sci* 282: 20150787.

Rudolph, D., A. Tafuri, P. Gass, G.J. Hammerling, B. Arnold, et al. (1998) Impaired fetal T cell development and perinatal lethality in mice lacking the cAMP response element binding protein. *Proc Natl Acad Sci U S A* 95: 4481–6.

Ruthenburg, A.J., C.D. Allis, and J. Wysocka (2007) Methylation of lysine 4 on histone H3: intricacy of writing and reading a single epigenetic mark. *Mol Cell* 25: 15–30.

Rutter, M., J. Silberg, T. O'Connor, and E. Simonoff (1999) Genetics and child psychiatry: II Empirical research findings. *J Child Psychol Psychiatry* 40: 19–55.

Samocha, K.E., E.B. Robinson, S.J. Sanders, C. Stevens, A. Sabo, et al. (2014) A framework for the interpretation of *de novo* mutation in human disease. *Nat Genet* 46: 944–50.

Samuels, B.A. and R. Hen (2011) Neurogenesis and affective disorders. *Eur J Neurosci* 33: 1152–9.

Sanders, S.J., A.G. Ercan-Sencicek, V. Hus, R. Luo, M.T. Murtha, et al. (2011) Multiple recurrent *de novo* CNVs, including duplications of the 7q11.23 Williams syndrome region, are strongly associated with autism. *Neuron* 70: 863–85.

Sanders, S.J., X. He, A.J. Willsey, A.G. Ercan-Sencicek, K.E. Samocha, et al. (2015) Insights into autism spectrum disorder genomic architecture and biology from 71 risk loci. *Neuron* 87: 1215–33.

Sanders, S.J., M.T. Murtha, A.R. Gupta, J.D. Murdoch, M.J. Raubeson, et al. (2012) *De novo* mutations revealed by whole-exome sequencing are strongly associated with autism. *Nature* 485: 237–41.

Sandnabba, N.K. (1996) Selective breeding for isolation-induced intermale aggression in mice: associated responses and environmental influences. *Behav Genet* 26: 477–88.

Sanger, F. and H. Tuppy (1951a) The amino-acid sequence in the phenylalanyl chain of insulin. 2. The investigation of peptides from enzymic hydrolysates. *Biochem J* 49: 481–90.

Sanger, F. and H. Tuppy (1951b) The amino-acid sequence in the phenylalanyl chain of insulin. 1. The identification of lower peptides from partial hydrolysates. *Biochem J* 49: 463–81.

Sartorius, N., A. Jablensky, and R. Shapiro (1977) Two-year follow-up of the patients included in the WHO International Pilot Study of Schizophrenia. *Psychol Med* 7: 529–41.

Sartorius, N., R. Shapiro, M. Kimura, and K. Barrett (1972) WHO international pilot study of schizophrenia. *Psychol Med* 2: 422–5.

Savage, J.E., P.R. Jansen, S. Stringer, K. Watanabe, J. Bryois, et al. (2018) Genome-wide association meta-analysis in 269,867 individuals identifies new genetic and functional links to intelligence. *Nat Genet* 50: 912–19.

Schadt, E.E., J. Lamb, X. Yang, J. Zhu, S. Edwards, et al. (2005) An integrative genomics approach to infer causal associations between gene expression and disease. *Nat Genet* 37: 710–17.

Schaub, M.A., A.P. Boyle, A. Kundaje, S. Batzoglou, and M. Snyder (2012) Linking disease associations with regulatory information in the human genome. *Genome Res* 22: 1748–59.

Schizophrenia Working Group of the Psychiatric Genomics Consortium (2014) Biological insights from 108 schizophrenia-associated genetic loci. *Nature* 511: 421–7.

Schmucker, D., J.C. Clemens, H. Shu, C.A. Worby, J. Xiao, et al. (2000) *Drosophila* Dscam is an axon guidance receptor exhibiting extraordinary molecular diversity. *Cell* 101: 671–84.

Schneider, J., J. Atallah and J.D. Levine (2017) Social structure and indirect genetic effects: genetics of social behaviour. *Biol Rev Camb Philos Soc* 92: 1027–38.

Schneider, K. (1958) *Psychopathic Personalities*. Cassell, London.

Schoenfelder, S., I. Clay, and P. Fraser (2010) The transcriptional interactome: gene expression in 3D. *Curr Opin Genet Dev* 20: 127–33.

Schumann, G., C. Liu, P. O'Reilly, H. Gao, P. Song, et al. (2016) *KLB* is associated with alcohol drinking, and its gene product β-Klotho is necessary for FGF21 regulation of alcohol preference. *Proc Natl Acad Sci U S A* 113: 14372–7.

Schwander, T., N. Lo, M. Beekman, B.P. Oldroyd, and L. Keller (2010) *Nature* versus nurture in social insect caste differentiation. *Trends Ecol Evol* 25: 275–82.

Sebat, J., B. Lakshmi, D. Malhotra, J. Troge, C. Lese-Martin, et al. (2007) Strong association of *de novo* copy number mutations with autism. *Science* 316: 445–9.

Sekar, A., A.R. Bialas, H. de Rivera, A. Davis, T.R. Hammond, et al. (2016) Schizophrenia risk from complex variation of complement component 4. *Nature* 530: 177–83.

Sexton, T., F. Bantignies, and G. Cavalli (2009) Genomic interactions: chromatin loops and gene meeting points in transcriptional regulation. *Semin Cell Dev Biol* 20: 849–55.

Shapiro, B.L. (1997) Whither Down syndrome critical region? *Hum Genet* 99: 421–3.

Sharpley, C.F., S.K. Palanisamy, N.S. Glyde, P.W. Dillingham, and L.L. Agnew (2014) An update on the interaction between the serotonin transporter promoter variant (5-HTTLPR), stress and depression, plus an exploration of non-confirming findings. *Behav Brain Res* 273: 89–105.

Shekhar, K., S.W. Lapan, I.E. Whitney, N.M. Tran, E.Z. Macosko, et al. (2016) Comprehensive classification of retinal bipolar neurons by single-cell transcriptomics. *Cell* 166: 1308–23.e30.

Sherrington, R., J. Brynjolfsson, H. Petursson, M. Potter, K. Dudleston, et al. (1988) Localization of a susceptibility locus for schizophrenia on chromosome 5. *Nature* 336: 164–7.

Shi, J., D.F. Levinson, J. Duan, A.R. Sanders, Y. Zheng, et al. (2009) Common variants on chromosome 6p22.1 are associated with schizophrenia. *Nature* 460: 753–7.

Shi, Y., F. Lan, C. Matson, P. Mulligan, J.R. Whetstine, et al. (2004) Histone demethylation mediated by the nuclear amine oxidase homolog LSD1. *Cell* 119: 941–53.

Shipman, S.L., J. Nivala, J.D. Macklis, and G.M. Church (2016) Molecular recordings by directed CRISPR spacer acquisition. *Science* 353: aaf1175.

Silberg, J., M. Rutter, M. Neale, and L. Eaves (2001) Genetic moderation of environmental risk for depression and anxiety in adolescent girls. *Br J Psychiatry* 179: 116–21.

Silva, A.J., R. Paylor, J.M. Wehner, and S. Tonegawa (1992a) Impaired spatial learning in a calcium-calmodulin kinase II mutant mice. Science 257: 206–11.

Silva, A.J., C.F. Stevens, S. Tonegawa and Y. Wang (1992b) Deficient hippocampal long-term potentiation in a-calcium-calmodulin kinase II mutant mice. *Science* 257: 201–6.

Silva, A.J., Y. Zhou, T. Rogerson, J. Shobe, and J. Balaji (2009) Molecular and cellular approaches to memory allocation in neural circuits. *Science* 326: 391–5.

Silventoinen, K., S. Sammalisto, M. Perola, D.I. Boomsma, B.K. Cornes, et al. (2003) Heritability of adult body height: a comparative study of twin cohorts in eight countries. *Twin Res* 6: 399–408.

Simmons, J.P., L.D. Nelson, and U. Simonsohn (2011) False-positive psychology: undisclosed flexibility in data collection and analysis allows presenting anything as significant. *Psychol Sci* 22: 1359–66.

Simpson, S.J., G.A. Sword, and N. Lo (2011) Polyphenism in insects. *Curr Biol* 21: R738–49.

Singer, J.B., A.E. Hill, L.C. Burrage, K.R. Olszens, J. Song, et al. (2004) Genetic dissection of complex traits with chromosome substitution strains of mice. *Science* 304: 445–8.

Singer, J.B., A.E. Hill, J.H. Nadeau, and E.S. Lander (2005) Mapping quantitative trait loci for anxiety in chromosome substitution strains of mice. *Genetics* 169: 855–62.

Singh, T., M.I. Kurki, D. Curtis, S.M. Purcell, L. Crooks, et al. (2016) Rare loss-of-function variants in *SETD1A* are associated with schizophrenia and developmental disorders. *Nat Neurosci* 19: 571–77.

Sittig, L.J., P. Carbonetto, K.A. Engel, K.S. Krauss, C.M. Barrios-Camacho, et al. (2016) Genetic background limits generalizability of genotype–phenotype relationships. *Neuron* 91: 1253–9.

Sjulson, L., D. Cassataro, S. DasGupta, and G. Miesenbock (2016) Cell-specific targeting of genetically encoded tools for neuroscience. *Annu Rev Genet* 50: 571–94.

Skarnes, W.C., B. Rosen, A.P. West, M. Koutsourakis, W. Bushell, et al. (2011) A conditional knockout resource for the genome-wide study of mouse gene function. *Nature* 474: 337–42.

Skuse, D.H., R.S. James, D.V. Bishop, B. Coppin, P. Dalton, et al. (1997) Evidence from Turner's syndrome of an imprinted X-linked locus affecting cognitive function. *Nature* 387: 705–8.

Smalley, S.L., R.F. Asarnow, and M.A. Spence (1988) Autism and genetics: a decade of research. *Arch Gen Psychiatry* 45: 953–61.

Smemo, S., J.J. Tena, K.H. Kim, E.R. Gamazon, N.J. Sakabe, et al. (2014) Obesity-associated variants within *FTO* form long-range functional connections with *IRX3*. *Nature* 507: 371–5.

Smiseth, P.T. and A.J. Moore (2004) Behavioral dynamics between caring males and females in a beetle with facultative biparental care. *Behav Ecol* 15: 621–8.

Smith, D.J., V. Escott-Price, G. Davies, M.E. Bailey, L. Colodro-Conde, et al. (2016) Genome-wide analysis of over 106 000 individuals identifies 9 neuroticism-associated loci. *Mol Psychiatry* 21: 749–57.

Smith, G.D. (2011) Epidemiology, epigenetics and the 'Gloomy Prospect': embracing randomness in population health research and practice. *Int J Epidemiol* 40: 537–62.

Smith, L.M., J.Z. Sanders, R.J. Kaiser, P. Hughes, C. Dodd, et al. (1986) Fluorescence detection in automated DNA sequence analysis. *Nature* 321: 674–9.

Smithies, O., R.G. Gregg, S.S. Boggs, M.A. Koralewski, and R.S. Kucherlapati (1985) Insertion of DNA sequences into the human chromosomal β-globin locus by homologous recombination. *Nature* 317: 230–4.

Smoller, J.W. and C.T. Finn (2003) Family, twin, and adoption studies of bipolar disorder. *Am J Med Genet C Semin Med Genet* 123C: 48–58.

Snell, G.D. (1979) Recent advances in histocompatibility immunogenetics. *Adv Genet* 20: 291–355.

So, H.C., M. Li, and P.C. Sham (2011) Uncovering the total heritability explained by all true susceptibility variants in a genome-wide association study. *Genet Epidemiol* 35: 447–56.

Sohail, M., R.M. Maier, A. Ganna, A. Bloemendal, A.R. Martin, et al. (2019) Polygenic adaptation on height is overestimated due to uncorrected stratification in genome-wide association studies. *Elife* 8: e39702.

Sokolowski, M.B. (1980) Foraging strategies of *Drosophila melanogaster*: a chromosomal analysis. *Behav Genet* 10: 291–302.

Sommer, B., M. Kohler, R. Sprengel, and P.H. Seeburg (1991) RNA editing in brain controls a determinant of ion flow in glutamate-gated channels. *Cell* 67: 11–19.

Spark Consortium (2018) SPARK: a US cohort of 50,000 families to accelerate autism research. *Neuron* 97: 488–93.

Sramek, A., M. Kriek, and F.R. Rosendaal (2003) Decreased mortality of ischaemic heart disease among carriers of haemophilia. *Lancet* 362: 351–4.

Stahl, E.A., G. Breen, A.J. Forstner, A. McQuillin, S. Ripke, et al. (2019) Genome-wide association study identifies 30 loci associated with bipolar disorder. *Nat Genet* 51: 793–803.

Stefansson, H., A. Meyer-Lindenberg, S. Steinberg, B. Magnusdottir, K. Morgen, et al. (2014) CNVs conferring risk of autism or schizophrenia affect cognition in controls. *Nature* 505: 361–6.

Stefansson, H., R.A. Ophoff, S. Steinberg, O.A. Andreassen, S. Cichon, et al. (2009) Common variants conferring risk of schizophrenia. *Nature* 460: 744–7.

Stefansson, H., D. Rujescu, S. Cichon, O.P. Pietilainen, A. Ingason, et al. (2008) Large recurrent microdeletions associated with schizophrenia. *Nature* 455: 232–6.

Stein, J.L., S.E. Medland, A.A. Vasquez, D.P. Hibar, R.E. Senstad, et al. (2012) Identification of common variants associated with human hippocampal and intracranial volumes. *Nat Genet* 44: 552–61.

Stern, D.L. (2013) The genetic causes of convergent evolution. *Nat Rev Genet* 14: 751–64.

Stern, S., C. Kirst, and C.I. Bargmann (2017) Neuromodulatory control of long-term behavioral patterns and individuality across development. *Cell* 171: 1649–62.e10.

Sterne, J.A. and G. Davey Smith (2001) Sifting the evidence-what's wrong with significance tests? *Br Med J* 322: 226–31.

Stoll, C., Y. Alembik, B. Dott, and J. Feingold (1994) Parental consanguinity as a cause of increased incidence of birth defects in a study of 131,760 consecutive births. *Am J Med Genet* 49: 114–17.

Stowers, L., T.E. Holy, M. Meister, C. Dulac, and G. Koentges (2002) Loss of sex discrimination and male–male aggression in mice deficient for TRP2. *Science* 295: 1493–500.

Stowers, L. and T.H. Kuo (2015) Mammalian pheromones: emerging properties and mechanisms of detection. *Curr Opin Neurobiol* 34: 103–9.

Strahl, B.D. and C.D. Allis (2000) The language of covalent histone modifications. *Nature* 403: 41–45.

Strenze, T. (2007) Intelligence and socioeconomic success: a meta-analytic review of longitudinal research. *Intelligence* 35: 401–26.

Strober, M., R. Freeman, C. Lampert, J. Diamond and W. Kaye (2000) Controlled family study of anorexia nervosa and bulimia nervosa: evidence of shared liability and transmission of partial syndromes. *Am J Psychiatry* 157: 393–401.

Styron, W. (1992) *Darkness Visible. A Memoir of Madness*. Open Road Integrated Media, New York.

Sullivan, P.F., K.S. Kendler, and M.C. Neale (2003) Schizophrenia as a complex trait: evidence from a meta-analysis of twin studies. *Arch Gen Psychiatry* 60: 1187–92.

Sullivan, P.F., M.C. Neale, and K.S. Kendler (2000) Genetic epidemiology of major depression: review and meta-analysis. *Am J Psychiatry* 157: 1552–62.

Sulston, J., Z. Du, K. Thomas, R. Wilson, L. Hillier, et al. (1992) The *C. elegans* genome sequencing project: a beginning. *Nature* 356: 37–41.

Sulston, J.E. (1976) Post-embryonic development in the ventral cord of *Caenorhabditis elegans*. *Philos Trans R Soc Lond B Biol Sci* 275: 287–97.

Sulston, J.E. and H.R. Horvitz (1977) Post-embryonic cell lineages of the nematode, *Caenorhabditis elegans*. *Dev Biol* 56: 110–56.

Sun, H., P.J. Kennedy, and E.J. Nestler (2013) Epigenetics of the depressed brain: role of histone acetylation and methylation. *Neuropsychopharmacology* 38: 124–37.

Surani, M.A., S.C. Barton, and M.L. Norris (1984) Development of reconstituted mouse eggs suggests imprinting of the genome during gametogenesis. *Nature* 308: 548–50.

Szasz, T.S. (1974) *The Myth of Mental Illness*. Harper Collins, New York.

Taher, A.T., D.J. Weatherall, and M.D. Cappellini (2018) Thalassaemia. *Lancet* 391: 155–67.

Takahashi, K., K. Tanabe, M. Ohnuki, M. Narita, T. Ichisaka, et al. (2007) Induction of pluripotent stem cells from adult human fibroblasts by defined factors. *Cell* 131: 861–72.

Talbot, C.J., A. Nicod, S.S. Cherny, D.W. Fulker, A.C. Collins, et al. (1999) High-resolution mapping of quantitative trait loci in outbred mice. *Nat Genet* 21: 305–8.

Taleb, N.N. (2008) *The Black Swan: The Impact of the Highly Improbable*. Penguin, London.

Tanay, A. and A. Regev (2017) Scaling single-cell genomics from phenomenology to mechanism. *Nature* 541: 331–8.

Tanzi, R.E. and L. Bertram (2001) New frontiers in Alzheimer's disease genetics. *Neuron* 32: 181–4.

Tarlungeanu, D.C., E. Deliu, C.P. Dotter, M. Kara, P.C. Janiesch, et al. (2016) Impaired amino acid transport at the blood brain barrier is a cause of autism spectrum disorder. *Cell* 167: 1481–94.e18.

Tasic, B., V. Menon, T.N. Nguyen, T.K. Kim, T. Jarsky, et al. (2016) Adult mouse cortical cell taxonomy revealed by single cell transcriptomics. *Nat Neurosci* 19: 335–46.

Taunton, J., C.A. Hassig, and S.L. Schreiber (1996) A mammalian histone deacetylase related to the yeast transcriptional regulator Rpd3p. *Science* 272: 408–11.

Taylor, S. and L. Campagna (2016) Avian supergenes. *Science* 351: 446–7.

Teasdale, T.W. and D.R. Owen (1984) Heredity and familial environment in intelligence and educational level—a sibling study. *Nature* 309: 620–22.

Teo, Y.Y., K.S. Small, and D.P. Kwiatkowski (2010) Methodological challenges of genome-wide association analysis in Africa. *Nat Rev Genet* 11: 149–60.

Thompson, P.M., T.D. Cannon, K.L. Narr, T. van Erp, V.P. Poutanen, et al. (2001) Genetic influences on brain structure. *Nat Neurosci* 4: 1253–8.

Thomson, J.P., P.J. Skene, J. Selfridge, T. Clouaire, J. Guy, et al. (2010) CpG islands influence chromatin structure via the CpG-binding protein Cfp1. *Nature* 464: 1082–6.

Tick, B., P. Bolton, F. Happe, M. Rutter, and F. Rijsdijk (2016) Heritability of autism spectrum disorders: a meta-analysis of twin studies. *J Child Psychol Psychiatry* 57: 585–95.

Tienari, P. (1991) Interaction between genetic vulnerability and family environment: the Finnish adoptive family study of schizophrenia. *Acta Psychiatr Scand* 84: 460–5.

Tillmann, W.A. and G.E. Hobbs (1949) The accident-prone automobile driver: a study of the psychiatric and social background. *Am J Psychiatry* 106: 321–31.

Timofeeva, M.N., R.J. Hung, T. Rafnar, D.C. Christiani, J.K. Field, et al. (2012) Influence of common genetic variation on lung cancer risk: meta-analysis of 14 900 cases and 29 485 controls. *Hum Mol Genet* 21: 4980–95.

Tinbergen, N. (1951) *The Study of Instinct*. Oxford University Press, Oxford, UK.

Tobacco and Genetics Consortium (2010) Genome-wide meta-analyses identify multiple loci associated with smoking behavior. *Nat Genet* 42: 441–7.

Toh, K.L., C.R. Jones, Y. He, E.J. Eide, W.A. Hinz, et al. (2001) An hPer2 phosphorylation site mutation in familial advanced sleep phase syndrome. *Science* 291: 1040–3.

Tolman, E.C. (1924) The inheritance of maze-learning ability in rats. *J Comp Psychol* 4: 1–18.

Toma, D.P., K.P. White, J. Hirsch, and R.J. Greenspan (2002) Identification of genes involved in *Drosophila melanogaster* geotaxis, a complex behavioral trait. *Nat Genet* 31: 349–53.

Tomida, S., T. Mamiya, H. Sakamaki, M. Miura, T. Aosaki, et al. (2009) *Usp46* is a quantitative trait gene regulating mouse immobile behavior in the tail suspension and forced swimming tests. *Nat Genet* 41: 688–95.

Trace, S.E., J.H. Baker, E. Penas-Lledo, and C.M. Bulik (2013) The genetics of eating disorders. *Annu Rev Clin Psychol* 9: 589–620.

Tramo, M.J., W.C. Loftus, T.A. Stukel, R.L. Green, J.B. Weaver, et al. (1998) Brain size, head size, and intelligence quotient in monozygotic twins. *Neurology* 50: 1246–52.

Tran, S.S., H.I. Jun, J.H. Bahn, A. Azghadi, G. Ramaswami, et al. (2019) Widespread RNA editing dysregulation in brains from autistic individuals. *Nat Neurosci* 22: 25–36.

Trible, W., L. Olivos-Cisneros, S.K. McKenzie, J. Saragosti, N.C. Chang, et al. (2017) *orco* mutagenesis causes loss of antennal lobe glomeruli and impaired social behavior in ants. *Cell* 170: 727–35.e10.

Trojer, P. and D. Reinberg (2006) Histone lysine demethylases and their impact on epigenetics. *Cell* 125: 213–17.

Trull, T.J. and T.A. Widiger (2013) Dimensional models of personality: the five-factor model and the DSM-5. *Dialogues Clin Neurosci* 15: 135–46.

Tryon, R.C. (1940) Genetic differences in mazelearning ability in rats. *Yearb Natl Soc Study Educ* 39: 111–19.

Tsankova, N., W. Renthal, A. Kumar, and E.J. Nestler (2007) Epigenetic regulation in psychiatric disorders. *Nat Rev Neurosci* 8: 355–67.

Tsien, J.Z., D.F. Chen, D. Gerber, C. Tom, E.H. Mercer, et al. (1996a) Subregion- and cell type-restricted gene knockout in mouse brain. *Cell* 87: 1317–26.

Tsien, J.Z., P.T. Huerta, and S. Tonegawa (1996b) The essential role of hippocampal CA1 NMDA receptor-dependent synaptic plasticity in spatial memory. *Cell* 87: 1327–38.

Turkheimer, E. (2000) Three laws of behavior genetics and what they mean. *Curr Dir Psychol Sci* 9: 160–4.

Turkheimer, E. (2011) Commentary: variation and causation in the environment and genome. *Int J Epidemiol* 40: 598–601.

Turkheimer, E. and M. Waldron (2000) Nonshared environment: a theoretical, methodological, and quantitative review. *Psychol Bull* 126: 78–108.

Turley, P., R.K. Walters, O. Maghzian, A. Okbay, J.J. Lee, et al. (2018) Multi-trait analysis of genome-wide association summary statistics using MTAG. *Nat Genet* 50: 229–37.

Turner, C.A., S.J. Watson, and H. Akil (2012) The fibroblast growth factor family: neuromodulation of affective behavior. *Neuron* 76: 160–74.

Turner, E.H., A.M. Matthews, E. Linardatos, R.A. Tell, and R. Rosenthal (2008) Selective publication of antidepressant trials and its influence on apparent efficacy. *N Engl J Med* 358: 252–60.

Tuttle, E.M. (2003) Alternative reproductive strategies in the white-throated sparrow: behavioral and genetic evidence. *Behav Ecol* 14: 425–32.

Tuttle, E.M., A.O. Bergland, M.L. Korody, M.S. Brewer, D.J. Newhouse, et al. (2016) Divergence and functional degradation of a sex chromosome-like supergene. *Curr Biol* 26: 344–50.

Tye, K.M. and K. Deisseroth (2012) Optogenetic investigation of neural circuits underlying brain disease in animal models. *Nat Rev Neurosci* 13: 251–66.

UK10K Consortium, K. Walter, J.L. Min, J. Huang, L. Crooks, et al. (2015) The UK10K project identifies rare variants in health and disease. *Nature* 526: 82–90.

Ustun, T.B., J.L. Ayuso-Mateos, S. Chatterji, C. Mathers, and C.J. Murray (2004) Global burden of depressive disorders in the year 2000. *Br J Psychiatry* 184: 386–92.

Valdar, W., L.C. Solberg, D. Gauguier, S. Burnett, P. Klenerman, et al. (2006a) Genome-wide genetic association of complex traits in heterogeneous stock mice. *Nat Genet* 38: 879–87.

Valdar, W., L.C. Solberg, D. Gauguier, W.O. Cookson, J.N. Rawlins, et al. (2006b) Genetic and environmental effects on complex traits in mice. *Genetics* 174: 959–84.

van Kesteren, R.E., A.B. Smit, R.P. de Lange, K.S. Kits, F.A. van Golen, et al. (1995) Structural and functional evolution of the vasopressin/oxytocin superfamily: vasopressin-related conopressin is the only member present in *Lymnaea*, and is involved in the control of sexual behavior. *J Neurosci* 15: 5989–98.

van Ravenzwaaij, D., P. Cassey, and S.D. Brown (2016) A simple introduction to Markov Chain Monte-Carlo sampling. *Psychon Bull Rev* 25: 143–54.

Vargha-Khadem, F., D.G. Gadian, A. Copp, and M. Mishkin (2005) *FOXP2* and the neuroanatomy of speech and language. *Nat Rev Neurosci* 6: 131–8.

Verhulst, B., M.C. Neale, and K.S. Kendler (2015) The heritability of alcohol use disorders: a meta-analysis of twin and adoption studies. *Psychol Med* 45: 1061–72.

Verkerk, A.J., M. Pieretti, J.S. Sutcliffe, Y.H. Fu, D.P. Kuhl, et al. (1991) Identification of a gene (*FMR-1*) containing a CGG repeat coincident with a breakpoint cluster region exhibiting length variation in fragile X syndrome. *Cell* 65: 905–14.

Visscher, P.M. (2008) Sizing up human height variation. *Nat Genet* 40: 489–90.

Visscher, P.M., M.E. Goddard, E.M. Derks and N.R. Wray (2012) Evidence-based psychiatric genetics, AKA the false dichotomy between common and rare variant hypotheses. *Mol Psychiatry* 17: 474–85.

Visscher, P.M., G. Hemani, A.A. Vinkhuyzen, G.B. Chen, S.H. Lee, et al. (2014) Statistical power to detect genetic (co)variance of complex traits using SNP data in unrelated samples. *PLoS Genet* 10: e1004269.

Visscher, P.M., J. Yang, and M.E. Goddard (2010) A commentary on 'common SNPs explain a large proportion of the heritability for human height' by Yang et al. (2010). *Twin Res Hum Genet* 13: 517–24.

Vissers, L.E., J. de Ligt, C. Gilissen, I. Janssen, M. Steehouwer, et al. (2010) A *de novo* paradigm for mental retardation. *Nat Genet* 42: 1109–12.

Vissers, L.E., C. Gilissen, and J.A. Veltman (2016) Genetic studies in intellectual disability and related disorders. *Nat Rev Genet* 17: 9–18.

Vitaterna, M.H., D.P. King, A.M. Chang, J.M. Kornhauser, P.L. Lowrey, et al. (1994) Mutagenesis and mapping of a mouse gene, *Clock*, essential for circardian behaviour. *Science* 264: 719–25.

von Frisch, K. (1967) *The Dance Language and Orientation of Bees*. Harvard University Press, Cambridge, MA.

Wacholder, S., S. Chanock, M. Garcia-Closas, L. El Ghormli, and N. Rothman (2004) Assessing the probability that a positive report is false: an approach for molecular epidemiology studies. *J Natl Cancer Inst* 96: 434–42.

Waddell, S. and W.G. Quinn (2001) Learning how a fruit fly forgets. *Science* 293: 1271–2.

Waddington, C.H. (1942) The epigenotype. *Endeavour* 1: 18–20.

Wagenmakers, E.J. (2007) A practical solution to the pervasive problems of *p* values. *Psychon Bull Rev* 14: 779–804.

Wagenmakers, E.J., R.D. Morey, and M.D. Lee (2016) Bayesian benefits for the pragmatic researcher. *Curr Dir Psychol Sci* 25: 169–76.

Walsh, J.J., D.J. Christoffel, B.D. Heifets, G.A. Ben-Dor, A. Selimbeyoglu, et al. (2018) 5-HT release in nucleus accumbens rescues social deficits in mouse autism model. *Nature* 560: 589–94.

Walsh, T., J.M. McClellan, S.E. McCarthy, A.M. Addington, S.B. Pierce, et al. (2008) Rare structural variants disrupt multiple genes in neurodevelopmental pathways in schizophrenia. *Science* 320: 539–43.

Wang, H., H. Yang, C.S. Shivalila, M.M. Dawlaty, A.W. Cheng, et al. (2013a) One-step generation of mice carrying mutations in multiple genes by CRISPR/Cas-mediated genome engineering. *Cell* 153: 910–18.

Wang, J., Y. Wurm, M. Nipitwattanaphon, O. Riba-Grognuz, Y.C. Huang, et al. (2013b) A Y-like social chromosome causes alternative colony organization in fire ants. *Nature* 493: 664–8.

Wang, Z., M. Gerstein, and M. Snyder (2009) RNA-Seq: a revolutionary tool for transcriptomics. *Nat Rev Genet* 10: 57–63.

Watson, J.D. and F.H. Crick (1953a) Genetical implications of the structure of deoxyribonucleic acid. *Nature* 171: 964–7.

Watson, J.D. and F.H. Crick (1953b) Molecular structure of nucleic acids: a structure for deoxyribose nucleic acid. *Nature* 171: 737–8.

Watson, H.J., Z. Yilmaz, L.M. Thornton, C. Hübel, J.R.I. Coleman, et al. (2019) Genome-wide association study identifies eight risk loci and implicates metabo-psychiatric origins for anorexia nervosa. *Nat Genet* 51: 1207–14.

Weaver, I.C., N. Cervoni, F.A. Champagne, A.C. D'Alessio, S. Sharma, et al. (2004) Epigenetic programming by maternal behavior. *Nat Neurosci* 7: 847–54.

Weaver, I.C., A.C. D'Alessio, S.E. Brown, I.C. Hellstrom, S. Dymov, et al. (2007) The transcription factor nerve growth factor-inducible protein a mediates epigenetic programming: altering epigenetic marks by immediate-early genes. *J Neurosci* 27: 1756–68.

Weber, J.N., B.K. Peterson and H.E. Hoekstra (2013) Discrete genetic modules are responsible for complex burrow evolution in *Peromyscus* mice. *Nature* 493: 402–5.

Weiner, D.J., E.M. Wigdor, S. Ripke, R.K. Walters, J.A. Kosmicki, et al. (2017) Polygenic transmission disequilibrium confirms that common and rare variation act additively to create risk for autism spectrum disorders. *Nat Genet* 49: 978–85.

Weinert, F. (2004) *The Scientist as Philosopher: Philosophical Consequences of Great Scientific Discoveries*. Springer, New York/Berlin/Heidelberg.

Weiss, I.C., C.R. Pryce, A.L. Jongen-Relo, N.I. Nanz-Bahr, and J. Feldon (2004) Effect of social isolation on stress-related behavioural and neuroendocrine state in the rat. *Behav Brain Res* 152: 279–95.

Weiss, J.M., C.H. West, M.S. Emery, R.W. Bonsall, J.P. Moore, et al. (2008a) Rats selectively-bred for behavior related to affective disorders: proclivity for intake of alcohol and drugs of abuse, and measures of brain monoamines. *Biochem Pharmacol* 75: 134–59.

Weiss, L.A., Y. Shen, J.M. Korn, D.E. Arking, D.T. Miller, et al. (2008b) Association between microdeletion and microduplication at 16p11.2 and autism. *N Engl J Med* 358: 667–75.

Weissman, M.M., R. Bland, P.R. Joyce, S. Newman, J.E. Wells, et al. (1993) Sex differences in rates of depression: cross-national perspectives. *J Affect Disord* 29: 77–84.

Weissman, M.M., R.C. Bland, G.J. Canino, C. Faravelli, S. Greenwald, et al. (1996) Cross-national epidemiology of major depression and bipolar disorder. *JAMA* 276: 293–9.

Weissman, M.M., E.S. Gershon, K.K. Kidd, B.A. Prusoff, J.F. Leckman, et al. (1984) Psychiatric disorders in the relatives of probands with affective disorders: the Yale University–National Institute of Mental Health Collaborative Study. *Arch Gen Psychiatry* 41: 13–21.

Weissman, M.M., P. Wickramaratne, Y. Nomura, V. Warner, D. Pilowsky, et al. (2006) Offspring of depressed parents: 20 years later. *Am J Psychiatry* 163: 1001–8.

Wellcome Trust Case Control Consortium (2007) Genome-wide association study of 14,000 cases of seven common diseases and 3,000 shared controls. *Nature* 447: 661–78.

Wheelwright, S., B. Auyeung, C. Allison, and S. Baron-Cohen (2010) Defining the broader, medium and narrow autism phenotype among parents using the Autism Spectrum Quotient (AQ). *Mol Autism* 1: 10.

White, J.G., E. Southgate, J.N. Thomson, and S. Brenner (1986) The structure of the nervous system of the nematode *C. elegans*. *Philos Trans R Soc Lond B Biol Sci* 314: 1–340.

White, J.K., A.K. Gerdin, N.A. Karp, E. Ryder, M. Buljan, et al. (2013) Genome-wide generation and systematic phenotyping of knockout mice reveals new roles for many genes. *Cell* 154: 452–64.

Wilcock, J. (1969) Gene action and behavior: an evaluation of major gene pleiotropism. *Psychol Bull* 72: 1–29.

Wilson, A.J., U. Gelin, M.C. Perron, and D. Reale (2009) Indirect genetic effects and the evolution of aggression in a vertebrate system. *Proc Biol Sci* 276: 533–41.

Wilson, A.J., M.B. Morrissey, M.J. Adams, C.A. Walling, F.E. Guinness, et al. (2011) Indirect genetics effects and evolutionary constraint: an analysis of social dominance in red deer, *Cervus elaphus*. *J Evol Biol* 24: 772–83.

Wilson, R., R. Ainscough, K. Anderson, C. Baynes, M. Berks, et al. (1994) 2.2 Mb of contiguous nucleotide sequence from chromosome III of *C. elegans*. *Nature* 368: 32–8.

Wing, J.K., J.L. Birley, J.E. Cooper, P. Graham, and A.D. Isaacs (1967) Reliability of a procedure for measuring and classifying "present psychiatric state". *Br J Psychiatry* 113: 499–515.

Winslow, J.T., N. Hastings, C.S. Carter, C.R. Harbaugh, and T.R. Insel (1993) A role for central vasopressin in pair bonding in monogamous prairie voles. *Nature* 365: 545–8.

Wolf, J.B. (2003) Genetic architecture and evolutionary constraint when the environment contains genes. *Proc Natl Acad Sci U S A* 100: 4655–60.

Wolpert, L. (2006) *Malignant Sadness. The Anatomy of Depression*. Faber & Faber, London.

Won, H., L. de la Torre-Ubieta, J.L. Stein, N.N. Parikshak, J. Huang, et al. (2016) Chromosome conformation elucidates regulatory relationships in developing human brain. *Nature* 538: 523–7.

World Health Organization (2017) *Depression and Other Common Mental Disorders: Global Health Estimates*. World Health Organization, Geneva, Switzerland.

Wray, N.R., S.H. Lee, and K.S. Kendler (2012) Impact of diagnostic misclassification on estimation of genetic correlations using genome-wide genotypes. *Eur J Hum Genet* 20: 668–74.

Wray, N.R., S. Ripke, M. Mattheisen, M. Trzaskowski, E.M. Byrne, et al. (2018) Genome-wide association analyses identify 44 risk variants and refine the genetic architecture of major depression. *Nat Genet* 50: 668–81.

Wu, Q., T. Wen, G. Lee, J.H. Park, H.N. Cai, et al. (2003) Developmental control of foraging and social behavior by the *Drosophila* neuropeptide Y-like system. *Neuron* 39: 147–61.

Wu, Z., A.E. Autry, J.F. Bergan, M. Watabe-Uchida, and C.G. Dulac (2014) Galanin neurons in the medial preoptic area govern parental behaviour. *Nature* 509: 325–30.

Xu, B., J.L. Roos, P. Dexheimer, B. Boone, B. Plummer, et al. (2011) Exome sequencing supports a *de novo* mutational paradigm for schizophrenia. *Nat Genet* 43: 864–8.

Yalcin, B., J. Flint, and R. Mott (2005) Using progenitor strain information to identify quantitative trait nucleotides in outbred mice. *Genetics* 171: 673–81.

Yalcin, B., J. Nicod, A. Bhomra, S. Davidson, J. Cleak, et al. (2010) Commercially available outbred mice for genome-wide association studies. *PLoS Genet* 6: e1001085.

Yalcin, B., S.A. Willis-Owen, J. Fullerton, A. Meesaq, R.M. Deacon, et al. (2004) Genetic dissection of a behavioral quantitative trait locus shows that *Rgs2* modulates anxiety in mice. *Nat Genet* 36: 1197–202.

Yalcin, B., K. Wong, A. Agam, M. Goodson, T.M. Keane, et al. (2011) Sequence-based characterization of structural variation in the mouse genome. *Nature* 477: 326–9.

Yang, C.F., M.C. Chiang, D.C. Gray, M. Prabhakaran, M. Alvarado, et al. (2013) Sexually dimorphic neurons in the ventromedial hypothalamus govern mating in both sexes and aggression in males. *Cell* 153: 896–909.

Yang, H., H. Wang, and R. Jaenisch (2014) Generating genetically modified mice using CRISPR/Cas-mediated genome engineering. *Nat Protoc* 9: 1956–68.

Yang, J., B. Benyamin, B.P. McEvoy, S. Gordon, A.K. Henders, et al. (2010) Common SNPs explain a large proportion of the heritability for human height. *Nat Genet* 42: 565–9.

Yang, J., T.A. Manolio, L.R. Pasquale, E. Boerwinkle, N. Caporaso, et al. (2011) Genome partitioning of genetic variation for complex traits using common SNPs. *Nat Genet* 43: 519–25.

Yang, J., J. Zeng, M.E. Goddard, N.R. Wray, and P.M. Visscher (2017) Concepts, estimation and interpretation of SNP-based heritability. *Nat Genet* 49: 1304–10.

Yin, J.C., J.S. Wallach, M. Del Vecchio, E.L. Wilder, H. Zhou, et al. (1994) Induction of a dominant negative CREB transgene specifically blocks long-term memory in *Drosophila*. *Cell* 79: 49–58.

Yonan, A.L., M. Alarcon, R. Cheng, P.K. Magnusson, S.J. Spence, et al. (2003) A genomewide screen of 345 families for autism-susceptibility loci. *Am J Hum Genet* 73: 886–97.

Young, L.J., R. Nilsen, K.G. Waymire, G.R. MacGregor and T.R. Insel (1999) Increased affiliative response to vasopressin in mice expressing the V1a receptor from a monogamous vole. *Nature* 400: 766–8.

Young, M.W. (1998) The molecular control of circadian behavioral rhythms and their entrainment in *Drosophila*. *Annu Rev Biochem* 67: 135–52.

Yu, S., M. Pritchard, E. Kremer, M. Lynch, J. Nancarrow, et al. (1991) Fragile X genotype characterized by an unstable region of DNA. *Science* 252: 1179–81.

Zabidi, M.A. and A. Stark (2016) Regulatory enhancer-core-promoter communication via transcription factors and cofactors. *Trends Genet* 32: 801–14.

Zeisel, A., A.B. Munoz-Manchado, S. Codeluppi, P. Lonnerberg, G. La Manno, et al. (2015) Cell types in the mouse cortex and hippocampus revealed by single-cell RNA-seq. *Science* 347: 1138–42.

Zelikowsky, M., M. Hui, T. Karigo, A. Choe, B. Yang, et al. (2018) The neuropeptide Tac2 controls a distributed brain state induced by chronic social isolation stress. *Cell* 173: 1265–79.e19.

Zeng, H. and J.R. Sanes (2017) Neuronal cell-type classification: challenges, opportunities and the path forward. *Nat Rev Neurosci* 18: 530–46.

Zerr, D.M., J.C. Hall, M. Rosbash, and K.K. Siwicki (1990) Circadian fluctuations of period protein immunoreactivity in the CNS and the visual system of *Drosophila*. *J Neurosci* 10: 2749–62.

Zhang, F., J. Vierock, O. Yizhar, L.E. Fenno, S. Tsunoda, et al. (2011) The microbial opsin family of optogenetic tools. *Cell* 147: 1446–57.

Zhou, Y., J. Won, M.G. Karlsson, M. Zhou, T. Rogerson, et al. (2009) CREB regulates excitability and the allocation of memory to subsets of neurons in the amygdala. *Nat Neurosci* 12: 1438–43.

Zhu, Z.H., F.T. Zhang, H. Hu, A. Bakshi, M.R. Robinson, et al. (2016) Integration of summary data from GWAS and eQTL studies predicts complex trait gene targets. *Nat Genet* 48: 481–7.

Index